Statistics and the Evaluation of Evidence for Forensic Scientists

STATISTICS IN PRACTICE

Founding Editor

Vic Barnett
Nottingham Trent University, UK

Statistics in Practice is an important international series of texts which provide detailed coverage of statistical concepts, methods and worked case studies in specific fields of investigation and study.

With sound motivation and many worked practical examples, the books show in down-to-earth terms how to select and use an appropriate range of statistical techniques in a particular practical field within each title's special topic area.

The books provide statistical support for professionals and research workers across a range of employment fields and research environments. Subject areas covered include medicine and pharmaceutics; industry, finance and commerce; public services; the earth and environmental sciences, and so on.

The books also provide support to students studying statistical courses applied to the above areas. The demand for graduates to be equipped for the work environment has led to such courses becoming increasingly prevalent at universities and colleges.

It is our aim to present judiciously chosen and well-written workbooks to meet everyday practical needs. Feedback of views from readers will be most valuable to monitor the success of this aim.

A complete list of titles in this series appears at the end of the volume.

Statistics and the Evaluation of Evidence for Forensic Scientists

Second Edition

Colin G.G. Aitken

School of Mathematics,
The University of Edinburgh, UK

Franco Taroni

Ecole des Sciences Criminelles and
Institut de Médecine Légale,
The University of Lausanne, Switzerland

John Wiley & Sons, Ltd

Other Wiley Editorial Offices

John Wiley & Sons Inc., 111 River Street, Hoboken, NJ 07030, USA

Jossey-Bass, 989 Market Street, San Francisco, CA 94103-1741, USA

Wiley-VCH Verlag GmbH, Boschstr. 12, D-69469 Weinheim, Germany

John Wiley & Sons Australia Ltd, 33 Park Road, Milton, Queensland 4064, Australia

John Wiley & Sons (Asia) Pte Ltd, 2 Clementi Loop #02-01, Jin Xing Distripark, Singapore 129809

John Wiley & Sons Canada Ltd, 22 Worcester Road, Etobicoke, Ontario, Canada M9W 1L1

Wiley also publishes its books in a variety of electronic formats. Some content that appears in print
may not be available in electronic books.

Library of Congress Cataloging in Publication Data

Aitken, C.G.G.
   Statistics and the evaluation of evidence for forensic scientists / Colin G.G. Aitken,
   Franco Taroni. — 2nd ed.
       p.   cm. — (Statistics in practice)
   Includes bibliographical references and index.
   ISBN 0-470-84367-5 (alk. paper)
   1. Forensic sciences—Statistical methods.   2. Forensic statistics.   3. Evidence, Expert.
   I. Taroni, Franco.   II. Title.   III. Statistics in practice (Chichester, England)

   HV8073.A55 2004
   363.23′01′5195—dc22

                                                                    2004048016

British Library Cataloguing in Publication Data

A catalogue record for this book is available from the British Library

ISBN 0-470-84367-5

Typeset in 10/12pt Photina by Integra Software Services Pvt. Ltd, Pondicherry, India
Printed and bound in Great Britain by TJ International, Padstow, Cornwall
This book is printed on acid-free paper responsibly manufactured from sustainable forestry
in which at least two trees are planted for each one used for paper production.

To

Liz, David and Catherine
and
Anna and Wanda

Contents

List of Tables

List of Figures

Foreword

The statistical evaluation of evidence is part of the scientific method, here applied to forensic circumstances. Karl Pearson was both perceptive and correct when he said that 'the unity of all science consists alone in its method, not in its material'. Science is a way of understanding and influencing the world in which we live. In this view it is not correct to say that physics is a science, whereas history is not: rather that the scientific method has been much used in physics, whereas it is largely absent from history. Scientific method is essentially a tool and, like any tool, is more useful in some fields than in others. If this appreciation of science as a method is correct, one might enquire whether the method could profitably be applied to the law, but before we can answer this it is necessary to understand something of what the scientific method involves. Books have been written on the topic, and here we confine ourselves to the essential ingredients of the method.

Two ideas dominate the scientific approach, observation and reasoning. Observation may be passive, as with a study of the motions of the heavenly bodies or the collection of medical records. Often it is active, as when an experiment is performed in a laboratory, or a controlled clinical trial is employed. The next stage is to apply reasoning to the observed data, usually to think of a theory that will account for at least some of the features seen in the data. The theory can then be used to predict further observations which are then made and compared with prediction. It is this see-saw between evidence and theory that characterises the scientific method and which, if successful, leads to a theory that accords with the observational material. Classic examples are Newton's theory to explain the motions of the planets, and Darwin's development of the theory of evolution. It is important to notice that, contrary to what many people think, uncertainty is present throughout any scientific procedure. There will almost always be errors in the measurements, due to variation in the material or limitations of the apparatus. A theory is always uncertain, and that is why it has to be rigorously tested. Only late in the cycle of movement between hard fact and mental activity is a theory admitted as being true. Even then the 'truth' is, in the long run, not absolute, as can be seen in the replacement of

Newton by Einstein. It is important to recognise the key role that uncertainty plays in the scientific method.

If the above analysis is correct, it becomes natural to see connections between the scientific method and legal procedures. All the ingredients are there, although the terminology is different. In a court of law, the data consist of the evidence pertinent to the case, evidence obtained by the police and other bodies, and presented by counsels for the defence and prosecution. There are typically only two theories, that the defendant is guilty or is innocent. As the trial proceeds, the see-saw effect is exhibited as evidence accumulates. Lawyers use an adversarial system, not openly present in scientific practice, but similar to the peer reviews that are employed therein. The most striking similarity between legal and scientific practice lies in the uncertainty that pervades both and the near-certainty that hopefully emerges at the end, the jurors oscillating as the evidence is presented. Indeed, it is striking that the law lightly uses the same term 'probability', as does the scientist, for expressing the uncertainty, for example in the phrase 'the balance of probabilities'.

The case for using the scientific method in a court of law therefore looks promising because the ingredients, in a modified form, are already present. Indeed, the method has had limited use, often by people who did not realise that they were acting in the scientific spirit, but it is only in the second half of the twentieth century that the method has been widely and successfully adopted. The major advances have taken place where the evidence itself is of the form that a scientist would recognise as material for study. Examples are evidence in the form of laboratory measurements on fragments of glass or types of blood, and, more recently and dramatically, DNA data. This book tells the story of this intrusion of science into law and, more importantly, provides the necessary machinery that enables the transition to be effected.

It has been explained how uncertainty plays important roles in both scientific method and courtroom procedure. It is now recognised that the only tool for handling uncertainty is probability, so it is inevitable that probability is to be found on almost every page of this book and must have a role in the courtroom. There are two aspects to probability: firstly, the purely mathematical rules and their manipulation; secondly, the interpretation of probability, that is, the connection between the numbers and the reality. Fortunately for legal applications, the mathematics is mostly rather simple – essentially it is a matter of appreciating the language and the notation, together with the use of the main tool, Bayes' theorem. Lawyers will be familiar with the need for specialist terms, so-called jargon, and will hopefully be appreciative of the need for a little mathematical jargon. Interpretation of probability is a more delicate issue, and difficulties here are experienced both by forensic scientists and lawyers. One miscarriage of justice was influenced by a scientist's flawed use of probability; another by a lack of legal appreciation of Bayes' theorem. A great strength of this book lies in the clear recognition of the interpretative problem and the inclusion of many examples of court cases, for example in Chapter 7. My personal view is that these problems are reduced by proper use of the mathematical notation and language, and by insisting that every statement of uncertainty is in the form of your probability of something, given clearly-stated assumptions. Thus the probability of the blood match, given that the defendant is innocent. Language that departs from this format can often lead to confusion.

The first edition (1995) of this book gave an admirable account of the subject as it was almost a decade ago. The current edition is much larger, and the enlargement reflects both the success of forensic science, by including recent cases, and also the new methods that have been used. A problem that arises in a courtroom, affecting both lawyers, witnesses and jurors, is that several pieces of evidence have to be put together before a reasoned judgement can be reached; as when motive has to be considered along with material evidence. Probability is designed to effect such combinations but the accumulation of simple rules can produce complicated procedures. Methods of handling sets of evidence have been developed; for example, Bayes nets in Chapter 14 and multivariate methods in Chapter 11. There is a fascinating interplay here between the lawyer and the scientist where they can learn from each other and develop tools that significantly assist in the production of a better judicial system. Another indication of the progress that has been made in a decade is the doubling in the size of the bibliography. There can be no doubt that the appreciation of some evidence in a court of law has been greatly enhanced by the sound use of statistical ideas, and one can be confident that the next decade will see further developments, during which time this book will admirably serve those who have cause to use statistics in forensic science.

D.V. Lindley
January 2004

Preface to the First Edition

In 1977 a paper by Dennis Lindley was published in *Biometrika* with the simple title 'A problem in forensic science'. Using an example based on the refractive indices of glass fragments, Lindley described a method for the evaluation of evidence which combined the two requirements of the forensic scientist, those of comparison and significance, into one statistic with a satisfactorily intuitive interpretation. Not unnaturally the method attracted considerable interest amongst statisticians and forensic scientists interested in seeking good ways of quantifying their evidence. Since then, the methodology and underlying ideas have been developed and extended in theory and application into many areas. These ideas, often with diverse terminology, have been scattered throughout many journals in statistics and forensic science and, with the advent of DNA profiling, in genetics. It is one of the aims of this book to bring these scattered ideas together and, in so doing, to provide a coherent approach to the evaluation of evidence.

The evidence to be evaluated is of a particular kind, known as transfer evidence, or sometimes trace evidence. It is evidence which is transferred between the scene of a crime and a criminal. It takes the form of traces — traces of DNA, traces of blood, of glass, of fibres, of cat hairs and so on. It is amenable to statistical analyses because data are available to assist in the assessment of variability. Assessments of other kinds of evidence, for example, eyewitness evidence, is not discussed.

The approach described in this book is based on the determination of a so-called likelihood ratio. This is a ratio of two probabilities, the probability of the evidence under two competing hypotheses. These hypotheses may be that the defendant is guilty and that he is innocent. Other hypotheses may be more suitable in certain circumstances and various of these are mentioned as appropriate throughout the book.

There are broader connections between statistics and matters forensic which could perhaps be covered by the title 'forensic statistics' and which are not covered here, except briefly. These might include the determination of a probability of guilt, both in the dicta 'innocent until proven guilty' and 'guilty beyond reasonable doubt'. Also, the role of statistical experts as expert witnesses presenting statistical assessments of data or as consultants preparing analyses for counsel is not discussed, nor is the possible involvement of statisticians as independent court assessors. A brief review of books on these other areas in the interface of statistics and the law is given in Chapter 1. There have also been two conferences on forensic statistics (Aitken, 1991, and Kaye, 1993a) with a third to be held in Edinburgh in 1996. These have included forensic science within their programme but have extended beyond this. Papers have also been presented and discussion sessions held at other conferences (e.g., Aitken, 1993, and Fienberg and Finkelstein, 1996).

The role of uncertainty in forensic science is discussed in Chapter 1. The main theme of the book is that the evaluation of evidence is best achieved through consideration of the likelihood ratio. The justification for this and the derivation of the general result is given in Chapter 2. A correct understanding of variation is required in order to derive expressions for the likelihood ratio and variation is the theme for Chapter 3 where statistical models are given for both discrete and continuous data. A review of other ways of evaluating evidence is given in Chapter 4. However, no other appears, to the author at least, to have the same appeal, both mathematically and forensically as the likelihood ratio and the remainder of the book is concerned with applications of the ratio to various forensic science problems. In Chapter 5, transfer evidence is discussed with particular emphasis on the importance of the direction of transfer, whether from the scene of the crime to the criminal or *vice versa*. Chapters 6 and 7 discuss examples for discrete and continuous data, respectively. The final chapter, Chapter 8, is devoted to a review of DNA profiling, though, given the continuing amount of work on the subject, it is of necessity brief and almost certainly not completely up to date at the time of publication.

In keeping with the theme of the Series, *Statistics in Practice*, the book is intended for forensic scientists as well as statisticians. Forensic scientists may find some of the technical details rather too complicated. A complete understanding of these is, to a large extent, unneccesary if all that is required is an ability to implement the results. Technical details in Chapters 7 and 8 have been placed in Appendices to these chapters so as not to interrupt the flow of the text. Statisticians may, in their turn, find some of the theory, for example in Chapter 1, rather elementary and, if this is the case, then they should feel free to skip over this and move on to the more technical parts of the later chapters.

The role of statistics in forensic science is continuing to increase. This is partly because of the debate continuing over DNA profiling which looks as if it will carry on into the foreseeable future. The increase is also because of increasing research by forensic scientists into areas such as transfer and persistence and because of increasing numbers of data sets. Incorporation of subjective probabilities will also increase, particularly through the role of Bayesian belief networks (Aitken and Gammerman, 1989) and knowledge-based systems (Buckleton and Walsh, 1991; Evett, 1993b).

Ian Evett and Dennis Lindley have been at the forefront of research in this area for many years. They have given me invaluable help throughout this time. Both made extremely helpful comments on earlier versions of the book for which I am grateful. I thank Hazel Easey for the assistance she gave with the production of the results in Chapter 8. I am grateful to Ian Evett also for making available the data in Table 7.3. Thanks are due to The University of Edinburgh for granting leave of absence and to my colleagues of the Department of Mathematics and Statistics in particular for shouldering the extra burdens such leave of absence by others entails. I thank also Vic Barnett, the Editor of the Series and the staff of John Wiley and Sons, Ltd for their help throughout the gestation period of this book.

Last, but by no means least, I thank my family for their support and encouragement.

Preface to the Second Edition

In the Preface to the first edition of this book it was commented that the role of statistics in forensic science was continuing to increase and that this was partly because of the debate continuing over DNA profiling which looked as if it would carry on into the foreseeable future. It now appears that the increase is continuing and perhaps at a greater rate than in 1995. The debate over DNA profiling continues unabated. We have left the minutiae of this debate to others, restricting ourselves to an overview of that particular topic. Instead, we elaborate on the many other areas in forensic science in which statistics can play a role.

There has been a tremendous expansion in the work in forensic statistics in the nine years since the first edition of this book was published. This is reflected in the increase in the size of the book. There are about 500 pages now, whereas there were only about 250 in 1995, and the bibliography has increased from 10 pages to 20 pages. The number of chapters has increased from 8 to 14. The title remains the same, yet there is more discussion of interpretation, in addition to new material on evaluation.

The first four chapters are on the same topics as in the first edition, though the order of Chapters 2 and 3 on evaluation and on variation has been reversed. The chapter on variation, the new Chapter 2, has been expanded to include many more probability distributions than mentioned in the first edition. As the subject has expanded so has the need for the use of more distributions. These have to be introduced sooner than before, hence the reversal of order with the chapter on evaluation. Chapter 4 has an additional section on the work of early twentieth-century forensic scientists as it has gradually emerged how far ahead of their time these scientists were in their ideas. Three new chapters have then been introduced before the chapter on transfer evidence. Bayesian inference has an increasing role to play in the evaluation of evidence, yet its use is still controversial and there have been some critical comments in the courts of some of its perceived uses in the legal process. Chapter 5 provides a discussion of Bayesian inference, somewhat separate from the main thrust of the book, in order to emphasise its particular relevance for evidence evaluation and interpretation. Appropriate sampling procedures are becoming ever more important. With scarce resources, sampling provides a means of achieving almost the same inferences but with the expenditure of considerably less resources. It is important, though, that correct inferences are drawn from the results obtained from a sample. In some jurisdictions and some types of crime, such as drug smuggling in the USA, the quantity of illicit material associated with the crime is a factor in the sentencing. Again, if only a sample has been taken from the initial consignment, then correct inferences about the quantities involved have to be made. These are the topics of Chapter 6. Chapter 7 is a consequence of the expansion of the book to consider interpretation. This includes a discussion of work on case assessment and interpretation done since the appearance of the first edition. Chapter 7 also includes brief comments on various evidence types for which statistical evaluation is beginning to be developed. This is in contrast to those areas such as glass, fibres and DNA profiling which are considerably more developed. Fibres and DNA each have chapters of their own, Chapters 12 and 13. Glass evaluation provides many examples throughout the book as it provides a context for much of what is discussed and it was felt this was better done at these places in the book rather than gathered together in a separate chapter. Chapters 8, 9 and 10 on transfer evidence, discrete data and continuous data are updated versions of chapters on the same topics in the first edition. Correct analysis of multivariate data is essential in forensic science as such data become more prevalent, for example, in the consideration of the elemental composition of glass or the chemical composition of drugs. Multivariate analysis is discussed in Chapter 11 with a worked analysis of a two-dimensional example. An appendix gives a brief description of the underlying mathematics of matrix algebra. Chapters 12 and 13 are the only chapters in the book which are specific to particular types of evidence, fibres and DNA profiling, respectively. Chapter 13 is a completely new chapter compared with the corresponding chapter in the first edition, such are the advances made in DNA profiling since the first edition appeared. It is still only a brief introduction to the topic. Other more specialised books, cited in Chapter 13, are recommended for the serious student of DNA profiling. The final chapter is an introduction to Bayesian networks, an exciting new topic for forensic science and evidence evaluation in general, predicted in the Preface to the first edition. A graphical representation, as provided by a Bayesian network, of the different types and pieces of evidence in a case aids considerably the understanding and analysis of the totality of the evidence. In addition to the bibliography and indexes, a list of notation has been provided at the end. It is hoped this will enable the reader to keep track of the symbolism which is a necessary part of ensuring clarity of exposition.

The role of Bayesian inference in forensic science remains controversial. In order to try to understand why this may be so, we can do no better than to quote from an eminent fibres expert, who wrote:

> There may be different reasons for the obvious reluctance or scepticism connected with the adoption of Bayesian theory for presenting fibres evidence. These may include:

- a lack of awareness of the explanatory literature available;
- difficulty in understanding the proposals therein;
- an antagonistic mind-set generated by an approach which is thought too complicated and too mathematical;
- not knowing how to apply Bayes' theorem in practical casework;
- criticism that case scenarios dealt with in the literature are over-simplified and not realistic.

(Grieve, 2000b)

We hope that this book goes some way towards overcoming such reluctance and scepticism.

Reference is made on occasion to probability values of statistical distributions. Rather than make reference to statistical packages and books of tables each time this is done, details of several packages are listed here instead:

- MINITAB. See http://www.minitab.com and Ryan *et al.* (2000).

- R. This is a language and environment for statistical computing and graphics which is available in source code form as Free Software under the terms of the Free Software Foundation's GNU General Public License. R can be considered as a different implementation of S. There are some important differences, but much code written for S runs unaltered under R. See http://www.r-project.org/ and Ihaka and Gentleman (1996).

- S-PLUS. See http://www.mathsoft.com/splus and Venables and Ripley (2002). See also http://lib.stat.cmu.edu/S/ for software and extensions.

In addition, for those who like paper a very useful book of statistical tables is Lindley and Scott (1995).

During the preparation of this book, two eminent forensic scientists, Barry Gaudette and Mike Grieve, died. Both did much to inspire our work in evidence evaluation, for which we will always be grateful.

Many people have helped in many ways in the preparation of this book. In particular, we acknowledge the assistance of Fred Anglada, Marc Augsburger, Luc Besson, Alex Biedermann, Christophe Champod, Pierre Esseiva, Paolo Garbolino, David Lucy, Willy Mazzella, Phil Rose and Bruce Weir. Whilst we have received much advice, we accept full responsibility for any errors of commission and omission. The Leverhulme Trust provided invaluable support for this work through the award of a research fellowship to one of us (CGGA). Dennis Lindley graciously agreed to write a foreword. He has been an inspiration to us throughout our careers, and we thank him most sincerely for the honour he has done us. We also thank the staff at John Wiley and Sons, Ltd, Lucy Bryan, Rob Calver, Siân Jones, Jane Shepherd, and a very assiduous copy-editor, Richard Leigh, for their help and support in bringing this project to fruition.

Last, but by no means least, we thank our families for their support and encouragement.

C.G.G. Aitken and F. Taroni
Edinburgh and Lausanne

1

Uncertainty in Forensic Science

1.1 INTRODUCTION

The purpose of this book is to discuss the statistical and probabilistic evaluation of scientific evidence for forensic scientists. For the most part the evidence to be evaluated will be so-called *transfer* or *trace* evidence.

There is a well-known principle in forensic science known as *Locard's principle* which states that every contact leaves a trace:

> tantôt le malfaiteur a laissé sur les lieux les marques de son passage, tantôt, par une action inverse, il a emporté sur son corps ou sur ses vêtements, les indices de son séjour ou de son geste. (Locard, 1920)

Iman and Rudin (2001) translate this as follows:

> either the wrong-doer has left signs at the scene of the crime, or, on the other hand, has taken away with him – on his person (body) or clothes – indications of where he has been or what he has done.

The principle was reiterated using different words in 1929:

> Les débris microscopiques qui recouvrent nos habits et notre corps sont les témoins muets, assurés et fidèles de chacun de nos gestes et de chacun de nos recontres. (Locard, 1929)

This may be translated as follows:

> Traces which are present on our clothes or our person are silent, sure and faithful witnesses of every action we do and of every meeting we have.

Transfer evidence and Locard's principle may be illustrated as follows. Suppose a person gains entry to a house by breaking a window and assaults the man of the house, during which assault blood is spilt by both victim and assailant. The criminal may leave traces of his presence at the crime scene in the form of bloodstains from the assault and fibres from his clothing. This evidence is said to be transferred from the criminal to the scene of the crime. The criminal may also take traces of the crime scene away with him. These could include

Statistics and the Evaluation of Evidence for Forensic Scientists: Second Edition
C.G.G. Aitken and F. Taroni © 2004 John Wiley & Sons, Ltd   ISBN: 0-470-84367-5

bloodstains from the assault victim, fibres of their clothes and fragments of glass from the broken window. Such evidence is said to be transferred to the criminal from the crime scene. A suspect is soon identified, at a time at which he will not have had the opportunity to change his clothing. The forensic scientists examining the suspect's clothing find similarities amongst all the different types of evidence: blood, fibres and glass fragments. They wish to *evaluate the strength of this evidence*. It is hoped that this book will enable them so to do.

Quantitative issues relating to the distribution of characteristics of interest will be discussed. However, there will also be discussion of qualitative issues such as the choice of a suitable population against which variability in the measurements of the characteristics of interest may be compared. Also, a brief history of statistical aspects of the evaluation of evidence is given in Chapter 4.

1.2 STATISTICS AND THE LAW

The book does not focus on the use of statistics and probabilistic thinking for legal decision making, other than by occasional reference. Also, neither the role of statistical experts as expert witnesses presenting statistical assessments of data nor their role as consultants preparing analyses for counsel is discussed. There is a distinction between these two issues (Fienberg, 1989; Tribe, 1971). The main focus of this book is on the assessment of evidence for forensic scientists, in particular for *identification* purposes. The process of addressing the issue of whether or not a particular item came from a particular source is most properly termed *individualisation*. 'Criminalistics is the science of individualisation' (Kirk, 1963), but established forensic and judicial practices have led to it being termed *identification*. The latter terminology will be used throughout this book. An *identification*, however, is more correctly defined as 'the determination of some set to which an object belongs or the determination as to whether an object belongs to a given set' (Kingston, 1965a). Further discussion is given in Kwan (1977) and Evett *et al.* (1998a).

For example, in a case involving a broken window, similarities may be found between the refractive indices of fragments of glass found on the clothing of a suspect and the refractive indices of fragments of glass from the broken window. The assessment of this evidence, in associating the suspect with the scene of the crime, is part of the focus of this book (and is discussed in particular in Section 10.4.2).

For those interested in the issues of statistics and the law beyond those of forensic science, in the sense used in this book, there are several books available and some of these are discussed briefly.

'*The Evolving Role of Statistical Assessments as Evidence in the Courts*' is the title of a report, edited by Fienberg (1989), by the Panel on Statistical Assessments as Evidence in the Courts formed by the Committee on National Statistics and the Committee on Research on Law Enforcement and the Administration of Justice of the USA, and funded by the National Science Foundation. Through the use of case studies the report reviews the use of statistics in selected areas of litigation, such as employment discrimination, antitrust litigation and environment law. One case study is concerned with identification in a criminal case. Such a matter is the concern of this book and the ideas relevant to this case study, which involves the evidential worth of similarities amongst human head hair samples, will be discussed in greater detail later (Sections 4.5.2 and 4.5.5). The report makes various recommendations, relating to the role of the expert witness, pre-trial discovery, the provision of statistical resources, the role of court-appointed experts, the enhancement of the capability of the fact-finder and statistical education for lawyers.

Two books which take the form of textbooks on statistics for lawyers are Vito and Latessa (1989) and Finkelstein and Levin (2001). The former focuses on the presentation of statistical concepts commonly used in criminal justice research. It provides criminological examples to demonstrate the calculation of basic statistics. The latter introduces rather more advanced statistical techniques and again uses case studies to illustrate such techniques.

The area of discrimination litigation is covered by a set of papers edited by Kaye and Aickin (1986). This starts by outlining the legal doctrines that underlie discrimination litigation. In particular, there is a fundamental issue relating to discrimination in hiring. The definition of the relevant market from which an employer hires has to be made very clear. For example, consider the case of a man who applies, but is rejected for, a secretarial position. Is the relevant population the general population, the representation of men amongst secretaries in the local labour force or the percentage of male applicants? The choice of a suitable reference population is also one with which the forensic scientist has to be concerned. This is discussed at several points in this book.

Another textbook, which comes in two volumes, is Gastwirth (1988a, b). The book is concerned with civil cases and 'is designed to introduce statistical concepts and their proper use to lawyers and interested policymakers' (1988a, p. xvii). Two areas are stressed which are usually given less emphasis in most statistical textbooks. The first area is concerned with measures of relative or comparative inequality. These are important because many legal cases are concerned with issues of fairness or equal treatment. The second area is concerned with the combination of results of several related statistical studies. This is important because existing administrative records or currently available studies often have to be used to make legal decisions and public policy; it is not possible to undertake further research. Gastwirth (2000) has also edited a collection of essays on statistical science in the courtroom, some of which are directly relevant to this current book and will be referred to as appropriate.

A collection of papers on *Statistics and Public Policy* has been edited by Fairley and Mosteller (1977). One issue in the book which relates to a particularly infamous case, the *Collins* case, is discussed in detail later (Section 4.4). Other articles concern policy issues and decision making.

The remit of this book is one which is not covered by these others in great detail. The use of statistics in forensic science in general is discussed in a collection of essays edited by Aitken and Stoney (1991). The remit of this book is to describe statistical procedures for the evaluation of evidence for forensic scientists. This will be done primarily through a modern Bayesian approach. This approach has its origins in the work of I.J. Good and A.M. Turing as code-breakers at Bletchley Park during World War II. A brief review of the history is given in Good (1991). An essay on the topic of probability and the weighing of evidence is Good (1950). This also refers to entropy (Shannon, 1948), the expected amount of information from an experiment, and Good remarks that the expected weight of evidence in favour of a hypothesis H and against its complement $\bar{H}$ (read as 'H-bar') is equal to the difference of the entropies assuming H and $\bar{H}$, respectively. A brief discussion of a frequentist approach and the problems associated with it is given in Section 4.6.

It is of interest to note that a high proportion of situations involving the formal objective presentation of statistical evidence uses the frequentist approach with tests of significance (Fienberg and Schervish, 1986). However, Fienberg and Schervish go on to say that the majority of examples cited for the use of the Bayesian approach are in the area of identification evidence. It is this area which is the main focus of this book and it is Bayesian analyses which will form the basis for the evaluation of evidence as discussed here. Examples of the applications of such analyses to legal matters include Cullison (1969), Fairley (1973), Finkelstein and Fairley (1970, 1971), Lempert (1977), Lindley (1977a, b), Fienberg and Kadane (1983) and Anderson and Twining (1998).

Another approach which will not be discussed here is that of Shafer (1976, 1982). This concerns so-called *belief functions* (see Section 4.1). The theory of belief functions is a very sophisticated theory for assessing uncertainty which endeavours to answer criticisms of both the frequentist and Bayesian approaches to inference. Belief functions are non-additive in the sense that belief in an event A (denoted Bel(A)) and belief in the opposite of A (denoted Bel($\bar{A}$)) do not sum to 1. See also Shafer (1978) for a historical discussion of non-additivity. Further discussion is beyond the scope of this book. Practical applications are few. One such, however, is to the evaluation of evidence concerning the refractive index of glass (Shafer, 1982).

It is very tempting when assessing evidence to try to determine a value for the probability of the so-called *probandum* of interest (or the ultimate issue) such as the guilt of a suspect, or a value for the odds in favour of guilt, and perhaps even to reach a decision regarding the suspect's guilt. However, this is the role of the jury and/or judge. It is not the role of the forensic scientist or statistical expert witness to give an opinion on this (Evett, 1983). It is permissible for the scientist to say that the evidence is 1000 times more likely, say, if the suspect is guilty than if he is innocent. It is not permissible to interpret this to say that, because of the evidence, it is 1000 times more likely that the suspect is guilty than innocent. Some of the difficulties associated with assessments of probabilities are discussed by Tversky and Kahneman (1974) and by Kahneman *et al.* (1982) and are further described in Section 3.3. An appropriate representation of probabilities is useful because it fits the analytic device most used by lawyers, namely the creation of a story. This is a narration of events 'abstracted from the evidence and arranged in a sequence to persuade the fact-finder that the story told is the most plausible account of "what really happened" that can be constructed from the evidence that has been or will be presented' (Anderson and Twining, 1998, p. 166). Also of relevance is Kadane and Schum (1996), which provides a Bayesian analysis of evidence in the *Sacco–Vanzetti* case (Sacco, 1969) based on subjectively determined probabilities and assumed relationships amongst evidential events. A similar approach is presented in Chapter 14.

1.3   UNCERTAINTY IN SCIENTIFIC EVIDENCE

Scientific evidence requires considerable care in its interpretation. Emphasis needs to be put on the importance of asking the question 'what do the results mean in this particular case?' (Jackson, 2000). Scientists and jurists have to abandon the idea of absolute certainty in order to approach the identification process in a fully objective manner. If it can be accepted that nothing is absolutely certain then it becomes logical to determine the degree of confidence that may be assigned to a particular belief (Kirk and Kingston, 1964).

There are various kinds of problems concerned with the random variation naturally associated with scientific observations. There are problems concerned with the definition of a suitable reference population against which concepts of rarity or commonality may be assessed. There are problems concerned with the choice of a measure of the value of the evidence.

The effect of the random variation can be assessed with the appropriate use of probabilistic and statistical ideas. There is variability associated with scientific observations. Variability is a phenomenon which occurs in many places. People are of different sexes, determination of which is made at conception. People are of different heights, weights and intellectual abilities, for example. The variation in height and weight is dependent on a person's sex. In general, females tend to be lighter and shorter than males. However, variation is such that there can be tall, heavy females and short, light males. At birth, it is uncertain how tall or how heavy the baby will be as an adult. However, at birth, it is known whether the baby is a boy or a girl. This knowledge affects the uncertainty associated with the predictions of adult height and weight.

People are of different blood groups. A person's blood group does not depend on the age or sex of the person but does depend on the person's ethnicity. The refractive index of glass varies within and between windows. Observation of glass as to whether it is window or bottle glass will affect the uncertainty associated with the prediction of its refractive index and that of other pieces of glass which may be thought to be from the same origin.

It may be thought that, because there is variation in scientific observations, it is not possible to make quantitative judgements regarding any comparisons between two sets of observations. The two sets are either different or they are not, and there is no more to be said. However, this is not so. There are many phenomena which vary, but they vary in certain specific ways. It is possible to represent these specific ways mathematically. Several such ways, including various probability distributions, are introduced in Chapter 2. It is then possible to assess differences quantitatively and to provide a measure of uncertainty associated with such assessments.

It is useful to recognise the distinction between statistics and probability. Probability is a deductive process which argues from the general to the particular. Consider a fair coin, that is, one for which, when tossed, the probability of a head landing uppermost equals the probability of a tail landing uppermost equals 1/2. A fair coin is tossed ten times. Probability theory enables a determination to be made of the probability that there are three heads and seven tails, say. The general concept of a fair coin is used to determine something about the outcome of the particular case in which it was tossed ten times.

On the other hand, statistics is an inductive process which argues from the particular to the general. Consider a coin which is tossed ten times and there are three heads and seven tails. Statistics enables the question as to whether the coin is fair or not to be addressed. The particular outcome of three heads and seven tails in ten tosses is used to determine something about the general case of whether the coin was fair or not.

Fundamental to both statistics and probability is uncertainty. Given a fair coin, the numbers of heads and tails in ten tosses is uncertain. The probability associated with each outcome may be determined but the actual outcome itself cannot be predicted with certainty. Given the outcome of a particular sequence of ten tosses, information is then available about the fairness or otherwise of the coin. For example, if the outcome were ten heads and no tails, one may believe that the coin is double-headed but it is not certain that this is the case. There is still a non-zero probability (1/1024) that ten tosses of a fair coin will result in ten heads. Indeed, this has occurred in the first author's experience. A class of some 130 students were asked to each toss a coin ten times. One student tossed ten consecutive heads from what it is safe to assume was a fair coin. The probability of this happening is $1 - (1 - \frac{1}{1024})^{130} = 0.12$.

1.3.1   The frequentist method

Consider a consignment of compact discs containing N discs. The consignment is said to be of size N. It is desired to make inferences about the proportion θ ($0 \leq \theta \leq 1$) of the consignment which is pirated. It is not practical to inspect the whole consignment so a sample of size n, where $n < N$, is inspected.

The frequentist method assumes that the proportion θ of the consignment which is pirated is unknown but fixed. The data, the number of discs in the sample which are pirated, are variable. A so-called *confidence interval* is calculated. The name *confidence* is used since no probability can be attached to the uncertain event that the interval contains θ. These ideas are discussed further in Chapter 5.

The frequentist approach derives its name from the relative frequency definition of probability. The probability that a particular event, A say, occurs is defined as the relative frequency of the number of occurrences of event A compared with the total number of occurrences of all possible events, over a long run of observations, conducted under identical condition of all possible events.

For example, consider tossing a coin n times. It is not known if the coin is fair. The outcomes of the n tosses are to be used to determine the probability of a head occurring on an individual toss. There are two possible outcomes, heads (H) and tails (T). Let $n(H)$ be the number of H and $n(T)$ be the number of T such that $n(H) + n(T) = n$. Then the probability of tossing a head on an individual toss of the coin is defined as the limit as $n \to \infty$ of the fraction $n(H)/n$. The frequentist approach relies on a belief in the long-run repetition of trials under identical conditions. This is an idealised situation, seldom, if ever, realised in practice. More discussion on the interpretation of such a result is given in Section 4.6.

The way in which statistics and probability may be used to evaluate evidence is the theme of this book. Care is required. Statisticians are familiar with variation, as are forensic scientists who observe it in the course of their work. Lawyers, however, prefer certainties. A defendant is found *guilty* or *not guilty* (or also, in Scotland, *not proven*). The scientist's role is to testify as to the worth of the evidence, the role of the statistician and this book is to provide the scientist with a quantitative measure of this worth. It is shown that there are few forms of evidence that are so definite that statistical treatment is neither needed nor desirable. It is up to other people (the judge and/or the jury) to use this information as an aid to their deliberations. It is for neither the statistician nor the scientist to pass judgement. The scientist's role in court is restricted to giving evidence in terms of what has been called *cognation* (Kind, 1994): the problem of whether or not the evidence from two places (e.g., the scene of the crime and the suspect) has the same origin.

The use of these ideas in forensic science is best introduced through the discussion of several examples. These examples will provide a constant theme throughout the book. Populations from which the criminal may be thought to have come, to which reference is made below, are considered in detail in Section 8.5, where they are called *relevant populations*. The value of evidence is measured by a statistic known as the *likelihood ratio*, and its logarithm. These are introduced in Sections 3.4 and 3.5.

1.3.2   Stains of body fluids

Example 1.1   A crime is committed. A bloodstain is found at the scene of the crime. All innocent explanations for the presence of the stain are eliminated.

A suspect is found. His DNA profile is identified and found to match that of the crime stain. What is the evidential value of this match? This is a very common situation, yet the answer to the question provides plenty of opportunity for discussion of the theme of this book.

Certain other questions need to be addressed before this particular one can be answered. Where was the crime committed, for example? Does it matter? Does the value of the evidence of the bloodstain change depending on where the crime was committed?

Apart from his DNA profile, what else is known about the criminal? In particular, is there any information, such as ethnicity, which may be related to his DNA profile? What is the population from which the criminal may be thought to have come? Could he be another member of the suspect's family?

Questions such as these and their effect on the interpretation and evaluation of evidence will be discussed in greater detail. First, consider only the evidence of the DNA profile in isolation and one particular locus, *LDLR*. Assume the crime was committed in Chicago and that there is eyewitness evidence that the criminal was a Caucasian. Information is available to the investigating officer about the genotypic distribution for the *LDLR* locus in Caucasians in Chicago and is given in Table 1.1. The information about the location of the crime and the ethnicity of the criminal is relevant. Genotypic frequencies vary across locations and among ethnic groups. A suspect is identified. For locus *LDLR* the genotype of the crime stain and that of the suspect match. The investigating officer knows a little about probability and works out that the probability of two people chosen at random and unrelated to the suspect having matching alleles, given the figures in Table 1.1

$$0.188^2 + 0.321^2 + 0.491^2 = 0.379 \qquad (1.1)$$

(see Section 4.5). He is not too sure what this result means. Is it high, and is a high value incriminating for the suspect? Is it low, and is a low value incriminating? In fact, a low value is more incriminating than a high value.

He thinks a little more and remembers that the genotypes not only match but also are both of type *BB*. The frequencies of genotypes *AA* and *AB* are not relevant. He works out the probability that two people chosen at random both have genotype *BB* as

$$0.321^2 = 0.103$$

Table 1.1   Genotypic frequencies for locus *LDLR* amongst Caucasians in Chicago based on a sample of size 200 (from Johnson and Peterson, 1999)

Genotype	AA	BB	AB
Frequency (%)	18.8	32.1	49.1

(see Section 4.5). He is still not too sure what this means but feels that it is more representative of the information available to him than the previous probability, since it takes account of the actual genotypes of the crime stain and the suspect.

The genotype of the crime stain for locus *LDLR* is *BB*. The genotype of the suspect is also *BB* (if it were not he would not be a suspect). What is the value of this evidence? The discussion above suggests various possible answers.

1. The probability that two people chosen at random have the same genotype for locus *LDLR*. This is 0.379.
2. The probability that two people chosen at random have the same, pre-specified, genotype. For genotype *BB* this is 0.103.
3. The probability that one person chosen at random has the same genotype as the crime stain. If the crime stain is of group *BB*, this probability is 0.321, from Table 1.1.

The phrase *at random* is taken to include the caveat that the people chosen are unrelated to the suspect.

The relative merits of these answers will be discussed in Section 4.5 for (1) and (2) and Section 9.2 for (3).

1.3.3 Glass fragments

The previous section discussed an example of the interpretation of the evidence of DNA profiling. Consider now an example concerning glass fragments and the measurement of the refractive index of these.

Example 1.2 As before, consider the investigation of a crime. A window has been broken during the commission of the crime. A suspect is found with fragments of glass on his clothing, similar in refractive index to the broken window. Several fragments are taken for investigation and their refractive index measurements taken.

Note that there is a difference here from Example 1.1 where it was assumed that the crime stain had come from the criminal and been transferred to the crime scene. In Example 1.2 glass is transferred from the crime scene to the criminal. Glass on the suspect need not have come from the scene of the crime; it may have come from elsewhere and by perfectly innocent means. This is an asymmetry associated with this kind of evidence. The evidence is known as *transfer evidence*, as discussed in Section 1.1, because evidence (e.g., blood or glass fragments) has been transferred from the criminal to the scene or vice versa. Transfer from the criminal to the scene has to be considered differently from evidence transferred from the scene to the criminal. A full discussion of this is given in Chapter 8.

Comparison in Example 1.2 has to be made between the two sets of fragments on the basis of their refractive index measurements. The evidential value of the outcome of this comparison has to be assessed. Notice that it is assumed that none of the fragments has any distinctive features and comparison is based only on the refractive index measurements.

Methods for evaluating such evidence were discussed in many papers in the late 1970s and early 1980s (Evett, 1977, 1978; Evett and Lambert, 1982, 1984, 1985; Grove, 1981, 1984; Lindley, 1977a; Seheult, 1978; Shafer, 1982). These methods will be described as appropriate in later chapters. Knowledge-based computer systems such as CAGE (Computer Assistance for Glass Evidence) and CAGE 2000 have been developed (Curran *et al.*, 2000; Hicks, 2004). See also Curran *et al.* (2000) for a review of current practice in the forensic interpretation of glass evidence.

Evett (1977) gave an example of the sort of problem which may be considered and developed a procedure for evaluating the evidence which mimicked the interpretative thinking of the forensic scientist of the time. The case is an imaginary one. Five fragments from a suspect are to be compared with ten fragments from a window broken at the scene of a crime. The values of the refractive index measurements are given in Table 1.2. The rather arbitrary and hybrid procedure developed by Evett is a two-stage one. It is described here briefly; more details are given in Chapter 4. While it follows the thinking of the forensic scientist, there are interpretative problems, which are described here, in attempting to provide due weight to the evidence. An alternative approach which overcomes these problems is described in Chapter 10.

The first stage is known as the *comparison* stage. The two sets of measurements are compared. The comparison takes the form of the calculation of a statistic, D say. This statistic provides a measure of the difference, known as a *standardised* difference, between the two sets of measurements which takes account of the natural variation there is in the refractive index measurements of glass fragments from within the same window. If the absolute value of D is less than (or equal to) some pre-specified value, known as a *threshold* value, then the two sets of fragments are deemed to be similar and the second stage is implemented. If the absolute value of D is greater than the threshold value then the two sets of fragments are deemed to be dissimilar. The two sets of fragments are then deemed to have come from different sources and the second stage is not implemented. (Note the use here of the word *statistic*, which in this context

Table 1.2   Refractive index measurements

Measurements	1.518 44	1.518 48	1.518 44	1.518 50	1.518 40
from the window	1.518 48	1.518 46	1.518 46	1.518 44	1.518 48
Measurements	1.518 48	1.518 50	1.518 48	1.518 44	1.518 46
from the suspect					

can be thought of simply as a function of the observations.) A classical example of such an approach is the use of the Student t-test or the modified Welch test for the comparison of means (Welch, 1937; Walsh *et al.*, 1996; Curran *et al.*, 2000); more details are given in Chapter 10.

The second stage is known as the *significance* stage. This stage attempts to determine the significance of the finding from the first stage that the two sets of fragments were similar. The significance is determined by calculating the probability of the result that the two sets of fragments were found to be similar, under the assumption that the two sets had come from different sources. If this probability is very low then this assumption is deemed to be false. The fragments are then assumed to come from the same source, an assumption which places the suspect at the crime scene.

The procedure can be criticised on two points. First, in the comparison stage the threshold provides a qualitative step which may provide very different outcomes for two different pairs of observations. One pair of sets of fragments may provide a value of D which is just below the threshold, whereas the other pair may provide a value of D just above the threshold. The first pair will proceed to the significance stage, the second stage will not. Yet, the two pairs may have measurements which are close together. The difference in the consequences is greater than the difference in the measurements' merits (such an approach is called a *fall-off-the-cliff* effect; see Robertson and Vignaux, 1995a). A better approach, which is described in Chapter 10, provides a measure of the value of the evidence which decreases as the distance between the two sets of measurements increases, subject, as explained later, to the rarity or otherwise of the measurements.

The second criticism is that the result is difficult to interpret. Because of the effect of the comparison stage, the result is not simply the probability of the evidence, assuming the two sets of fragments came from different sources. A reasonable interpretation, as will be explained in Section 3.5, of the value of the evidence is the effect that it has on the odds in favour of guilt of the suspect. In the two-stage approach this effect is difficult to measure. The first stage discards certain sets of measurements which may have come from the same source and may not discard other sets of measurements which have come from different sources. The second stage calculates a probability, not of the evidence but of that part of the evidence for which D was not greater than the threshold value, assuming the two sets came from different sources. It is necessary to compare this probability with the probability of the same result, assuming the two sets came from the same source. There is also an implication in the determination of the probability in the significance stage that a small probability for the evidence, assuming the two sets came from different sources, means that there is a large probability that the two sets came from the same source. This implication is unfounded; see Section 3.3.1.

A review of the two-stage approach and the development of a Bayesian approach is provided by Evett (1986).

As with DNA profiling, there are problems associated with the definition of a suitable population from which probability distributions for refractive measurements may be obtained; see, for example, Walsh and Buckleton (1986).

These examples have been introduced to provide a framework within which the evaluation of evidence may be considered. In order to evaluate evidence, something about which there is much uncertainty, it is necessary to establish a suitable terminology and to have some method for the assessment of uncertainty. First, some terminology will be introduced, followed by the method for the assessment of uncertainty. This method is probability. The role of uncertainty, as represented by probability, in the assessment of the value of scientific evidence will form the basis of the rest of this chapter. A commentary on so-called *knowledge management*, of which this is one part, has been given by Evett (1993b).

1.4   TERMINOLOGY

It is necessary to have clear definitions of certain terms. The crime scene and suspect materials have fundamentally different roles. Determination of the probability of a match between two randomly chosen sets of materials is not the important issue. One set of materials, crime scene or suspect, can be assumed to have a known source. It is then required to assess the probability of the corresponding item, suspect or crime scene, matching in some sense, the known set of materials, under two competing propositions. Examples 1.1 and 1.2 serve to illustrate this.

Example 1.1 (continued)   A crime is committed. A bloodstain is found at the scene of the crime. All innocent explanations for the presence of the stain are eliminated. A suspect is found. His DNA profile is found to match that of the crime stain. The crime scene material is the DNA profile of the crime stain. The suspect material is the DNA profile of the suspect.

Example 1.2 (continued)   As before, consider the investigation of a crime. A window has been broken during the commission of the crime. Several fragments are taken for investigation and measurements made of their refractive indices. These fragments, as their origin is known, are sometimes known as control fragments and the corresponding measurements are known as control measurements. A suspect is found. Fragments of glass are found on his person and measurements of the refractive indices of these fragments are made. These fragments (measurements) are sometimes known as recovered fragments (measurements). Their origin is not known. They may have come from the window broken at the crime scene but need not necessarily have done so.

The crime scene material is the fragments of glass and the measurements of refractive index of these at the scene of the crime. The suspect material is the fragments of glass found on the suspect and their refractive index measurements.

Evidence such as this where the source is known and is of a bulk form will be known as *source* or *bulk form* evidence. These fragments of glass will be known as source or bulk fragments and the corresponding measurements will be known as source or bulk measurements, as their source is known and they have been taken from a bulk form of glass, namely a window (Stoney, 1991a). In general, only the term *source* will be used when referring to this type of evidence.

A suspect is found. Fragments of glass are found on his person and measurements of the refractive indices of these fragments are made. Evidence such as this where the evidence has been received and is in particulate form will be known as *receptor* or *transferred particle* evidence. These fragments (measurements) of glass in this example will be known as *receptor* or *transferred particle* fragments (measurements). Their origin is not known. They have been 'received' from somewhere by the suspect. They are particles which have been transferred to the suspect from somewhere. They may have come from the window broken at the crime scene but need not necessarily have done so.

There will also be occasion to refer to the location at which, or the person on which, the evidence was found. Evidence found at the scene of the crime will be referred to as *crime* evidence. Evidence found on the suspect's clothing or in the suspect's natural environment, such as his home, will be referred to as *suspect* evidence. Note that this does not mean that the evidence itself is of a suspect nature!

Locard's principle (see Section 1.1) is that every contact leaves a trace. In the examples above the contact is that of the criminal with the crime scene. In Example 1.1, the trace is the bloodstain at the crime scene. In Example 1.2, the trace is the fragments of glass that would be removed from the crime scene by the criminal (and, later, hopefully, be found on his clothing).

The evidence in both examples is *transfer evidence* (see Section 1.1) or sometimes *trace evidence*. Material has been transferred between the criminal and the scene of the crime. In Example 1.1 blood has been transferred *from* the criminal *to* the scene of the crime. In Example 1.2 fragments of glass *may* have been transferred *from* the scene of the crime *to* the criminal. The direction of transfer in these two examples is different. Also, in the first example the blood at the crime scene has been identified as coming from the criminal. Transfer is known to have taken place. In the second example it is not known that glass has been transferred from the scene of the crime to the criminal. The suspect has glass fragments on his clothing but these need not necessarily have come from the scene of the crime. Indeed, if the suspect is innocent and has no connection with the crime scene the fragments will not have come from the crime scene.

Other terms have been suggested and these suggestions provide a potential source of confusion. For example, the term *control* has been used to indicate the material whose origin is known. This can be either the bulk (source) form of the material or the transferred particle form, however. Similarly, the term *recovered* has been used to indicate the material whose origin is unknown.

Again, this can be either the bulk (source) form or the transferred particle form, depending on which has been designated the control form. Alternatively, *known* has been used for 'control' and *questioned* has been used for 'recovered'. See, for example, Brown and Cropp (1987). Also Kind *et al.* (1979) used *crime* for material known to be associated with a crime and *questioned* for material thought to be associated with a crime. All these terms are ambiguous. The need to distinguish the various objects or persons associated with a crime was pointed out by Stoney (1984a).

Consider the recovery of a jumper of unknown origin from a crime scene (Stoney, 1991a). A suspect is identified and fibres similar in composition to those on the jumper are found at his place of residence. The two parts of the transfer evidence are the jumper found at the crime scene and the fibres found at the suspect's place of residence. The bulk (source) form of the material is the jumper. The transferred particle form of the material is the fibres. However, the jumper may not be the control evidence. It is of unknown origin. The fibres, the transferred particle form, could be considered the control as their source is known, in the sense that they have been found at the suspect's place of residence. They are associated with the suspect in a way the jumper, by itself, is not. Similarly, the fibres have a known origin, and the jumper has a questioned origin.

Definitions given in the context of fibres evidence are provided by Champod and Taroni (1999). The object or person on which traces have been recovered is defined as the *receptor* and the object or person that could be the source (or one of the sources) of the traces, and which is the origin of the material defined as *known material*, is defined as the *known source*.

Material will be referred to as source form where appropriate and to receptor or transferred particle form where appropriate. This terminology conveys no information as to which of the two forms is of known origin. There are two possibilities for the origin of the material which is taken to be known: the scene of the crime and the suspect. One or other is taken to be known, the other to be unknown. The two sets of material are compared by determining two probabilities, both of which depend on what is assumed known and provide probabilities for what is assumed unknown. The two possibilities for the origin of the material which is taken to be known are called *scene-anchored* and *suspect-anchored*, where the word 'anchored' refers to that which is assumed known (Stoney, 1991a). The distinction between scene-anchoring and suspect-anchoring is important when determining so-called *correspondence probabilities* (Section 8.2); it is not so important in the determination of likelihood ratios. Reference to form (source or receptor, bulk or transferred particle) is a reference to one of the two parts of the evidence. Reference to anchoring (scene or suspect) is a reference to a perspective for the evaluation of the evidence.

Other terms such as control, known, recovered, questioned, will be avoided as far as possible. However, it is sometimes useful to refer to a sample of material found at the scene of a crime as the *crime sample* and to a sample of material found on or about a suspect as the *suspect sample*. This terminology reflects the site at which the material was found. It does not indicate the kind of material (bulk or transferred particle form) or the perspective (scene- or suspect-anchored) by which the evidence will be evaluated.

1.5   TYPES OF DATA

A generic name for observations which are made on items of interest, such as bloodstains or refractive indices of glass, is *data*. There are different types of data and some terminology is required to differentiate amongst them. For example, consider the *ABO* blood grouping system. The observations of interest are the blood groups of the crime stain and of the suspect. These are not quantifiable. There is no numerical significance which may be attached to these. The blood group is a qualitative characteristic. As such, it is an example of so-called *qualitative data*. The observation of interest is a quality, the blood group, which has no numerical significance. The different blood groups are sometime known as *categories*. The assignation of a person to a particular category is called a *classification*. A person may be said to be classified into one of several categories (see the discussion on the definition of *identification* in Section 1.2).

It is not possible to order blood groups and say that one is larger or smaller than another. However, there are other qualitative data which do have a natural ordering, such as the level of burns on a body. There is not a numerical measure of this but the level of burns may be classified as first, second or third degree, for example. Qualitative data which have no natural ordering are known as *nominal* data. Qualitative data to which a natural ordering may be attached are known as *ordinal* data. An ordinal characteristic is one in which there is an underlying order even though it is not quantifiable. Pain is one such characteristic; level of trauma may be ordered as none, slight, mild, severe, very severe. The simplest case of nominal data arises when an observation (e.g., of a person's blood group) may be classified into one of only two possible categories. For example, consider the old methodology of blood typing, such as the Kell genetic marker system where a person may be classified as either Kell+ or Kell−. Such data are known as *binary*. Alternatively, the variable of interest, here the Kell system, is known as *dichotomous*.

Other types of data are known as *quantitative* data. These may be either counts (known as *discrete* data, since the counts take discrete, integer, values) or measurements (known as *continuous* data, since the measurements may take any value on a continuous interval).

A violent crime involving several people, victims and offenders, may result in much blood being spilt and many stains from each of several DNA profiles being identified. Then the numbers of stains for each of the different profiles are examples of discrete, quantitative data.

The refractive indices and elemental concentrations of glass fragments are examples of continuous measurements. In practice, variables are rarely truly continuous because of the limits imposed by the sensitivity of the measuring instruments. Refractive indices may be measured only to a certain accuracy.

Observations, or data, may thus be classified as qualitative or quantitative. Qualitative data may be classified further as nominal or ordinal and quantitative data may be classified further as discrete or continuous.

1.6   PROBABILITY

1.6.1   Introduction

The interpretation of scientific evidence may be thought of as the assessment of a comparison. This comparison is that between evidential material found at the scene of a crime (denote this by M_c) and evidential material found on a suspect, a suspect's clothing or around his environment (denote this by M_s). Denote the combination by $M = (M_c, M_s)$. As a first example, consider the bloodstains of Example 1.1. The crime stain is M_c, the receptor or transferred particle form of the evidential material; M_s is the genotype of the suspect and is the source form of the material. From Example 1.2, suppose glass is broken during the commission of a crime. M_c would be the fragments of glass (the source form of the material) found at the crime scene, M_s would be fragments of glass (the receptor or transferred particle form of the material) found on the clothing of a suspect, and M would be the two sets of fragments.

Qualities (such as the genotypes) or measurements (such as the refractive indices of the glass fragments) are taken from M. Comparisons are made of the source form and the receptor form. Denote these by E_c and E_s, respectively, and let $E = (E_c, E_s)$ denote the combined set. Comparison of E_c and E_s is to be made and the assessment of this comparison has to be quantified. The totality of the evidence is denoted Ev and is such that $Ev = (M, E)$.

Statistics has developed as a subject, one of whose main concerns is the quantification of the assessments of comparisons. The performance of a new treatment, drug or fertiliser has to be compared with that of an old treatment, drug or fertiliser, for example. It seems natural that statistics and forensic science should come together. Two samples, the crime and the suspect sample, are to be compared. Yet, apart from the examples discussed in Chapter 4, it is only recently that this has happened. As discussed in Section 1.2, there have been several books describing the role of statistics in the law. Until the first edition of this book there had been none concerned with statistics and the evaluation of scientific evidence. Two factors may have been responsible for this.

First, there was a lack of suitable data from relevant populations. There was a consequential lack of a baseline against which measures of typicality of any characteristics of interest might be determined. One exception are the reference data which have been available for many years on blood group frequencies among certain populations. Not only has it been possible to say that the suspect's blood group matched that of a stain found at the scene of a crime but also that this group is only present in, say, 0.01% of the population. Now these have been superseded by databases of DNA profiles. Also, data collections exist for the refractive index of glass fragments found at random on clothing and for transfer and persistence parameters linked to glass evidence (Curran *et al.*, 2000). Contributions towards estimating the frequency of fibre types have also been published (Grieve and Biermann, 1997; Grieve, 2000a, b; Grieve *et al.*, 2001). There is also much information about the frequency of characteristics in DNA profiles. Announcements of population data are published regularly in peer-reviewed journals such as *Forensic Science International* and the *Journal of Forensic Sciences*.

Secondly, the approach adopted by forensic scientists in the assessment of their evidence has been difficult to model. The approach has been one of comparison and significance. Characteristics of the crime and suspect samples are compared. If the examining scientists believe them to be similar, the typicality, and hence the significance of the similarity, of the characteristics is then assessed. This approach is what has been modelled by the two-stage approach of Evett (1977), described briefly in Section 1.3.3 and in fuller detail in Chapter 4. However, interpretation of the results provided by this approach is difficult.

Then, in a classic paper, Lindley (1977a) described an approach which was easy to justify, to implement and to interpret. It combined the two parts of the two-stage approach into one statistic and is discussed in detail in Section 10.2. The approach compares two probabilities, the probability of the evidence, assuming one proposition about the suspect to be true (that he is guilty, for example) and the probability of the evidence, assuming another proposition about the suspect to be true (that he is innocent, for example). (Note: some people use the term *hypothesis* rather than *proposition*; the authors will endeavour to use the term *proposition* as they believe this reduces the risk of confusion of their ideas with the ideas of *hypothesis testing* associated with the alternative term.) This approach implies that it is not enough for a prosecutor to show that evidence is unlikely if a suspect is innocent. The evidence has also to be more likely if the suspect is guilty. Such an approach had a good historical pedigree (Good, 1950; see also Good, 1991, for a review) yet it had received very little attention in the forensic science literature, even though it was clearly proposed at the beginning of the twentieth century (Taroni *et al.*, 1998). It is also capable of extension beyond the particular type of example discussed by Lindley, as will be seen by the discussion throughout this book, for example in Section 10.4.

However, in order to proceed it is necessary to have some idea about how uncertainty can be measured. This is best done through probability (Lindley, 1991, 1998).

1.6.2   A standard for uncertainty

An excellent description of probability and its role in forensic science has been given by Lindley (1991). Lindley's description starts with the idea of a standard for uncertainty. He provides an analogy using the concept of balls in an urn. Initially, the balls are of two different colours, black and white. In all other respects – size, weight, texture etc. – they are identical. In particular, if one were to pick a ball from the urn, without looking at its colour, it would not be possible to tell what colour it was. The two colours of balls are in the urn in proportions b and w for black and white balls, respectively, such that $b + w = 1$. For example, if there were 10 balls in the urn of which 6 were black and 4 were white, then $b = 0.6, w = 0.4$ and $b + w = 0.6 + 0.4 = 1$.

The urn is shaken up and the balls thoroughly mixed. A ball is then drawn from the urn. Because of the shaking and mixing it is assumed that each ball, regardless of colour, is equally likely to be selected. Such a selection process, in which each ball is equally likely to be selected, is known as a random selection, and the chosen ball is said to have been chosen *at random*.

The ball, chosen at random, can be either black, an event which will be denoted B, or white, an event which will be denoted W. There are no other possibilities; one and only one of these two events has to occur. The uncertainty of the event B, the drawing of a black ball, is related to the proportion b of black balls in the urn. If b is small (close to zero), B is unlikely. If b is large (close to 1), B is likely. A proportion b close to $1/2$ implies that B and W are about equally likely. The proportion b is referred to as the probability of obtaining a black ball on a single random draw from the urn. In a similar way, the proportion w is referred to as the probability of obtaining a white ball on a single random draw from the urn.

Notice that on this simple model probability is represented by a proportion. As such it can vary between 0 and 1. A value of $b = 0$ occurs if there are no black balls in the urn and it is, therefore, impossible to draw a black ball from the urn. The probability of obtaining a black ball on a single random draw from the urn is zero. A value of $b = 1$ occurs if all the balls in the urn are black. It is certain that a ball drawn at random from the urn will be black. The probability of obtaining a black ball on a single random draw from the urn is one. All values between these extremes of 0 and 1 are possible (by considering very large urns containing very large numbers of balls).

A ball has been drawn at random from the urn. What is the probability that the selected ball is black? The event B is the selection of a black ball. Each ball has an equal chance of being selected. The colours black and white of the balls are in the proportions b and w. The proportion, b, of black balls corresponds to the probability that a ball, drawn in the manner described (i.e., at random) from the urn is black. It is then said that the probability a black ball is drawn from the urn, when selection is made at random, is b. Some notation is needed to denote the probability of an event. The probability of B, the drawing of a black ball, is denoted $Pr(B)$, and similarly $Pr(W)$ denotes the probability of the drawing of a white ball. Then it can be written that $Pr(B) = b, Pr(W) = w$. Note that

$$Pr(B) + Pr(W) = b + w = 1.$$

This concept of balls in an urn can be used as a reference for considering uncertain events. Let R denote the uncertain event that the England football team will win the next European football championship. Let B denote the uncertain event that a black ball will be drawn from the urn. A choice has to be made between R and B and this choice has to be ethically neutral. If B is chosen and a black ball is indeed drawn from the urn then a prize is won. If R is chosen and England do win the championship the same prize is won. The proportion b of black balls in the urn is known in advance. Obviously, if $b = 0$ then R is the better choice, assuming, of course, that England do have some non-zero chance of winning the championship. If $b = 1$ then B is the better choice. Somewhere in the interval $[0, 1]$, there is a value of b, b_0 say, where the choice does not matter. One is indifferent as to whether R or B is chosen. If B is chosen, $Pr(B) = b_0$. Then it is said that $Pr(R) = b_0$, also. In this way the uncertainty in relation to any event can be measured by a probability b_0, where b_0 is the proportion of black balls which leads to indifference between the two choices, namely the choice of drawing a black ball from the urn and the choice of the uncertain event in whose probability one is interested.

Notice, though, that there is a difference between these two probabilities. By counting, the proportion of black balls in the urn can be determined precisely. Probabilities of other events such as the outcome of the toss of a coin or the roll of a die are also relatively straightforward to determine, based on assumed physical characteristics such as fair coins and fair dice. Let H denote the event that when a coin is tossed it lands head uppermost. Then, for a fair coin, in which the outcomes of a head H and a tail T at any one toss are equally likely, the probability the coin comes down head uppermost is $1/2$. Let F denote the event that when a die is rolled it lands 4 uppermost. Then, for a fair die, in which the outcomes $1, 2, \ldots, 6$ at any one roll are equally likely, the probability the die lands 4 uppermost is $1/6$.

Probabilities relating to the outcomes of sporting events, such as football matches or championships or horse races, or to the outcome of a civil or criminal trial, are rather different in nature. It may be difficult to decide on a particular value for b_0. The value may change as evidence accumulates such as the results of particular matches and the fitness or otherwise of particular players, the fitness of horses, the identity of the jockey or the going of the race track. Also, different people may attach different values to the probability of a particular event. These kinds of probability are sometimes known as *subjective* or *personal* probabilities; see de Finetti (1931), Good (1959), Savage (1954) and DeGroot (1970). Another term is *measure of belief*, since the probability may be thought to provide a measure of one's belief in a particular event. Despite these difficulties, the arguments concerning probability still hold. Given an uncertain event R, the probability of $R, Pr(R)$, is defined as the proportion of balls b_0 in the urn such that if one had to choose between B (the event that a black ball was chosen) where $Pr(B) = b_0$ and R then one would be indifferent as to which one was chosen. There are difficulties, but the point of importance is that a standard for probability exists. A comment on subjective probabilities is given in Section 9.5.5. A use of probability as a measure of belief is described in Section 9.5, where it is used to represent *relevance*. The differences and similarities in the two kinds of probability discussed above and their ability to be combined has been referred to as a *duality* (Hacking, 1975).

It is helpful also to consider two quotes concerning the relationship amongst probability, logic and consistency, both from Ramsey (1931).

> We find, therefore, that a precise account of the nature of partial beliefs reveals that the laws of probability are laws of consistency, an extension to partial beliefs of formal logic, the logic of consistency. They do not depend for their meaning on any degree of belief in a proposition being uniquely determined as the rational one; they merely distinguish those sets of beliefs which obey them as consistent ones. (p. 182)

> We do not regard it as belonging to formal logic to say what should be a man's expectation of drawing a white or black ball from an urn; his original expectations may within the limits of consistency be any he likes; all we have to point out is that if he has certain expectations he is bound in consistency to have certain others. This is simply bringing probability into line with ordinary formal logic, which does not criticize premises but merely declares that certain conclusions are the only ones consistent with them. (p. 189)

1.6.3 Events

The outcome of the drawing of a ball from the urn was called an event. If the ball was black, the event was denoted B. If the ball was white, the event was denoted W. It was not certain which of the two events would happen: would the ball be black, event B, or white, event W? The degree of uncertainty of the event (B or W) was measured by the proportion of balls of the appropriate colour (B or W) in the urn and this proportion was called the probability of the event (B or W). In general, for an event R, $Pr(R)$ denotes the probability that R occurs.

Events can be events which may have happened (past events), events which may be relevant at the present time (present events) and events which may happen in the future (future events). There is uncertainty associated with each of these. In each case, a probability may be associated with the event.

- Past event: a crime is committed and a bloodstain of a particular type is found at the crime scene. A suspect is found. The event of interest is that the suspect left the stain at the crime scene. Though the suspect either did or did not leave the stain, the knowledge of it is uncertain and hence the event can have a probability associated with it.

- Present event: a person is selected. The event of interest is that he is of blood group *O*. Again, before the result of a blood test is available, this knowledge is uncertain.

- Future event: The event of interest is that it will rain tomorrow.

All of these events are uncertain and have probabilities associated with them. Notice, in particular, that even if an event has happened, there can still be uncertainty associated with it. The probability the suspect left the stain at the crime scene requires consideration of many factors, including the possible location of the suspect at the crime scene and the properties of transfer of blood from a person to a site. With reference to the blood group, consideration has to be given to the proportion of people in some population of blood group *O*. Probabilistic predictions are common with weather forecasting. Thus, it may be said, for example, that the probability it will rain tomorrow is 0.8 (though it may not always be obvious what this means).

1.6.4 Subjective probability

In forensic science, it is often emphasised that there is a real paucity of numerical data, so that the calculation of likelihood ratios (Section 3.4.1) is sometimes very difficult. Examples of this difficulty are the numerical assessments of parameters such as transfer or persistence probabilities (see Chapter 8) or even the relevance of a piece of evidence (see Section 9.5). The Bayesian approach considers probabilities as measures of belief (also called subjective probabilities) since such probabilities may be thought of as measures of one's belief in the occurrence of a particular event. The approach enables the combination of the objective probabilities, based on data, and subjective probabilities, for which the certified knowledge and experience of the forensic scientist may assist in the provision of estimates. Jurists are also interested in probability calculations using subjective probabilities, notably probabilities associated with the credibility of witnesses and the conclusions that might be drawn from their testimony.

From a formal point of view, the frequentist definition of probability involves a long sequence of repetitions of a given situation, under identical conditions. Consider a sequence of N repetitions in which an event E occurs X times, where X is some value greater than or equal to 0 and less than or equal to N. The relative frequency X/N could vary in different sequences of N repetitions, but it is supposed that, in a sequence where the number N of repetitions grows indefinitely under identical conditions, the relative frequency tends to a definite limiting value. In a frequentist framework the probability of event E, denoted $Pr(E)$, is defined to be that limiting value. Note, as with the balls in the urn, that this is a value between 0 and 1. In reality, it is difficult, if not impossible, to maintain identical conditions between trials. Therefore, in anything other than idealised situations, such a definition of probability proves unworkable. For example, consider prediction of the unemployment rate for the following year. It is not possible to use the frequentist definition to determine the probability that the rate will lie between 3% and 4% of the workforce, since it is not possible to consider unemployment as a sequence of repetitions under identical conditions. Unemployment in the following year is a unique, one-time event (Berger, 1985).

Frequentist probabilities are objective probabilities. They are objective in the sense that there is a well-defined set of circumstances for the long-run repetition of the trials, such that the corresponding probabilities are well defined and that one's personal or subjective views will not alter the value of the probabilities. Each person considering these circumstances will provide the same values for the probabilities. The frequency model relates to a relative frequency obtained in a long sequence of trials, assumed to be performed in an identical manner, physically independent of each other. Such a circumstance has certain difficulties. If taken strictly, this point of view does not allow a statement of probability for any situation that does not happen to be embedded, at least conceptually, in a long sequence of events giving equally likely outcomes.

The underlying idea of subjective probability is that the probability an event happens reflects a measure of personal belief in the occurrence of the event. For example, a person may have a personal feeling that the unemployment rate will be between 3% and 4%, even though no frequency probability can be assigned to the event. There is nothing surprising about this. It is common to think in terms of personal probabilities all the time, such as when betting on the outcome of a football game or when stating the probability of rain tomorrow.

In many situations, law being a prime example, we cannot assume equally likely outcomes any more than we can count past occurrences of events to determine relative frequencies. The reason is that the events of interest, if they have occurred, have done so only once (Schum, 2000). So, a subjective probability is defined as 'a degree of belief (as actually held by someone based on his whole knowledge, experience, information) regarding the truth of a statement, or event E (a fully specified single event or statement whose truth or falsity is, for whatever reason, unknown to the person)' (de Finetti, 1968, p. 45). There are three factors to consider for this probabilistic assessment. First, it depends on the available information. Second, it may change as the information changes. Third, it may vary amongst individuals because different individuals may have different information or assessment criteria. The only constraint in such an assessment is that it must be what is known as *coherent*. Coherence may be understood through consideration of subjective probability in terms of betting, in particular on a horse race. For the probabilities on winning for each horse in a race to be coherent, the sum of the probabilities over all the horses in the race must be 1 (Taroni *et al.*, 2001).

Under either definition (frequentist or Bayesian), probability takes a value between 0 and 1. Events or parameters of interest, in a wide range of academic fields (such as history, theology, law, forensic science), are usually not the result of repetitive or replicable processes. These events are singular, unique or one of a kind. It is not possible to repeat the events under identical conditions and tabulate the number of occasions on which some past event actually occurred. The use of subjective probabilities allows us to consider probability for events in situations such as these.

For a historical and philosophical definition of subjective probabilities and a commentary on the work of statisticians, de Finetti and Savage, working in this field in the middle of the twentieth century, see Lindley (1980) and Taroni *et al.* (2001).

1.6.5   Laws of probability

There are several laws of probability which describe the values which probability may take and how probabilities may be combined. These laws are given here, first for events which are not conditioned on any other information and then for events which are conditioned on other information.

The first law of probability has already been suggested implicitly in the context of proportions.

First law of probability   Probability can take any value between 0 and 1, inclusive, and only those values. Let R be any event and let $Pr(R)$ denote the probability that R occurs. Then $0 \leq Pr(R) \leq 1$. For an event which is known to be impossible, the probability is zero. Thus if R is impossible, $Pr(R) = 0$. This law is sometimes known as the *convexity rule* (Lindley, 1991).

Consider the hypothetical example of the balls in the urn of which a proportion b are black and a proportion w white, with no other colours present, such that $b + w = 1$. Proportions lie between 0 and 1; hence $0 \leq b \leq 1, 0 \leq w \leq 1$. For any event $R, 0 \leq Pr(R) \leq 1$. Consider B, the drawing of a black ball. If this event is impossible then there are no black balls in the urn and $b = 0$. This law is sometimes strengthened to say that a probability can *only* be 0 when the associated event is known to be impossible.

The first law concerns only one event. The next two laws, sometimes known as the second and third laws of probability, are concerned with combinations of events. Events combine in two ways. Let R and S be two events. One form of combination is to consider the event 'R and S', the event that occurs if and only if R and S both occcur. This is known as the *conjunction* of R and S.

Consider the roll of a six-sided fair die. Let R denote the throwing of an odd number. Let S denote the throwing of a number greater than 3 (i.e., a 4, 5 or 6). Then the event 'R and S' denotes the throwing of a 5.

Secondly, consider rolling two six-sided fair die. Let R denote the throwing of a six with the first die. Let S denote the throwing of a six with the second die. Then the event 'R and S' denotes the throwing of a double six.

The second form of combination is to consider the event 'R or S', the event that occurs if R or S (or both) occurs. This is known as the *disjunction* of R and S.

Consider again the roll of a single six-sided fair die. Let R, the throwing of an odd number (1, 3 or 5), and S, the throwing of a number greater than 3 (4, 5 or 6), be as before. Then 'R or S' denotes the throwing of any number other than a 2 (which is both even and less than 3).

Secondly, consider drawing a card from a well-shuffled pack of 52 playing cards, such that each card is equally likely to be drawn. Let R denote the event that the card drawn is a spade. Let S denote the event that the card drawn is a club. Then the event 'R or S' is the event that the card drawn is from a black suit.

The second law of probability concerns the disjunction 'R or S' of two events. Events are called *mutually exclusive* when the occurrence of one excludes the occurrence of the other. For such events, the conjunction 'R and S' is impossible. Thus $Pr(R$ and $S) = 0$.

Second law of probability   If R and S are mutually exclusive events, the probability of their disjunction 'R or S' is equal to the sum of the probabilities of R and S. Thus, for mutually exclusive events,

$$Pr(R \text{ or } S) = Pr(R) + Pr(S). \tag{1.2}$$

Consider the drawing of a card from a well-shuffled pack of cards with R defined as the drawing of a spade and S the drawing of a club. Then $Pr(R) = 1/4, Pr(S) = 1/4, Pr(R$ and $S) = 0$ (a card may be a spade, a club, neither but not both). Thus, the probability that the card is drawn from a black suit, $Pr(R$ or $S)$ is $1/2$, which equals $Pr(R) + Pr(S)$.

Consider the earlier example, the rolling of a single six-sided fair die. Then the events R and S are not mutually exclusive. In the discussion of conjunction it was noted that the event 'R and S' denoted the throwing of a 5, an event with probability $1/6$. The general law, when $Pr(R$ and $S) \neq 0$, is

$$Pr(R \text{ or } S) = Pr(R) + Pr(S) - Pr(R \text{ and } S).$$

This rule can be easily verified in this case where $Pr(R) = 1/2, Pr(S) = 1/2$, $Pr(R$ and $S) = 1/6, Pr(R$ or $S) = 5/6$.

The third law of probability concerns the conjunction of two events. Initially, it will be assumed that the two events are what is known as *independent*. By independence is meant that knowledge of the occurrence of one of the two events does not alter the probability of occurrence of the other event. Thus two events which are mutually exclusive cannot be independent. As a simple example of independence, consider the rolling of two six-sided fair dice, A and B say. The outcome of the throw of A does not affect the outcome of the throw of B. If A lands '6' uppermost, this result does not alter the probability that B

will land '6' uppermost. The same argument applies if one die is rolled two or more times. Outcomes of earlier throws do not affect the outcomes of later throws.

Third law of probability   Let R and S be two independent events. Then

$$Pr(R \text{ and } S) = Pr(R) \times Pr(S). \tag{1.3}$$

This relationship is sometimes used as the definition of independence. Thus, two events R and S are said to be independent if $Pr(R \text{ and } S) = Pr(R) \times Pr(S)$. There is symmetry in this definition. Event R is independent of S and S is independent of R. This law may be generalised to more than two events. Consider n events $S_1, S_2, \ldots, S_n$. If they are mutually independent, then

$$Pr(S_1 \text{ and } S_2 \text{ and } \ldots \text{ and } S_n) = Pr(S_1) \times Pr(S_2) \times \cdots \times Pr(S_n) = \prod_{i=1}^{n} Pr(S_i).$$

1.6.6   Dependent events and background information

Not all events are independent. Consider one roll of a fair die, with R the throwing of an odd number as before, and S the throwing of a number greater than 3 as before. Then, $Pr(R) = 1/2, Pr(S) = 1/2, Pr(R) \times Pr(S) = 1/4$ but $Pr(R \text{ and } S) = Pr(\text{throwing a} 5) = 1/6$.

Events which are not independent are said to be *dependent*. The third law of probability for dependent events was first presented by Thomas Bayes (1763), see also Barnard (1958), Pearson and Kendall (1970) and Poincaré (1912). It is the general law for the conjunction of events. The law for independent events is a special case. Before the general statement of the third law is made, some discussion of dependence is helpful.

It is useful to consider that a probability assessment depends on two things: the event R whose probability is being considered and the information I available when R is being considered. The probability $Pr(R \mid I)$ is referred to as a *conditional probability*, acknowledging that R is conditional or dependent on I. Note the use of the vertical bar $\mid$. Events listed to the left of it are events whose probability is of interest. Events listed to the right are events whose outcomes are known and which may affect the probability of the events listed to the left of the bar, the vertical bar having the meaning 'given' or 'conditional on'.

Consider a defendant in a trial who may or may not be guilty. Denote the event that he is guilty by G. The uncertainty associated with his guilt, the probability that he is guilty, may be denoted by $Pr(G)$. It is a subjective probability. The uncertainty will fluctuate during the course of a trial. It will fluctuate as evidence is presented. It depends on the evidence. Yet neither the notation, $Pr(G)$, nor the language, the probability of guilt, makes mention of this dependence. The probability of guilt at any particular time depends on the knowledge (or information) available at that time. Denote this information by I. It is then possible to speak of the probability of guilt given, or conditional on, the information available at that time. This is written as $Pr(G \mid I)$. If additional evidence E is presented this then becomes, along with I, part of what is known. What is taken as known is then 'E and I', the conjunction of E and I. The revised probability of guilt is $Pr(G \mid E$ and $I)$.

All probabilities may be thought of as conditional probabilities. Personal experience informs judgements made about events. For example, judgement concerning the probability of rain the following day is conditioned on personal experiences of rain following days with similar weather patterns to the current one. Similarly, judgement concerning the value of evidence or the guilt of a suspect is conditional on many factors. These include other evidence at the trial but may also include a factor to account for the reliability of the evidence. There may be eyewitness evidence that the suspect was seen at the scene of the crime, but this evidence may be felt to be unreliable. Its value will then be lessened.

The value of scientific evidence will be conditioned on the background data relevant to the type of evidence being assessed. Evidence concerning frequencies of different DNA profiles will be conditioned on information regarding ethnicity of the people concerned for the values of these frequencies. Evidence concerning distributions of the refractive indices of glass fragments will be conditioned on information regarding the type of glass from which the fragments have come (e.g., building window, car headlights). The existence of such conditioning events will not always be stated explicitly. However, they should not be forgotten. As stated above, all probabilities may be thought of as conditional probabilities. The first two laws of probability can be stated in the new notation, for events R, S and information I as follows:

First law

$$0 \leq Pr(R \mid I) \leq 1 \tag{1.4}$$

and $Pr(\text{not } I \mid I) = 0$. If I is known, the event 'not I' is impossible and thus $Pr(I \mid I) = 1$.

Second law

$$Pr(R \text{ or } S \mid I) = Pr(R \mid I) + Pr(S \mid I) - Pr(R \text{ and } S \mid I). \tag{1.5}$$

Third law for independent events

If R and S are independent, then, conditional on I,

$$Pr(R \text{ and } S \mid I) = Pr(R \mid I) \times Pr(S \mid I). \tag{1.6}$$

Notice that the event I appears as a conditioning event in *all* the probability expressions. The laws are the same as before but with this simple extension. The third law for dependent events is given later by (1.7).

As an example of the use of the ideas of independence, consider a diallelic system in genetics in which the alleles are denoted A and a, with $Pr(A) = p$, $Pr(a) = q$ and $Pr(A) + Pr(a) = p + q = 1$. This gives rise to three genotypes that, assuming Hardy–Weinberg equilibrium to hold, are expected to have the following probabilities:

- p^2 (homozygotes for allele A),
- $2pq$ (heterozygotes),
- q^2 (homozygotes for allele a).

The genotype probabilities are calculated by simply multiplying the two allele probabilities together on the assumption that the allele inherited from one's father is independent of the allele inherited from one's mother. The factor 2 arises for heterozygotes case because two cases must be considered, that in which allele A was contributed by the mother and allele a by the father, and vice versa. Each of these cases has probability pq because of the assumption of independence (see Table 1.3). The particular locus under consideration is said to be in Hardy–Weinberg equilibrium when the two parental alleles are independent.

Suppose now that events R and S are dependent – that the knowledge that R has occurred affects the probability that S will occur, and vice versa. For example, let R be the outcome of drawing a card from a well-shuffled pack of 52 playing cards. This card is not replaced in the pack, so there are now only 51 cards in the pack. Let S be the draw of a card from this reduced pack of cards. Let R be the event 'an ace is drawn'. Thus $Pr(R) = 4/52 = 1/13$. (Note here the conditioning information I that the pack is well shuffled, with its implication that each of the 52 cards is equally likely to be drawn, has been omitted for simplicity of notation; explicit mention of I will be omitted in many cases but its existence should never be forgotten.) Let S be the event 'an ace is drawn' also. Then $Pr(S \mid R)$ is the probability that an ace was drawn at the second draw, given that an ace was drawn at the first draw (and given everything

Table 1.3 Genotype probabilities, assuming Hardy–Weinberg equilibrium, for a diallelic system with allele probabilities p and q

Allele from mother	Allele from father	
	$A(p)$	$a(q)$
$A(p)$	p^2	pq
$a(q)$	pq	q^2

else that is known, in particular that the first card was not replaced). There are 51 cards at the time of the second draw, of which 3 are aces. (Remember that an ace was drawn the first time, which is the information contained in R.) Thus $Pr(S \mid R) = 3/51$. It is now possible to formulate the third law of probability for dependent events:

Third law of probability for dependent events

$$Pr(R \text{ and } S \mid I) = Pr(R \mid I) \times Pr(S \mid R \text{ and } I). \tag{1.7}$$

Thus in the example of the drawing of the aces from the pack, the probability of drawing two aces is

$$Pr(R \text{ and } S \mid I) = Pr(R \mid I) \times Pr(S \mid R \text{ and } I) = \frac{4}{52} \times \frac{3}{51}.$$

Note that if the first card had been replaced in the pack and the pack shuffled again, the two draws would have been independent and the probability of drawing two aces would have been $4/52 \times 4/52$.

Example 1.3   Consider the genetic markers Kell and Duffy. For both of these markers a person may be either positive ($+$) or negative ($-$). In a relevant population 60% of the people are Kell+ and 70% are Duffy+. An individual is selected at random from this population; that is, an individual is selected using a procedure such that each individual is equally likely to be selected.

Consider the Kell marker first in relation to the urn example. Let those people with Kell+ correspond to black balls and those with Kell− correspond to white balls. By analogy with the urn example, the probability a randomly selected individual (ball) is Kell+ (black) corresponds to the proportion of Kell+ people (black balls) in the population, namely 0.6. Let R be the event that a randomly selected individual is Kell+. Then $Pr(R) = 0.6$. Similarly, let S be the event that a randomly selected individual is Duffy+. Then $Pr(S) = 0.7$.

If Kell and Duffy markers were independent, it would be calculated that the probability that a person selected at random were both Kell+ and Duffy+, namely $Pr(R \text{ and } S)$, was given by

$$Pr(R \text{ and } S) = Pr(R) \times Pr(S) = 0.6 \times 0.7 = 0.42.$$

However, it may not necessarily be the case that 42% of the population is both Kell+ and Duffy+. Nothing in the information currently available justifies the assumption of independence used to derive this result. It is perfectly feasible that, say, 34% of the population are both Kell+ and Duffy+, that is, $Pr(R \text{ and } S) = 0.34$. In such a situation where $Pr(R \text{ and } S) \neq Pr(R) \times Pr(S)$ it can be said that the Kell and Duffy genetic markers are not independent. (See also Chapter 3.)

Table 1.4 The proportion of people in a population who fall into the four possible categories of genetic markers

Kell	Duffy		Total
	+	−	
+	34	26	60
−	36	4	40
Total	70	30	100

The information about the Kell and Duffy genetic markers can be represented in a tabular form, known as a 2×2 (two-by-two) table as in Table 1.4, the rows of which refer to Kell (positive or negative) and the columns of which refer to Duffy (positive or negative). It is possible to verify the third law of probability for dependent events (1.7) using this table. Of the 60% of the population who are Kell+ (R), 34/60 are Duffy+ (S). Thus $Pr(S \mid R) = 34/60 = 0.68$. Also, $Pr(R) = 60/100$ and

$$Pr(R \text{ and } S) = Pr(R) \times Pr(S \mid R) = \frac{60}{100} \times \frac{34}{60} = 0.34,$$

as may be derived from the table directly.

This example also illustrates the symmetry of the relationship between R and S as follows. Of the 70% of the population who are Duffy+ (S), 34/70 are Kell+ (R). Thus $Pr(R \mid S) = 34/70$. Also $Pr(S) = 70/100$ and

$$Pr(R \text{ and } S) = Pr(S) \times Pr(R \mid S) = \frac{70}{100} \times \frac{34}{70} = 0.34.$$

Thus, for dependent events, R and S, the third law of probability, (1.7) may be written as

$$Pr(R \text{ and } S) = Pr(S \mid R) \times Pr(R) = Pr(R \mid S) \times Pr(S), \qquad (1.8)$$

where the conditioning on I has been omitted. The result for independent events follows as a special case with $Pr(R \mid S) = Pr(R)$ and $Pr(S \mid R) = Pr(S)$.

1.6.7 Law of total probability

This is sometimes known as the extension of the conversation (Lindley, 1991). Events $S_1, S_2, \ldots, S_n$ are said to be *mutually exclusive and exhaustive* if one of them has to be true and only one of them can be true; they exhaust the possibilities and the occurrence of one excludes the possibility of any other.

Alternatively, they are called a *partition*. The event $(S_1$ or $\ldots$ or $S_n)$ formed from the conjunction of the individual events $S_1, \ldots, S_n$ is certain to happen since the events are exclusive. Thus, it has probability 1 and

$$Pr(S_1 \text{ or } \ldots \text{ or } S_n) = Pr(S_1) + \cdots + Pr(S_n) = 1, \tag{1.9}$$

a generalisation of the second law of probability (1.5) for exclusive events. Consider allelic distributions at a locus, for example locus *LDLR* for Caucasians in Chicago (Johnson and Peterson, 1999). The three alleles A, B and C are mutually exclusive and exhaustive.

Consider $n = 2$. Let R be any other event. The events 'R and S_1' and 'R and S_2' are exclusive. They cannot both occur. The event ' "R and S_1" *or* "R and S_2" ' is simply R. Let S_1 be male, S_2 be female, R be left-handed. Then 'R and S_1' denotes a left-handed male, while 'R and S_2' denotes a left-handed female. The event ' "R and S_1" or "R and S_2" ' is the event that a person is a left-handed male or a left-handed female, which implies the person is left-handed (R). Thus,

$$\begin{aligned} Pr(R) &= Pr(R \text{ and } S_1) + Pr(R \text{ and } S_2) \\ &= Pr(R \mid S_1)Pr(S_1) + Pr(R \mid S_2)Pr(S_2). \end{aligned}$$

The argument extends to any number of mutually exclusive and exhaustive events to give the law of total probability.

Law of total probability   If $S_1, S_2, \ldots, S_n$ are n mutually exclusive and exhaustive events,

$$Pr(R) = Pr(R \mid S_1)Pr(S_1) + \cdots + Pr(R \mid S_n)Pr(S_n). \tag{1.10}$$

An example for blood types and paternity cases is given by Lindley (1991). Consider two possible groups, S_1 (Rh−) and S_2 (Rh+) for the father, so here $n = 2$. Assume the relative frequencies of the two groups are p and $1 - p$, respectively. The child is Rh− (event R) and the mother is also Rh− (event M). The probability of interest is the probability a Rh− mother will have a Rh− child, in symbols $Pr(R \mid M)$. This probability is not easily derived directly but the derivation is fairly straightforward if the law of total probability is invoked to include the father:

$$Pr(R \mid M) = Pr(R \mid M \text{ and } S_1)Pr(S_1 \mid M) + Pr(R \mid M \text{ and } S_2)Pr(S_2 \mid M). \tag{1.11}$$

This is a generalisation of the law to include information M. If both parents are Rh−, event 'M and S_1', then the child is Rh− with probability 1, so $Pr(R \mid M \text{ and } S_1) = 1$. If the father is Rh+ (the mother is still Rh−), event S_2, then $Pr(R \mid M \text{ and } S_2) = 1/2$. Assume that parents mate at random with respect to the rhesus quality. Then $Pr(S_1 \mid M) = p$, the relative frequency of Rh− in the population, independent of M. Similarly, $Pr(S_2 \mid M) = 1 - p$, the relative frequency of Rh+ in the population. These probabilities can now be inserted in (1.11) to obtain

$$Pr(R \mid M) = 1(p) + \frac{1}{2}(1 - p) = (1 + p)/2$$

for the probability that a Rh− mother will have a Rh− child. This result is not intuitively obvious, unless one considers the approach based on the law of total probability.

An example using DNA profiles is given in Evett and Weir (1998). According to the 1991 census, the New Zealand population consists of 83.47% Caucasians, 12.19% Maoris and 4.34% Pacific Islanders; denote the event that a person chosen at random from the 1991 New Zealand population is Caucasian, Maori or Pacific Islander as *Ca*, *Ma* and *Pa*, respectively. The probabilities of finding the same *YNH24* genotype g (event G), as in a crime sample, for a Caucasian, Maori or Pacific Islander are 0.012, 0.045 and 0.039, respectively. These values are the assessments for the following three conditional probabilities: $Pr(G \mid Ca)$, $Pr(G \mid Ma)$, $Pr(G \mid Pa)$. Then the probability of finding the *YNH24* genotype, G, in a person taken at random from the whole population of New Zealand is

$$
\begin{aligned}
Pr(G) &= Pr(G \mid Ca)Pr(Ca) + Pr(G \mid Ma)Pr(Ma) + Pr(G \mid Pa)Pr(Pa) \\
&= 0.012 \times 0.8347 + 0.045 \times 0.1219 + 0.039 \times 0.0434 \\
&= 0.017.
\end{aligned}
$$

A further extension of this law to consider probabilities for combinations of genetic marker systems in a racially heterogeneous population has been given by Walsh and Buckleton (1988). Let C and D be two genetic marker systems. Let S_1 and S_2 be two mutually exclusive and exhaustive sub-populations such that a person from the population belongs to one and only one of S_1 and S_2. Let $Pr(S_1)$ and $Pr(S_2)$ be the probabilities that a person chosen at random from the population belongs to S_1 and to S_2, respectively. Then $Pr(S_1) + Pr(S_2) = 1$. Within each sub-population C and D are independent so that the probability an individual chosen at random from one of these sub-populations is of type CD is simply the product of the individual probabilities. Thus

$$
\begin{aligned}
Pr(CD \mid S_1) &= Pr(C \mid S_1) \times Pr(D \mid S_1), \\
Pr(CD \mid S_2) &= Pr(C \mid S_2) \times Pr(D \mid S_2).
\end{aligned}
$$

However, such a so-called *conditional independence* result does not imply unconditional independence (i.e., that $Pr(CD) = Pr(C) \times Pr(D)$). The probability that an individual chosen at random from the population is CD, without regard to his sub–population membership, may be written as

$$Pr(CD) = Pr(CDS_1) + Pr(CDS_2)$$
$$= Pr(CD \mid S_1) \times Pr(S_1) + Pr(CD \mid S_2) \times Pr(S_2)$$
$$= Pr(C \mid S_1) \times Pr(D \mid S_1) \times Pr(S_1) + Pr(C \mid S_2) \times Pr(D \mid S_2) \times Pr(S_2).$$

This is not necessarily equal to $Pr(C) \times Pr(D)$ as is illustrated in the following example. Let $Pr(C \mid S_1) = \gamma_1, Pr(C \mid S_2) = \gamma_2, Pr(D \mid S_1) = \delta_1, Pr(D \mid S_2) = \delta_2,$ $Pr(S_1) = \theta$ and $Pr(S_2) = 1 - \theta$. Then

$$Pr(CD) = \gamma_1 \delta_1 \theta + \gamma_2 \delta_2 (1 - \theta),$$
$$Pr(C) = \gamma_1 \theta + \gamma_2 (1 - \theta),$$
$$Pr(D) = \delta_1 \theta + \delta_2 (1 - \theta).$$

The product of $Pr(C)$ and $Pr(D)$ is not necessarily equal to $Pr(CD)$. Suppose, for example that $\theta = 0.40, \gamma_1 = 0.10, \gamma_2 = 0.20, \delta_1 = 0.15$ and $\delta_2 = 0.05$. Then

$$Pr(CD) = \gamma_1 \delta_1 \theta + \gamma_2 \delta_2 (1 - \theta)$$
$$= 0.10 \times 0.15 \times 0.4 + 0.20 \times 0.05 \times 0.6$$
$$= 0.0120,$$
$$Pr(C) = \gamma_1 \theta + \gamma_2 (1 - \theta) = 0.04 + 0.12 = 0.16,$$
$$Pr(D) = \delta_1 \theta + \delta_2 (1 - \theta) = 0.06 + 0.03 = 0.09,$$
$$Pr(C) \times Pr(D) = 0.0144 \neq 0.0120 = Pr(CD).$$

1.6.8 Updating of probabilities

Notice that the probability of guilt is a subjective probability, as mentioned before (Section 1.6.4). Its value will change as evidence accumulates. Also, different people will have different values for it. The following examples, adapted from similar ones in DeGroot (1970), illustrate how probabilities may change with increasing information. The examples have several parts and each part has to be considered in turn without information from a later part.

Example 1.4

(a) Consider four events S_1, S_2, S_3 and S_4. Event S_1 is that the area of Lithuania is no more than $50\,000\,\text{km}^2$, S_2 is the event that the area of Lithuania is greater than $50\,000\,\text{km}^2$ but no more than $75\,000\,\text{km}^2$, S_3 is the event that the area of Lithuania is greater than $75\,000\,\text{km}^2$ but no more than $100\,000\,\text{km}^2$, and S_4 is the event that the area of Lithuania is greater than $100\,000\,\text{km}^2$. Assign probabilities to each of these four events. Remember that these are four mutually exclusive events and that the four probabilities should add up to 1. Which do you consider the most probable and what probability do you assign to it? Which do you consider the least probable and what probability do you assign to it?

(b) Now consider the information that Lithuania is the 25th largest country in Europe (excluding Russia). Use this information to reconsider your probabilities in part (a).

(c) Consider the information that Estonia, which is the 30th largest country in Europe, has an area of $45\,000\,\text{km}^2$ and use it to reconsider your probabilities from the previous part.

(d) Consider the information that Austria, which is the 21st largest country in Europe, has an area of $84\,000\,\text{km}^2$ and use it to reconsider your probabilities from the previous part.

The area of Lithuania is given at the end of the chapter.

Example 1.5

(a) Imagine you are on a jury. The trial is about to begin but no evidence has been led. Consider the two events: S_1, the defendant is guilty; S_2, the defendant is innocent. What are your probabilities for these two events?

(b) The defendant is a tall Caucasian male. An eyewitness says he saw a tall Caucasian male running from the scene of the crime. What are your probabilities now for S_1 and S_2?

(c) A bloodstain at the scene of the crime was identified as coming from the criminal. A partial DNA profile has been obtained, with proportion 2% in the local Caucasian population. What are your probabilities now for S_1 and S_2?

(d) A window was broken during the commission of the crime. Fragments of glass were found on the defendant's clothing of a similar refractive index to that of the crime window. What are your probabilities now for S_1 and S_2?

(e) The defendant works as a demolition worker near to the crime scene. Windows on the demolition site have refractive indices similar to the crime window. What are your probabilities now for S_1 and S_2?

This example is designed to mimic the presentation of evidence in a court case. Part (a) asks for a prior probability of guilt before the presentation of any evidence. It may be considered as a question concerning the understanding of the dictum 'innocent until proven guilty'. See Section 3.5.5 for further discussion of this with particular reference to the logical problem created if a prior probability of zero is assigned to the event that the suspect is guilty.

Part (b) involves two parts. First, the value of the similarity in physical characteristics between the defendant and the person running from the scene of the crime, assuming the eyewitness is reliable, has to be assessed. Secondly, the assumption that the eyewitness is reliable has to be assessed.

In part (c) it is necessary to check that the defendant has the same DNA profile. It is not stated that he has, but if he has not he should never have been a defendant. Secondly, is the local Caucasian population the correct population? The evaluation of evidence of the form in (c) is discussed in Chapter 9.

The evaluation of refractive index measurements mentioned in (d) is discussed in Chapter 10. Variation both within and between windows has to be considered. Finally, how information about the defendant's lifestyle may be considered is discussed in Chapter 8.

It should be noted that the questions asked initially in Example 1.5 are questions which should be addressed by the judge and/or jury. The forensic scientist is concerned with the evaluation of his evidence, not with probabilities of guilt or innocence. These are the concern of the jury. The jury combines the evidence of the scientist with all other evidence and uses its judgement to reach a verdict. The theme of this book is the evaluation of evidence. Discussion of issues relating to guilt or otherwise of suspects will not be very detailed.

(As a tail piece to this chapter, the area of Lithuania is $65\,301\,km^2$.)

2

Variation

So far, events and the probability of their occurrence have been discussed. These ideas may be extended to consider counts and measurements about which there may be some uncertainty or randomness. In certain fairly general circumstances the way in which probability is distributed over the possible numbers of counts or over the possible values for the measurements can be represented mathematically, by functions known as probability distributions. Distributions for counts and for measurements will be described in Sections 2.3 and 2.4, respectively. Further details of all the distributions mentioned here and others may be found in Evans *et al.* (2000). Before probability distributions can be discussed here, however, certain other concepts have to be introduced.

2.1 POPULATIONS

'Who is "random man"?' This is the title of a paper by Buckleton *et al.* (1991). In order to evaluate evidence it is necessary to have some idea of the variability or distribution of the evidence under consideration within some population. This population will be called the *relevant* population (and a more formal definition will be given later in Section 8.5) because it is the population which is deemed relevant to the evaluation of the evidence. Variability is important because if the suspect did not commit the crime and is, therefore, assumed innocent it is necessary to be able to determine the probability of associating the evidence with him when he is innocent. Surveys of populations are required in order to obtain this information. Three such surveys for glass fragments are Pearson *et al.* (1971), Harrison *et al.* (1985) and McQuillan and Edgar (1992). There are many reports on the distribution of blood groups (e.g., Gaensslen *et al.*, 1987a, b, c) and DNA frequencies (e.g., in *Forensic Science International* and the *Journal of Forensic Sciences*) and Forensic Science Communications (http://www.fbi.gov/hq/lab/fsc/current/backissu.htm). Surveys for other materials are discussed in Chapter 7.

Care has to be taken when deciding how to choose the relevant population. Buckleton *et al.* (1991) describe two situations and explain how the relevant population is different for each.

Statistics and the Evaluation of Evidence for Forensic Scientists: Second Edition
C.G.G. Aitken and F. Taroni © 2004 John Wiley & Sons, Ltd   ISBN: 0-470-84367-5

The first situation is one in which there is transfer from the criminal to the crime scene as in Example 1.1 and discussed in greater detail in Section 8.3.1. In this situation, the details of any suspect are irrelevant under H_d, the hypothesis that the suspect was not present at the scene of the crime. Consider a bloodstain at the crime scene which, from the background information I, it is possible to assume is blood from the criminal. If the suspect was not present, then clearly some other person must have left the stain. There is no reason to confine attention to any one group of people. In particular, attention should not be confined only to any group (e.g., ethnic group) to which the suspect belongs. However, if there is some information which might cause one to reconsider the choice of population then that choice may be modified. Such information may come from an eyewitness, for example, who is able to provide information about the offender's ethnicity. This would then be part of the background information I. In general, though, information about blood group frequencies would be required from a survey which is representative of all possible offenders. For evidence of blood groups it is known that age and sex are not factors affecting a person's blood group, but that ethnicity is. It is necessary to consider the racial composition of the population of possible offenders (not suspects). Normally, it is necessary to study a general population since there will be no information available to restrict the population of possible criminals to any particular ethnic group or groups.

The second situation considered by Buckleton *et al.* (1991) is possible transfer from the crime scene to the suspect or criminal, discussed further in Section 8.3.2. The details of the suspect are now relevant, even assuming he was not present at the crime scene. Consider the situation where there is a deceased victim who has been stabbed numerous times. A suspect, with a history of violence, has been apprehended with a heavy bloodstain on his jacket which is not of his own blood. What is the evidential value in itself, and not considering possible DNA evidence, of the existence of such a heavy bloodstain, not of the suspect's blood? The probability of such an event (the existence of a heavy bloodstain) if the suspect did not commit the crime needs to be considered.

The suspect may offer an alternative explanation. The jury can then assign a probability to the occurrence of the evidence, given the suspect's explanation. The two propositions to be considered would then be

- H_p, the blood was transferred during the commission of the crime,

- H_d, the suspect's explanation is true,

and the jury would assess the evidence of the existence of transfer under these two propositions. Evaluation of the evidence of the blood group frequencies would be additional to this. The two parts could then be combined using the technique described in Section 5.1.3.

In the absence of an explanation from the suspect, the forensic scientist could conduct a survey of persons as similar as possible to the suspect in whatever are the key features of his behaviour or lifestyle. The survey would be conducted with respect to the suspect since it is of interest to learn about the transfer of bloodstains for people of the suspect's background. In a particular case, it may be that a survey of people of a violent background is needed. Results may be adequately provided by Briggs (1978) in which 122 suspects who were largely vagrants, alcoholics and violent sexual deviants were studied. The nature and lifestyle of the suspect determined the type of population to survey. Buckleton *et al.* (1991) reported also the work of Fong and Inami (1986) in which clothing items from suspects, predominantly in offences against the person, were searched exhaustively for fibres that were subsequently grouped and identified.

The idea of a relevant population is a very important one and is discussed further in Chapter 8. Consider the example of offender profiling, one which is not strictly speaking forensic science but which is still pertinent. Consider its application to rape cases. Suppose the profiler is asked to comment on the offender's lifestyle – age, marital status, existence and number of previous convictions, etc. – which the profiler may be able to do. However, it is important to know something about the distribution of these in some general population. The question arises here, as in Buckleton *et al.* (1991) described above, as to what is the relevant population. In rape cases, it may not necessarily be the entire male population of the local community. It could be argued that it might be the population of burglars, not so much because rapists are all burglars first but rather because burglars are a larger group of people who commit crimes involving invasion of someone else's living space. Information from control groups is needed regarding both the distribution of observed traits among the general non-offending population and the distribution of similar offences amongst those without the observed traits.

2.2 SAMPLES AND ESTIMATES

For any particular type of evidence the distribution of the characteristic of interest is important. This is so that it may be possible to determine the rarity or otherwise of any particular observation. For a genetic marker system, the relative frequencies of each of the groups within the system is important. For the refractive index of glass, the distribution of the refractive index measurements is important. In practice, these distributions are not known exactly. They are estimated. The allelic frequencies for locus *LDLR* in Chicago exist but are not known. Instead, they are estimated from a sample (see Table 1.1). Similarly, the distribution of the refractive indices of glass exists but is not known. Instead it may be estimated from a sample (see Table 10.5, extracted from Lambert and Evett, 1984, in which the number of fragments was 2269). Table 10.5 refers to float glass from buildings. Other types of glass discussed by Lambert and Evett (1984) include 'not float glass' from buildings, glass from vehicles (divided into windows and others, such as headlamps and driving mirrors) and container glass. In each of these situations, a sample has been observed (e.g., blood groups of people or measurements of the refractive index of glass fragments). These samples are assumed to be representative, in some sense, of some population (e.g., all Caucasians in Chicago (Johnson and Peterson, 1999) or all float glass from buildings (Lambert and Evett, 1984)).

A characteristic of interest from the population is known as a *parameter*. The corresponding characteristic from the corresponding sample is known as an *estimate*. For example, the proportion γ_{AA} of Caucasians in Chicago with allele *AA* at locus *LDLR* is a parameter. The proportion of Caucasians with allele *AA* at locus *LDLR* in the sample of 200 people studied by Johnson and Peterson (1999) is an estimate of γ_{AA}. Conventionally a ˆ symbol (read as 'hat') is used to denote an estimate. Thus $\hat{\gamma}_{AA}$ (read as 'gamma-hat AA') denotes an estimate of γ_{AA}. From Table 1.1, $\hat{\gamma}_{AA} = 0.188$.

It is hoped that an estimate will be a good estimate in some sense. Different samples from the same population may produce different estimates. The proportion of people with an *AA* allele at locus *LDLR* in a second sample of 200 Caucasians from Chicago may have produced a different number of people with *AA* alleles and hence a different value for $\hat{\gamma}_{AA}$. Different results from different samples do not mean that some are wrong and others are right. They merely indicate the natural variability in the distribution of allelic frequencies amongst people.

An estimate may be considered good if it is *accurate* and *precise*. Accuracy may be thought of as a measure of closeness of an estimate to the true value of the parameter of interest. In the example above it is obviously desirable that $\hat{\gamma}_{AA}$ be close to γ_{AA}. Precision may be thought of as a measure of the variability of the estimates, whether or not they are close to the true value (Kendall and Buckland, 1982, pp. 3, 152). If different samples lead to very different estimates of the same characteristic then the variability is great and the estimates are not very precise. For example, if the variability in the estimation procedure is large then a second estimate, from a different sample of people, of γ_{AA} may produce an estimate very different from 0.188 ($\hat{\gamma}_{AA}$).

The importance of allowing for variability is illustrated by the following hypothetical example from a medical context. The reaction times of two groups of people, group *A* and group *B* say, are measured. Both groups have the same median reaction time, 0.20 seconds, but group *A*'s times vary from 0.10 to 0.30 seconds whereas group *B*'s range from 0.15 to 0.25 seconds. Samples of equal numbers of people from each group are then given a drug designed to reduce reaction times. In both cases, the reaction times of the samples of people given the drug range from 0.11 to 0.14 seconds. For group *A*, this is within the range of previous knowledge and there is a little, but not very strong, evidence to suggest that the drug is effective in reducing reaction times. For group *B*, however, the result is outwith the range of previous knowledge and there is very strong evidence to suggest that the drug is effective. Note that both group *A* and group *B* had the same initial median reaction time. The drug produced the same range of reaction times in samples from both groups. The distinction in the interpretation of the results between the two groups arises because of the difference in the range, or variability, of the results for the whole of each of the two groups.

Later, in Section 2.4.2, it will be seen that when measurements are standardised, variation is accounted for by including a measure of the variation, known as the *standard deviation*, in the process.

Another notational convention, as well as the ˆ notation, uses Latin letters for functions evaluated from measurements from samples and Greek letters for the corresponding parameters from populations. Thus a sample mean may be denoted $\bar{x}$ and the corresponding population mean μ. A sample standard deviation may be denoted s and the corresponding population standard deviation σ. The square of the standard deviation is known as the *variance*; a sample variance may be denoted s^2, and the corresponding population variance σ^2.

The concept of a *random variable* (or *random quantity*; or *uncertain quantity*; Lindley, 1991) also needs some explanation. A *random variable*, in a rather circular definition, is something which varies at random. For example, the number of sixes in four rolls of a die varies randomly amongst the five numbers {0, 1, 2, 3, 4} as the die is rolled for several sets of four rolls. Similarly, the refractive index of a fragment of glass varies over the set of all fragments of glass. The variation to be considered in the refractive index of glass, is, however, of a more complicated structure than the number of sixes in four rolls of a die. There is variation in refractive index within a window and between windows. This requires parameters to measure two standard deviations, one for each type of variation, and this problem is discussed in greater detail in Chapter 10.

Notation is useful in the discussion of random variables. Rather than write out in long-hand phrases such as 'the number of sixes in four rolls of a die' or 'the refractive index of a fragment of glass' the phrases may be abbreviated to a single upper-case Latin letter. For example, let X be short for 'the number of sixes in four rolls of a die'. It then makes sense to write mathematically $Pr(X = 3)$, which may be read as 'the probability that the number of sixes in four rolls of a die equals 3'. More generally still, the 3 may be replaced by a lower-case Latin letter to give $Pr(X = x)$, say, where x may then be substituted by one of the permissible values {0, 1, 2, 3, 4} as required.

Similarly, X may be substituted for 'the refractive index of a fragment of glass' and the phrase 'the probability that the refractive index of a fragment of glass is less than 1.5185' may be written as $Pr(X < 1.5185)$, or more generally as $Pr(X < x)$ for a general value x of the refractive index. For reasons which are explained later (Section 2.4.2) it is not possible to evaluate $Pr(X = x)$ for a random variable representing a continuous measurement.

The mean of a random variable is the corresponding population mean. In the examples above this would be the mean number of sixes in the conceptual population of all possible sets of four rolls of a die (and here note that the population need not necessarily exist except as a concept) or the mean refractive index of the population of all fragments of glass (again, effectively, a conceptual population). The mean of a random variable is given a special name, the *expectation*, and for a random variable, X say, it is denoted $E(X)$. Similarly, the variance of a random variable is the corresponding population variance. For a random variable X, it is denoted $Var(X)$.

A *statistic* is a function of the data. Thus, the sample mean and the sample variance are statistics. A particular value of a statistic which is determined to estimate the value of a parameter is known as an *estimate*. The corresponding random variable is known as an *estimator*. An estimator, X say, of a parameter, θ say, which is such that $E(X) = \theta$ is said to be unbiased. If $E(X) \neq \theta$, the estimator is said to be biased.

The applications of these concepts are now discussed in the context of probability distributions for counts and for measurements.

2.3 COUNTS

2.3.1 Probabilities

Suppose a fair six-sided die is to be rolled four times. Let the event of interest be the number of occurrences of a six being uppermost; denote this by X. Then X can take one of five different integer values, 0, 1, 2, 3 or 4. Over a sequence of groups of four rolls of the die, X will vary *randomly* over this set of five integers. Outcomes of successive rolls are independent. For any one group of four rolls of the die, X takes a particular value, one of the integers $\{0, 1, 2, 3, 4\}$. Denote this particular value by x.

There is a formula which enables this probability to be evaluated fairly easily. Notice that in any one roll, the probability of throwing a six is $1/6$; the probability of not throwing a six is $5/6$, as these are complementary events. Then

$$Pr(X = x) = \binom{4}{x}\left(\frac{1}{6}\right)^x \left(\frac{5}{6}\right)^{4-x}, \qquad x = 0, 1, \ldots, 4;$$

an example of the binomial distribution (Section 2.3.3). The term $(1/6)^x$ corresponds to the x sixes, each occurring independently with probability $1/6$. The term $(5/6)^{4-x}$ corresponds to the $4 - x$ non-sixes, each occurring independently with probability $5/6$. The term $\binom{4}{x}$ is the *binomial coefficient*

$$\binom{4}{x} = \frac{4!}{x!(4-x)!},$$

where $x! = x(x - 1)(x - 2) \cdots 1$, known as x-factorial and, conventionally, $0! = 1$. The binomial coefficient here is the number of ways in which x sixes and $4 - x$ non-sixes may be selected from 4 rolls, without attention being paid to the order in which the sixes occur.

Table 2.1 Probabilities for the number of sixes, X, in four rolls of a fair six-sided die

Number of sixes (x)	0	1	2	3	4	Total
$Pr(X = x)$	0.4823	0.3858	0.1157	0.0154	0.0008	1.0000

Suppose $x = 1$, that is, there is one six and three non-sixes; then

$$Pr(X = 1) = \binom{4}{1} \left(\frac{1}{6}\right)^1 \left(\frac{5}{6}\right)^3.$$

Now

$$\binom{4}{1} = \frac{4!}{1!3!} = \frac{4 \times 3 \times 2 \times 1}{1 \times 3 \times 2 \times 1} = 4,$$

$$\left(\frac{1}{6}\right)^1 = \frac{1}{6},$$

$$\left(\frac{5}{6}\right)^3 = \frac{125}{216},$$

$$Pr(X = 1) = 4 \times \frac{1}{6} \times \frac{125}{216} = 0.3858.$$

The probabilities for the five possible outcomes relating to the number of sixes in four rolls of the die are given in Table 2.1. Notice that the sum of the probabilities is 1 since the five possible outcomes 0, 1, 2, 3 and 4 are mutually exclusive and exhaustive.

2.3.2 Summary measures

It is possible to determine a value for the mean of the number of sixes in four rolls of the die; this is the expectation of the number of sixes in four rolls of the die. Consider 10 000 groups of four rolls of the die. The probabilities in Table 2.1 may be considered as the expected proportion of times in which each of 0, 1, 2, 3 and 4 sixes would occur. Thus it would be expected that there would be no sixes 4823 times, one six 3858 times, two sixes 1157 times, three sixes 154 times, and four sixes 8 times. The total number of sixes expected is thus

$$(0 \times 4823) + (1 \times 3858) + (2 \times 1157) + (3 \times 154) + (4 \times 8) = 6666.$$

In any one group of four rolls the expected number $E(X)$ of sixes is then $6666/10,000 = 0.6666$. Notice that this is not an achievable number (0, 1, 2, 3 or 4) but is justified by the calculations. (In a similar way, an average family size of 2.4 children is not an achievable family size.) There is a formula for its calculation:

$$E(X) = 0 \times Pr(X = 0) + \cdots + 4 \times Pr(X = 4)$$

$$= \sum_{x=0}^{4} x \, Pr(X = x).$$

This can be further shortened by denoting $Pr(X = x)$ by p_x so that

$$E(X) = \sum_{x=0}^{4} x \, p_x.$$

In general, for $n + 1$ outcomes $\{0, 1, \ldots, n\}$ with associated probabilities $p_0, p_1, \ldots, p_n$,

$$E(X) = \sum_{x=0}^{n} x \, p_x.$$

Note the use of the Greek capital letter $\sum$ to denote summation (S for *summation*). The expression below $\sum$ (when it is displayed) or as a subscript (when in the body of the text) denotes the symbol over which the summation is being made and the starting point of the sum. The finishing point of the sum is above $\sum$ (when it is displayed) or as a superscript (when in the body of the text). This symbol should be compared with the Greek capital letter $\prod$ to denote product (P for *product*) where the same convention for indexing the product is used. The first example of the use of $\prod$ is in Section 4.5.4.

The expectation is a well-known statistic. Not so well known is the variance which measures the variability in a set of observations. The number of sixes which occurs in any group of four rolls of the die varies from group to group over the integers 0, 1, 2, 3 and 4.

Consider the square of the difference, $d(X)^2 = \{X - E(X)\}^2$ between an outcome X and the expectation. This squared difference is itself a random variable and as such has an expectation. The expectation of $d(X)^2$, $E\{d(X)^2\}$, for a set of $n + 1$ outcomes $\{0, 1, \ldots, n\}$ with associated probabilities $p_0, p_1, \ldots, p_n$ is

$$\sum_{x=0}^{n} \{x - E(X)\}^2 p_x,$$

and is the *variance* of X. The square root of the variance is the *standard deviation*. Another, quicker, method of evaluation of the variance is to calculate

$$Var(X) = \sum_{x=0}^{n} x^2 p_x - \left(\sum_{x=0}^{n} x p_x \right)^2.$$

Table 2.2   Intermediate calculations for the variance of the number of sixes, x, in four rolls of a fair six-sided die

x	0	1	2	3	4
d	−0.6666	0.3334	1.3334	2.3334	3.3334
d^2	0.4444	0.1112	1.7780	5.4448	11.1116
p_x	0.4823	0.3858	0.1157	0.0154	0.0008
x^2	0	1	4	9	16

The variance may be worked out for the example of the number of sixes in four rolls of the die as follows, where $E(X) = 0.6666$. The variance of X may then be calculated as

$$Var(X) = \sum_{x=0}^{4} \{x - E(X)\}^2 p_x = \sum_{x=0}^{4} d^2 p_x = 0.5557.$$

The quicker way is to evaluate

$$Var(X) = \sum_{x=0}^{4} x^2 p_x - \left(\sum_{x=0}^{4} x p_x\right)^2 = 1.0000 - 0.6666^2 = 0.5556.$$

The intermediate calculations are given in Table 2.2.

This example of four rolls of a fair six-sided die may be generalised. Consider each roll of the die as a *trial*, in a statistical sense. At each trial there will be one of only two outcomes, a six or a non-six. Conventionally, in general terms, these may be known as *success* and *failure*. The trials are independent of each other. The probability of each of the outcomes is constant from trial to trial (the probability of a six is 1/6 for each roll). Such a set of trials is known as a set of Bernoulli trials (after the Swiss mathematician, Jacob Bernoulli, 1654–1705). The conditions to be met are:

- a fixed number of trials;
- independent trials;
- two and only two outcomes, conventionally denoted success and failure or positive and negative;
- a constant probability of success from trial to trial.

2.3.3   Binomial distribution

Let n denote the number of trials. Let X denote the number of successes. Let p denote the probability of a success in any individual trial and let $q (= (1 - p))$ denote the probability of a failure in any individual trial. Let the probability, $Pr(X = x)$, that X equals x be denoted by p_x, $x = 0, 1, \ldots, n$. This probability is dependent on n and p and more correctly should be written as $Pr(X = x \mid n, p)$.

The situation described above is a very common one. Examples include the number of heads in ten tosses of a fair coin ($n = 10$, $p = 1/2$), the number of sixes in five rolls of a fair die ($n = 5$, $p = 1/6$), the number of people with genotype $(11, 12)$, at the *FES* locus in a sample of size 50 from a relevant population ($n = 50$, p may be estimated from previous population data). The distribution of the probabilities (*probability distribution*) over the set of possible outcomes is known as the *binomial distribution*. The function which gives the formula for the probabilities $Pr(X = x)$ is known as a *probability function*. For the binomial distribution $Pr(X = x)$ is given by

$$Pr(X = x) = \binom{n}{x} p^x (1 - p)^{n-x} = \binom{n}{x} p^x q^{n-x}, \qquad x = 0, 1, \ldots, n, \qquad (2.1)$$

where

$$\binom{n}{x} = \frac{n!}{x!(n-x)!}, \qquad (2.2)$$

the binomial coefficient. The distribution of X can be denoted in short-hand as

$$X \sim Bin(n, p),$$

where $\sim$ indicates *is distributed as*, the first term n in the parentheses denotes the number of trials and the second term p denotes the probability of success. For example, if X is the number of sixes in 10 throws of a fair die then this can be denoted as

$$X \sim Bin(10, 1/6).$$

It can be shown that

$$E(X) = np, \ Var(X) = npq.$$

(Verification of these formulae can be made by reference to the numerical results in the example above.) Note that $E(X/n) = E(X)/n = np/n = p$. Thus, X/n, the sample proportion of successes, is an unbiased estimator of p, the probability of success in an individual trial.

2.3.4   Multinomial distribution

The multinomial distribution is a generalisation of the binomial distribution. The binomial distribution models a situation in which there is a sequence of independent trials in each of which there are only two possible mutually exclusive outcomes. The multinomial distribution models a situation in which there is a sequence of independent trials in each of which there are k possible mutually exclusive outcomes ($k \geq 2$). Denote the probabilities for the k outcomes $\theta_1, \ldots, \theta_k$, with $\sum_{i=1}^{k} \theta_i = 1$. Consider n trials in which the observed number of occurrences of each of the k outcomes is $x_1, \ldots, x_k$, with $\sum_{i=1}^{k} x_i = n$. The corresponding random variables are denoted $X_1, \ldots, X_k$, where X_i is short-hand for the phrase 'the number of occurrences of outcome i'. The probability of observing $\{X_1 = x_1, \ldots, X_k = x_k\}$ is then

$$Pr(X_1 = x_1, \ldots, X_k = x_k) = \frac{n!}{x_1! x_2! \ldots x_k!} \theta_1^{x_1} \ldots \theta_k^{x_k},$$

where $\sum_{i=1}^{k} x_i = n$, $\sum_{i=1}^{k} \theta_i = 1$. This distribution may be used to model allele frequencies at loci where there are more than two possible alleles and to model drug frequencies in consignments of tablets in which there are more than two possible drug types. When $k = 3$ and there are three mutually exclusive outcomes, the distribution is also known as a *trinomial* distribution.

2.3.5 Hypergeometric distribution

For the binomial and multinomial distributions the probability of a particular outcome in any one trial is assumed constant. Thus the probability of a six in a throw of a fair die is assumed equal to $1/6$ regardless of the number of throws of the die. The probability of a particular allelic type is assumed constant, regardless of the number of other people who have been observed with or without that type. The population from which these observations have been taken (all throws of a fair die, all people in the population) is sufficiently large that the observation of a particular outcome does not alter the probability of that outcome in future trials. In a sense, it may be considered that once that outcome from the population has been observed it is then returned to the population and may be selected for observation again. The selection (or sampling) of observations from the population is said to be *with replacement*. However, there are instances when the population is not large and the observation of the outcome of a particular trial does change the probability of that outcome in future. For example, consider sampling from a consignment of N white tablets to determine the proportion which are illicit, an example for which further details are given in Chapter 6. The tablets are assumed indistinguishable by size, colour, weight and texture, but each is assumed to be either licit or illicit. A sample of size m is taken from the consignment. The true, but unknown, number of illicit tablets is R and the true, but unknown, number of licit tablets is $N - R$. The first tablet examined is either illicit (with probability R/N, the proportion of illicit tablets in the consignment) or licit (with probability $(N - R)/N$). After examination it is put to one side; it is not placed back in the consignment. A second tablet is then examined.

Assume the first tablet was illicit. The second tablet is either illicit or licit. The probability that it is illicit is $(R - 1)/(N - 1)$, the proportion of illicit tablets remaining in the consignment. The probability that it is licit is $(N-R)/(N-1)$, the proportion of licit tablets remaining in the consignment.

Assume the first tablet was licit. The second tablet is either illicit or licit. The probability that it is illicit is $R/(N-1)$, the proportion of illicit tablets remaining in the consignment. The probability that it is licit is $(N-R-1)/(N-1)$, the proportion of licit tablets remaining in the consignment. After examination, the second tablet is put to one side. A third tablet is examined. There are three possibilities for the probability it is illicit. These depend on the outcomes of the first two examinations, in which there may zero, one or two illicit tablets.

Sampling in this context where N, the consignment size, is small is said to be *without replacement*. The distribution which models the probability of the number X of illicit tablets in a sample of size m from a consignment of size N in which R are illicit and $N-R$ are licit is the hypergeometric distribution. The probability function is given by

$$Pr(X=x) = \frac{\binom{R}{x}\binom{N-R}{m-x}}{\binom{N}{m}}. \tag{2.3}$$

Another example of the use of the hypergeometric distribution in a forensic context is given by Bates and Lambert (1991). The problem is to decide how many transferred particle fragments of the same generic type are needed to make a comparison with a source sample.

Example 2.1   Suppose there are 20 transferred particle fragments of glass found on the clothing of a suspect of which 10 match the source and 10 do not. A sample of 6 of these 20 fragments is selected at random for examination, so that each of the 20 fragments has an equal probability of being selected. The probability that 3 are found to match the source can be determined. Here, $N=20$, $R=10$, $m=6$. Then from (2.3)

$$Pr(X=3) = \frac{\binom{10}{3}\binom{10}{3}}{\binom{20}{6}}.$$

Since

$$\binom{10}{3} = 120 \text{ and } \binom{20}{6} = 38\,760,$$

it can be calculated that

$$Pr(X=3) = 120 \times 120/38\,760 = 0.37.$$

Further examples are given in Bates and Lambert (1991). With $N = 10$, $R = 2$ and $m = 5$, the probability that $X = 0$ is 0.22 ($> 1/5$). Thus, with 10 transferred particle items of which 2 match the source (and 8 do not), if a sample of 5 items is examined there is a probability greater than 1 in 5 that no matches will be found. Conversely, in practice, what is known is that $N = 10$, $m = 5$ and $x = 0$. There is a reasonably large probability (0.22) that no matching items will be found in a sample of size 5, when, in fact, there are two. A conclusion from such an examination that there were no matching items, and hence that this evidence did not associate the suspect with the crime scene, would be wrong.

With 50 transferred particle items, 10 of which match and 40 which do not ($N = 50$, $R = 10$, the same proportions as in the previous paragraph) and a sample of size $m = 5$, then $Pr(X = 0) = 0.31$. There is about a 1 in 3 probability that none of the five items selected randomly will match and a similar erroneous conclusion to the one above would be made.

Another example of the use of the hypergeometric distribution is in sampling for the proportion of a consignment of homogeneous tablets which contain illicit drugs. Consider a consignment of N tablets which are homogeneous in nature (colour, texture, type of logo) where it is desired to know the proportion which are illicit. Let R denote the number, out of N, which are illicit. Then $N - R$ are licit. A random sample of size m is taken from N, and x are found to be illicit and $m - x$ licit. The probability of this event (x tablets found to be illicit when m tablets are chosen at random from N tablets of which R are illicit) is given by the hypergeometric distribution (2.3).

Examples of the use of the hypergeometric distribution for sampling in drugs-related cases are given in Aitken (1999), Colón *et al.* (1993), Coulson *et al.* (2001b), Frank *et al.* (1991) and Tzidony and Ravreboy (1992). The hypergeometric distribution is also recommended by the United Nations in this context (United Nations, 1998). An application to fibres for the determination of the optimal sample size is given by Faber *et al.* (1999). Further details are provided in Chapter 6. Further discussion in the context of fibres is provided in Chapter 12.

Probability values for the hypergeometric distribution may be obtained from statistical software. Also, tables of the hypergeometric distribution have been published. See, for example, Lindley and Scott (1995) for values of the probability distribution for $N \leq 17$. For larger values of N with m relatively small compared to N (e.g., $m < N/20$), probabilities may be calculated using the binomial distribution (2.1) with $p = R/N$, $n = m$. For example, with $N = 50$, $R = 10$, $m = 5$, $p = 0.2$,

$$Pr(X = 0) = \binom{5}{0} 0.2^0 0.8^5 = 0.33,$$

which should be compared with an exact probability of 0.31. Notice that $m > N/20$ so that the approximation is not too good.

Suppose that $N = 120$, $R = 80$, $m = 5$, $p = 2/3$. The hypergeometric probability

$$Pr(X = 2) = \frac{\binom{80}{2}\binom{40}{3}}{\binom{120}{5.}} = 0.164.$$

Using the binomial distribution,

$$Pr(X = 2) = \binom{5}{2}\left(\frac{2}{3}\right)^2\left(1 - \frac{2}{3}\right)^3 = 0.165,$$

and a very good approximation is obtained.

It is possible to extend the hypergeometric distribution to the situation where there are more than two categories. This is analogous to the extension of the binomial to the multinomial. No further discussion of this extension is given here.

2.3.6   Poisson distribution

This distribution is named after the French mathematician, S.D. Poisson (1781–1840). It is generally used to describe the number of events which occur randomly in a specified period of time or of space. It is characterised by a single parameter, the mean of the distribution (in unit time or space). This parameter is then multiplied by the period of time or space under consideration to give the mean number of events within that period. As an example in time, consider the emission of radioactive particles from a radioactive source, as measured by a Geiger counter. Take the unit of time to be 1 second. Denote the mean number of particles emitted in 1 second by λ, a number greater than 0. The mean number of particles emitted in t seconds is then λt, where t can take any value in the range $(0, \infty)$. As an example in space, consider the number of characteristics of a particular kind in a piece of handwriting. Take the unit of space to be one character in the handwriting. Again, denote the mean by λ, which is now the mean number of the particular kind of characteristics, and it would be expected that λ is very much less than 1. The mean number of characteristics of the particular kind in a length of handwriting of s characters is then λs. Note that the parameter λ has units 'per unit time' or 'per unit interval of space'. Thus, when considering the distribution of the number of events it is important to specify the length of time or space which is being considered.

Let X denote the number of events in a period of time t, which is considered as a random variable with mean λt, that follows a Poisson distribution. Then, the probability that X takes a particular value x (a non-negative integer) is given by the equation

$$Pr(X = x) = \frac{(\lambda t)^x}{x!}\exp(-\lambda t). \tag{2.4}$$

where x can take values $0, 1, 2, \ldots$ and $\exp\{\cdots\}$ denotes e, the base of Napierian logarithms ($e = 2.718\ldots$) and $\exp(-\lambda t)$ denotes $e^{-\lambda t}$. Equation (2.4) is sometimes written, more conveniently, as

$$Pr(X = x) = \frac{(\lambda t)^x}{x!} e^{-\lambda t}. \tag{2.5}$$

A characteristic of the Poisson distribution is that the variance equals the mean. Thus, the variance of the number of events in a period of time t which have a Poisson distribution with mean λt is also λt.

An application to forensic science concerns the estimation of transfer probabilities. Experiments which have been carried out on the transfer and persistence of glass fragments suggest that the persistence of glass fragments on clothing can be described as a mixture of exponential decay curves. A possible model for the probability distribution for the number of fragments remaining at time t after the event which caused the transfer is then a Poisson distribution (Evett *et al.*, 1995). Note that in this example, however, it is not possible to have a mean number of fragments transferred per unit time as in the Geiger counter example. This is because of the nature of the exponential decay; the number of fragments left after a particular time decays, rather than increases. A Poisson model may still be used. The mean number of fragments remaining after time t is still a function of time t but not a linear function. The mean in this context will be denoted μ_t, where the subscript is used to denote the dependence on t without the implication that the mean is a linear function of t. Let X here denote the random variable corresponding to the number of fragments remaining after time t. Then, from (2.5) with μ_t replacing λt, the probability that X takes the value x, where x is a non-negative integer $(0, 1, 2, \ldots)$, is

$$Pr(X = x) = \frac{\mu_t^x}{x!} e^{-\mu_t}. \tag{2.6}$$

For practical application of this distribution, an estimate has to be obtained for μ_t. Imagine, for example, that, before examination of the clothing in the case under consideration, the expert was asked how many fragments would be expected to be found on the clothing of the suspect which matched (in some sense) a window which had been broken in the commission of the crime. Before answering this question, the expert would need background information about the circumstances surrounding the apprehension of the suspect. This information would include details of the clothing that the suspect was wearing when examined and an estimate of the time since the crime was committed and the suspect was apprehended. Given such information, suppose the expert replies 'about '4'. This value can then be substituted for μ_t in (2.6) and values for $Pr(X = x)$ can then be calculated. The values for $x = 0, 1, 2, 3$ and 4 are given in Table 2.3.

The final probability $Pr(X > 4) = 0.371$ is obtained from (1.9) where the events $X = 0$, $X = 1$, $X = 2$, $X = 3$, $X = 4$ and $X > 4$ form a partition and thus

Table 2.3 Probabilities that a random variable X with a Poisson distribution with mean 4 takes values $0, 1, 2, 3, 4$ and greater than 4

Value	Probability
0	0.018
1	0.073
2	0.147
3	0.195
4	0.195
>4	0.371

the probabilities of the six events sum to 1(actually 0.999 in Table 2.3 because of rounding errors). If the probability of what has been observed is very different from what is expected (4) then what has been observed is supportive of the proposition that the suspect is not the criminal. How one may measure support is a major component of this book.

In a method analogous to this, the approach has been used in the context of examinations of fibres to assess the value for the probability of the presence of extraneous fibres by chance on a receptor (Champod and Taroni, 1997). Support for the use of a Poisson distribution can be investigated using data from previous surveys to obtain estimates for the mean number of fibres transferred innocently under certain circumstances and, hence, the so-called background probabilities $\{b_i; i = 0, 1, 2, \ldots\}$ where b_i denotes the probability of an innocent transfer of i fibres of the type in question to clothing of a suspect or to a crime scene. (See also Chapters 10, 12 and 14 for further references to background probabilities.)

Another example of the use of a Poisson distribution is the count of the number of consecutive matching striations (CMS) on fired bullets. This count may be used as a criterion for the identification of a particular firearm as the one which fired a particular bullet. Two data sets can be examined, one for bullets fired from the same barrel and one for those fired from different barrels. These data sets can be used to characterise the within- and between-source populations, respectively. A hypothetical histogram is presented by Bunch (2000). The Poisson distribution is appropriate for the consideration of data under a probability distribution characterised by parameters θ (the maximum CMS count) and λ (the weighted average maximum CMS count, details of the calculation of which are in Section 7.3.2).

In the context of the transfer of glass fragments, available data indicate that the Poisson distribution is a plausible model, except that it tends to overestimate b_1, the probability of an innocent transfer of a single glass fragment of the type in question to clothing of a suspect or to a crime scene. Note that the assumption of a Poisson distribution is a simplification which is made for the purpose of exploring concepts. The deviation of b_1 from that predicted by the Poisson assumption is acceptable as the deviation provides a value for the likelihood ratio which is more supportive of the defence proposition (a conservative value) than the value for b_1 provided by the Poisson model (Curran *et al.*, 1998a).

2.3.7 Beta-binomial distribution

Consider again the example of a consignment of tablets, a proportion of which are suspected to be drugs. For large consignments, the probability distribution of the proportion θ which are drugs can be modelled with a beta distribution, which treats the proportion θ as a variable which is continuous over the interval $(0, 1)$. For small consignments, say $N < 50$, a more accurate distribution, which recognises the discrete nature of the possible values of the proportion, may be used.

Assume there are n units in that part of the consignment which has not been inspected such that $m + n = N$, the total consignment size. Let Y ($\leq n$ and unknown) be the number of units in the remainder of the consignment which contain drugs. The total number of units in the consignment which contain drugs is then $z + y$ ($\leq N$) where z is the number of units in the inspected part which contain drugs. Then ($Y \mid m, n, z, \alpha, \beta$) has a Bayesian predictive distribution known as the beta-binomial distribution (Bernardo and Smith, 1994) with

$$Pr(Y = y \mid m, n, z, \alpha, \beta)$$

$$= \frac{\Gamma(m + \alpha + \beta) \binom{n}{y} \Gamma(y + z + \alpha) \Gamma(m + n - z - y + \beta)}{\Gamma(z + \alpha) \Gamma(m - z + \beta) \Gamma(m + n + \alpha + \beta)}, (y = 0, 1, \ldots, n),$$

$$(2.7)$$

where

$$\Gamma(x + 1) = x! \quad \text{for integer } x > 0,$$

$$\Gamma(1/2) = \sqrt{\pi},$$

$$(2.8)$$

is the gamma function.

The derivation of this distribution requires a beta prior (Section 2.4.4) and a binomial model for the data (m, z). This gives a posterior distribution for the proportion. This is then combined with a binomial model for the uninspected portion (n, y) of the consignment to give the beta-binomial distribution above. Further details are given in Chapter 6 and Aitken (1999).

The beta-binomial distribution may be generalised to consider more than two categories, and the corresponding distribution is known as the *Dirichlet-multinomial* distribution.

For the example of a consignment of tablets, various proportions may be suspected of being various types of drugs. For large consignments, the probability distribution of the proportions $\{\theta_i, \ i = 1, \ldots, k\}$ of the various kinds of drugs can be modelled with a Dirichlet distribution, which treats the proportions θ_i as variables which are continuous over the interval $(0, 1)$.

As before, consider a consignment of tablets. A sample of size m has been inspected and z_i are found to be of type $i, i = 1, \ldots, k$, such that $\sum_{i=1}^{k} z_i = m$. Assume there are n units in that part of the consignment which has not been inspected such that $m + n = N$, the total consignment size. Let $(Y_i, i = 1, \ldots, k)$ be the numbers (unknown) of tablets in the remainder of the consignment which contain drugs.

The total number of tablets in the consignment of type i is then $z_i + y_i \ (\leq N)$. Then $(Y_i \mid m, n, z_1, \ldots, z_k, \alpha_1, \ldots, \alpha_k)$ has the Bayesian predictive distribution known as the Dirichlet-multinomial distribution (Bernardo and Smith, 1994) with

$$Pr(Y_1 = y_1, \ldots, Y_k = y_k \mid m, n, z_1, \ldots, z_k, \ \alpha_1, \ldots, \alpha_k)$$

$$= \frac{\Gamma(m + \sum_{i=1}^{k} \alpha_i) \dfrac{n!}{y_1! \cdots y_k!} \prod_{i=1}^{k} \Gamma(y_i + z_i + \alpha_i)}{\prod_{i=1}^{k} \Gamma(z_i + \alpha_i) \ \Gamma(m + n + \sum_{i=1}^{k} \alpha_i)}, 0 \leq y_i \leq n; \sum_{i=1}^{k} y_i = n.$$

$$(2.9)$$

The derivation of this distribution requires a Dirichlet prior (Section 2.4.5) and a multinomial model (Section 2.3.4) for the data $(m, z_1, \ldots, z_k)$. This gives a posterior distribution for the proportions of each of the k types. This distribution is then combined with a multinomial model for the uninspected portion $(n, y_1, \ldots, y_k)$ of the consignment to give the Dirichlet-multinomial distribution above. A brief further reference is given in Chapter 6.

2.4   MEASUREMENTS

2.4.1   Summary statistics

Consider a population of continuous measurements with mean μ and standard deviation σ.

Given sample data, $(x_1, x_2, \ldots, x_n)$, of measurements from this population, μ and σ may be estimated from the sample data as follows. The sample mean, denoted $\bar{x}$, is defined by

$$\bar{x} = \sum_{i=1}^{n} x_i / n. \tag{2.10}$$

The sample standard deviation, denoted s, is defined as the square root of the sample variance, s^2, which is itself defined by

$$s^2 = \sum_{i=1}^{n} (x_i - \bar{x})^2 / (n - 1). \tag{2.11}$$

Expression (2.11) can also be calculated as

$$s^2 = \left\{ \sum_{i=1}^{n} x_i^2 - \left(\sum_{i=1}^{n} x_i \right)^2 \middle/ n \right\} \middle/ (n-1).$$ (2.12)

As an example of the calculations, consider the following five measurements of the medullary widths, in microns, of cat hairs ($n = 5$):

x_1	x_2	x_3	x_4	x_5
17.767	18.633	19.067	19.300	19.933

Then, from (2.10)

$$\bar{x} = \sum_{i=1}^{n} x_i/n = 94.700/5 = 18.9400.$$

From (2.12),

$$s^2 = \left\{ \sum_{i=1}^{n} x_i^2 - \left(\sum_{i=1}^{n} x_i \right)^2 \middle/ n \right\} \middle/ (n-1) = (1796.220 - 94.700^2/5)/4 = 0.6505,$$

and the sample standard deviation is

$$s = \sqrt{0.6505} = 0.8065.$$

Note that the sample mean and standard deviation are quoted to one more significant figure than the original measurements.

2.4.2 Normal distribution

When considering data in the form of counts, the variation in the possible outcomes can be represented by a function known as a *probability function*. The variation in measurements, which are continuous, may also be represented mathematically by a function known as a *probability density function*. Probability functions and probability density functions are both examples of *probability models*.

As an example of a probability model for a continuous measurement, consider the estimation of the quantity of alcohol in blood. From experimental results, it has been determined that there is variation in the measurements, x (in g/kg), provided by a certain procedure. The variation is such that it may be represented by a probability density function which in this case is unimodal, symmetric and bell-shaped. The particular function which is used here is the *Normal* or *Gaussian* probability density function (named after the German mathematician Carl Friedrich Gauss, 1777–1855).

The binomial distribution required the number of trials and the probability of a success to be known in order that the probability function could be defined. Two characteristics (or parameters) of the measurement are required to define the Normal probability density function. These are the mean, or expectation, θ, and the standard deviation, σ. The mean may be thought of as a *measure of location* to indicate the size of the measurements. The standard deviation may be thought of as a *measure of dispersion* to indicate the variability in the measurements. The square of the standard deviation, the variance, is denoted σ^2. Given these parameters, the Normal probability density function for x, $f(x \mid \theta, \sigma^2)$, is given by

$$f(x \mid \theta, \sigma^2) = \frac{1}{\sqrt{2\pi\sigma^2}} \exp\left\{-\frac{(x-\theta)^2}{2\sigma^2}\right\}. \qquad (2.13)$$

The function $f(x \mid \theta, \sigma^2)$ is a function which is symmetric about θ. It takes its maximum value when $x = \theta$, it is defined on the whole real line for $-\infty < x < \infty$ and is always positive. The area underneath the function is 1, since x has to lie between $-\infty$ and ∞.

In some countries if the alcohol level in blood is estimated to be greater than 0.8 g/kg a person is considered to be under the influence of alcohol. The variability inherent in a measurement, x, of alcohol quantity is known from previous experiments to be such that it is Normally distributed about the true value θ with variance, σ^2, of 0.005. Consider a person whose quantity, θ, of alcohol in the blood is 0.7 g/kg. The probability density function $f(x \mid \theta, \sigma^2)$ for the measurement of the quantity of alcohol in the blood is then obtained from (2.13) with the substitution of 0.7 for θ and 0.005 for σ^2. The function is illustrated in Figure 2.1. Note the labelling of the ordinate as 'probability density'. The reasoning for this is described later in this section. In particular, it is possible for the probability density function to take values greater than 1.

There is a special case of zero mean ($\theta = 0$) and unit variance ($\sigma^2 = 1$). The Normal probability density function is then

$$f(z \mid 0, 1) = \frac{1}{\sqrt{2\pi}} \exp\left(-\frac{z^2}{2}\right), \qquad (2.14)$$

where z is used instead of x to denote the special nature of parameter values of zero mean and unit variance. The Normal probability density function is so common that it has special notation. If a random variable Z is Normally distributed with mean 0 and variance 1, it is denoted

$$Z \sim N(0, 1),$$

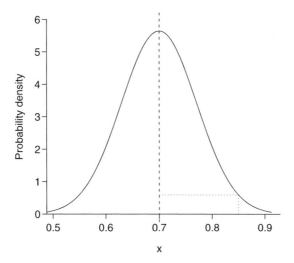

Figure 2.1 Probability density function for a Normal distribution, with mean 0.7 and variance 0.005.

where Z is the random variable corresponding to z and the conditioning on $\theta = 0$ and $\sigma^2 = 1$ on the left-hand-side has been omitted for clarity. This distribution is known as a standard Normal distribution. In general, a Normally distributed measurement, X say, with mean θ and variance σ^2, may be said to be such that

$$(X \mid \theta, \sigma^2) \sim N(\theta, \sigma^2).$$

The first symbol within parentheses on the right conventionally denotes the mean, the second conventionally denotes the variance. It is not always necessary for the notation to make explicit the dependence of X on θ and σ^2. The distributional statement may then be denoted

$$X \sim N(\theta, \sigma^2),$$

and such abbreviated notation will be used often.

The determination of probabilities associated with Normally distributed random variables is made possible by a process known as *standardisation*, whereby a general Normally distributed random variable is transformed into one which has a standard Normal distribution. Let

$$Z = (X - \theta)/\sigma.$$

Then $E(Z) = 0$ and $Var(Z) = 1$ and the random variable Z has a standard Normal distribution. Notice that standardisation requires variability, as represented by σ, to be taken into account. For example, the division by σ ensures the resulting statistic is dimensionless.

Consider the following numerical example of blood alcohol measurements using the parameter values above. Let X be the random variable of measurements of blood alcohol for a particular person, with x denoting the value of a particular measurement. Suppose the true, unknown, level of alcohol in the person's blood is $\theta = 0.7\,\text{g/kg}$ and the standard deviation in the measurement is σ, the square root of 0.005, which equals 0.07 (to two decimal places). The measurement x of the blood alcohol quantity recorded by the measuring apparatus is $0.85\,\text{g/kg}$ which is over the permitted limit of $0.8\,\text{g/kg}$. Note that θ is unknown in this case. The variance σ^2 is assumed known as it has been estimated from many previous experiments with the measuring apparatus and it is assumed to be a constant, independent of θ. Substitution of $x = 0.85$ g/kg, $\theta = 0.7$ g/kg and $\sigma^2 = 0.005$ into (2.13) gives

$$ f(x) = f(0.85) = \frac{1}{\sqrt{0.01\pi}} \exp\left(- \frac{(0.85 - 0.7)^2}{0.01} \right) = 0.60; \qquad (2.15) $$

see Figure 2.1. In practice, what is of interest is the probability that the true blood alcohol level is greater than $0.8\,\text{g/kg}$, when the instrument provides a measurement of $0.85\,\text{g/kg}$. This requires consideration of a prior distribution for θ and is discussed in detail in Section 5.5.

Consider the continuous case in more detail. The function modelling the variation is known as a probability density function, not a probability function as it does not measure probabilities. An intuitive understanding of the terminology can be gained by considering the following analogy. A cylindrical rod, with circular cross-section, has a density which varies along its length according to some function, f say. Then its weight over any particular part of its length is the integral of this function f over that part. In the same way, with a probability density function, the probability of a random variable lying in a certain interval is the integral of the corresponding density function over the interval. Thus the probability of the measurement of the blood alcohol quantity x lying within a certain interval would be the integral of $f(x)$ over this interval. Note, however, the following theoretical detail. A cross-section of zero thickness of the rod would have zero weight since its volume would be zero. Similarly, the probability of a continuous random variable taking a particular value is zero. In practice, measuring instruments are not sufficiently accurate to measure to an infinite number of decimal places and this problem does not arise so long as one determines the probability of a measurement lying within a particular interval and does not attempt to calculate the probability of a measurement taking a particular value. (See Section 4.5.5 for an application of this idea.)

The Normal probability cannot be determined analytically and reference has to be made to tables of probabilities of the standard Normal distribution or to statistical packages.

Let Z be a random variable with a standard Normal distribution, thus

$$ Z \sim N(0, 1). $$

The probability that Z is less than a particular value z, $Pr(Z < z)$ is denoted $\Phi(z)$. Certain values of z are used commonly in the discussion of significance probabilities, particularly those values for which $1 - \Phi(z)$ is small, and some of these are tabulated in Table 2.4. Corresponding probabilities for absolute values of Z may be deduced from the tables by use of the symmetry of the Normal distribution. By symmetry,

$$\Phi(-z) = Pr(Z < -z) = 1 - Pr(Z < z) = 1 - \Phi(z).$$

Thus

$$
\begin{aligned}
Pr(|Z| < z) &= Pr(-z < Z < z) \\
&= Pr(Z < z) - Pr(Z < -z) \\
&= \Phi(z) - \Phi(-z) \\
&= 2\Phi(z) - 1.
\end{aligned}
$$

Particular, commonly used, values of z with the corresponding probabilities for the absolute values of z are given in Table 2.5. Figure 2.2 illustrates the probabilities for the following events: (a) $Pr(Z > 1) = 0.159$; (b) $Pr(Z > 2) = 0.023$; (c) $Pr(|Z| < 2) = Pr(-2 < Z < 2) = 0.954$; and (d) $Pr(Z > 2.5) = 0.006$.

Table 2.4 Values of cumulative distribution function $\Phi(z)$ and its complement $1 - \Phi(z)$ for the standard Normal distribution for given values of z

z	$\Phi(z)$	$1 - \Phi(z)$
1.6449	0.950	0.050
1.9600	0.975	0.025
2.3263	0.990	0.010
2.5758	0.995	0.005

Table 2.5 Probabilities for absolute values from the standard Normal distribution function

z	$\Phi(z)$	$Pr(|Z| < z) = 2\Phi(z) - 1$	$Pr(|Z| > z)$
1.6449	0.950	0.90	0.10
1.9600	0.975	0.95	0.05
2.3263	0.990	0.98	0.02
2.5758	0.995	0.99	0.01

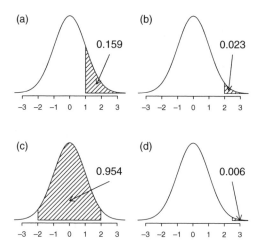

Figure 2.2 Selected tail area probabilities for a standard Normal random variable Z: (a) $Pr(Z > 1)$, (b) $Pr(Z > 2)$, (c) $Pr(|Z| < 2)$, (d) $Pr(Z > 2.5)$.

An interval, known as a confidence interval, for the mean θ of a random variable X with an $N(\theta, \sigma^2)$ distribution may be determined from the expression

$$Pr\left(-d_{\alpha/2} < \frac{X - \theta}{\sigma} < d_{\alpha/2}\right) = 1 - \alpha, \tag{2.16}$$

where $Pr(Z > d_{\alpha/2}) = Pr(Z < -d_{\alpha/2}) = \alpha/2$ and $Z = (X - \theta)/\sigma \sim N(0, 1)$. Rearrangement of (2.16) shows that

$$Pr\left(X - d_{\alpha/2}\sigma < \theta < X + d_{\alpha/2}\sigma\right) = 1 - \alpha. \tag{2.17}$$

Replacement of the random variable X with an observation x gives the interval $x - d_{\alpha/2}\sigma < \theta < x + d_{\alpha/2}\sigma$, which is said to be a $100(1 - \alpha)\%$ confidence interval for θ. For example, if $\alpha = 0.05$, the 95% confidence interval for θ is

$$(x - 1.96\sigma, x + 1.96\sigma),$$

where the figure 1.96 is taken from Table 2.5. Note that a 95% confidence interval for θ means that if an experiment is repeated many times under identical conditions, 95% of the confidence intervals estimated will cover the true value of the parameter of interest. This does not means that there is a 95% probability that the true value is in the estimated interval. For further details, see Kaye (1987) and Section 5.6.

Some variables, including blood alcohol level, can only take positive values. If the mean is sufficiently far away from zero, in units of standard deviations, then the probability the variable takes a value less than zero can effectively be discounted. In some cases, the distribution may be *positively skewed* in the sense that the right-hand tail of the distribution is much longer than the left-hand tail and the distribution is asymmetric (e.g., the mean is greater than the median). (A distribution in which the left-hand tail of the distribution is much longer than the right-hand tail and the mean is less than the median is said to be *negatively skewed*.) For positively skewed distributions, a transformation to the logarithm of the variable of interest will often produce a variable which is more symmetric than the original variable and for which inferences based on the Normal distribution may be used. Care must then be taken to transform the results back to the original measurements for the final summary.

Normal approximations to the binomial and Poisson distributions

One of the advantages of the Normal distribution is that it can be used as an approximation to other distributions in situations where it may be impractical or just simply tedious (such as in the absence of a suitable statistical package to do the sums) to use the other distributions. Two examples are the binomial and Poisson distributions. These are two discrete distributions where it is tedious to evaluate exact probabilities for large numbers of events. For example, while possible, it is tedious to evaluate exactly the probability of 530 or fewer heads in 1000 tosses of a fair coin.

Let X be a random variable with a binomial distribution with n trials and success probability θ so that $E(X) = n\theta$ and $Var(X) = n\theta(1 - \theta)$. For n large and θ not too close to 0 or 1, the distribution of X may be approximated by a Normal distribution with the same mean and variance. Thus, approximately,

$$X \sim N\left(n\theta, n\theta(1 - \theta)\right).$$

Also, $E(X/n) = \theta, Var(X/n) = \theta(1 - \theta)/n$ and

$$X/n \sim N(\theta, \theta(1 - \theta)/n). \tag{2.18}$$

In answer to the question posed immediately above, let X denote the number of heads in 1000 tosses of a fair coin. Then

$$Pr(X \leqslant 530 \mid n = 1000, \theta = 0.5) = \sum_{x=0}^{530} \binom{n}{x} 0.5^{1000},$$

a sum which can be evaluated, given time. Alternatively, the Normal approximation is the following

$$Pr(X \leqslant 530 \mid n = 1000, \theta = 0.5) \simeq \Phi\left((530.5 - 500)/\sqrt{250}\right)$$
$$= \Phi(1.929) = 0.9731,$$

where 0.5 is added to 530 to allow for the approximation of a discrete distribution by a continuous distribution. For the discrete distribution, X takes integer values only ($\ldots$, 529, 530, 531, $\ldots$) whereas for the continuous distribution, X can take any value. In this example, the value 530.5 is chosen as being the value midway between 530 and the value 531 immediately above it. The exact probability evaluated by statistical software is 0.9732. The approximation is excellent.

Let $y_1, \ldots, y_n$ be n observations for a Poisson distribution with mean λ. Let $\bar{Y}$ be the random variable corresponding to the sample mean. Then

$$\bar{Y} \sim N(\lambda, \lambda/n). \tag{2.19}$$

2.4.3   Student's *t*-distribution

In practice, the standard deviation σ of data from a Normal distribution is rarely known and it is estimated from the data by the sample standard deviation s. Consider n independent, identically distributed Normal random variables $X_1, X_2, \ldots, X_n$ such that

$$X_i \sim N(\theta, \sigma^2), \qquad i = 1, \ldots, n.$$

It can be shown that the random variable $\bar{X}$ corresponding to the mean of $X_1, \ldots, X_n$, and given by

$$\bar{X} = \sum_{i=1}^{n} X_i/n$$

has itself a Normal distribution, such that

$$\bar{X} \sim N(\theta, \sigma^2/n).$$

The transformed, standardised variable Z, defined as

$$Z = (\bar{X} - \theta)/(\sigma/\sqrt{n})$$

has a standard Normal $N(0, 1)$ distribution.

If the standard deviation, σ, is not known and it is replaced by the random variable S corresponding to its estimate s, the resulting statistic is

$$(\bar{X} - \theta)/(S/\sqrt{n}). \tag{2.20}$$

This statistic does not have a standard Normal distribution. It is the ratio of two random variables $\bar{X}$ and S. The distribution is known as *Student's t-distribution* and the corresponding statistic is known as a *t*-statistic. ('Student' was the pseudonym of W.S. Gosset, 1876–1937.) The distribution is symmetric about zero. The extra uncertainty induced by replacing σ with an estimate s leads to the t-distribution having greater dispersion than the standard Normal distribution. Also, the distribution depends on the sample size, n. In particular, if the standard deviation s is estimated from a sample of n observations $x_1, \ldots, x_n$ for use in the statistic (2.20) then the value $n-1$ is known as the number of *degrees of freedom* associated with the t-statistic. The degrees of freedom are determined from the denominator of the expression (2.11) used to derive s. Informally, the number of degrees of freedom may be considered as the number of observations free to estimate s after 1 has been deducted from n to estimate $\bar{x}$. Given the values of $n-1$ observations and the value of the mean $\bar{x}$ on n observations, it is possible to determine the value of the nth observation from the expression

$$x_n = n\bar{x} - \sum_{i=1}^{n-1} x_i.$$

As n increases, the t-distribution approaches the standard Normal distribution. As with the standard Normal distribution, the probability cannot be determined analytically and reference has to be made to statistical software or tables of probabilities of the t-distribution.

Some probabilities for the t-distribution are given in Table 2.6 where $t_{(n-1)}(P)$ is the value of t from a t-distribution with $n-1$ degrees of freedom (denoted $t_{(n-1)}$) such that the probability the random variable T (with a $t_{(n-1)}$ distribution) is greater than $t_{(n-1)}(P)$ is $P/100$. For example, when the sample size n is 20, the probability that T is greater than 2.539 is $1/100$ or 0.01.

There is a more general form of the t-distribution which is not centred about zero, known as the non-central t-distribution. There are three parameters, which will be denoted μ, λ and ν. If X has such a non-central t-distribution then the transformed variable $Y = (X - \mu)/\lambda$ has a (central) t-distribution with ν degrees of freedom. An example of the use of this distribution is given in Section 10.6 to determine the value of the numerator in the evaluation of glass fragments, where μ is a control mean and λ is an estimate of the standard deviation of the

Table 2.6   Percentage points $t_{(n-1)}(P)$ for the t-distribution for given values of sample size n, degrees of freedom $(n-1)$ and P, and the corresponding point $z(P)$ for the standard Normal distribution

$P\%$	$(100-P)\%$	n	$(n-1)$	$t_{(n-1)}(P)$	$z(P)$
95	5	10	9	1.833	1.645
95	5	20	19	1.729	1.645
99	1	10	9	2.821	2.326
99	1	20	19	2.539	2.326
99.5	0.5	10	9	3.250	2.576
99.5	0.5	20	19	2.861	2.576

refractive index of the population of glass fragments from which the recovered fragments have come. The value of the numerator is the value of the central t-density at the appropriate point, with an adjustment by multiplication of the density value by a factor of $1/\lambda$ to allow for the standardisation.

2.4.4 Beta distribution

Consider an example in which it is desired to know the proportion of a consignment which is illicit drugs. This example has been discussed in Section 2.3.5 with reference to a consignment of tablets. The number of tablets which are illicit is R and the consignment size is N. The proportion of illicit tablets is then R/N, which takes a finite number of values, depending on the value of R, ranging from $0/N$ to N/N in steps of $1/N$. As N increases, this proportion becomes closer to a continuous measurement, over the interval $(0, 1)$. The variability in a continuous random variable which is a proportion is modelled by the beta distribution. Denote the random variable by θ. For a consignment of drugs, assume that it is representative of a super-population of drugs in which the proportion of illicit tablets is $\theta (0 < \theta < 1)$. See Smith and Charrow (1975) and Finney (1977) for comments about super-populations and also Section 3.3.5. For example, the consignment may be known to have come from a particular location and θ is the proportion of units in the super-population which contain drugs. In order to make probability statements about θ it is necessary to have a probability distribution for θ to represent the variability in θ. This variability may simply be uncertainty in one's knowledge of the exact value of θ, uncertainty which may arise because the consignment is considered as a random sample from a super-population. The Bayesian philosophy permits this uncertainty to be represented as a probability distribution. The most common distribution for θ is the so-called beta distribution, characterised by two parameters, α and β, with probability density function

$$f(\theta \mid \alpha, \beta) = \frac{\theta^{\alpha-1}(1-\theta)^{\beta-1}}{B(\alpha, \beta)}, \qquad 0 < \theta < 1, \qquad (2.21)$$

denoted $Be(\alpha, \beta)$, where

$$B(\alpha, \beta) = \frac{\Gamma(\alpha)\Gamma(\beta)}{\Gamma(\alpha+\beta)},$$

and Γ is the gamma function (2.8). The mean is $\alpha/(\alpha+\beta)$ and the variance is $\alpha\beta/(\alpha+\beta)^2(\alpha+\beta+1)$. The function $B(\alpha, \beta)$ is known as the *beta function*.

The use of the beta distribution in this context is described in Aitken (1999). Values for α and β may be chosen subjectively to represent the scientist's prior beliefs before inspection about the proportion of the units in the consignment (as a random sample from the super-population) which contain drugs. A large value of α relative to β would imply a belief that θ was high. Larger values of α and β would correspond to higher certainty about the value of θ. A detailed discussion is given in Aitken (1999) and summarised in Chapter 6. In many cases, the scientist will not wish to quantify his prior beliefs and will wish to remain neutral. This can be done by choosing $\alpha = \beta = 1$. Also, as shown in Aitken (1999), for variations in α and β, when both are small, the evidence from the sample will soon reduce the effect of the values of α and β considerably. This is intuitively reasonable: little prior information is soon subsumed by the data.

A variation of the beta density function for θ when θ has a range $(0, n)$ is

$$f(\theta \mid \alpha, \beta, n) = \frac{1}{B(\alpha, \beta)} \frac{\theta^{\alpha-1}(n-\theta)^{\beta-1}}{n^{\alpha+\beta-1}}, \qquad 0 < \theta < n. \qquad (2.22)$$

2.4.5 Dirichlet distribution

The example in which it is desired to know the proportion of a consignment which is illicit drugs may be generalised to a situation in which there may be several (say, k) drug types and it is desired to know the proportions in each type. Consider again a consignment of tablets of size N. Denote the number of tablets in each of the k types as $R_i, i = 1, \ldots, k$. The proportion of tablets of type i is then R_i/N, which takes a finite number of values, depending on the value of R_i, ranging from $0/N$ to N/N in steps of $1/N$. As N increases, these proportions become closer to a continuous measurement, over the interval $(0, 1)$. The variability in a set of random variables which are proportions and for which the sum is 1 is modelled by a generalisation of the beta distribution, known as the Dirichlet distribution. This generalisation is analogous to the generalisation of the binomial distribution (Section 2.3.3) by the multinomial distribution (Section 2.3.4) .

Denote the set of random variables by $\theta_i, i = 1, \ldots, k$, which are such that $\sum_{i=1}^{k} \theta_i = 1$. The beta distribution is the case described here for which $k = 2$ and, conventionally, θ_1 is denoted θ and $\theta_2 = 1 - \theta_1 = 1 - \theta$. For a consignment of drugs, assume, as before, that it is representative of a super-population of drugs in which the proportion of tablets in each of the k categories is $\theta_i, i = 1, \ldots, k; 0 < \theta_i < 1; \sum_{i=1}^{k} \theta_i = 1$). For example, the consignment may be known to have come from a particular location and the set $\{\theta_i; i = 1, \ldots, k\}$ are the proportions of units in the super-population which fall into the k categories. In order to make probability statements about $\{\theta_i\}$ it is necessary to have a probability distribution for $\{\theta_i\}$ to represent the variability in $\{\theta_i\}$. This variability may simply be uncertainty in one's knowledge of the exact values of $\{\theta_i\}$, uncertainty which may arise because the consignment is considered as a random sample from a super-population. The Dirichlet distribution is the most common distribution for $\{\theta_i; i = 1, \ldots, k\}$ with probability density function

$$f(\theta_1, \ldots, \theta_k \mid \alpha_1, \ldots, \alpha_k) = \frac{\theta_1^{\alpha_1 - 1} \cdots \theta_k^{\alpha_k - 1}}{B(\alpha_1, \ldots, \alpha_k)},$$

$$0 < \theta_i < 1, i = 1, \ldots, k, \ \sum_{i=1}^{k} \theta_i = 1, \quad (2.23)$$

where

$$B(\alpha_1, \ldots, \alpha_k) = \frac{\Gamma(\alpha_1) \cdots \Gamma(\alpha_k)}{\Gamma(\alpha_1 + \cdots + \alpha_k)}.$$

The mean $E(\theta_i)$ of θ_i is $\alpha_i / (\sum_{i=1}^{k} \alpha_i)$ and the variance of θ_i is $E(\theta_i)(1 - E(\theta_i))/(1 + \sum_{i=1}^{k} \alpha_i)$. The $\{\theta_i\}$ sum to 1 so they are correlated. The covariance $Cov(\theta_i, \theta_j)$, $i \neq j$, between θ_i and θ_j is given by $Cov(\theta_i, \theta_j) = -E(\theta_i)E(\theta_j)/(1 + \sum_{i=1}^{k} \alpha_i)$. Note that this is negative; given the value of θ_i, the range of values for θ_j is reduced from $(0,1)$ to $(0, 1 - \theta_i)$.

The Dirichlet distribution is characterised by k parameters, $\{\alpha_1, \ldots, \alpha_k\}$. Values for $\{\alpha_1, \ldots, \alpha_k\}$ may be chosen subjectively to represent the scientist's prior beliefs before inspection about the proportions of the units in the consignment (as a random sample from the super-population) for each of the k categories.

Consider a single-locus probe for DNA profiling. Let (X_1, X_2) be the sample frequencies of the two bands of the locus found on a crime scene profile. The overall sample size, from which X_1 and X_2 are obtained, is n. Let $X_3 = n - X_1 - X_2$. The corresponding population relative frequencies are θ_1, θ_2 and θ_3 with $\sum_{i=1}^{3} \theta_i = 1$. The Dirichlet distribution provides a convenient prior distribution for the $\{\theta_i\}$, with three categories, $k = 3$.

Further details of the above example and inferences which may be drawn from study of the crime scene profile are given in Balding (1995). An example of the use of the Dirichlet distribution as a prior for a multinomial likelihood for blood grouping data is given in Leonard and Hsu (1999, pp. 195–196). Applications to forensic match probabilities are described in Lange (1995).

2.4.6 Multivariate Normal and correlation

Often, more than one characteristic is of interest, for example, the refractive index, the density and various elemental compositions for window glass. The data (measurements of these characteristics) are referred to as *multivariate data* and in the special case where only two characteristics are measured they are known as *bivariate data*. Let the vector of measurements be denoted **x**. (A vector in mathematics is a list of more than one characteristic associated with a unit of interest. A scalar is used to denote one characteristic.) There is a notational convention as to how a vector is written. A vector **x** is conventionally written in bold script and expanded as a column (illustrated here with p variables)

$$\mathbf{x} = \begin{pmatrix} x_1 \\ x_2 \\ \vdots \\ x_p \end{pmatrix}.$$

The T notation denotes the transpose of the column vector to a row vector such that $\mathbf{x}^T = (x_1, x_2, \ldots, x_p)$.) For bivariate data $p = 2$. In the example of window glass, x_1 would be the value of the refractive index, x_2 the value of the density, and the elemental compositions would be denoted $x_3, x_4, \ldots$. For continuous data, the vector **x** has a probability density function, just as the individual characteristics have.

If the characteristics are independent (Section 1.6.5) then the joint probability density function $f(\mathbf{x})$ is the product of the individual probability density functions. Thus

$$f(\mathbf{x}) = f(x_1, \ldots, x_p) = \prod_{i=1}^{p} f(x_i), \qquad (2.24)$$

which may be thought of as an extension of the third law of probability for independent events (1.6).

If the characteristics are not independent, however, such an approach is not possible. Assume the measurements of these characteristics are Normally distributed and dependent. The measurements are said to be *correlated*. A multivariate analogue of the Normal distribution may be obtained. The multivariate mean $\boldsymbol{\theta}$ is the vector formed by the means of the individual variables. Instead of a variance σ^2 there is a square $(p \times p)$ symmetric matrix $\boldsymbol{\Sigma}$ of variances and *covariances*. (See Section 11.9 for a brief introduction to matrices and the associated terminology). The matrix $\boldsymbol{\Sigma}$ is known as the *covariance matrix*. Covariance is a measure of the association between a pair of characteristics and is the product of the individual standard deviations and a factor which measures the *correlation* (degree of association) between the two characteristics. The variances of the p variables are located on the diagonal of $\boldsymbol{\Sigma}$. The covariances are the off-diagonal terms so that the (i, j)th cell of $\boldsymbol{\Sigma}$ contains the covariance between X_i and X_j (the covariance of X_i and X_i is simply the variance of X_i). The correlation between two variables is a parameter which measures the amount of linear association between the variables. It takes values between -1 and 1. Two variables which have a perfect linear relationship with a positive slope (as one increases so does the other) have a correlation of 1. Two variables which have a perfect linear relationship with a negative slope (as one increases, the other decreases) have a correlation of -1. A correlation of 0 implies that there is no linear association between the two variables. Notice that this does not mean there is no association between the variables, only that the association is not linear.

Denote the variance of X_i by σ_i^2, $i = 1, \ldots, p$. The correlation between X_i and X_j is denoted by the correlation coefficient ρ_{ij} ($i = 1, \ldots, p$; $j = 1, \ldots, p$; $i \neq j$). Then the covariance between X_i and X_j is given by

$$\mathrm{Cov}(X_i, X_j) = \rho_{ij}\sigma_i\sigma_j, \qquad i = 1, \ldots, p; \ j = 1, \ldots, p.$$

Denote the determinant of $\boldsymbol{\Sigma}$ by $|\boldsymbol{\Sigma}|$ and the inverse by $\boldsymbol{\Sigma}^{-1}$. Then the probability density function of $\mathbf{X}$ is given by

$$f(\mathbf{x}) = (2\pi)^{-\frac{1}{2}p} \, |\boldsymbol{\Sigma}|^{-\frac{1}{2}} \, \exp\left\{-\frac{1}{2}(\mathbf{x} - \boldsymbol{\theta})^T \boldsymbol{\Sigma}^{-1}(\mathbf{x} - \boldsymbol{\theta})\right\}. \qquad (2.25)$$

This may be written in short-hand, equivalent to the univariate case, as

$$(\mathbf{X} \mid \boldsymbol{\theta}, \boldsymbol{\Sigma}) \sim \mathrm{N}(\boldsymbol{\theta}, \boldsymbol{\Sigma}). \qquad (2.26)$$

Consider the special case of $p = 2$. The multivariate Normal distribution is then called the bivariate Normal distribution. The vector parameters may be written out in full:

$$\boldsymbol{\theta} = \begin{pmatrix} \theta_1 \\ \theta_2 \end{pmatrix},$$

$$\boldsymbol{\Sigma} = \begin{pmatrix} \sigma_1^2 & \rho\sigma_1\sigma_2 \\ \rho\sigma_1\sigma_2 & \sigma_2^2 \end{pmatrix}, \qquad (2.27)$$

$$\boldsymbol{\Sigma}^{-1} = \frac{1}{1 - \rho^2} \begin{pmatrix} \sigma_1^{-2} & -\rho/\sigma_1\sigma_2 \\ -\rho/\sigma_1\sigma_2 & \sigma_2^{-2} \end{pmatrix},$$

$$|\boldsymbol{\Sigma}|^{\frac{1}{2}} = \sigma_1\sigma_2\sqrt{(1 - \rho^2)}.$$

The matrix $\boldsymbol{\Sigma}^{-1}$ is sometimes denoted $\boldsymbol{\Omega}$ (see Section 4.5.5). Notice that

$$(\mathbf{x} - \boldsymbol{\theta})^T \boldsymbol{\Sigma}^{-1}(\mathbf{x} - \boldsymbol{\theta}) = \left\{\frac{(x_1 - \theta_1)^2}{\sigma_1^2} - 2\rho\frac{(x_1 - \theta_1)(x_2 - \theta_2)}{\sigma_1\sigma_2} + \frac{(x_2 - \theta_2)^2}{\sigma_2^2}\right\} \Big/ (1 - \rho^2).$$

The bivariate Normal density function may then be written as

$$f(x_1, x_2) = \frac{1}{2\pi\sigma_1\sigma_2\sqrt{(1 - \rho^2)}}$$

$$\times \exp\left[-\frac{1}{2(1 - \rho^2)} \left\{\frac{(x_1 - \theta_1)^2}{\sigma_1^2} - 2\rho\frac{(x_1 - \theta_1)(x_2 - \theta_2)}{\sigma_1\sigma_2} + \frac{(x_2 - \theta_2)^2}{\sigma_2^2}\right\}\right].$$

For the special case in which $\theta_1 = \theta_2 = 0$,

$$f(x_1, x_2) = \frac{1}{2\pi\sigma_1\sigma_2\sqrt{(1 - \rho^2)}} \exp\left\{-\frac{1}{2(1 - \rho^2)}\left(\frac{x_1^2}{\sigma_1^2} - 2\rho\frac{x_1 x_2}{\sigma_1\sigma_2} + \frac{x_2^2}{\sigma_2^2}\right)\right\}.$$

Another special case is when $\rho = 0$. The bivariate Normal density function is then

$$f(x_1, x_2) = \frac{1}{2\pi\sigma_1\sigma_2} \exp\left[-\frac{1}{2}\left\{\frac{(x_1 - \theta_1)^2}{\sigma_1^2} + \frac{(x_2 - \theta_2)^2}{\sigma_2^2}\right\}\right],$$

which may be written as

$$\frac{1}{2\sqrt{\pi\sigma_1^2}} \exp\left[-\frac{1}{2}\left\{\frac{(x_1 - \theta_1)^2}{\sigma_1^2}\right\}\right] \times \frac{1}{2\sqrt{\pi\sigma_2^2}} \exp\left[-\frac{1}{2}\left\{\frac{(x_2 - \theta_2)^2}{\sigma_2^2}\right\}\right].$$

This is the product of the probability density functions for two Normal distributions, one with mean θ_1 and variance σ_1^2 and one with mean θ_2 and variance σ_2^2. It is an example of (2.24) with $p = 2$.

Applications are given in Chapter 11 and Sections 4.5.5 and 4.6.3.

3

The Evaluation of Evidence

3.1 ODDS

3.1.1 Complementary events

There is a measure of uncertainty, known as *odds* (also known as *betting quotients*) which will be familiar to people who know about gambling. Bookmakers quote odds in sporting events such as horse races or football matches. For example, a particular horse may be given odds of '6 to 1 against' it winning a race or a particular football team may be quoted at odds of '3 to 2 on' to win a match or, equivalently, '3 to 2 in favour' of winning the match. Odds are equivalent to probability. The above phrases can be related directly to probability statements about the probability of the horse winning its race or the football team winning its match.

First, an event, known as the negation, or *complement*, of another event has to be introduced and given some notation. Let R be an event. Then the negation or complement of R, denoted $\bar{R}$, is the event which is true when R is false and false when R is true. The events R and $\bar{R}$ are known as *complementary events*. Often in this book comparison will be made of the probability of the evidence under two competing propositions, that put forward by the prosecution and that put forward by the defence. The proposition put forward by the prosecution will be denoted H_p. The proposition put forward by the defence will be denoted by H_d. The subscripts p and d denote *prosecution* and *defence*, respectively. The letter H denotes *hypothesis* and has stuck despite the more common usage now of the term *proposition*. These two propositions may denote complementary events, such as 'Guilty' and 'Not guilty'. However, there will be occasions on which the events denoted are not complementary, such as 'Suspect A and one unknown person were present at the crime scene' and 'Two unknown people were present at the crime scene'. There are many other events not covered by these two propositions, such as the naming of the two individuals at the crime scene and the consideration of less than, or more than, two people at the crime scene.

Statistics and the Evaluation of Evidence for Forensic Scientists: Second Edition
C.G.G. Aitken and F. Taroni © 2004 John Wiley & Sons, Ltd   ISBN: 0-470-84367-5

3.1.2   Examples

1. A coin is tossed. Let R be the event it lands heads. Then $\bar{R}$ is the event that it lands tails. If the coin is fair, $Pr(R) = 1/2, Pr(\bar{R}) = 1/2$.
2. A die is rolled. Let R be the event that a six is face uppermost. Then $\bar{R}$ is the event that a 1, 2, 3, 4 or 5 is rolled. If the die is fair, $Pr(R) = 1/6, Pr(\bar{R}) = 5/6$.
3. A person is checked to see if he is Kell+ or Kell−. Let R be the event that he is Kell+. Then $\bar{R}$ is the event that he is Kell−. Using the data in Table 1.4, $Pr(R) = 0.6, Pr(\bar{R}) = 0.4$.
4. A person is charged with a crime. Let G be the event that he is guilty. Then $\bar{G}$ is the event that he is not guilty.

Notice that the event 'R or $\bar{R}$', formed from the disjunction of R and its complement $\bar{R}$, is certain. Thus it has probability 1. Also, since R and $\bar{R}$ are mutually exclusive,

$$1 = Pr(R \text{ or } \bar{R}) = Pr(R) + Pr(\bar{R})$$

and hence

$$Pr(\bar{R}) = 1 - Pr(R). \tag{3.1}$$

In general, for complementary events R and $\bar{R}$,

$$Pr(R) + Pr(\bar{R}) = 1. \tag{3.2}$$

It is now possible to define odds.

3.1.3   Definition

If an event R has probability $Pr(R)$ of occurring, the *odds* against R are

$$Pr(\bar{R})/Pr(R).$$

From (3.1), the odds against R are

$$\frac{1 - Pr(R)}{Pr(R)}.$$

The odds in favour of R are

$$\frac{Pr(R)}{1 - Pr(R)}.$$

Given a probability for an event it is possible to derive the odds against the event.

Given a value for the odds against an event it is possible to determine the probability of that event occurring. Thus, if the horse has odds of 6 to 1 against it winning the race and R is the event that it wins the race, then

$$\frac{1 - Pr(R)}{Pr(R)} = 6,$$

where '6 to 1' is taken as the ratio $6/1$ and written as 6. Then

$$1 - Pr(R) = 6 \times Pr(R)$$
$$1 = \{6 \times Pr(R)\} + Pr(R)$$
$$= 7 \times Pr(R).$$

Thus $Pr(R) = 1/7$.

The phrases 'odds on' and 'odds in favour of' are equivalent and are used as the reciprocal of 'odds against'. Consider the football team which is '3 to 2 on' to win its match. The phrase '3 to 2' is taken as the ratio $3/2$ as this is odds *on*. The relationship between odds and probability is written as

$$\frac{Pr(R)}{1 - Pr(R)} = \frac{3}{2}.$$

Thus

$$2 \times Pr(R) = 3\{1 - Pr(R)\}$$
$$5 \times Pr(R) = 3$$
$$Pr(R) = 3/5.$$

The general result may be derived as follows. Let O denote the odds against the occurrence of an event R. Then

$$O = \frac{1 - Pr(R)}{Pr(R)}$$
$$O \times Pr(R) = 1 - Pr(R)$$
$$(O + 1) \times Pr(R) = 1$$
$$Pr(R) = \frac{1}{O + 1}.$$

This can be verified directly for the horse whose odds were 6 to 1 against it winning (with $O = 6$). For the football team with odds of 3 to 2 on this can be taken as 2 to 3 against ($O = 2/3$) and the result follows. Odds equal to 1 are known as *evens*.

The concept of odds is an important one in the evaluation of evidence. Evidence is evaluated for its effect on the probability of a certain supposition about a suspect (before he comes to trial) or defendant (while a trial is in progress). This supposition may be that the suspect was present at the crime scene, and it is this supposition which will be most discussed in this book. Initially, however, the discussion will be in terms of the effect of evidence on the probabilities of the guilt (H_p) and the innocence (H_d) of a suspect. These are two complementary events. The ratio of the probabilities of these two events, $Pr(H_p)/Pr(H_d)$, is the odds against innocence or the odds in favour of guilt. Notice, also, that the events are that the suspect is truly guilty or truly innocent, not that he is judged to be guilty or innocent. The same principles concerning odds also apply for conditional probabilities. Given background information I, the ratio $Pr(H_p \mid I)/Pr(H_d \mid I)$ is the odds in favour of guilt, given I. Much of this book will be concerned with the effect of the evidence E under consideration on the odds in favour of a supposition about the suspect.

3.2   BAYES' THEOREM

Bayes' theorem is an important part of the process of the consideration of the odds. In fact, the theorem permits the revision, based on new information, of a measure of uncertainty about the truth or otherwise of an outcome or issue (such as a hypothesis or proposition). This perspective is common to numerate scientific fields where data are combined with prior or background information to give posterior probabilities for a particular outcome or issue. The essential feature of Bayesian inference is that it permits the move from prior (initial or pre-test) to posterior (final or post-test) probabilities on the basis of data.

3.2.1   Statement of the theorem

Consider the last two parts of the third law of probability as given in (1.8), namely that for events R and S,

$$Pr(S \mid R) \times Pr(R) = Pr(R \mid S) \times Pr(S).$$

If $Pr(R) \neq 0$, it is possible to divide by $Pr(R)$ and obtain:

Bayes' theorem   For two events, R and S,

$$Pr(S \mid R) = \frac{Pr(R \mid S) \times Pr(S)}{Pr(R)} \qquad (3.3)$$

assuming $Pr(R) \neq 0$.

3.2.2   Examples

A numerical verification of this result is available from Table 1.4 and Example 1.3. Let R denote Kell+, S denote Duffy+. Then, as before, $Pr(R \mid S) = 34/70, Pr(S) = 70/100, Pr(R) = 60/100$ and

$$\frac{Pr(R \mid S) \times Pr(S)}{Pr(R)} = \frac{\dfrac{34}{70} \times \dfrac{70}{100}}{\dfrac{60}{100}} = \frac{34}{60} = Pr(S \mid R).$$

The importance of Bayes' theorem is that it links $Pr(S)$ with $Pr(S \mid R)$. The uncertainty about S as given by $Pr(S)$ on the right-hand side of (3.3) is altered by the knowledge about R to give the uncertainty about S as given by $Pr(S \mid R)$ on the left-hand side of (3.3). Note that the connection between $Pr(S)$ and $Pr(S \mid R)$ involves $Pr(R \mid S)$ and $Pr(R)$.

An important example of such reasoning is found in medical diagnosis. Consider the following example where a doctor in a clinic is interested in the proposition 'This patient has disease S'. By regarding the patient as a random member of a large collection (population) of patients presenting themselves in the clinic (for a discussion on the population characteristics relevant to the case, see Section 8.5), the doctor associates a probability with the proposition of interest: this probability is the prior (or pre-test) probability. It is the probability that a person has the disease S, before any test results or new observations are taken. Suppose the doctor then carries out a test (e.g., a blood test) which gives a positive result; call this event R. After that, the doctor is now interested in assessing the new probability that the patient has disease S. This new value is the posterior or post-test probability because it refers to a new situation, as expressed by equation (3.3). Note that the probability of a positive blood test could be expanded using the *extension of the conversation* (Section 1.6.7). A positive blood test result could be considered under two competing situations: first, the patient has disease S (event S); and second, the patient does not have disease S (event $\bar{S}$). So $Pr(R)$ becomes

$$Pr(R \mid S) \times Pr(S) + Pr(R \mid \bar{S}) \times Pr(\bar{S}),$$

and the posterior probability

$$Pr(S \mid R) = \frac{Pr(R \mid S) \times Pr(S)}{Pr(R \mid S) \times Pr(S) + Pr(R \mid \bar{S}) \times Pr(\bar{S})}. \tag{3.4}$$

Reconsider the previous simple example where a doctor is interested in the probability the patient has disease S given the positive blood result R. For the quantitative assessment of conditional probabilities, it is important that in an earlier stage (i.e. before the analysis), the blood test used by the doctor is evaluated using two groups of patients with and without the disease. The groups are classified using a reference test (a so-called *gold standard*) to obtain a two-by-two contingency table, Table 3.1. There are n patients in total. The subscripts refer to the number of patients in the category identified by the subscripts. The sums of pairs of numbers in rows or columns in the body of the table are the values in the margins (bottom row and right-hand column). Thus, for example $n_{RS} + n_{R\bar{S}} = n_R$, the number of patients with a positive blood test. In medical terminology it is common to refer to the *sensitivity* and *specificity* of a test. Sensitivity is the probability of a positive result in the blood test given that the patient has disease S. It is estimated by the ratio of n_{RS} to n_S, the proportion of positive patients in the diseased group. Specificity is the probability of a negative result in the blood test given that the patient does not have the disease. It is estimated by the ratio of $n_{\bar{R}\bar{S}}$ to $n_{\bar{S}}$, the proportion of negative patients in the non-diseased group. Sensitivity and specificity provide a measure of the quality of a test, with high values implying high quality. Thus, in (3.3), $Pr(S)$ represents the prior probability that the patient has disease S (in medical terms, this probability is also called *prevalence*); $Pr(\bar{S})$ equals $1 - Pr(S)$. $Pr(R \mid S)$ is the sensitivity of the test. $Pr(R \mid \bar{S})$, which is $1 - Pr(\bar{R} \mid \bar{S})$, is known as the *false positive* rate and is estimated by the ratio of $n_{R\bar{S}}$ to $n_{\bar{S}}$. The *false negative* rate is $Pr(\bar{R} \mid S)$ and is estimated by the ratio of $n_{\bar{R}S}$ to n_S.

Table 3.1 can also be presented using probabilities instead of frequencies (see, for example, Leonard, 2000, pp. 35–40).

In Table 3.2, $Pr(S)$ is the prior probability or prevalence of the disease in the relevant population and $Pr(S, R) = Pr(R \mid S) \times Pr(S)$ is assessed using the sensitivity of the test. $Pr(S, \bar{R})$ may then be calculated by subtraction, $Pr(S) - Pr(S, R)$. An analogous procedure is adopted for the column $\bar{S}$.

The distinction between $Pr(S \mid R)$ and $Pr(R \mid S)$ is very important and needs to be recognised. In $Pr(S \mid R)$, R is known or given, S is uncertain. In Example 1.3 the individual is assumed to be Kell+, and it is uncertain whether he is Duffy+ and or Duffy−. In the medical example, the result of the blood test is known, the disease status is unknown. In $Pr(R \mid S)$, S is known or given, R is uncertain. In Example 1.3 the individual is assumed to be Duffy+, and it is uncertain whether

Table 3.1 Two-by-two contingency table for frequencies for the tabulation of patients with or without a disease (S or $\bar{S}$) given a blood test positive or negative (R or $\bar{R}$)

	S	$\bar{S}$	Total
R	n_{RS}	$n_{R\bar{S}}$	n_R
$\bar{R}$	$n_{\bar{R}S}$	$n_{\bar{R}\bar{S}}$	$n_{\bar{R}}$
Total	n_S	$n_{\bar{S}}$	n

Table 3.2 Two-by-two contingency table for probabilities for the tabulation of patients with or without a disease (S or $\bar{S}$) given a blood test positive or negative (R or $\bar{R}$)

	S	$\bar{S}$	Total
R	$Pr(S, R)$	$Pr(\bar{S}, R)$	$Pr(R)$
$\bar{R}$	$Pr(S, \bar{R})$	$Pr(\bar{S}, \bar{R})$	$Pr(\bar{R})$
Total	$Pr(S)$	$Pr(\bar{S})$	1

he is Kell+ or Kell−. In the medical example, the disease status is known, the result of the blood test is uncertain.

Further examples will emphasise the difference between these two conditional probabilities.

Example 3.1 First, let S be the event 'I have two arms and two legs' and let R be the event 'I am a monkey'. Then $Pr(S \mid R) = 1$, whereas $Pr(R \mid S) \neq 1$. The first probability is equivalent to saying that 'If I am a monkey then I have two arms and two legs'. The second probability is equivalent to saying that 'If I have two arms and two legs, I need not be a monkey.' Similarly, in the previous medical example, a patient is more interested in the probability of not having the disease, given that the test has a positive result, than in the probability of a positive test given that he does not have the disease. The latter probability is the false positive rate, $Pr(R \mid \bar{S})$, the former is a posterior probability, $Pr(\bar{S} \mid R)$. For a discussion on this very important point, see Saks and Koehler (1991).

Example 3.2 This example is from Lindley (1991). Consider the following two statements.

1. The death rate amongst men is twice that amongst women.
2. In the deaths registered last month there were twice as many men as women.

Let M denote male, F denote female, so that M and F are complementary events ($M \equiv \bar{F}, F \equiv \bar{M}$). Let D denote the event of death. Then statements (1) and (2) may be written as

1. $Pr(D \mid M) = 2Pr(D \mid F)$,
2. $Pr(M \mid D) = 2Pr(F \mid D)$.

Notice, also that $Pr(M \mid D) + Pr(F \mid D) = 1$ since M and F are complementary events. Equation (3.2) generalises to include a conditioning event (D in this case). Thus from statement (2),

$$1 - Pr(F \mid D) = 2Pr(F \mid D)$$

Table 3.3 Hypothetical results for deaths amongst a population

	Male	Female	Total
Dead	2	1	3
Alive	98	99	197
Total	100	100	200

and

$$Pr(F \mid D) = 1/3, Pr(M \mid D) = 2/3.$$

It is not possible to make any similar inferences from statement (1). Table 3.3 illustrates the point numerically.There are 100 males of whom 2 died, and 100 females of whom 1 died. Thus $Pr(D \mid M) = 0.02, Pr(D \mid F) = 0.01$, satisfying (1). There were 3 deaths in total, of whom 2 were male and 1 female, satisfying (2).

Example 3.3 Consider the problem of determining which of three sub-populations (Ψ_1, Ψ_2, Ψ_3) an individual belongs to, based on observations of genotypes at several loci and knowledge of genotype frequencies in each of the sub-populations (Shoemaker *et al.*, 1999).

The context may be that of a bloodstain found at a crime scene and the question is to determine which of three populations, Caucasian, Maori or Western Polynesian, the contributor of the stain belongs (assuming that attention can be restricted to these three sub-populations).

The New Zealand census in 1991 reported that the population in that country had the following composition: 81.9% Caucasian, 13.7% Maori and 4.4% Western Polynesian. The probability of the observed genotypes (a DNA forensic profile) X of the individual can be calculated. For this example, suppose the three probabilities $Pr(X \mid \Psi_1)$, $Pr(X \mid \Psi_2)$, $Pr(X \mid \Psi_3)$ have been calculated to be $3.96 \times 10^{-9}, 1.18 \times 10^{-8}, 1.91 \times 10^{-7}$. The prior probabilities $Pr(\Psi_i)$ for the three sub-populations are 0.819, 0.137 and 0.044. Then

$$
\begin{aligned}
Pr(&\Psi_i \mid X) \\
&= \frac{Pr(X \mid \Psi_i) \times Pr(\Psi_i)}{Pr(X \mid \Psi_1) \times Pr(\Psi_1) + Pr(X \mid \Psi_2) \times Pr(\Psi_2) + Pr(X \mid \Psi_3) \times Pr(\Psi_3)} \\
&= \frac{3.96 \times 10^{-9} \times 0.819}{3.96 \times 10^{-9} \times 0.819 + 1.18 \times 10^{-8} \times 0.137 + 1.91 \times 10^{-7} \times 0.044} \\
&= 0.245.
\end{aligned}
$$

Note that the probability of the stain being a Caucasian has dropped from a prior probability of 0.819 to a posterior probability of 0.245. This is because of the relative rarity of the profile of X in the Caucasian population. It can be checked that

$$Pr(\Psi_2 \mid X) = 0.121, \quad Pr(\Psi_3 \mid X) = 0.634.$$

Thus, for the Western Polynesian sub-population, the prior probability has increased from 0.044 to 0.634. This is because the profile X is comparatively common in the Western Polynesian sub-population.

Example 3.4 Another example to illustrate the difference between the two probability statements $Pr(S \mid R)$ and $Pr(R \mid S)$ has been provided by Darroch (1987). Consider a town in which a rape has been committed. There are 10 000 men of suitable age in the town, of whom 200 work underground at a mine. Evidence is found at the crime scene from which it is determined that the criminal is one of the 200 mineworkers. Such evidence may be traces of minerals which could only have come from the mine. A suspect is identified and traces of minerals similar to those found at the crime scene are found on some of his clothing. How might this evidence be assessed?

Denote the evidence by E: the event that 'mineral traces have been found on clothing of the suspect which are similar to mineral traces found at the crime scene'. Denote the proposition that the suspect is guilty by H_p and the proposition that he is innocent by H_d (these are complementary propositions: one and only one is true).

A proposition may be thought of in a similar way to an event, if subjective probabilities are considered. Events may be measurements of characteristics of interest, such as concentrations of certain minerals within the traces. There may be a well-specified model representing the randomness in such measurements. However, the guilt or innocence of the suspect is something for which there is no well-specified model but for which it is perfectly reasonable for an individual to represent with a probability his state of uncertainty about the truth or otherwise of the propositions.

(Note the use of the word *proposition*. As stated before, in Section 1.6.1, this word is being used here in preference to the word *hypothesis* in order to avoid confusion with the statistical process known as *hypothesis testing*. Propositions may be complementary in the same way as events are said to be complementary. One and only one can be true and together they exhaust all possibilities.)

Assume that all people working underground at the mine will have mineral traces similar to those found at the crime scene on some of their clothing. This assumption is open to question, but the point about conditional probabilities will still be valid. The probability of finding the evidence on an innocent person may then be determined as follows. There are 9 999 innocent men in the town of whom 199 work underground at the mine. These 199 men will, as a result of their work, have this evidence on their clothing, under the above assumption. Thus $Pr(E \mid H_d) = 199/9\,999 \simeq 200/10\,000 = 0.02$, a small number. Does this imply that a man who is found to have the evidence on him is innocent with probability 0.02? Not at all. There are 200 men in the town with the evidence (E) or them of whom 199 are innocent (H_d). Thus $Pr(H_d \mid E) = 199/200 = 0.995$. The equation of $Pr(E \mid H_d)$ with the probability $Pr(H_d \mid E)$ is known as the *fallacy of the transposed conditional* (Diaconis and Freedman, 1981) and is discussed in more detail later in Sections 3.3.1 and 3.5.5.

3.3   ERRORS IN INTERPRETATION

A large part of the controversy over scientific evidence is due to the way in which the evidence is classically presented. At trial, it is already a complex operation to ensure that judge and jury members understand the scientific evidence; additional difficulties are added when the forensic scientist gives the court an evaluation to illustrate the convincing force of the results (see, for example, the misunderstanding in *R. v. Adams, D.J.*). The assessment of the value of the analytical results is associated with probabilities as measures of uncertainty. Hence experts' statements are associated with uncertainty. It is important to ensure that this uncertainty is measured accurately and represented correctly to avoid the so-called 'fallacies' or pitfalls of intuition (Saks and Koehler, 1991; Fienberg and Kaye, 1991). Psychological research has demonstrated that intuition is a bad substitute for the laws of probability in evaluating uncertainty (Bar-Hillel and Falk, 1982; Koehler, 1992; Piattelli-Palmarini, 1994) and the presentation of scientific argument at trial can create confusion: victims of this are both jurists and experts (Koehler, 1993b; Reinstein, 1996). Thus, statistical reasoning will help the forensic scientists and members of the jury in reaching conclusions. Since the beginning of the twentieth century, some scientists and jurists have plainly been conscious of the lack of intuition in dealing with the calculation of chances. The French mathematician Henri Poincaré gave a remarkable example of these limits in his course taught during 1895 under the title 'The problem of the three caskets' (Poincaré, 1896). Around one hundred years later, the same problem creates large controversies (see Selvin, 1975; Engel and Venetoulias, 1991; Morgan *et al.*, 1991; Falk, 1992).

Consider three boxes A, B and C, within one of which there is a prize. There is a competitor, K, and a compère, L. The compère knows the identity of the box which contains the prize. The prior probability $Pr(i)$ that a particular box contains the prize is $1/3$ for $i = A, B, C$. K has to choose a box. If the chosen box contains the prize, K keeps the prize. The choice is made in two stages. K chooses a box but it is not opened. L then opens one of the two other boxes and shows K that it is empty (remember L knows which box contains the prize). K is then allowed to remain with the first choice of box or to change to the other unopened box. What should K do in order to maximise the probability of winning the prize?

Suppose, without loss of generality, that the competitor initially chooses C. The compère, L, opens one of A or B which is empty. Assume, again without loss of generality, that A is opened. If the prize is in C, then $Pr(L$ opens $A) = Pr(L$ opens $B) = 1/2$. If the prize is in B, then $Pr(L$ opens $A) = 1$. If the prize is in A, then $Pr(L$ opens $A) = 0$.

The probability of interest is $Pr($true box is $C \mid L$ opens A and K chooses C). Denote this as $Pr(C \mid L_A, K_C)$. As K_C is common to what follows it will be omitted for ease of notation. Let L_B and L_C denote the events that the compère opens B and C, respectively. The probability of interest $Pr(C \mid L_A, K_C)$ may then be written as

$$\frac{Pr(L_A \mid C)Pr(C)}{Pr(L_A \mid A)Pr(A) + Pr(L_A \mid B)Pr(B) + Pr(L_A \mid C)Pr(C)}$$

which is equal to

$$\frac{\frac{1}{2} \times \frac{1}{3}}{0 + \left(1 \times \frac{1}{3}\right) + \left(\frac{1}{2} \times \frac{1}{3}\right)} = \frac{1}{3}.$$

The prior probabilities for the box which contained the prize were $Pr(A) = Pr(B) = Pr(C) = 1/3$. The competitor chose C. The compère then opened A and showed that it did not contain the prize. The calculations above show that the posterior probability that C contains the prize is still $1/3$. However, A has been eliminated. Thus, the posterior probability that B contains the prize is $2/3$. The odds are 2 to 1 in favour of B containing the prize. The competitor should thus change his choice of box.

More subtle forms of pitfalls of intuition have been described analysing court reports and scientists' statements on evidence. For example, case reports on DNA evidence have been studied and submitted to practitioners (forensic scientists, lawyers, advocates) and students to investigate their understanding of measures of uncertainty (Taroni and Aitken, 1998b, c). Results suggested improvements for the presentation of scientific evidence and for the education of future lawyers and forensic scientists. Examples of these fallacies abound in the literature and are presented in this section (see also Goodman, 1992; R. v. *Deen*, *The Times*, 10 January, 1994; Matthews, 1994; Dickson, 1994; Balding and Donnelly, 1994b).

3.3.1 Fallacy of the transposed conditional

Examples of this fallacy abound in judicial and forensic literature. References to it in the courts include R. v. *Adams, D.J.*, R. v. *Doheny and Adams, G.*, R. v. *Clark* and *Wilson v. Maryland*. Consider the following example, from Gaudette and Keeping (1974). The authors conducted a lengthy experiment to attempt to determine the ability of hair samples to distinguish between different people. Multiple comparisons were made of hairs from many different people. In one experiment, nine hairs, selected to be mutually dissimilar, from one source were compared, one by one, with a single hair from another source. It was estimated, from the results of many such comparisons, that the probability that in at least one of the nine comparisons the two hairs examined, from different sources, would be indistinguishable would be 1/4500. The authors concluded that 'it is estimated that if one human scalp hair found at the scene of a crime is indistinguishable from at least one of a group of about nine dissimilar hairs from a given source the probability that it could have originated from another source is very small, about 1 in 4500' (Gaudette and Keeping, 1974, p. 605).

Let R denote the event that 'one human scalp hair found at the scene of a crime is indistinguishable from at least one of a group of about nine dissimilar hairs from a given source'. Let S denote the event that 'the nine dissimilar hairs come from a different source from the single hair'. Then the authors' experiment gives a value for $Pr(R \mid S)$, but the authors' summarising statement gives a value for $Pr(S \mid R)$.

From a historical point of view, it is interesting to recall the Dreyfus case and the probabilistic testimony offered by Alphonse Bertillon. Bertillon failed with the same problem of intuition (for a full description of the judicial case and comments on experts' conclusions, see Champod *et al.*, 1999). According to Bertillon, Dreyfus was the author of the bordereau (document). To increase the credibility of his allegations, Bertillon submitted a calculation of probability. If the individual probability for one coincidence were set to 0.2, then the probability of observing four coincidence is $0.2^4 = 0.0016$, and generally for N coincidences the probability will be 0.2^N. Considering the four coincidences observed by Bertillon, this probability was considered so remote that it demonstrated the forgery. Even if we admitted that the probabilistic value p provided by Bertillon was correct (for a comment on this point, see Darboux *et al.*, 1908; Champod *et al.*, 1999), he claimed (indirectly) that it was possible to deduce from p that the probability that the questioned document was a forgery was $1 - p$. This latter probability was sufficiently close to 1 that it constituted an unequivocal demonstration with a reasonable degree of scientific certainty that Dreyfus was the author. Bertillon's statement is fallacious because he seemed to argue that $Pr(H_d \mid E) = p$, hence $Pr(H_p \mid E) = 1 - p$, whereas p only represents $Pr(E \mid H_d)$.

Other examples of the fallacy of the transposed conditional are given by Thompson and Schumann (1987) who called it the *prosecutor's fallacy*. It has also been called the *inversion fallacy* (Kaye, 1993b). For example:

> There is a 10% chance that the defendant would have the crime blood type if he were innocent. Thus there is a 90% chance that he is guilty.

or

> The blood test is highly relevant. The suspect has the same blood type as the attacker. This blood type is found in only 1% of the population so there is only a 1% chance that the blood found at the scene came from someone other than the

suspect. Since there is a 1% chance that someone else committed the crime there is a 99% chance that the suspect is guilty. (Thompson and Schumann, 1987, p. 177)

In general, let E denote the evidence and H_d the proposition that a suspect is innocent. A value is determined for $Pr(E \mid H_d)$, the probability of the evidence if the suspect is innocent. The interpretation of the value calculated can cause considerable confusion. Two special cases of the fallacy of the transposed conditional in which $Pr(E \mid H_d)$ is confused with

(a) the probability the suspect is not the source of the evidence (*source probability error*), and
(b) the probability the suspect is not guilty (*ultimate issue error*)

are discussed by Koehler (1993a), Balding and Donnelly (1994b), Evett (1995) and Redmayne (1995, 1997).

3.3.2 Source probability error

A crime is committed. Trace evidence is found which is considered to have come from the criminal. Let H_d be the proposition that the suspect was not the source of the evidence.

For instance, the evidence E may be that a DNA match has been found between blood from a murder victim and blood recovered from the clothing of a suspect. A scientist determines a value for $Pr(E \mid H_d)$ as 1 in 7 million. Consider the following possible statement of the value of the evidence (based on *Wike v. State*, transcript, pp. 147–148, given in Koehler, 1993a).

> With probability 1 in 7 million it can be said that the blood on the clothing of the suspect could be that of someone other than the victim.

Other possibilities are (where the first four examples are not taken from any particular case but they and the figures quoted are for illustrative purposes only):

- the probability that the DNA that was found at the scene of the crime came from anyone else other than the suspect is 1 in 7 million;
- the probability of this DNA profile occurring at random is 1 in 18 billion; thus the likelihood that the DNA belongs to someone other than the suspect is 1 in 18 billion;
- the probability of finding the evidence on an innocent person is 0.01% (1 in 10 000) thus the likelihood that the suspect is guilty is 99.99%;
- the trace evidence has the same DNA profile as the suspect, thus the trace evidence has been left by the suspect;

- After conducting DNA testing on the vaginal swab samples taken from the victim and Ross' [the suspect's] blood samples, the DNA expert stated that Ross was the source of the seminal fluid (*Ross v. State of Indiana*);
- The expert offered probability statistics on whether the DNA sample found on the victim came from someone other than defendant (*State of Vermont v. T. Streich*).

None of the conclusions in these statements is justified by the value given by $Pr(E \mid H_d)$. All give a mistaken probability for the source of the evidence.

3.3.3   Ultimate issue error

The source probability error may be extended to an error known as the ultimate issue error (Koehler, 1993a). This extends the hypothesis that the suspect is the source of the evidence to the hypothesis that the suspect is guilty. The case of *People v. Collins* (see Chapter 4) is a particular example of this. Consider a case in which $Pr(E \mid H_d)$ is 1 in 5 million, say, where, as before, H_d is the proposition that the suspect was not the source of the evidence E. The ultimate issue error would interpret this as a probability of 1 in 5 million that the suspect was innocent.

3.3.4   Defender's fallacy

As well as the fallacy of the transposed conditional there is also a *defender's fallacy* (Thompson and Schumann, 1987). Consider the following illustrative statement from a defence lawyer:

> The evidence for blood types has very little relevance for this case. Only 1% of the population has the rare blood type found at the crime scene and in the suspect. However, in a city, like this one in which the crime occurred, with a population of 200 000 people who may have committed the crime, this blood type would be found in approximately 2000 people. The evidence merely shows that the suspect is one of 2000 people in the city who might have committed the crime. The blood test evidence has provided a probability of guilt of 1 in 2000. Such a small probability has little relevance for proving the suspect is guilty.

Strictly speaking (from an inferential point of view) the defence lawyer is correct. However, before the evidence of the blood test was available, the suspect had a probability of only 1 in 200 000 of being guilty (not accounting for any other evidence which may have been presented). The effect of the blood test evidence is to increase this probability by a factor of 100. The evidence is 100 times more likely if the suspect is *guilty* than if he is *innocent*. This may be thought to show that the blood test evidence is quite compelling in support of a hypothesis of guilt. Of course, on its own, this evidence is unlikely to be sufficient for a verdict of guilty to be returned.

Two other errors discussed by Koehler (1993a) are the probability (another match) error and the numerical conversion error.

3.3.5 Probability (another match) error

As in Example 1.1, a crime is committed. Evidence E of a bloodstain with profile Γ is found at the scene and identified as belonging to the criminal. A suspect is identified. Let H_d be the proposition that the evidence E was not left by the suspect. Suppose the frequency of the profile of the stain is γ amongst the relevant population (details of such estimation are given in Chapter 13) and so $Pr(E \mid H_d) = \gamma$. Then the probability that a person selected at random from the population does not match this profile is $1 - \gamma$. Let N be the size of the population. Then, assuming independence amongst the members of the population with respect to E, the probability of no matches with the crime stain profile amongst the N members is $(1 - \gamma)^N$ (a generalisation to N events of the third law of probability for independent events, (1.3)). The complement of no matches is at least one match. Hence, the probability of at least one match is $\theta = 1 - (1 - \gamma)^N$. Two numerical examples are given in Table 3.4. Let $N = 1$ million and take γ as a value which has been estimated from some larger population, assumed similar to the relevant population with regard to profile random match probabilities (see Chapter 13). As in Section 2.4.4, see Smith and Charrow (1975) and Finney (1977) for comments about super-populations. Thus it is possible for γ to be less than $1/N$. The two probabilities for θ in Table 3.4 are considerably larger than the corresponding values for γ. The probability (another match) error arises when the two probabilities, γ and θ, are equated. In other words, a small value of γ is taken to imply a small value for the probability that at least one other person has the same matching evidence. The results in Table 3.4 illustrate why this implication is false. So, a random match probability of 1 in 1 million (if we assume that the value is derived correctly and if we assume that there is no possibility of error, lying, or misinterpretation of the data), means that there is 1 chance in 1 million that a single randomly selected person would share the observed characteristics. In other words, assuming that the data and its interpretation are infallible, we would expect to see this DNA profile in approximately 1 out of every million people. Notice that this is not identical to the probability that there exists someone else who shares the observed profile. Although it may be

Table 3.4 Probability θ of at least one match, given a frequency of the trace evidence of γ, in a population of size 1 million

γ:	1 in 1 million	1 in 10 million
θ:	0.632	0.095

extremely unlikely that a single randomly selected person would share a DNA profile with another person, it may be quite likely that others share this profile (Koehler, 1996). There is only 1 chance in 1 million that a random person shares a DNA profile that is common to one in every million people, but there is a 63.2% chance that there are others in a population of size 1 million people (see Table 3.4) who share the profile.

3.3.6 Numerical conversion error

Let γ be the random match probability (RMP, see details in Chapter 13) of the crime stain as in Section 3.3.5. A match between the crime stain and the profile of a suspect has been made. Let n be the number of people who would have to be tested before there is another match. It may be thought that the significance of the value of γ can be measured by equating $1/\gamma$ with n. A small value of γ implies a large value of n. It is fairly straightforward to calculate n, given γ and given some value for the probability of another match occurring, $Pr(M)$ say. The numerical conversion error claims that n equals $1/\gamma$, but this is not so.

Suppose $\gamma = 0.01$. There is a probability of 0.01 that a randomly selected individual would match the evidence E. The numerical conversion error would claim that 100 people need to be tested before another match might be expected, but this is not the case. The error is exposed by consideration of $Pr(M)$. Suppose, initially, that $Pr(M)$ is taken to be equal to 0.5. A value of n greater than the value determined using a value of $Pr(M)$ of 0.5 would imply that a match is more likely to happen than not if n people were tested.

Let n be the number of people who are to be tested before the probability of finding a match is greater than 0.5. The probability that a randomly selected individual does not match the evidence is $1 - \gamma$. For n independent randomly selected individuals, the probability none matches the evidence is $(1 - \gamma)^n$. The probability there is at least one match is thus $1 - (1 - \gamma)^n$. Thus, for a match to be more likely than not with n individuals, $1 - (1 - \gamma)^n$ has to be greater than 0.5 and so

$$(1 - \gamma)^n < 0.5.$$

This inequality may then be written as $n \log(1 - \gamma) < \log 0.5$. Thus, $n > \log 0.5 / \log(1 - \gamma) = \psi_5$, say (remembering that here $1 - \gamma$ is less than 1 and so its logarithm is negative). A similar argument shows that if $Pr(M)$ is taken to be greater than 0.9 then $n > \log 0.1 / \log(1 - \gamma) = \psi_9$, say. Values of ψ and n are given in Table 3.5 for $\gamma = 0.1, 0.01$ and 0.001 and for values of $Pr(M)$ equal to 0.5 and 0.9.

It is also worth noting that if $n' = 1/\gamma$ people were tested this does not make the probability of a match certain. If n' people are tested, the probability of at least one match is $\theta' = 1 - (1 - \gamma)^{n'}$; see Table 3.6 for examples. Notice that as $\gamma \to 0$, $\theta' \to 1 - e^{-1} = 0.632\ldots$. Notice also, that $n_5 < n'$. Thus the

Table 3.5 Evidence occurs with *RMP* γ. Smallest number ψ of people to be observed before a match with the evidence occurs with a given probability, $Pr(M) = 0.5, 0.9$; $\psi_5 = \log 0.5 / \log(1 - \gamma)$, $\psi_9 = \log 0.1 / \log(1 - \gamma)$, n_5 is the smallest integer greater than ψ_5, n_9 is the smallest integer greater than ψ_9

γ	$Pr(M) = 0.5$		$Pr(M) = 0.9$	
	ψ_5	n_5	ψ_9	n_9
0.1	6.6	7	21.9	22
0.01	69.0	69	229.1	230
0.001	692.8	693	2301.4	2302

Table 3.6 The probability, θ', of at least one match with the evidence, which occurs with *RMP* γ, when $n' = 1/\gamma$ people are tested

γ	n'	θ'
0.1	10	0.651
0.01	100	0.634
0.001	1000	0.632

numerical conversion error, based on $Pr(M) = 0.5$, exaggerates the number of people that need to be tested before a match may be expected. For illustrative purposes, consider the following case. The random match probability equates 1 in 209 100 000 and the expert said that he has a database of blood samples from all over the country and he asked the question 'How many people would we have to look at before we saw another person like this ?' The answer given is 209 100 000 (*Ross v. State*). This exaggerates the probative strength of a match and favours the prosecution.

3.3.7 False positive fallacy

Consider the possibility of a misclassification error for a scientific sample so that the sample is classified as positive when in fact it is negative. An example of this would be the misclassification, as a match, of two DNA profiles from different sources which did not match. Serious errors of interpretation can occur through ignorance or underestimation of the potential for a false positive. A low value for the probability of a false positive does not imply that the probability of a false match is low in every case. A forensic scientist who thinks there is only a probability of 0.01 of declaring, falsely, a match between the samples in a case if they really do not match may assume that there is, necessarily, a probability of 0.99 that the reported match is a true match.

Let M be the event that the suspect and the perpetrator have matching DNA profiles and R be the report of a match. The false positive probability is $Pr(R \mid \bar{M})$, the probability of reporting a match when the samples do not have matching profiles. The probability $Pr(M \mid R)$ is the probability of a true match, given that a true match has been reported. The fallacy of the false positive is to equate $Pr(M \mid R)$ with $1 - Pr(R \mid \bar{M})$.

The fallacy is a version of the prosecutor's fallacy. Further details are given in Section 3.5.5, Chapter 13 and Thompson *et al.* (2003).

3.3.8   Uniqueness

The identification (or better the 'individualization') statement (also known as 'uniqueness' or 'individuality') is regularly used in fields such as fingerprint, shoeprint, toolmarks, firearms, earprints, and speaker recognition (e.g., Simons, 1997) and regularly discussed in the forensic literature (e.g., Champod, 2000). DNA evidence is not immune from this tendency (Smith and Budowle, 1998; Budowle *et al.*, 2000; Kaye, 1997b; Zeisel and Kaye, 1997; Robertson and Vignaux, 1998; Balding, 1999). In fact, the Federal Bureau of Investigation (FBI) have announced that their experts will be permitted to testify that DNA from blood, semen, or other biological evidence recovered at a scene of crime came from a specific person (Holden, 1997). The FBI officially said that the new policy states that if the likelihood of a random match is less than 1 in 260 billion, the examiner can testify that the samples are an exact match, and that such a scientific certainty invokes that we are identifying one individual to the exclusion of all others. So, some experts will give opinions that DNA genotypes are unique, and that there is absolutely no doubt two samples have come from the same human being – for a discussion on the pertinence of the statistic presented (1 in 260 billion) see Evett and Weir (1998) and Buckleton *et al.* (2004). The establishment of an identification is an opinion on the issue itself. This (illogical) conclusion came from a statement of probability expressing that the chance of observing on earth another person presenting the same characteristic is zero. There is sufficient uniqueness within the observed characteristics to eliminate all other possible donors in the world. No contrary evidence (even alibi) can change the expert's certainty. The movement from a probability statement to one of certainty represents a 'leap of faith' rather then a logical consequence (Stoney, 1991b) and such a conclusion points out a misinterpretation of the role of forensic scientists and of the court in scientific inference procedure, and of the role of statistics in forensic science. For further comments, see Buckleton (1999), Champod (1999), Taroni and Margot (2000) and Buckleton *et al.* (2004). In particular, the last authors comment that 'on balance (they) cannot see much positive coming from a policy of declaring common source' (Buckleton *et al.*, 2004). It is not the role of the expert to qualify the acceptable level of reasonable doubt using an illegitimate relevant population (the size is set at its maximum to consider the possibility that all persons on earth could be at the origin of the trace). Jurists should deal with the legal standard 'beyond reasonable doubt' which represents a threshold regarding the issue of identification.

3.3.9 Other difficulties

The forensic scientist has to assess the value of the evidence; this means that he has to evaluate the strength of the link between, for example, a recovered trace and a suspect. Therefore, it seems important to point out that forensic evidence evaluation has – by its nature – a close link to statistical assessment (results are associated with probabilities as measures of uncertainty). But there is a potential for misinterpretation of the value of statistical evidence when, routinely, such evidence supports a scientific argument in the adversarial (or inquisitorial) system of the trial process. In this subsection the meaning of some concluding statements given by European DNA experts is discussed. Note that the probabilistic meaning of these statements is presented assuming that the the relative frequency was derived correctly and that there is no possibility of error or misinterpretation of the data (Taroni *et al.*, 2002).

Relative frequency of occurrence

The method of the relative frequency, γ, seems to be inappropriate when describing a match between two samples of DNA. A typical way of expressing the result, used by several laboratories, is 'The DNA profile in question occurs in about one person in 100 000 of the population'. There are four main objections to this approach:

1. If the population in question is considerably greater in size than $1/\gamma$ (for example, 3 million), then it might be reasonable for a court to consider that about $3\,000\,000 \times 1/100\,000$ people in the population would have the same profile. The court could then perhaps use this DNA evidence as prior odds when evaluating the remaining evidence in the case. If, however, the population is much smaller in size than $1/\gamma$ (e.g., a laboratory reports 1 person in 2.5×10^9) we would not expected to find anyone else in the population who possesses the profile, and it seems impossible logically to combine DNA evidence with the other evidence in the case which may provide support for the defence hypothesis, for example, convincing evidence of an alibi.
2. A more serious objection to 'frequency of occurrence' occurs when the scientist is considering the alternative hypothesis that the DNA has originated from a close relative of the suspect. It does not make any sense to say 'The DNA profile in question occurs in about one brother in 400', so the scientist has to find a different way of expressing this result, with more confusion for the court.

3. In considering the numerator and denominator of the likelihood ratio we are evaluating two probabilities, that is, we are evaluating the probability of observing a match between the crime stain and the suspect, given that the stain has come from the suspect, and given that the stain has come from someone else. Also, when we are considering the proposition that the stain has come from someone else, we are conditioning on the result that we have obtained from the suspect. Quoting a relative frequency may be appropriate when considering a sample of DNA from the scene of an undetected crime, but is inappropriate when matching DNA has been obtained from a suspect.

4. There is a range of cases – for example, missing persons, paternity and, in particular, cases where mixed profiles have been obtained – where it is not possible to use a relative frequency to express the value of the DNA evidence. The simple reason for this is that in general the numerator of the likelihood ratio is less than 1, and hence the value of the DNA evidence is not given by $1/\gamma$.

Note that point (3) above introduces the notion of the *random match probability* which represents an acceptable way of expressing the value of DNA evidence; it can be used as an alternative to the likelihood ratio in simple cases where matching profiles have been obtained. For example, the scientist may set out his interpretation as: 'I have considered two propositions with respect to the matching DNA profiles: H_p, the semen has come from the suspect; and H_d, the semen has come from an unknown person who is unrelated to the suspect. These findings are what I would have expected if the first proposition is true. I have assessed the probability of obtaining matching profiles as 1 in a million if the second proposition is true.'

It is also possible to express a match probability without explicitly stating the propositions: 'The genetic profiles of the semen traces are identical to the suspect. The probability of an unrelated person of European origin presenting by chance the same genetic profile as the semen stain is about 1 in 1 billion' or 'The DNA profile of the semen matched that of the blood of the suspect. The chance of a person chosen at random in the population who is not related to the suspect sharing this profile is less than one in ten million.' Details on the calculation of a random match probability are given in Chapter 13.

'Could be' approach

Several scientists have prefaced an estimate of a relative frequency with phrases like 'The semen stain could have come from Mr X, the suspect', 'Sample A could have come from the donor of sample B', 'The semen stain may originate from the suspect', 'Based on the results of the DNA analysis, it can be concluded that this semen may originate from the suspect', 'According to the results of the DNA analysis the bloodstain may originate from the victim', or 'According to the results of the DNA analysis the bloodstain could originate from the person in question'.

It may be helpful to the investigator to spell out what may seem obvious, namely that if the DNA from the crime stain matches the suspect, then the suspect could be the source of the DNA. If this is followed by a statement about the relative frequency of the profile, it is not clear what message is given about the strength of the evidence. A fruitful discussion on this topic is presented in Evett and Weir (1998) and Evett *et al.* (2000a).

A 'could have come from the suspect' statement could be seen as a transposed conditional as it is expressing a view about the probability of the proposition. If this type of explanation is considered necessary then it would be preferable to use a form of words such as 'The DNA profile from the bloodstain matches that obtained from the suspect. Therefore the suspect, or anyone else with the same DNA profile, could be the donor of the bloodstain.' This could be read as providing further explanation of the matching profiles, rather than as a probability statement about a proposition.

'Cannot be excluded' approach

Among the possible statements, a frequently used conclusion is that 'the defendant cannot be excluded as the stain donor'. Such a statement is close to the previous one ('could be' approach) for its vagueness. But it is also related to a statement typically used in paternity cases where the concept of exclusion is presented in a numerical form using a 'probability of exclusion'. For example, if the characteristic is shared by 0.1% of the population, then the probability of exclusion is 0.999. As clearly explained by Robertson and Vignaux (1992), such a probability tells the scientist what proportion of the population the test would exclude, regardless of who is the father (the donor) of the child (stain). Therefore, this estimate is a measure of the efficacy of the test, because it answers the question 'how likely is this test to exclude Mr X if he is not the father (the donor of the stain)?' However, the court is interested in another question: 'how much more likely is the evidence if Mr X is the father (the donor) of the child (stain) than if some randomly selected person were? The probability of exclusion is not relevant to the answer to this question.

Care is required in choosing a form of words which avoids ambiguity. Phrases like 'the probability of finding another person who has the same genetic profile is 1 in 1 million in the population' could be interpreted as saying that if a DNA profile was obtained from every member of the population, then the probability of finding another person with the same profile is 1 in 1 million. Clearly, if the population is about 50 million, then there is a very high probability of finding someone else with the same profile! A similar example of ambiguous wording is 'this genotype can be found in 4.07×10^{-10} people in the reference population'.

3.3.10   Empirical evidence of errors in interpretation

Scientific evidence is often presented in a numerical way. Such an evaluation inevitably uses probabilities as measures of uncertainty. Judges are concerned that scientific evidence may overwhelm or mislead the jury, especially when its presentation by an expert may appear to give the evidence greater probative value to a layman than it would to another expert. This concerns the legal community when a decision rests on resolving the differences between opposing experts, with seemingly conflicting testimony. In such circumstances, the potential for basing a decision on misunderstanding is considerable when the uncertainty of the scientific evidence is not understood (Fienberg *et al.*, 1996). Statistical forms of scientific evidence have probably yielded the greatest confusion and concern for the courts in the application of DNA (Kaye, 1993b; National Research Council, 1996). The confusion is not surprising, given that the courts have little expertise in genetics and statistics. Scientists have also provided sources of misinterpretation in their reports and statements (Koehler, 1993a). Moreover, it has also been argued that the presentation at trial of the evidence value in the form of a likelihood ratio could be very prejudicial in the decision-making process (Koehler, 1996). Koehler noted that '[e]ven when likelihood ratios are properly conveyed, there is little reason to believe that jurors will understand what they mean and how they should be used. Although they have scientific merit, likelihood ratios – which are the ratios of conditional probabilities – are not easy to understand.' Psychological research has emphasised the fallacious way in which people reason in managing uncertainty and probabilities, especially with conditional probabilities (for a review of these studies, see Kaye and Koehler, 1991; Fienberg and Finkelstein, 1996). Empirical research has been carried out in the last decade. Following from the results, methods of improving the reporting of the evidence and the trial presentation have been proposed in order to assist with the questions:

- Is the evidence correctly interpreted?
- Can the way in which the evidence is presented at trial influence a verdict?
- Can the way in which the evidence is presented at trial influence an update of the probability of guilt?
- Can the explanation of the evidence be misunderstood?

Cases have been studied where scientific evidence has been presented in court, and, using information gained from such study, a series of problems have been developed for, and given to, law and forensic science students, to practitioners (advocates and forensic scientists) and to mock jurors to investigate their understanding of uncertainty. Problems associated with the presentation of scientific evidence at trial were investigated using the responses of students and practitioners. Research studied the interpretation of numbers related to scientific evidence, numbers which had been used by experts to explain the value of the evidence. For example, the impact of different ways of presentation, and the value of similarities between a DNA recovered trace and a DNA control material on the verdict (guilty or not guilty) and on the update of the probability of guilt were studied. The results showed an underestimate of the value as expressed by the posterior probabilities computed using Bayes' theorem. The fact that these posterior assessments were substantially below those computed from Bayes' theorem confirmed results of previous studies (for details, see Taroni and Aitken, 1998b; Koehler, 1996). For a review of the previous studies, see Fienberg and Finkelstein (1996). The results also suggested that subjects did not treat different methods of presentation of the evidence (percentage of exclusion, relative frequency, likelihood ratio and posterior odds) similarly but that there was an association between the assessment of the posterior probabilities and the verdicts. Moreover, subjects did not seem capable of distinguishing between the magnitude of the difference in values between scenarios and the effect of error rate, if reported (Koehler *et al.*, 1995; Koehler, 2001a). Even if the results have to be treated with caution, because of the limits of sample size and geographical areas studied, they show a clear problem in dealing with measures of uncertainty (Koehler, 2001b). Studies are described in which

> DNA match statistics that target the individual suspect and that are presented as probabilities (i.e., 'the probability that the suspect would match the blood drops if he were not their source is 0.1%') are more persuasive than mathematically equivalent presentations that target a broader reference group and that are framed as frequencies (i.e., 'one in one thousand people in Houston would also match the blood drops'). (Koehler, 2001b)

Other empirical research has approached the problem of the possible pitfalls of intuition related to the presentation of scientific evidence in numerical ways. The available DNA data (as collected in population studies) allow the scientist to offer to the court a number that should quantify the strength of the link generally established between a suspect and a trace recovered on a victim or on a crime scene. This number generally represents the relative frequency of the matching characteristic in some relevant population (or the *random match probability*, Section 13.3). Is the use of these numbers prejudicial? Conclusions based solely on the relative frequency of the matching trait could have dangerous consequences, as already discussed in Sections 3.3.1–3.3.6, where the scientists' statements use subtle forms of reasoning. Research has tried to measure the extent of the misunderstanding of the meaning of the value of the statistical evidence. Generally, the experts' statements in criminal trials were presented to survey participants. Experts gave different explanations of the meaning of the statistical evidence presented. Participants were asked to detect which statements were correct and which ones were erroneous. Where an expert's explanation was thought to be erroneous, participants were asked to explain their reasons. A correct answer was taken to be one in which the respondent said the expert was correct when the expert was indeed correct or the respondent said the expert was wrong when the expert was indeed wrong (all details of cases and analyses can be found in Taroni and Aitken, 1998b, 1999a). The concern here is with principles. For example, in *Ross v. State of Indiana* (Section 3.3.2), *R. v. Gordon, M.* and *R. v. Deen* (see also Section 3.3), the problem was the fact that the expert gave an opinion on an issue (the suspect is the source of some evidence, or the suspect is the rapist). These are examples of the transposed conditional. Participants in the surveys were confused by the statements and unfortunately accepted them. In *R. v. Montella*, the expert presented the value of the evidence in a Bayesian form expressing the value of the evidence as a likelihood ratio. This logical method of assessing and presenting the evidence unfortunately created considerable confusion in the comprehension of the expert's statement. There were comments from members of all groups of participants (students and practitioners) that they believed the explanation to be wrong, confusing and too difficult to understand. The results supported the viewpoint of Koehler (2001b) above. The aim of this book is to reduce, or even eliminate, the incidence of these problems. In *U.S. v. Jakobetz*, the expert fell into a *source probability error* (Section 3.3.2). This equates the frequency of the trait with the probability that someone other than the defendant was the source of the trace evidence. The case *Ross v. State* presents another statistical problem. This fallacy is an example of the numerical conversion error (Section 3.3.6) because it may be thought that the significance of the value of the relative frequency can be measured by equating the reciprocal of the frequency with the number of people who would have to be tested before there is another match. Generally, results showed that participants believed that experts were right in their explanation of the evidence. In this last situation, the participating members of the Faculty of Advocates of Scotland correctly identified the statistical meaning of the evidence presented stating that 'just because the odds (*sic!*) are 1 to 209,100,000 it doesn't mean that you have to look at 209,100,000 people before finding a match – the next sample could be the next match'. From a research point of view, studies with mock jurors and/or students have been essentially focused on management of two distinct pitfalls of intuition: the prosecutor's fallacy and the defender's fallacy (Sections 3.3.1 and 3.3.4; Thompson and Schumann, 1987; Thompson, 1989; and Carracedo *et al.*, 1996). These studies involved mock trial scripts rather than actual trials. Other research submitted real criminal cases, where the statistical evidence had been explained, to specialist groups such as students (who represent future judges, lawyers and forensic scientists) and as practitioners (forensic scientists and advocates). Results in both kinds of research showed that the great majority of participants failed to detect the error in the arguments exposed by experts at trial. The tendency to draw erroneous conclusions from fallacious descriptions of the meaning of the evidence is troubling. It demonstrates a lack of knowledge of conditional probability which is required to assess correctly the value of the evidence and to appreciate correctly the meaning of this value.

Consider fibres and glass evidence. As empirical research has supported, risks of misconception exist when experts use relative frequencies to support their analytical results. In fact, the relative frequency is only one of many parameters that should be considered in a complete perspective of evidence evaluation (see Chapter 8). The use of the likelihood ratio constitutes, for the expert, an interesting subject for reflection on scientific proof, because the expert must choose the relevant questions considering the physical evidence from two opposite points of view. Case surveys represent an attempt to study the evaluation framework for different scenarios involving fibres, blood and glass fragments (Taroni and Aitken, 1998b). These scenarios were chosen deliberately in order to illustrate the assessments and the evolution of the different parameters in different situations. Subjects' reactions to a situation in which more than one foreign group of fibres was recovered at a crime scene and determined to have been left by the aggressors were studied. Only one group of fibres was compatible with an article of clothing associated with a suspect. Scientific literature demonstrates that, in evaluating the match between recovered fibres and clothing from a suspect, the scientist must consider similar and dissimilar elements, as already suggested for bloodstain evidence (Evett, 1987b). Therefore, it is important not only to focus on the fibres that match the suspect's garments, but also to consider other groups of fibres compatible with the facts, which potentially could have been left by the offender (who is not the suspect!). Two scenarios, in which the value of the fibre evidence was different in the two scenarios, were described to participants who were asked for assessments (Evett, 1983; and Buckleton and Evett, 1989). It was found that subjects did not change their assessment of the evidence according to the difference between the scenarios. Subjects did not take into account, in their assessment of the value of the evidence, the number of groups of fibres which were compatible with the issue in question. These failings produced an overestimate of the value of the evidence in the case where there was more than one distinct group of fibres (details of the likelihood ratio development are given in Sections 3.4 and 3.5). When two individuals (or an individual and an object) have contact during a criminal action, a reciprocal transfer of material (e.g., fibres or glass) is involved. Where this happens, the two sets of recovered traces have to be considered as dependent. If a transfer has occurred in one direction, and the expert has recovered traces characterising this transfer, then the expert would expect to find trace evidence characterising the transfer in the other direction. The presence of one set of transfer evidence gives information about the likelihood of the presence of the other set of transfer evidence (details of such a situation are presented in Chapter 8). The ability of participants to distinguish the scenario where the two sets of recovered traces are dependent from the scenario where the two sets of recovered traces are independent was investigated, with particular reference to the reaction to new technical information about the presence or the absence of cross-evidence involved in the criminal contact. The absence of the expected presence of some trace evidence which would show a reciprocal exchange of material has to be taken into account in the assessment of the value of the recovered matching evidence. The results obtained in the surveys supported previous results and emphasised the subjects' inability to take into account technical information in the assessment of the real value of a link detected between two persons (or objects) (Taroni and Aitken, 1998b). The assessment by the participants of the probative force of an entire aggregation of evidence was also investigated.

In one scenario, participants were asked to make an aggregated judgement which concerned a large collection of evidence in a criminal case involving glass evidence presented by the experts for the prosecution and for the defence. In a second scenario, for comparison with the first, the participants made assessments for two subsets of the evidence and these were combined to provide an overall judgement. This required more assessments, but each one was made with reference to a smaller and more specific body of evidence. A comparison was made of the assessments of posterior probabilities made by the participants in each of the two scenarios. The posterior probability of guilt arrived at by participants who received the entire body of evidence was significantly smaller than that arrived at by the participants who updated their probability twice. The results indicate that arguments tend to provide smaller posterior probability assessments if the body of evidence is not decomposed. Note also that, in both scenarios, smaller values for the probability of guilt were obtained than would have been if the laws of probability had been followed.

In general, results have shown that methods of assessment used by participants are insufficient to obtain a correct value of the scientific evidence. Dangers of underestimation and overestimation of the real value of the evidence still remain. Forensic scientists endeavour to give the court an accurate evaluation to illustrate the true worth of their results. Unfortunately, judging from the results of surveys, the evaluations of the scientists fail to consider all the parameters involved in the scenarios proposed. Furthermore, comments on the calculation of the posterior probabilities show that these have been based upon 'a subjective decision' instead of being in accordance with the probability rules. As stated earlier, the studies were designed to answer four questions:

- Is the evidence correctly interpreted?
- Can the way in which the evidence is presented at trial influence a verdict?
- Can the way in which the evidence is presented at trial influence an update of the probability of guilt?
- Can the explanation of the evidence be misunderstood?

The studies were particularly pertinent as the National Research Council of the USA had only recently commented that 'there is a lack of research into how jurors react to different ways of presenting statistical information' and that 'no research has as yet tested the reactions of triers of fact to the detailed presentations of evidence on DNA profiling that are encountered in the courtroom' (National Research Council, 1996).

The answers to the questions were all undesirable: the evidence may not be correctly interpreted, the way in which the evidence is presented at trial may influence a verdict, the way in which the evidence is presented at trial may influence an update of the probability of guilt and the explanation of the evidence may be misunderstood. This book presents the view that the likelihood ratio (in the Bayesian framework) should be used by experts because it allows them to take into consideration the evidence under two alternative propositions and it enables the consideration of other relevant factors in the calculation of the value of the evidence (as will be presented in the following chapters). Jurists should also appreciate the approach, because it clarifies the roles of the expert and of the judge or members of the jury: the latter take the decision on an issue, the former compares the likelihoods of the evidence under two proposed propositions.

It is important to realise that in the evaluation of evidence the probability of the evidence has to be considered under two propositions, separately. The following quotes illustrate this point.

> The concept of a match is gratuitous.
>
> The factfinder's task is to assess the relative probability of two hypotheses – that the samples came from a common source, and that they did not.
>
> The evidence is the two profiles revealed by the samples.
>
> A factfinder can ask how likely it is that the evidence would have arisen, given each of the competing hypotheses, without asking whether the evidence satisfies an arbitrary defined match standard. (Friedman, 1996, p. 1826)

and

> This concern might have some theoretical force, in the context of DNA evidence, when a prosecutor presents evidence that two samples do not match because the disparity between the measurements is so great. Such a conclusion tells the factfinder that the evidence would be unlikely to arise given the hypothesis that the samples had a common origin, but it does not combine easily with other evidence because it does not tell the factfinder how likely the evidence would arise given an alternative hypothesis. (Friedman, 1996, p. 1827)

3.4 THE ODDS FORM OF BAYES' THEOREM

3.4.1 Likelihood ratio

Replace S by $\bar{S}$ in (3.3) and the equivalent version of Bayes' theorem is

$$Pr(\bar{S} \mid R) = \frac{Pr(R \mid \bar{S})Pr(\bar{S})}{Pr(R)} \tag{3.5}$$

$(Pr(R) \neq 0)$. Dividing equation (3.3) by (3.5) gives:

Odds form of Bayes' theorem

$$\frac{Pr(S \mid R)}{Pr(\bar{S} \mid R)} = \frac{Pr(R \mid S)}{Pr(R \mid \bar{S})} \times \frac{Pr(S)}{Pr(\bar{S})}. \tag{3.6}$$

The left-hand side is the odds in favour of S, given R has occurred. The right-hand side is the product of two terms,

$$\frac{Pr(R \mid S)}{Pr(R \mid \bar{S})} \text{ and } \frac{Pr(S)}{Pr(\bar{S})}.$$

The latter of these is the odds in favour of S, without any information about R. The former is a ratio of probabilities but it is not in the form of odds. The conditioning events, S and $\bar{S}$, are different in the numerator and denominator, whereas the event R, the probability of which is of interest, is the same.

In the odds form of Bayes' theorem given here, the odds in favour of S are changed on receipt of information R by multiplication by the ratio $\{Pr(R \mid S)/Pr(R \mid \bar{S})\}$. This ratio is important in the evaluation of evidence and is given the name *likelihood ratio* or *Bayes factor*. It is the ratio of two probabilities, the probability of R when S is true and the probability of R when S is false. Thus, to consider the effect of R on the odds in favour of S, that is, to change $Pr(S)/Pr(\bar{S})$ to $Pr(S \mid R)/Pr(\bar{S} \mid R)$, the former is multiplied by the likelihood ratio. The odds $Pr(S)/Pr(\bar{S})$ are known as the *prior odds* in favour of S, that is, odds prior to receipt of R. The odds $Pr(S \mid R)/Pr(\bar{S} \mid R)$ are known as the *posterior odds* in favour of S, that is, odds posterior to receipt of R. With similar terminology, $Pr(S)$ is known as the *prior probability* of S and $Pr(S \mid R)$ is known as the *posterior probability* of S. Notice that to calculate the change in the odds on S, it is probabilities of R that are needed. The difference between $Pr(R \mid S)$ and $Pr(S \mid R)$, as explained before (Section 3.3), is vital. Consider two examples.

1. $Pr(R \mid S)/Pr(R \mid \bar{S}) = 3$; the event R is three times more likely if S is true than if S is false. The prior odds in favour of S are multiplied by a factor of 3.
2. $Pr(R \mid S)/Pr(R \mid \bar{S}) = 1/3$; the event R is three times more likely if S is false than if S is true. The prior odds in favour of S are reduced by a factor of 3.

When considering the effect of R on S it is necessary to consider both the probability of R when S is true *and* the probability of R when S is false. It is a frequent mistake (the fallacy of the transposed conditional, Section 3.3.1, again) to consider that an event R which is unlikely if $\bar{S}$ is true thus provides evidence in favour of S. For this to be so, it is required additionally that R is not so unlikely when S is true. The likelihood ratio is then greater than 1 and the posterior odds are greater than the prior odds.

Notice that the likelihood ratio is a ratio of probabilities. It is greater than zero (except when $Pr(R \mid S) = 0$, in which case it is zero also) but has no theoretical upper limit. Probabilities take values between 0 and 1 inclusive; the likelihood ratio takes values between 0 and ∞. It is not an odds statistic, however. Odds are the ratio of the probabilities of two complementary events, perhaps conditioned on some other event. The likelihood ratio is the ratio of the probability of the same event conditioned upon two exclusive events, though they need not necessarily be complementary. Thus, $Pr(R \mid S)/Pr(R \mid \bar{S})$ is a likelihood ratio; $Pr(S \mid R)/Pr(\bar{S} \mid R)$ is an odds statistic.

Notice that (3.6) holds also if S and $\bar{S}$ are propositions rather than events. Propositions may be complementary, such as presence (H_p) or absence (H_d) of a suspect from a crime scene, but need not necessarily be so (see Example 3.6 and also Section 8.1.3 where more than two propositions are compared). In general, the two propositions to be compared will be known as competing propositions. The odds $Pr(S)/Pr(\bar{S})$ in such circumstances should be explicitly stated as odds in favour of S, relative to $\bar{S}$. In the special case in which the propositions are mutually exclusive and exhaustive, they are complementary. The odds may then be stated as the odds in favour of S, where the relationship to the complementary proposition is implicit.

Example 3.5   This hypothetical example is taken from Walsh and Buckleton (1991). Consider three events:

- Let X_A refer to Kell phenotype: $A = \text{Kell}+$, $\bar{A} = \text{Kell}-$.
- Let X_B refer to Duffy phenotype: $B = \text{Duffy}+$, $\bar{B} = \text{Duffy}-$.
- Let X_C refer to colour: $C = \text{pink}$, $\bar{C} = \text{blue}$.

Table 3.7 gives frequency counts in an area of 100 people. These frequency counts can be transformed into relative frequencies by dividing throughout by 100. These relative frequencies will now be considered as probabilities for the purpose of illustration. In practice, relative frequencies from a sample will only provide estimates of probabilities; the effects of such considerations are not discussed here.

Table 3.7 Frequency of Kell and Duffy types by colour in a hypothetical area. (Reproduced by permission of The Forensic Science Society)

Kell	Duffy		Total
	+	−	
Pink			
+	32	8	40
−	8	2	10
Total	40	10	50
Blue			
+	2	8	10
−	8	32	40
Total	10	40	50
Combined			
+	34	16	50
−	16	34	50
Total	50	50	100

From Table 3.7, the following probabilities, among others, may be derived.

$$Pr(A) = Pr(\text{Kell}+) = 50/100 = 0.5.$$
$$Pr(\bar{B}) = Pr(\text{Duffy}-) = 50/100 = 0.5.$$

$$Pr(A) \times Pr(\bar{B}) = 0.5 \times 0.5 = 0.25.$$

However,

$$Pr(A\bar{B}) = Pr(\text{Kell}+, \text{Duffy}-) = 16/100 = 0.16.$$

Thus

$$Pr(A\bar{B}) \neq Pr(A) \times Pr(\bar{B}).$$

The Kell and Duffy phenotypes are not, therefore, independent.

As well as the above joint probability, conditional probabilities can also be determined. Fifty people are Kell+ (A). Of these, 16 are Duffy− ($\bar{B}$). Thus, $Pr(\bar{B} \mid A) \times Pr(A) = 0.32 \times 0.5 = 0.16 = Pr(A\bar{B})$ which provides verification of the third law of probability (1.8). Similarly, $Pr(A \mid \bar{B}) = 16/50 = 0.32$. Equation (3.5) may be verified by noting that $Pr(\bar{B}) = 0.5$, and so

$$Pr(A \mid \bar{B}) = \frac{Pr(\bar{B} \mid A) \times Pr(A)}{Pr(\bar{B})} = \frac{0.32 \times 0.5}{0.5} = 0.32.$$

Now, consider only 'pink' (C) people:

$$Pr(A \mid C) = Pr(\text{Kell}+ \mid \text{pink}) = 40/50 = 0.8,$$
$$Pr(B \mid CA) = Pr(\text{Duffy}+ \mid \text{pink and Kell}+) = 32/40 = 0.8.$$

Thus,

$$Pr(AB \mid C) = Pr(B \mid AC) \times Pr(A \mid C) = 0.8 \times 0.8 = 0.64 = 32/50$$

from (1.7).

The odds version of Bayes' theorem,

$$\frac{Pr(C \mid A)}{Pr(\bar{C} \mid A)} = \frac{Pr(A \mid C) \times Pr(C)}{Pr(A \mid \bar{C}) \times Pr(\bar{C})}$$
$$= \frac{Pr(A \mid C)}{Pr(A \mid \bar{C})} \times \frac{Pr(C)}{Pr(\bar{C})},$$

which is of particular relevance to the evaluation of evidence, may also be verified. From Table 3.7, when $C = $ 'pink', $\bar{C} = $ 'blue', then $Pr(C) = Pr(\bar{C}) = 0.5$.

For someone of Kell+ (A) phenotype, $Pr(A \mid C) = 40/50 = 0.8, Pr(A \mid \bar{C}) = 10/50 = 0.2$. Thus

$$\frac{Pr(A \mid C)}{Pr(A \mid \bar{C})} \times \frac{Pr(C)}{Pr(\bar{C})} = \frac{0.8 \times 0.5}{0.2 \times 0.5} = 4. \tag{3.7}$$

Also,

$$Pr(C \mid A) = 40/50,$$
$$Pr(\bar{C} \mid A) = 10/50.$$

The ratio of these two probabilities is $Pr(C \mid A)/Pr(\bar{C} \mid A) = 4$, equal to (3.7). The odds form of Bayes' theorem has been verified numerically.

Consider this example as an identification problem. Before any information about Kell types, say, is available a person is equally likely to be 'pink' or 'blue'. The odds in favour of one colour or the other is evens. It is then discovered that the person is Kell+. The odds are now changed, using the procedure in the example, to being 4 to 1 on in favour of 'pink'. The probability the person is 'pink' is $4/5 = 0.8$. The effect or value of the evidence has been to multiply the prior odds by a factor of 4. The value of the evidence is 4.

3.4.2   Logarithm of the likelihood ratio

Odds and the likelihood ratio take values between 0 and ∞. Logarithms of these statistics take values on $(-\infty, \infty)$. Also, the odds form of Bayes' theorem involves a multiplicative relationship. If logarithms are taken, the relationship becomes an additive one:

$$\log \left\{ \frac{Pr(S \mid R)}{Pr(\bar{S} \mid R)} \right\} = \log \left\{ \frac{Pr(R \mid S)}{Pr(R \mid \bar{S})} \right\} + \log \left\{ \frac{Pr(S)}{Pr(\bar{S})} \right\}.$$

The idea of evaluating evidence by adding it to the logarithm of the prior odds is very much in keeping with the intuitive idea of weighing evidence in the scales of justice. The logarithm of the likelihood ratio has been called the *weight of evidence* (Good, 1950; see also Peirce, 1878). A likelihood ratio with a value greater than 1, which leads to an increase in the odds in favour of S, has a positive weight. A likelihood ratio with a value less than 1, which leads to a decrease in the odds in favour of S, has a negative weight. A positive weight may be thought to tip the scales of justice one way, a negative weight may be thought to tip the scales of justice the other way. A likelihood ratio with a value equal to 1 leaves the odds in favour of S and the scales unchanged.

The evidence is logically relevant only when the probability of finding that evidence given the truth of some proposition at issue in the case differs from the probability of finding the same evidence given the falsity of the proposition at issue; that is, the log-likelihood ratio is not zero (Kaye, 1986). The logarithm of the likelihood ratio (sometimes called the *relevance ratio* or *weight*) provides an equivalent measure of relevance. This method is advantageous, because it equates the relevance of evidence offered by both the prosecution and the defence (Lempert, 1977). The log-likelihood ratio also has qualities of symmetry and additivity that other measures lack (Edwards, 1986). Lyon and Koehler (1996) believe the simplicity and intuitive appeal of the relevance ratio make it a good candidate for heuristic use by judges. The mathematical symmetry between the weight of evidence for the prosecution proposition and the weight of evidence for the defence proposition can be maintained by inverting the weight of evidence when considering the defence proposition.

A verbal scale based on logarithms has been developed by Aitken and Taroni (1998).

Example 3.6   Consider two propositions regarding a coin.

- S, the coin is double-headed; if S is true, the probability of tossing a head equals 1, the probability of tossing a tail equals 0.

- $\bar{S}$, the coin has two heads and is fair; if $\bar{S}$ is true the probability of tossing a head equals the probability of tossing a tail and both equal $1/2$.

Notice that these are not complementary propositions.

The coin is tossed ten times and the outcome of any one toss is assumed independent of the others. The result R is ten heads. Then $Pr(R \mid S) = 1$, $Pr(R \mid \bar{S}) = (\frac{1}{2})^{10}$. The likelihood ratio is

$$\frac{Pr(R \mid S)}{Pr(\bar{R} \mid \bar{S})} = \frac{1}{(1/2)^{10}} = 2^{10} = 1024.$$

The evidence is 1024 times more likely if the coin is double-headed. The weight of the evidence is $10 \log(2)$. Each toss which yields a head contributes a weight $\log(2)$ to the hypothesis S that the coin is double-headed. Suppose, however, that the outcome of one toss is a tail (T). Then $Pr(T \mid S) = 0, Pr(T \mid \bar{S}) = 1/2$. The likelihood ratio $Pr(T \mid S)/Pr(T \mid \bar{S}) = 0$ and the posterior odds in favour of S relative to $\bar{S}$ equals 0. This is to be expected. A double-headed coin cannot produce a tail. If a tail is the outcome of a toss then the coin cannot be double-headed.

A brief history of the use of the weight of evidence is given by Good (1991). Units of measurement are associated with it. When the base of the logarithms is 10, Turing suggested that the unit should be called a *ban* and one-tenth of this would be called a *deciban*, abbreviated to db.

3.5 THE VALUE OF EVIDENCE

3.5.1 Evaluation of forensic evidence

Consider the odds form of Bayes' theorem in the forensic context of assessing the value of some evidence. The initial discussion is in the context of the guilt or otherwise of the suspect. This may be the case, for example, in the context of Example 1.1, if all innocent explanations for the bloodstain have been eliminated. Later, greater emphasis will be placed on propositions that the suspect was, or was not, present at the scene of the crime. At present, replace event S by a proposition H_p, that the suspect (or defendant if the case has come to trial) is truly guilty. Event $\bar{S}$ is replaced by proposition H_d, that the suspect is truly innocent. Event R is replaced by event Ev, the evidence under consideration. This may be written as $(M, E) = (M_c, M_s, E_c, E_s)$, the type of evidence and observations of it as in Section 1.6.1. The odds form of Bayes' theorem then enables the prior odds (i.e., prior to the presentation of Ev) in favour of guilt to be updated to posterior odds given Ev, the evidence under consideration. This is done by multiplying the prior odds by the likelihood ratio which, in this context, is the ratio of the probabilities of the evidence assuming guilt and assuming innocence of the suspect. With this notation, the odds form of Bayes' theorem may be written as

$$\frac{Pr(H_p \mid Ev)}{Pr(H_d \mid Ev)} = \frac{Pr(Ev \mid H_p)}{Pr(Ev \mid H_d)} \times \frac{Pr(H_p)}{Pr(H_d)}.$$

Explicit mention of the background information I is omitted in general from probability statements for ease of notation. With the inclusion of I the odds form of Bayes' theorem is

$$\frac{Pr(H_p \mid Ev, I)}{Pr(H_d \mid Ev, I)} = \frac{Pr(Ev \mid H_p, I)}{Pr(Ev \mid H_d, I)} \times \frac{Pr(H_p \mid I)}{Pr(H_d \mid I)}.$$

Notice the important point that in the evaluation of the evidence Ev it is two probabilities that are necessary: the probability of the evidence if the suspect is guilty and the probability of the evidence if the suspect is innocent. For example, it is not sufficient to consider only the probability of the evidence if the suspect is innocent and to declare that a small value of this is indicative of guilt. The probability of the evidence if the suspect is guilty has also to be considered.

Similarly, it is not sufficient to consider only the probability of the evidence if the suspect is guilty and to declare that a high value of this is indicative of guilt. The probability of the evidence if the suspect is innocent has also to be considered. An example of this is the treatment of the evidence of a bite mark in the Biggar murder in 1967–68 (Harvey *et al.*, 1968), an early example of odontology in forensic science. In that murder a bite mark was found on the breast of the victim, a young girl, which had certain characteristic marks, indicative of the conformation of the teeth of the person who had bitten her. A 17-year-old boy was found with this conformation and became a suspect. Such evidence would help towards the calculation of $Pr(Ev \mid H_p)$. However, there was no information available about the incidence of this conformation among the general public. Examination was made of 342 boys of the suspect's age. This enabled an estimate – albeit an intuitive one – of $Pr(E \mid H_d)$ to be obtained and to show that the particular conformation found on the suspect was not at all common.

Consider the likelihood ratio $Pr(Ev \mid H_p)/Pr(Ev \mid H_d)$ further, where explicit mention of I has again been omitted. This equals

$$\frac{Pr(E \mid H_p, M)}{Pr(E \mid H_d, M)} \times \frac{Pr(M \mid H_p)}{Pr(M \mid H_d)}.$$

The second ratio in this expression, $Pr(M \mid H_p)/Pr(M \mid H_d)$, concerns the type and quantity of evidential material found at the crime scene and on the suspect. It may be written as

$$\frac{Pr(M_s \mid M_c, H_p)}{Pr(M_s \mid M_c, H_d)} \times \frac{Pr(M_c \mid H_p)}{Pr(M_c \mid H_d)}.$$

The value of the second ratio in this expression may be taken to be 1. The type and quantity of material at the crime scene are independent of whether the suspect is the criminal or someone else is. The value of the first ratio, which concerns the evidential material found on the suspect given the evidential material found at the crime scene and the guilt or otherwise of the suspect, is a matter for subjective judgement and it is not proposed to consider its determination further here. Instead, consideration will be concentrated on

$$\frac{Pr(E \mid H_p, M)}{Pr(E \mid H_d, M)}.$$

In particular, for notational convenience, M will be subsumed into I and omitted, for clarity of notation. Then

$$\frac{Pr(M \mid H_p)}{Pr(M \mid H_d)} \times \frac{Pr(H_p)}{Pr(H_d)},$$

which equals

$$\frac{Pr(H_p \mid M)}{Pr(H_d \mid M)},$$

will be written as

$$\frac{Pr(H_p)}{Pr(H_d)}.$$

Thus

$$\frac{Pr(H_p \mid Ev)}{Pr(H_d \mid Ev)} = \frac{Pr(H_p \mid E, M)}{Pr(H_d \mid E, M)}$$

will be written as

$$\frac{Pr(H_p \mid E)}{Pr(H_d \mid E)},$$

and

$$\frac{Pr(Ev \mid H_p)}{Pr(Ev \mid H_d)} \times \frac{Pr(H_p)}{Pr(H_d)}$$

will be written as

$$\frac{Pr(E \mid H_p)}{Pr(E \mid H_d)} \times \frac{Pr(H_p)}{Pr(H_d)}.$$

The full result is then

$$\frac{Pr(H_p \mid E)}{Pr(H_d \mid E)} = \frac{Pr(E \mid H_p)}{Pr(E \mid H_d)} \times \frac{Pr(H_p)}{Pr(H_d)}, \tag{3.8}$$

or, if I is included,

$$\frac{Pr(H_p \mid E, I)}{Pr(H_d \mid E, I)} = \frac{Pr(E \mid H_p, I)}{Pr(E \mid H_d, I)} \times \frac{Pr(H_p \mid I)}{Pr(H_d \mid I)}. \tag{3.9}$$

The likelihood ratio is the ratio

$$\frac{Pr(H_p \mid E, I)/Pr(H_d \mid E, I)}{Pr(H_p \mid I)/Pr(H_d \mid I)} \tag{3.10}$$

of posterior odds to prior odds. It is the factor which converts the prior odds in favour of guilt to the posterior odds in favour of guilt. The representation in (3.9) also emphasises the dependence of the prior odds on background information. Previous evidence may also be included here; see, for example, Section 8.1.3.

It is often the case that another representation may be appropriate. Sometimes it may not be possible to consider the effect of the evidence on the guilt or innocence of the suspect. However, it may be possible to consider the effect of the evidence on the possibility that there was contact between the suspect and the crime scene. For example, a bloodstain at the crime scene may be of the same type as that of the suspect. This, considered in isolation, would not necessarily be evidence to suggest that the suspect was guilty, only that the suspect was at the crime scene. Consider the following two complementary propositions:

- H_p, the suspect was at the crime scene;
- H_d, the suspect was not at the crime scene.

The odds form of Bayes' theorem is then

$$\frac{Pr(H_p \mid E, I)}{Pr(H_d \mid E, I)} = \frac{Pr(E \mid H_p, I)}{Pr(E \mid H_d, I)} \times \frac{Pr(H_p \mid I)}{Pr(H_d \mid I)}, \tag{3.11}$$

identical to (3.9) but with different definitions for H_p and H_d. The likelihood ratio converts the prior odds in favour of H_p into the posterior odds in favour of H_p.

The likelihood ratio may be thought of as the value of the evidence. Evaluation of evidence, the theme of this book, will be taken to mean the determination of a value for the likelihood ratio. This value will be denoted V.

Definition 3.1   Consider two competing propositions, H_p and H_d, and background information I. The *value* V of the evidence E is given by

$$V = \frac{Pr(E \mid H_p, I)}{Pr(E \mid H_d, I)}, \tag{3.12}$$

the likelihood ratio which converts prior odds $Pr(H_p \mid I)/Pr(H_d \mid I)$ in favour of H_p relative to H_d into posterior odds $Pr(H_p \mid E, I)/Pr(H_d \mid E, I)$ in favour of H_p relative to H_d.

An illustration of the effect of evidence with a value V of 1000 on the odds in favour of H_p, relative to H_d, is given in Table 3.8.

This is not a new idea. Consider the following quotes (Kaye, 1979):

> 'That approach does not ask the jurors to produce any number, let alone one that can qualify as a probability. It merely shows them how a 'true' prior probability would be altered, if one were in fact available. It thus supplies the jurors with

Table 3.8   Effect on prior odds in favour of H_p relative to H_d of evidence E with value V of 1 000. Reference to background information I is omitted

Prior odds $Pr(H_p)/Pr(H_d)$	V	Posterior odds $Pr(H_p \mid E)/Pr(H_d \mid E)$
1/10 000	1 000	1/10
1/100	1 000	10
1 (evens)	1 000	1 000
100	1 000	100 000

as precise and accurate an illustration of the probative force of the quantitative data as the mathematical theory of probability can provide. Such a chart, it can be maintained, should have pedagogical value for the juror who evaluates the entire package of evidence solely by intuitive methods, and who does not himself attempt to assign a probability to the 'soft' evidence.

[A] more fundamental response is that there appears to be no reason in principle why a juror could not generate a prior probability that could be described in terms of the objective, relative-frequency sort of probability. One could characterize the juror's prior probability as an estimate of the proportion of cases in which a defendant confronted with the same pattern of nonquantitative evidence as exists in the case at bar would in fact turn out to have stabbed the deceased.

This practical difficulty does not undercut the conceptual point.

3.5.2   Summary of competing propositions

Initially, when determining the value of evidence, the two competing propositions were taken to be that the suspect is guilty and that the suspect is innocent. However, these are not the only ones possible, as the discussion at the end of the previous subsection has shown.

Care has to be taken in a statistical discussion about the evaluation of evidence as to the purpose of the analysis. Misunderstandings can arise. For example, it has been said that 'the statistician's objective is a final, composite probability of guilt' (Kind, 1994). This may be an intermediate objective of a statistician on a jury. A statistician in such a position would then have to make a decision as to whether to find the defendant guilty or not guilty. Determination of the guilt, or otherwise, of a defendant is the duty of a judge and/or jury. The objective of a statistician advising a scientist on the worth of the scientist's evidence is rather different. It is to assess the value of the evidence under two competing propositions. The evidence which is being assessed will often be transfer evidence. The propositions may be guilt or innocence. However, in many cases they will not be. An illustration of the effect of evidence on the odds in favour of a proposition H_p relative to a proposition H_d has been given in Table 3.8.

However, it has to be emphasised that the determination of the prior odds is also a vital part of the equation. That determination is part of the duty of the judge and/or jury.

Several suggestions for competing propositions, including those of guilt and innocence, are given below.

1. H_p, the suspect is guilty;
   H_d, the suspect is innocent.
2. H_p, there was contact between the suspect and the crime scene;
   H_d, there was no contact between the suspect and the crime scene.
3. H_p, the crime sample came from a Caucasian;
   H_d, the crime sample came from an Afro-Caribbean (Evett *et al.*, 1992a).

4. H_p, the alleged father is the true father of the child;
   H_d, the alleged father is not the true father of the child (Evett *et al.*, 1989a).
5. H_p, the two crime samples came from the suspect and one other man;
   H_d, the two crime samples came from two other men (Evett *et al.*, 1991).
6. H_p, the suspect was the person who left the crime stain;
   H_d, the suspect was not the person who left the crime stain (Evett *et al.*, 1989b).
7. H_p, paint on the injured party's vehicle originated from the suspect's vehicle;
   H_d, paint on the injured party's vehicle originated from a random vehicle (McDermott *et al.*, 1999).

In general, the two propositions can be referred to as the prosecutor's and defence propositions, respectively. The prosecutor's proposition is the one to be used for the determination of the probability in the numerator, the defence proposition is the one to be used for the determination of the probability in the denominator.

At a particular moment in a trial, the context is restricted to two competing propositions. Consider a rape case in which the victim reported to police that she had been raped by a former boyfriend. A T-shirt belonging to the boyfriend is examined and foreign fibres are collected from it. The propositions offered may include:

- H_p, the suspect is the offender;

- H_{d1}, the suspect is not the offender and has not seen the victim during the past three weeks;

- H_{d2}, the suspect is not the offender but on the night of the alleged rape he went dancing with the victim.

The evidence includes the attributes of the foreign fibres found on the boyfriend's T-shirt and the attributes of fibres taken from the victim's garments. The value of this evidence will change as the propositions offered by the prosecution and defence change. The propositions are defined taking into account also the background information I.

Example 3.7 (Example 1.1 Continued)   The following discussion is simplistic in the context of DNA profiling and is provided simply to illustrate the use of the likelihood ratio for the evaluation of evidence. More realistic examples for DNA profiling are given in Chapter 13.

A crime has been committed. A bloodstain is found at the scene of the crime. All innocent sources of the stain are eliminated and the criminal is determined as the source of the stain. The purpose of the investigation is to determine the identity of the criminal; it is at present unknown. For the *LDLR* locus, the bloodstain is of genotype Γ, with a frequency γ in the relevant population. A suspect has been identified with the same genotype for locus *LDLR* as the crime stain. These two facts, the genotype of the crime stain (E_c) and the

Table 3.9   Value of the evidence for each genotype

Genotype	(Γ)	AA	BB	AB
Value	($1/\gamma$)	5.32	3.12	2.04

genotype of the suspect (E_s) together form the observations for the evidence $E(= (E_c, E_s))$. The likelihood ratio can be evaluated as follows. If the suspect is guilty (H_p), the match between the two genotypes is certain and $Pr(E \mid H_p) = 1$. If the suspect is innocent (H_d), the match is coincidental. The criminal is of genotype Γ. The probability that the suspect would also have group Γ is γ, the frequency of Γ in the relevant population. Thus, $Pr(E \mid H_d) = \gamma$. The likelihood ratio is $Pr(E \mid H_p)/Pr(E \mid H_d) = 1/\gamma$. The odds in favour of G are multiplied by $1/\gamma$.

Consider the following numerical illustration of this, for the *LDLR* locus. The three genotypes and their frequencies γ for a Caucasian Chicago population (Johnson and Peterson, 1999) were given in Table 1.1.

The effect $(1/\gamma)$ on the odds in favour of H_p is given in Table 3.9 for each genotype. Verbal interpretations may be given to these figures. For example, if the crime stain were of genotype *AA*, it could be said that 'the evidence of the genotype of the crime stain matching the suspect's genotype is about five times more likely if the suspect is guilty than if he is innocent'. If the crime stain were of type *AB*, it could be said that 'the evidence . . . is two times more likely if the suspect is guilty than if he is innocent'.

3.5.3   Qualitative scale for the value of the evidence

This quantitative value has been given a qualitative interpretation (Jeffreys, 1983; Evett, 1987a, 1990; Evett *et al.*, 2000a). Consider two competing propositions H_p and H_d and a value V for a piece of evidence. The qualitative scale suggested by Evett *et al* (2000a) is given Table 3.10.

Note that this scale works with reciprocal values for $V < 1$ in support of the defence proposition. For DNA in which there are very large likelihood ratios, the verbal scale becomes inadequate. However, it has become accepted practice

Table 3.10   Qualitative scale for reporting the value of the support of the evidence for H_p against H_d (Evett *et al.*, 2000a)

1	$< V \leq$	10	Limited evidence to support
10	$< V \leq$	100	Moderate evidence to support
100	$< V \leq$	1000	Moderately strong evidence to support
1000	$< V \leq$	10000	Strong evidence to support
10000	$< V$		Very strong evidence to support

(Evett *et al.*, 2000a) to use the phrase *extremely strong* for likelihood ratios of 1 million or more. A comment on the extremely high values of V obtained with DNA evidence is presented in Evett *et al.* (2000c). To face this problem, Aitken and Taroni (1998) proposed an approach based on the logarithm of the likelihood ratio.

Interestingly, Fienberg (1989) provides a quote from a nineteenth-century jurist, Jeremy Bentham, which appears to anticipate the Jeffreys–Evett scale, though perhaps in application to the strength of the belief in the hypothesis of guilt rather than in the strength of the evidence.

> The scale being understood to be composed of ten degrees – in the language applied by the French philosophers to thermometers, a decigrade scale – a man says, My persuasion is at 10 or 9 *etc.* affirmative, or at least 10 *etc.* negative . . . (Bentham, 1827; quoted in Fienberg, 1989)

Notice, also, what is not said in the verbal interpretation. Consider blood of profile *AB* for locus *LDLR* again. It is not claimed that the evidence is such that the suspect is twice as likely to be guilty as he would have been if the evidence had not been presented. It is the evidence which is twice as likely, not the proposition of guilt. The largest value for γ in Example 3.7 is 5.32, a value which would be said to provide limited support using the scale in Table 3.10. At present, there is no general agreement amongst jurists concerning the association of verbal and numeric scales. Until there is such agreement, the verbal description for a numeric value will have to remain a matter of personal judgement.

This consideration of support for a proposition is illustrated in the following case from 1998:

> We note and we follow and accept unreservedly Dr Evett's evidence to us and his strictures to us that we cannot look at one hypothesis, we must look at two and we must test each against the other . . . what is the probability of the evidence if the Respondent's hypothesis is correct? what is the probability of the evidence if the Appellant's hypothesis is correct?
>
> Dr Evett tells us (and we follow it) that if the answer to the first question is greater than the answer to the second question, then the Respondent's hypothesis is supported by the evidence. (*Johannes Pruijsen v. H.M. Customs & Excise*)

Statements concerning the probability of guilt require knowledge of the prior odds in favour of guilt, something which is not part of the scientist's knowledge. A similar interpretation is given in Royall (1997, p.13):

> a likelihood ratio of k corresponds to evidence strong enough to cause a k-fold increase in a prior probability ratio, regardless of whether a prior ratio is actually available in a specific problem or not.

Note that this is a measure of the value of the evidence. The implications for a particular value of evidence will vary according to the context (Evett, 1998).

Other possibilities include a scale based on logarithms (Kass and Raftery, 1995) which provides a conversion from the logarithm of prior odds to a logarithm of posterior odds.

The use of logarithms transforms the relationship between prior and posterior odds and the likelihood ratio into an additive one, and a useful discussion of this is given in Schum (1994). Thus

$$\log Pr(H_p|E)/Pr(H_d|E) = \log Pr(E|H_p)/Pr(E|H_d) + \log Pr(H_p)/Pr(H_d).$$

Logarithms provide a good way of comprehending the magnitude of large numbers. Consider logarithms to base 10. A relative frequency of 1 in 10 million has a logarithm of -7, and a relative frequency of 1 in 1 million has a logarithm of -6. The corresponding reciprocals have logarithms of 7 and 6. These numbers, 7 and 6, are much more meaningful to the statistical layman and the difference between them is much more comprehensible. This provides a measure of the influence on the judge's assessment of the ultimate issue.

Logarithms are also used to measure effects in other areas with which many people are familiar. The Richter scale for measuring the strength of earthquakes is a logarithmic scale. Sound is measured in decibels, another logarithmic scale. The pH scale for measuring acidity/basicity is also a logarithmic scale.

Consider a case in which the likelihood ratio is 500 million in favour of the prosecution's proposition. Prior odds in favour of innocence of 1 million to 1 will be converted to posterior odds of 500 to 1 in favour of guilt.

In the verbal scale proposed by Aitken and Taroni (1998), odds in favour of innocence of 1 million to 1 would relate to a city of 1 million people in which there were one guilty person. A person is selected at random from this city. The probability he is the guilty person is approximately 1 in 1 million.

Consider now posterior odds of 500 to 1 in favour of guilt. Imagine a large street with about 500 inhabitants. All except one are guilty. A person is selected at random from the street. The probability he is guilty is $499/500$ and the odds are approximately 500 to 1 in favour of guilt.

Different people will have different prior odds and thus different posterior odds. Tables can be provided, such as Table 3.8, to show how the prior odds are converted to posterior odds. Table 3.11 from Aitken and Taroni (1998), converts log prior odds to log posterior odds.

As an example of how the table may be used, consider a case in which the prior odds against the defendant are 1 million to 1, 10 million to 1 or 100 million to 1. The posterior odds in favour of the guilt of the defendant are then 7 million to 1, 70 million to 1 or 700 000 to 1.

The interpretation of these odds can be in the context of a community scale. (Calman, 1996; Calman and Royston, 1997). Consider odds of 1 million to 1 against guilt and consider a city of 1 million people (a little bigger than Glasgow). A criminal is one person in this city. All others are innocent. A person is selected at random from the city. The odds are a million to 1 against that

Table 3.11 The values of the logarithm of the posterior odds in favour of an issue determined from the values of the logarithm of the prior odds in favour of guilt (log(Prior odds)) and the logarithm of the likelihood ratio (Aitken and Taroni, 1998). The values in the body of the table are obtained by adding the appropriate row and column values. Logarithms are taken to base 10. The verbal description is taken from Calman and Royston (1997). (Reproduced by permission of The Forensic Science Society)

Verbal description	log (Prior odds)	Logarithm of the likelihood ratio												
		−2	−1	0	1	2	3	4	5	6	7	8	9	10
Individual	0	−2	−1	0	1	2	3	4	5	6	7	8	9	10
Family	−1	−3	−2	−1	0	1	2	3	4	5	6	7	8	9
Street	−2	−4	−3	−2	−1	0	1	2	3	4	5	6	7	8
Village	−3	−5	−4	−3	−2	−1	0	1	2	3	4	5	6	7
Small town	−4	−6	−5	−4	−3	−2	−1	0	1	2	3	4	5	6
Large town	−5	−7	−6	−5	−4	−3	−2	−1	0	1	2	3	4	5
City	−6	−8	−7	−6	−5	−4	−3	−2	−1	0	1	2	3	4
Province/country	−7	−9	−8	−7	−6	−5	−4	−3	−2	−1	0	1	2	3
Large country	−8	−10	−9	−8	−7	−6	−5	−4	−3	−2	−1	0	1	2
Continent	−9	−11	−10	−9	−8	−7	−6	−5	−4	−3	−2	−1	0	1
World	−10	−12	−11	−10	−9	−8	−7	−6	−5	−4	−3	−2	−1	0

person being the criminal. The advantage of such an approach is that it offers an image associated with a number.

3.5.4   Misinterpretations

The above examples are appropriate summaries of the evidence. However, misinterpretations still occur in which the evidence is summarized as a comment about the truth or otherwise of the prosecution's proposition. Note that the use of verbal scales about the truth or otherwise of a proposition are accepted in many scientific fields (e.g., medicine, weather forecasting). Here, in contrast to forensic science, the scientist plays a different role and uses a different amount of information. Therefore, posterior scales seem acceptable in a way they are not in forensic science. An 11-point subjective posterior scale of scientific uncertainty based on legally defined standards of proof has been proposed (Weiss, 2003), motivated by discussions within the Intergovernmental Panel on Climate Change (2001).

In a response to a survey conducted by Taroni and Aitken (2000) on fibres evidence, the comment was made that the strength of the evidence was categorised in terms of the probability of the prosecution's proposition, that it was

- beyond reasonable doubt,
- most probable,
- probable,
- quite possible, or
- possible

that matching evidence associated with the defendant comes from the same source as that found at the crime scene. In this survey, laboratories generally commented on the truthfulness or otherwise of the proposition proposed by the prosecution. This was instead of the value of the evidence.

Also, in the context of human hair comparisons, Gaudette (2000) gives a scale for the questioned hairs originating or not from the same person as the known sample. There is a match and the scale interprets this as follows.

Strong positive:	Questioned hairs originated from same person as the known sample.
Normal positive:	Questioned hairs are consistent with the known sample.
Inconclusive:	No conclusion can be given.
Normal negative:	Questioned hairs are not consistent with the known sample.
Strong negative:	Questioned hairs could not have originated from the known sample.

However, this is making a judgement about the source of the hair without prior knowledge of the background information of the case. Similar problems have been discussed in the context of shoeprint examinations (Champod *et al.*, 2000; Taroni and Margot, 2001; Katterwe, 2002a, b; Taroni and Buckleton, 2002; Champod and Jackson, 2002) and speaker recognition (Champod and Evett, 2000).

3.5.5 Explanation of transposed conditional and defence fallacies

Using the odds form of Bayes' theorem (3.11), insight into the fallacy of the transposed conditional (or prosecutor's fallacy), the false positive fallacy and the defence fallacy can be gained.

Fallacy of the transposed conditional

As before, a crime has been committed. A bloodstain has been found at the scene which has been identified as coming from the criminal. A suspect is identified and his blood group is the same as the crime stain. Let E denote the evidence that the suspect's blood group is the same as that of the crime stain. Let H_p denote the proposition that the suspect is guilty and its complement H_d denote the proposition that the suspect is innocent.

Consider the following two statements:

- The blood group is found in only 1% of the population.
- There is a 99% chance that the suspect is guilty.

The second statement does not follow from the first without an unwarranted assumption about the prior odds in favour of guilt. The first statement that the blood group is found in only 1% of the population is taken to mean that the probability that a person selected at random from the population has the same blood group as the crime stain is 0.01. Thus $Pr(E \mid H_d) = 0.01$. Also, $Pr(E \mid H_p) = 1$. The value of V is 100.

The second statement is taken to mean that the posterior probability in favour of guilt (posterior to the presentation of E) is 0.99, that is, $Pr(H_p \mid E) = 0.99$. Thus $Pr(H_d \mid E) = 0.01$ since H_p and H_d are complementary propositions. The posterior odds are then 0.99/0.01 or 99, which is approximately equal to 100. However, V also equals 100. From (3.11), the prior odds are approximately equal to 1:

$$Pr(H_p) \simeq Pr(H_d).$$

The second statement of the fallacy of the transposed conditional follows from the first only if $Pr(H_p) \simeq Pr(H_d)$. In other words, the suspect is just as likely to be guilty as innocent. This is not in accord with the dictum that a person is innocent until proven guilty. The prosecutor's conclusion, therefore, does not follow from the first statement unless this unwarranted assumption is made.

The use of prior odds of 1 is also advocated for shoeprint examination (Katterwe, 2003). A population of N shoes is postulated as the one to which the shoe that made the print at a crime scene belongs. There is one shoe which may be considered as the suspect shoe. In the absence of any other information, the probability that this shoe made the crime print is $1/N$. The probability that another shoe from this population made the print is $(N-1)/N$. The prior odds that the suspect shoe made the print are $1/(N-1)$. The argument advanced by Katterwe (2003), however, is that, apart from the suspect shoe and in the absence of any other information, there is a population of only 1 shoe that could also have made the print. The defence proposition is that only one other shoe made the print. Therefore the relevant population is of size 2. This argument is analogous to the probability of hitting a target, for example in a game of darts. The dart may hit the bull's-eye or it may not. There are only two possibilities. In the absence of other information, the probabilities of 'hit' or 'miss' are equal at $1/2$ each. However, in reality, there is always other information. For the dart player, there is information about the area of the bull's-eye compared with the area of the rest of the dartboard and the wall on which the dartboard hangs. For shoeprint examination there is information on the number of shoes in the world which could have made the print. Similar arguments hold for other evidence types, such as paternity (Section 9.8).

The false positive fallacy

It is important to have accurate information about both the random match probability (see Chapter 13 for details) and the false positive probability when evaluating DNA evidence. Ignorance of, or an underestimation of the potential for, a false positive can lead to serious errors of interpretation, particularly when the other evidence against the suspect (apart from the DNA evidence) is weak.

It is considered essential to have valid scientific data on the random match probability but, paradoxically, it is thought unnecessary to have valid data on the false positive probability. The explanation for this lies partly in a common logical fallacy which may be called the *false positive fallacy*. It is assumed, mistakenly, that if the false positive probability is low then the probability of a false match must also be low in every case. For example, a forensic scientist who thinks that there is only a 1% chance of falsely declaring a match between samples in a case if they really do not match, might assume that there is, necessarily, a 99% chance that the reported match is a true match. This assumption is fallacious. The fallacy arises from mistaken equation of the conditional probability of a match being reported when the samples do not match (the false positive probability) with the probability that the samples do not match when a match has been reported. These two probabilities are not the same.

The false positive probability is the probability of a match being reported under a specified condition (no match). It does not depend on the probability of the occurrence of that condition. By contrast, the probability that the samples do not match when a match has been reported depends on both the probability of a match being reported under the specified condition (no match) and the prior probability that that condition will occur. Consequently, the probability that a reported match is a true match or a false match cannot be determined from the false positive probability alone. In formal terms, the fallacious assumption is that $Pr(M \mid R) = 1 - Pr(R \mid \bar{M})$, where M is the event that the suspect and the perpetrator have matching DNA profiles, $\bar{M}$ is the event that they do not have matching profiles, and $Pr(R \mid \bar{M})$ is the false positive probability; that is, the probability of a match being reported given that the samples do not have matching profiles. This assumption is fallacious because it ignores the prior odds that the suspect's profile matches the sample profile. Let the prior odds, $Pr(M)/Pr(\bar{M})$, equal $1/k$ where k is large. Then

$$\frac{Pr(M \mid R)}{Pr(\bar{M} \mid R)} = \frac{Pr(R \mid M)}{Pr(R \mid \bar{M})} \times \frac{1}{k}.$$

Assume $Pr(R \mid M) = 1$, that is, there are no false negatives. Then

$$Pr(M \mid R) = \frac{1}{1 + kPr(R \mid \bar{M})}$$

which can be much lower than $1 - Pr(R \mid \bar{M})$ when k is large.

For example, suppose that the prior odds the suspect will match are $1/1000$ because the suspect is selected through a large DNA dragnet and appears, initially, to be an unlikely perpetrator. Suppose, further, that a DNA match is reported and that the false positive probability is 0.01. The probability that this reported match is a true match is, therefore, $1/(1 + 1000 \times 0.01) = 0.0909$. In other words, the probability that this reported match is a true match is not 0.99, as the false positive fallacy would suggest; it is less than 0.1.

During database searches, true matches are expected to be rare. Therefore, the probability in a particular case that a non-match will mistakenly be reported as a match, even if low, may approach or even exceed the probability that the suspect truly matches. The false positive fallacy is similar in form to the prosecutor's fallacy (Thompson and Schumann, 1987), but differs somewhat in content. Victims of the false positive fallacy mistakenly assume that $Pr(M \mid R) = 1 - Pr(R \mid \bar{M})$. Victims of the prosecutor's fallacy mistakenly assume that $Pr(S \mid M) = 1 - Pr(M \mid \bar{S})$ where

- S is the proposition that the specimen came from the suspect, and
- $\bar{S}$ is the proposition that the specimen did not come from the suspect

(Thompson and Schumann, 1987). Both fallacies arise from failure to take account of prior probabilities (or odds) when evaluating new evidence; both can lead to significant overestimation of the posterior probability when the prior probability is low. The prosecutor's fallacy is an erroneous way of estimating the probability that the suspect is the source of a sample based on evidence of a matching characteristic; the false positive fallacy is an erroneous way of estimating the probability of a true match based on a reported match.

Defender's fallacy

Assume the likelihood ratio is 100, as immediately above. Consider the relevant population to contain 200 000 people. The defence says that there are 2000 people with the same blood group as the defendant, the probability the defendant is guilty is $1/2000$ and thus the evidence has very little value in showing this particular person guilty. As before $Pr(E \mid H_p) = 1, Pr(E \mid H_d) = 0.01$ and $V = 100$. Also, $Pr(H_p \mid E) = 1/2000$ and so $Pr(H_d \mid E) = 1999/2000$. The posterior odds in favour of H_p are

$$\frac{1/2000}{1999/2000} = \frac{1}{1999} \simeq \frac{1}{2000}.$$

The prior odds are equal to the ratio of the posterior odds to V:

$$\frac{Pr(H_p)}{Pr(H_d)} = \frac{Pr(H_p \mid E)}{Pr(H_d \mid E)} \Big/ V \simeq \left(\frac{1}{2000}\right) \Big/ 100 = \frac{1}{200\,000}.$$

Thus, $Pr(H_p) = 1/200001, Pr(H_p) = 200\,000/200\,0001$. The prior probability of guilt is $1/200\,001$. The denominator is the size of the relevant population (of innocent people) plus one for the criminal. The implication is that everybody is equally likely to be guilty. This does seem in accord with the dictum of innocent until proven guilty. Colloquially, it could be said that the defendant is just as likely to be guilty as anyone else. The defence fallacy is not really a fallacy. It is misleading, though, to claim that the evidence has little relevance for proving the suspect is guilty. Evidence which increases the odds in favour of guilt from $1/200\,000$ to $1/2000$ is surely relevant. On the Jeffreys–Evett scale (Table 3.10) the likelihood ratio expresses moderate evidence in support of the proposition of guilt.

Notice that it is logically impossible to equate the dictum *innocent until proven guilty* with a prior probability of guilt of zero, $Pr(H_p) = 0$. If $Pr(H_p) = 0$, then $Pr(H_p \mid E) = 0$, from (3.8), no matter how overwhelming the evidence, no matter how large the value of V. The probability of guilt may be exceedingly small, so long as it is not zero. So long as the prior probability of guilt is greater than zero, it will be possible, given sufficiently strong evidence, to produce a posterior probability of guilt sufficiently large to secure a conviction. It could be argued that the defendant is as likely to be as guilty as anyone else; his prior probability of guilt would then be the reciprocal of the size of the population defined by 'anyone else'. Comments on this point are presented in Robertson and Vignaux (1994).

3.5.6   The probability of guilt

Another legal dictum is that the jury is told to find the suspect 'guilty' if it is persuaded that the prosecution has shown 'beyond reasonable doubt' that the suspect committed the crime (Robertson and Vignaux, 1991). However, the meaning of 'beyond reasonable doubt' is not clear. Lord Denning has said that 'there may be degrees of proof within that standard' (Eggleston, 1983). The more serious the crime, the greater the presumption of innocence. There is evidence that this is how people behave in the results of a questionnaire administered to judges, jurors and students of sociology (Simon and Mahan, 1971) and summarised in Table 3.12. This gives the value for the probability of guilt which would be taken as proof beyond reasonable doubt by those groups of people for various crimes. The variability amongst these figures is surprisingly small. For judges, the range is from 0.92 to 0.87 and for jurors from 0.86 to 0.74. These probabilities may be used to determine odds in favour of guilt. For example, for judges the odds in favour of guilt for a charge of murder to be proved 'beyond reasonable doubt' are 0.92 to 0.08 or approximately 12 to 1. An alternative interpretation is that 1 out of 13 people convicted of murder by that standard would be innocent. For a charge of petty larceny the odds are approximately 7 to 1. These odds (7 to 1 and 12 to 1) are remarkably close when the different levels of punishment are considered. The results need careful interpretation. For example, it is unlikely that judges in America in 1971 were convicting, wrongly, 1 out of 13 people convicted of murder. It is far more likely that the judges do not have a good intuitive feel for the meaning of probability figures. As the value of the probability of guilt required to find a person guilty increases more protection is given to the innocent but at the expense of making it more difficult to convict the guilty. For example, Jaynes (2003) comments that

> If 1000 guilty men are set free, we know from only too much experience that 200 or 300 of them will proceed immediately to inflict still more crimes upon society, and their escaping justice will encourage 100 more to take up crime. So it is clear that the damage to society as a whole caused by allowing 1000 guilty men to go free is far greater than that caused by falsely convicting one innocent man.

Table 3.12 Probability of guilt required for proof beyond reasonable doubt (Simon and Mahan, 1971)

Crime	Mean of persons surveyed		
	Judges	Jurors	Students
Murder	0.92	0.86	0.93
Forcible rape	0.91	0.75	0.89
Burglary	0.89	0.79	0.86
Assault	0.88	0.75	0.85
Petty larceny	0.87	0.74	0.82

If you have an emotional reaction against this statement, I ask you to think: if you were a judge, would you rather face one man whom you had convicted falsely or 100 victims of crime that could have been prevented?

Consider the following illustration of the effect of changes in probability at the upper end of the scale on the odds. Suppose that a probability value of 0.999 for the probability of guilt was thought sufficient to prove someone guilty 'beyond reasonable doubt'. The odds in favour of guilt are then 999 to 1. A change in the probability of guilt to 0.99 reduces the odds to 99 to 1. A reduction by a factor of 10 in the odds has been brought about by a change in the probability of less than 0.01.

Another interpretation of 'beyond reasonable doubt' may be given by consideration of (3.10) and the interpretation of the likelihood ratio as the factor which converts the prior odds in favour of guilt to the posterior odds in favour of guilt. Suppose the prior odds for the dictum 'innocent until proven guilty' are taken to be 1/1000, though in practice they may be a lot smaller. Suppose the posterior odds for the dictum 'proof beyond reasonable doubt' are taken to be 1000 to 1, though in practice they may be a lot bigger. Then the likelihood ratio, the factor which converts the prior odds into the posterior odds, has to take the value 1000/(1/1000), which is 1 million. In other words, the evidence has to be 1 million times more likely if the suspect is guilty than if he is innocent. Such evidence is very compelling and may reasonably be said to provide very strong support for the hypothesis that the suspect is guilty.

Determination of the probability of guilt has been a long-standing problem. The problem has often been approached by consideration of a finite population such as may be found on an island, of size $N + 1$, say. A crime is committed and evidence of a characteristic (e.g., a bloodstain of group Γ, with frequency γ amongst some population) is found at the scene of the crime. A suspect is found who possesses this characteristic. What is the probability he is guilty? Such a problem has been dubbed the *island problem* (Eggleston, 1983; Yellin, 1979; Lindley, 1987). Various solutions have been proposed. These include

- $\left\{ 1 - (1 - \gamma)^{N+1} \right\} / \{(N+1)\gamma\}$ (Yellin, 1979);
- $1/(1 + N\gamma)$, (Lindley, in correspondence with Eggleston – see Eggleston, 1983).

Balding and Donnelly (1994a) and Dawid (1994), in the context of discussions of wider issues, show that the second of these is theoretically correct in the mathematical constraints of the problem as specified. It is, though, of limited practical value.

Many important aspects of evidence evaluation are illustrated by the island problem. A crime has been committed on an island. Consider a single stain left at the crime scene. The island has a population of $N + 1$ individuals. In connection with this crime, a suspect has been identified by other evidence. The genotype G_s of the suspect and the genotype G_c of the crime stain are the same.

The probability that a person selected at random from the island population has this genotype is γ. The two propositions are

- H_p, the suspect has left the crime stain;
- H_d, some other person left the crime stain.

There is considerable debate as to how prior odds may be determined. This is related to a debate about the meaning of the phrase 'innocent until proven guilty'. Note that, from the basic equation, a value of zero for the prior odds means that the posterior odds will also be zero, regardless of the value of the likelihood ratio. Thus, if 'innocent until proven guilty' is taken to mean the prior probability $Pr(H_p) = 0$, then the posterior probability $Pr(H_p \mid E)$ will equal zero, regardless of the value of the likelihood ratio.

A more realistic assessment of the prior odds is to say, before any evidence is presented, that the suspect is as likely to be guilty as anyone else in the relevant population. In the context of this example, this implies that $Pr(H_p \mid I) = 1/(N+1)$ (Robertson and Vignaux, 1995a). The prior odds are then $1/N$. With some simplifying assumptions, such as that $P(G_c \mid G_s, H_p, I) = 1$, it can be shown that the posterior odds are then $1/N\gamma$. Values for the posterior odds provide valuable information for the debates as to what is meant by 'proof beyond reasonable doubt'. Further details are given in Section 8.5.

3.6 SUMMARY

Odds have been defined and Bayes' theorem, relating conditional probabilities, presented. Various possible errors in the interpretation of probabilistic measures of the value of evidence have been discussed. The likelihood ratio and the odds version of Bayes' theorem have been defined. The role of the likelihood ratio as the value of the evidence has been defined as the factor which converts prior odds in favour of a proposition to posterior odds in favour of the proposition. This has enabled the fallacies of the transposed conditional and of the defence to be explained. The previous chapter discussed how probability may be evaluated in certain situations and how it is distributed over the set of possible outcomes. Given these general probability distributions, it is then possible to determine procedures for evaluating evidence within certain general frameworks. The next three chapters review the historical development of evidence evaluation, introduce the ideas of Bayesian inference and discuss issues associated with sampling of evidence. Following these, the role of Bayes' theorem in the evaluation of evidence will be discussed in greater detail.

4

Historical Review

4.1 EARLY HISTORY

The earliest use of probabilistic reasoning in legal decision making, albeit in a somewhat rudimentary form, appears to have been over 18 centuries ago by Jewish scholars in Babylon and Israel writing in the *Talmud* (Zabell, 1976, in a review of Rabinovitch, 1973). For example, if nine stores in a town sold kosher meat and one sold non-kosher meat then a piece of meat which is found, at random, in the town is presumed to be kosher and thus ritually permissible to eat since it is assumed to have come from one of the shops in the majority (Rabinovitch, 1969). However, consider the following quotation from the *Talmud*:

> All that is stationary (fixed) is considered half and half . . . . If nine shops sell ritually slaughtered meat and one sells meat that is not ritually slaughtered and he bought in one of them and does not know which one, it is prohibited because of the doubt; but if meat was found in the street, one goes after the majority. (Kethuboth 15a, quoted in Rabinovitch, 1969)

The reasoning seems to be as follows. If the question arises at the source of the meat (i.e., in the shops) the odds in favour of kosher meat are not really 9 to 1. The other nine shops are not considered – the piece of meat certainly did not come from any of them. There are, hence, only two possibilities: the meat is either kosher or it is not. The odds in favour of it being kosher are evens. However, if meat is found outwith the shops (e.g., in the street) the probability that it came from any one of the ten shops is equal for each of the ten shops. Thus, the probability that it is kosher is 0.9.

The works of Cicero (*De Inventione* and *Rhetorica ad Herennium*) and Quintillian (*Institutio Oratoria*), among others, are cited by Garber and Zabell (1979). Garber and Zabell also quote an example from Jacob Bernoulli's *Ars Conjectandi* (1713, Part 4, Chapter 2) which is of interest given the examples in Section 1.6.8 of updating evidence. One person, Titius, is found dead on the road. Another,

Statistics and the Evaluation of Evidence for Forensic Scientists: Second Edition
C.G.G. Aitken and F. Taroni © 2004 John Wiley & Sons, Ltd   ISBN: 0-470-84367-5

Maevius, is accused of committing the murder. There are various pieces of evidence in support of this accusation:

1. It is well known that Maevius regarded Titius with hatred. (This is evidence of motive: hatred could have driven Maevius to kill.)
2. On interrogation, Maevius turned pale and answered apprehensively. (This is evidence of effect: the paleness and apprehension could have come from his own knowledge of having committed a crime.)
3. A bloodstained sword was found in Maevius' house. (This is evidence of a weapon.)
4. On the day Titius was slain, Maevius travelled over the road. (This is evidence of opportunity.)
5. A witness, Gaius, alleges that on the day before the murder he had interceded in a dispute between Titius and Maevius.

Later (Chapter 3 of Part 4 of *Ars Conjectandi*) Bernoulli (1713) discusses how to calculate *numerically* the weight which should be afforded a piece of evidence or proof.

> The degree of certainty or the probability which this proof generates can be computed from these cases by the method discussed in the first part (i.e., the ratio of favourable to total cases) just as the fate of the gamblers in games of chance [is] accustomed to be investigated. (Garber and Zabell, 1979, p. 44)

Garber and Zabell (1979) then go on to say:

> What is new in the '*Ars Conjectandi*' is not its notation of evidence – which is based on the rhetorical treatment of circumstantial evidence – but its attempt to *quantify* such evidence by means of the newly developed calculus of chances. (p. 44)

Thus it is that over two hundred years ago consideration was being given to methods of evaluating evidence numerically.

A long discussion of *Ars Conjectandi*, Part 4, is given by Shafer (1978). The distinction is drawn between pure and mixed arguments. A *pure argument* is one which proves a thing in certain cases in such a way as to prove nothing positively in other cases. A *mixed argument*, on the other hand, is one which proves a thing in some cases in such a way that they prove the contrary in the remaining cases. Shafer discusses an example of this from Part 4 of *Ars Conjectandi*.

A man is stabbed with a sword in the midst of a rowdy mob. It is established by the testimony of trustworthy men who were standing at a distance that the crime was committed by a man in a black cloak. It is found that one person, by the name of Gracchus, and three others in the crowd were wearing cloaks of that colour. This is an argument that the murder was committed by Gracchus but it is a *mixed* argument. In one case it proves his guilt, in three cases his innocence, according to whether the murder was perpetrated by himself or one of the other three. If one of those three perpetrated the murder then Gracchus is supposed innocent.

However, if at a subsequent hearing, Gracchus went pale, this is a *pure* argument. If the change in his pallor arose from a guilty conscience it is indicative of his guilt. If it arose otherwise it does not prove his innocence; it could be that Gracchus went pale for a different reason but that he is still the murderer.

Shafer (1978) draws an analogy between these two kinds of argument and his mathematical theory of evidence (Shafer, 1976) and belief functions (see Section 1.2). In that theory a probability p is assigned to a proposition and a probability q to its negation, or complement, such that $0 \leq p \leq 1$, $0 \leq q \leq 1$, $p + q \leq 1$. It is not necessarily the case that $p + q = 1$, in contradiction to (3.2). There are then three possibilities:

- $p > 0$, $q = 0$ implies the presence of evidence in favour of the proposition and the absence of evidence against it;

- $p > 0$, $q > 0$ implies the presence of evidence on both sides, for and against the proposition;

- $p > 0$, $q > 0$, $p + q = 1$ (additivity) occurs only when there is very strong evidence both for and against the proposition.

Only probabilities which satisfy the additivity rule (3.2) are considered in this book. Sheynin (1974) comments that:

> According to Sambursky (*Treatise on divine government*, Q. 105, art. 7, *Great Books*, vol. 9, p. 544), Socrates held that in law-courts 'men care nothing about truth but only about conviction, and this is based on probability.'

Sheynin also quotes Aristotle (*Rhetorica*, 1376a, 19) as saying

> If you have no witnesses . . . you will argue that the judges must decide from what is probable . . . . If you have witnesses, and the other man has not, you will argue that probabilities cannot be put on their trial, and that we could do without the evidence of witnesses altogether if we need do no more than balance the pleas advanced on either side.

Sheynin (1974) also mentions that probability in law was discussed by Thomas Aquinas (*Treatise on Law*, Question 105, Article 2, *Great Books*, volume 20, p. 314) who provides a comment on collaborative evidence:

> In the business affairs of men, there is no such thing as demonstrative and infallible proof and we must contend with a certain conjectural probability . . . . Consequently, although it is quite possible for two or three witnesses to agree to a falsehood, yet it is neither easy nor probable that they succeed in so doing; therefore their testimony is taken as being true.

Jacob Bernoulli (1713) gave a probabilistic analysis of the cumulative force of circumstantial evidence. His nephew, Nicholas Bernoulli (1709), applied the calculus of probabilities to problems including the presumption of death, the value of annuities, marine insurance, the veracity of testimony and the probability of innocence; see Fienberg (1989).

The application of probability to the verdicts by juries in civil and criminal trials was discussed by Poisson (1837), and there is also associated work by Condorcet (1785), Cournot (1838) and Laplace (1886). The models developed by Poisson have been put in a modern setting by Gelfand and Solomon (1973).

Two early examples of the use of statistics to query the authenticity of signatures on wills were given by Mode (1963). One of these, the Howland will case from the 1860s, has also been discussed by Meier and Zabell (1980). This case is probably the earliest instance in American law of the use of probabilistic and statistical evidence. The evidence was given by Professor Benjamin Peirce, Professor of Mathematics at Harvard University, and by his son Charles, then a member of staff of the United States Coast Survey. The evidence related to the agreement of 30 downstrokes in a contested signature with those of a genuine signature. It was argued that the probability of this agreement if the contested signature were genuine was extremely small; the probability of observing two spontaneous signatures with the number of overlaid strokes observed in those two signatures was $(1/5)^{30}$. Hence the contested signature was a forgery. Comments on this case pointed out the now famous *prosecutor's fallacy* (Section 3.3.1). Good (1983) argues that Charles Peirce (1878) – in a pre-Bayesian statistical model – considered only two hypotheses with implicit initial odds of 1, thereby excluding some alternatives that might have had a prior probability greater than zero. Interestingly, in the same article, Good (1983) comments:

> It might be better to call ... 'hypothesis testing' *hypothesis determination*, as in a court of law where a judge or jury 'determines' that an accused person is innocent or guilty and where stating a numerical probability might even be regarded as contempt of court.

More recent attempts to evaluate evidence are reviewed here in greater detail.

4.2   THE DREYFUS CASE

This example concerns the trial of Dreyfus in France at the end of the last century. Dreyfus, an officer in the French Army assigned to the War Ministry, was accused in 1894 of selling military secrets to the German military attaché. Part of the evidence against Dreyfus centred on a document called the *bordereau*, admitted to have been written by him, and said by his enemies to contain cipher messages. This assertion was made because of the examination of the position of words in the bordereau. In fact, after reconstructing the bordereau and tracing on it with 4 mm interval vertical lines, Alphonse Bertillon showed that four pairs of polysyllabic words (among 26 pairs) had the same relative position with respect to the grid. Then, with reference to probability theory, Bertillon stated that the coincidences described could not be attributed to normal handwriting. Therefore, the bordereau was a forged document. Bertillon submitted probability calculations to support his conclusion. His statistical argument can be expressed as follows: if the probability for one coincidence equals 0.2, then the probability of observing N coincidences is 0.2^N. Bertillon calculated that the four coincidences observed by him had, then, a probability of 0.2^4, or $1/625$, a value that was so small as to demonstrate that the bordereau was a forgery (Charpentier, 1933). However, this value of 0.2 was chosen purely for illustration and had no evidential foundation; for a comment on this point, see Darboux *et al.* (1908).

Bertillon's deposition included not only this simple calculation but also an extensive argument to identify Dreyfus as the author of the bordereau on the basis of other measurements and a complex construction of hypotheses. (For an extensive description of the case, see the literature quoted in Taroni *et al.*, 1998, p. 189.)

As noted in Section 3.3.1 and using a Bayesian perspective, it is not difficult to see where Bertillon's logic had failed in his conclusion on the forgery. It seems that Bertillon argued that $Pr(H_d \mid E, I) = p = 1/625 = 0.0016$ and hence that $Pr(H_p \mid E, I) = 1 - p = 0.9984$. However, p represents $Pr(E \mid H_d, I)$. This seems to be an early example of the prosecutor's fallacy (Section 3.3.1).

The reliability of Bertillon's approaches was discussed at a retrial. Notably, Darboux, Appell and Poincaré, mathematicians and members of the French Academy of Sciences, offered their opinions. They commented that the probabilistic assessment proposed by Bertillon had no sound mathematical basis. In fact, the value of 0.0016 is the probability of observing four independent coincidences out of four comparisons (with the probability, θ, of one coincidence being 0.2), whereas Darboux, Appell and Poincaré are quoted as determining the probability of observing four coincidences out of 26 comparisons to be quite different, namely 0.7, or 400 times greater ($0.7/0.0016 = 437.5$) (Moras, 1906; Darboux *et al.*, 1908).

It is not clear how this figure of 0.7 was derived. The binomial expression $\binom{26}{4}0.2^4 0.8^{22} = 0.176$, and the probability of four or more coincidences out of 26 comparisons is approximately 0.8. It is not possible to choose a value of θ for which $\binom{26}{4}\theta^4(1 - \theta)^{22} = 0.7$. The value of θ for which the probability of four or more coincidences, out of 26, is 0.7 is $\theta = 0.18$. Further comments on Bertillon's calculations are given in Section 7.3.6.

Another assertion by Dreyfus' enemies was that the letters of the alphabet did not occur in the documents in the proportions in which they were known to occur in average French prose. The proportions observed had a very small probability of occurring (see Tribe, 1971). Though it was pointed out to the lawyers that the most probable proportion of letters was itself highly improbable, this point was not properly understood. A simple example from coin tossing will suffice to explain what is meant by the phrase 'the most probable proportion of letters was itself highly improbable'. Consider a fair coin, that is, one in which the probabilities of a head and of a tail are equal at $1/2$. If the coin is tossed $10\,000$ times, the expected number of heads is 5000 (see Section 2.3.3 with $n = 10\,000$, $p = 1/2$) and this is also the most probable outcome. However, the probability of 5000 heads, as distinct from 4999 or 5001 or any other number, is $\simeq 0.008$ or 1 in 125, which is a very low probability. The most probable outcome is itself improbable. The situation, of course, would be considerably enhanced given all the possible choices of combinations of letters in French prose in Dreyfus' time. This idea may be expressed in mathematical symbols as follows. If Dreyfus were innocent (H_d) the positions of the words (E) which he had used would be extremely unlikely; $Pr(E \mid H_d)$ would be very small. The prosecuting lawyers concluded that Dreyfus must have deliberately chosen the letters he did as a cipher and so must be a spy; $Pr(H_d \mid E)$ must be very small. The lawyers did not see that any other combination of letters would also be extremely unlikely and that the particular combination used by Dreyfus was of no great significance. This is another example of the prosecutor's fallacy.

Darboux, Appell and Poincaré also expressed a more fundamental point: the nature of the inferential process they used to reach the conclusion. They stated that the case under consideration was a classical problem of *probability of the causes* and not a problem of *probability of the effects*. The difference between the two statistical concepts (and inferences) could be illustrated by the following example (see also Poincaré, 1992). If you draw a ball from an urn containing 90 white balls and 10 black balls, the probability of drawing a black ball is $1/10$ and it corresponds to the probability of the effect. Suppose now you are facing two identical urns. Urn 1 contains black and white balls in the proportion 90:10. The second urn contains black and white balls in the proportion 10:90. You choose an urn (at random, each urn is equally likely to be chosen) and pick up a ball. It is white. What is the probability that you have picked up a ball from urn 1? In this example, the effect is known, but it is the cause which is uncertain.

This is another example concerning urns (Section 1.6.2). In order to infer something about a possible cause from an observation of an effect, two assessments are needed: the probabilities *a priori* of the causes under examination (i.e., forgery or not in the Dreyfus case), and the probabilities of the observed effect for each possible cause (the coincidences observed by Bertillon). A more detailed description of this kind of reasoning applied in forensic science was proposed by Poincaré and his colleagues; it will be presented in Section 4.8 of this chapter.

For a complete analysis of the questioned document examination in the Dreyfus case, see Champod *et al.* (1999).

4.3 STATISTICAL ARGUMENTS BY EARLY TWENTIETH-CENTURY FORENSIC SCIENTISTS

A review of the forensic science literature suggests that, for the evaluation of forensic evidence, early forensic scientists recognised that adequate data and consideration of the case as a whole should be used to reach a decision. Their points of view were generally compatible with a Bayesian framework (Taroni *et al.*, 1998).

Despite his argument in the Dreyfus case, Bertillon wrote that experts must be prepared to present evidence in a numerical form which was more demanding than that generally required of expert opinions. He proposed that reports should be concluded in the following form:

> this writing characterized by the set of unique features the expert enumerated, can only be encountered in one individual among a hundred, among a thousand, among ten thousand or among a million individuals. (Bertillon, 1897/1898)

Moreover, Bertillon argued that the only way to accept a conclusion of the final issue (e.g., the identification of a writer) was to consider not only the statistical evidence provided by the examination of the document, but also other information pertaining to the inquiry. Bertillon considered the presentation of results without such information as a methodological error. The value of the comparison results, even if not absolute, could supply sufficient information to allow a conviction when the case is considered as a whole. The same approach – more clearly expressed in a numerical way – was proposed in 1934, for typewritten documents, by William Souder (1934/1935):

> Suppose the report does not establish an extremely remote possibility of recurrence (of characteristics or agreements between the questioned and known writings). Suppose the final fraction for recurrence of the typed characteristics had come out as only 1 in 100. Is such a report of value? Yes, if the number of typewriters upon which the document could have been written can be limited to 100 or less, the report is vital. Similarly, in handwriting we do not have to push the tests until we get a fraction represented by unity divided by the population of the world. Obviously the denominator can always be reduced to those who can write and further to those having the capacity to produce the work in question. In a special case, it may be possible to prove that one of three individuals must have produced the document. Our report, even though it shows a mathematical probability of only 1 in 100, would then irresistibly establish the conclusion.

The same idea that a final issue could only be assessed if the case is considered as a whole is often reiterated in judicial literature:

> Forensic science alone cannot identify the probability that O.J. Simpson – or any other criminal defendant – is or is not the source of the recovered genetic evidence. Non-genetic considerations must be factored into any equation that purports to identify the chance that someone is the source of genetic sample. (Koehler, 1997a)

One of the oldest examples of the use of a probabilistic inference is fingerprint identification, and notably the works of Balthazard, a French legal examiner. Balthazard's works influenced rules regarding standards for the establishment of a fingerprint identification such as the rule for 17 concordant minutiae expressed in Italian jurisprudence since 1954 (Balthazard, 1911). For general comments on identification rules for fingerprints, see Champod (1996).

Despite the weakness of Balthazard's hypotheses and assumptions used to perform his simple calculation, which have been extensively challenged in the scientific literature (see comments in Champod, 1996), it is important to note that part of Balthazard's text is in agreement with the Bayesian framework:

> In medico-legal work, the number of corresponding minutiae can be lowered to eleven or twelve if you can be certain that the population of potential criminals is not the entire world population but it is restricted to an inhabitant of Europe, a French citizen, or an inhabitant of a city, or of a village, etc. (Balthazard, 1911, p. 1964)

So, as years later Souder proposed for questioned documents, here Balthazard stated that prior assessment (based on inquiry information and reducing the size of the suspect population) has to be associated with a statistical value of the evidence to allow the decision maker to judge on an identification (a posterior assessment). Unfortunately, ninety years later, the discussion on fingerprint identification and the use of probabilistic models remains open. See, for example, Taroni and Margot (2000), Champod and Evett (2001) and Friedman *et al.* (2002).

4.4   PEOPLE v. COLLINS

The Dreyfus case is a rather straightforward abuse of probabilistic ideas, though the fallacy of which it is an example still occurs. It is easy now to expose the fallacy through consideration of the odds form of Bayes' theorem (see Section 3.4). At the time, however, the difficulty of having the correct reasoning accepted had serious and unfortunate consequences for Dreyfus. A further example of the fallacy occurred in a case which has achieved a certain notoriety in the probabilistic legal literature, namely that of *People v. Collins* (Kingston, 1965a, b, 1966; Fairley and Mosteller, 1974, 1977). In this case, probability values, for which there was no objective justification, were quoted in court.

Briefly, the crime was as follows. An old lady, Juanita Brooks, was pushed to the ground in an alleyway in the San Pedro area of Los Angeles by someone whom she neither saw nor heard. According to Mrs Brooks, a blond-haired woman wearing dark clothing grabbed her purse and ran away. John Bass, who lived at the end of the alley, heard the commotion and saw a blond-haired woman wearing dark clothing run from the scene. He also noticed that the woman had a ponytail and that she entered a yellow car driven by a black man who had a beard and a moustache (Koehler, 1997a).

A couple answering this description were eventually arrested and brought to trial. The prosecutor called as a witness an instructor of mathematics at a state college in an attempt to bolster the identifications. This witness testified to the product rule for multiplying together the probabilities of independent events (the third law of probability (1.3), Section 1.6.5). The third law may be extended to a set of n independent events and also to take account of conditional probabilities. Thus, if $E_1, E_2, \ldots, E_n$ are mutually independent pieces of evidence and H_d denotes the hypothesis of innocence then

$$Pr(E_1 E_2 \cdots E_n \mid H_d) = Pr(E_1 \mid H_d) Pr(E_2 \mid H_d) \cdots Pr(E_n \mid H_d). \qquad (4.1)$$

In words, this states that, for mutually independent events, the probability that they all happen is the product of the probabilities of each individual event happening.

The instructor of mathematics then applied this rule to the characteristics as testified to by the other witnesses. Values to be used for the probabilities of the individual characteristics were suggested by the prosecutor without any justification, a procedure which would not now go unchallenged. The jurors were invited to choose their own values but, naturally, there is no record of whether they did or not. The individual probabilities suggested by the prosecutor are given in Table 4.1. Using the product rule for independent characteristics, the prosecutor calculated the probability that a couple selected at random from a population would exhibit all these characteristics as 1 in 12 million ($10 \times 4 \times 10 \times 3 \times 10 \times 1000 = 12\,000\,000$).

The accused were found guilty. This verdict was overturned on appeal for two statistical reasons:

1. The statistical testimony lacked an adequate foundation both in evidence and in statistical theory.
2. The testimony and the manner in which the prosecution used it distracted the jury from its proper function of weighing the evidence on the issue of guilt.

Table 4.1 Probabilities suggested by the prosecutor for various characteristics of the couple observed in the case of *People v. Collins*

Evidence	Characteristic	Probability
E_1	Partly yellow automobile	1/10
E_2	Man with moustache	1/4
E_3	Girl with ponytail	1/10
E_4	Girl with blond hair	1/3
E_5	Negro man with beard	1/10
E_6	Interracial couple in car	1/1000

The first reason refers to the lack of justification offered for the choice of probability values and the assumption that the various characteristics were independent. As an example of this latter point, an assumption of independence implies that the propensity of a man to have a moustache does not affect his propensity to have a beard. Moreover, the computation implicitly assumed that the six reported characteristics were true and were accurately reported. It made no allowance for the possibility of disguise (e.g., dyed hair).

The second reason still has considerable force today. When statistical evidence is presented, great care has to be taken that the jury is not distracted from its proper function of weighing the evidence on the issue of guilt.

The fallacy of the transposed conditional is also evident. The evidence is $(E_1, E_2, \ldots, E_6)$ and $Pr(E_1 E_2 \ldots E_6 \mid H_d)$ is extremely small (1 in 12 million). The temptation for a juror to interpret this figure as a probability of innocence is very great.

Lest it be thought that matters have improved, consider the case of *R. v. Clark*. Sally Clark's first child, Christopher, died unexpectedly at the age of about 3 months when Clark was the only other person in the house. The death was initially treated as a case of sudden infant death syndrome (SIDS). Her second child, Harry, was born the following year. He died in similar circumstances. Sally Clark was arrested and charged with murdering both her children. At trial, a professor of paediatrics quoted from a report (Fleming *et al.*, 2000) that, in a family like the Clarks, the probability that two babies would both die of SIDS was around 1 in 73 million. This was based on a study which estimated the probability of a single SIDS death in such a family as 1 in 8500, and then squared this to obtain the probability of two deaths, a mathematical operation which assumed the two deaths were independent. In an open letter to the Lord Chancellor, copied to the President of the Law Society of England and Wales, the Royal Statistical Society expressed its concern at the statistical errors which can occur in the courts, with particular reference to Clark. To quote from the letter:

> One focus of the public attention was the statistical evidence given by a medical expert witness, who drew on a published study [Confidential Enquiry into Stillbirths and Deaths in Infancy] to obtain an estimate [1 in 8543] of the frequency of sudden infant death syndrome (SIDS, or 'cot death') in families having some of the characteristics of the defendant's family. The witness went on to square this estimate to obtain a value of 1 in 73 million for the frequency of two cases of SIDS in such a family . . . .
>
> Some press reports at the time stated that this was the chance that the deaths of Sally Clark's two children were accidental. This (mis-)interpretation is a serious error of logic known as the Prosecutor's Fallacy. (*Royal Statistical Society News*, March 2002)

A similar American case is that of *Wilson v. Maryland*. Note that the prosecutor's fallacy equates a small probability of finding the evidence on an innocent person with the probability that the person is innocent.

4.5 DISCRIMINATING POWER

4.5.1 Derivation

How good is a method at distinguishing between two samples of material from different sources? If a method fails to distinguish between two samples, how strong is this as evidence that the samples come from the same source? Questions like these and the answers to them were of considerable interest to forensic scientists in the late 1960s and in the 1970s; see, for example, theoretical work in Parker (1966, 1967), Jones (1972) and Smalldon and Moffat (1973). Experimental attempts to answer these questions are described in Tippett *et al.* (1968) for fragments of paint, in Gaudette and Keeping (1974) for human head hairs, in Groom and Lawton (1987) for shoeprints, in Massonnet and Stoecklein (1999) for paint and in Adams (2003) for dental impressions.

Two individuals are selected at random from some population. The probability that they are found to match with respect to some characteristic (e.g., blood profile, paint fragments on clothing, head hairs) is known as the *probability of non-discrimination* or the *probability of a match*, PM. The complementary probability, the probability they are found not to match with respect to this characteristic, is known as the *probability of discrimination* (Jones, 1972) or *discriminating power*, DP (Smalldon and Moffat, 1973). The idea was first applied to problems concerning ecological diversity (Simpson, 1949) and later to blood group genetics (Fisher, 1951). See also Jeffreys *et al.* (1987) for an application to DNA profiles.

Consider a population and locus in which there are k genotypes, labelled $1, \ldots, k$, and in which the jth genotype has relative frequency p_j, such that $p_1 + \cdots + p_k = 1$ from the second law of probability for mutually exclusive and exhaustive events (Section 1.6.6). Two people are selected at random from this population such that their genotypes may be assumed independent. What is the probability of a match of genotypes between the two people at this locus?

Let the two people be called C and D. Let C_1 and D_1 be the events that C and D, respectively, are of genotype labelled 1. Then

$$Pr(C_1) = Pr(D_1) = p_1,$$

and the probability of C_1 and of D_1 is given by

$$Pr(C_1 D_1) = Pr(C_1) \times Pr(D_1) = p_1^2$$

by the third law of probability (1.3) applied to independent events. Thus, the probability they are both of genotype 1 is p_1^2. In general, let C_j, D_j be the events that C, D are of genotype j $(j = 1, \ldots, k)$. The probability that the individuals selected at random match on genotype j, is given by

$$Pr(C_j D_j) = Pr(C_j) \times Pr(D_j) = p_j^2.$$

The probability of a match on *any* genotype is the disjunction of k mutually exclusive events, the matches on genotypes $1, \ldots, k$, respectively. Let Q be the probability PM of a match. Then

$$
\begin{aligned}
Q &= Pr(C_1 D_1 \text{ or } C_2 D_2 \text{ or} \ldots \text{ or } C_k D_k) \\
&= Pr(C_1 D_1) + Pr(C_2 D_2) + \cdots + Pr(C_k D_k) \\
&= p_1^2 + p_2^2 + \cdots + p_k^2,
\end{aligned} \tag{4.2}
$$

by the second and third laws of probability ((1.3) and (1.2)); see also (1.1) of Example 1.1. The discriminating power, or probability of discrimination, is $1 - Q$.

Example 4.1   Consider the frequencies from Table 1.1, where $k = 3$. Then

$$
Q = 0.188^2 + 0.321^2 + 0.491^2 = 0.379.
$$

The discriminating power $DP = 1 - Q = 0.621$.

4.5.2   Evaluation of evidence by discriminating power

The approach using discriminating power has implications for the assessment of the value of forensic evidence. If two samples of material (e.g., two blood stains, two sets of paint fragments, two groups of human head hairs) are found to be indistinguishable, it is of interest to know if this is forensically significant. If a system has a high value of Q it implies that a match between samples of materials from two different sources under this system is quite likely. For example, if there were only one category, no discrimination would be possible. In such a case, $k = 1$, $p_1 = 1$ and $Q = p_1^2 = 1$. It is intuitively reasonable that a match under such a system will not be very significant. Conversely, if a system has a very low value of Q a match will be forensically significant. Blood grouping data from the Strathclyde region of Scotland for which there is a discriminating power of 0.602 are given by Gettinby (1984). He interprets this, in the context of blood grouping, to mean that 'in 100 cases where two blood samples come from different people then, on average, 60 will be identifiable as such'.

Limits for Q may be determined (Jones, 1972). First, note that $p_1 + \cdots + p_k = 1$ and that $0 \le p_j \le 1$ ($j = 1, \ldots, k$) from the first law of probability (1.4). Thus $p_j^2 \le p_j$ ($j = 1, \ldots, k$) and so $Q = p_1^2 + \cdots + p_k^2 \le 1$; that is, Q can never be greater than 1 (and is equal to 1 if and only if one of the p_j is 1, and hence the rest are 0, as illustrated above). A value of Q equal to 1 implies that all members of the population fall into the same category; the discriminating power is zero.

Now, consider the lower bound. It is certainly no less than zero. Suppose the characteristic of interest (h_0, say) divides the system into k classes of equal probability $1/k$, so that $p_j = p_{0j} = 1/k$ ($j = 1, \ldots, k$). Then

$$
\begin{aligned}
Q = Q_0 &= p_{01}^2 + p_{02}^2 + \cdots + p_{0k}^2 \\
&= \frac{1}{k^2} + \frac{1}{k^2} + \cdots + \frac{1}{k^2} \\
&= \frac{1}{k}.
\end{aligned}
$$

Consider another characteristic (h_1, say) which divides the system into k classes of unequal probability such that

$$
p_j = p_{1j} = \frac{1}{k} + \epsilon_j; \qquad j = 1, \ldots, k.
$$

Since $\sum_{j=1}^{k} p_j = 1$ it can be inferred that $\sum_{j=1}^{k} \epsilon_j = 0$. Thus, for h_1,

$$
\begin{aligned}
Q = Q_1 &= p_{11}^2 + p_{12}^2 + \cdots + p_{1k}^2 \\
&= \sum_{j=1}^{k} \left(\frac{1}{k} + \epsilon_j \right)^2 \\
&= \sum_{j=1}^{k} \left(\frac{1}{k^2} + \frac{2\epsilon_j}{k} + \epsilon_j^2 \right) \\
&= \frac{1}{k} + \frac{2}{k} \sum_{j=1}^{k} \epsilon_j + \sum_{j=1}^{k} \epsilon_j^2 \\
&= \frac{1}{k} + \sum_{j=1}^{k} \epsilon_j^2 \quad \text{since} \quad \sum_{j=1}^{k} \epsilon = 0 \\
&\geq Q_0
\end{aligned}
$$

since $\sum_{j=1}^{k} \epsilon_j^2$ is never negative (and equals zero if and only if $\epsilon_1 = \epsilon_2 = \cdots = \epsilon_k = 0$; i.e., if and only if $p_{0j} = p_{1j}$ ($j = 1, \ldots, k$)). Thus Q takes values between $1/k$ and 1 where k is the number of categories in the system. The probability of a match is minimised, and the discriminating power is maximised when the class probabilities are all equal. This is confirmation of a result which is intuitively reasonable, namely that if a choice has to be made among several techniques as to which to implement, then techniques with greater variability should be preferred over those with lesser variability.

As an example of the application of this result, note that for the *LDLR* locus in Table 1.1, $k = 3$ and $1/k = 0.33$. Thus, Q cannot be lower than 0.33 for this locus and the discriminating power cannot be greater than 0.67. Notice also that the minimum value $(1/k)$ of Q decreases as k increases. Discriminating power increases as the number of categories into which an item can be classified increases, a result which is intuitively attractive.

The above calculations assume that the total population size (N say) of interest is very large and that for at least one p_j, p_j^2 is much greater than $1/N$. Failure of these assumptions can lead to misleading results as described by Jones (1972) with reference to the results of Tippett *et al.* (1968) on paint fragments; see Examples 4.2 and 4.3.

The experiment described by Tippett *et al.* (1968) compared pairwise 2000 samples of paint fragments. For various reasons, the number of samples was reduced to 1969, all from different sources. These were examined by various tests and only two pairs of samples from different sources were found to be indistinguishable. The total number of pairs which can be picked out at random is $\frac{1}{2} \times 1969 \times 1968 = 1\,937\,496$. Two pairs of samples were found to agree with each other. The probability of picking a pair of fragments at random which are found to be indistinguishable is thus determined empirically as $2/1\,937\,496 = 1/968\,478$.

This probability is an estimate of the probability of a match (Q). The method by which it was determined is extremely useful in situations such as the one described by Tippett *et al.* (1968) in which frequency probabilities are unavailable and, indeed, for which a classification system has not been devised. The extremely low value $(1/968\,748)$ of Q demonstrates the high evidential value of the methods used by the authors. The conclusion from this experiment is that these methods are very good at differentiating between paints from different sources. Low values of Q were also obtained in work on head hairs (Gaudette and Keeping, 1974) and footwear (Groom and Lawton, 1987).

The equivalence of the theoretical and empirical approaches to the determination of Q can be verified numerically using the *LDLR* locus with genotypic frequencies as given in Table 1.1. Assume there is a sample of 1000 people with genotypic frequencies in the proportions in Table 1.1. All possible pairs of people in this sample are considered and their genotypes compared. There are $\frac{1}{2}(1000 \times 999) = 499\,500 \,(= P$, say) different pairings. Of these pairings there are the following numbers of matches for each genotype:

- *AA*, $188 \times 187/2 = 17\,578$;
- *BB*, $321 \times 320/2 = 51\,360$;
- *AB*, $491 \times 490/2 = 120\,295$.

There are, thus, $M = \{(188 \times 187) + (321 \times 320) + (491 \times 490)\}/2 = 189\,233$ pairings of people who have the same genotype. The probability of a match, by this numerical method, is then $M/P = 189\,233/499\,500 = 0.3788$. The probability of a match is $Q = p_1^2 + \cdots + p_3^2 = 0.188^2 + 0.321^2 + 0.491^2 = 0.3795$. The approximate equality of these two values is not a coincidence, as a study of the construction of the ratio M/P shows.

The probability Q is sometimes called an *average probability* (Aitken and Robertson, 1987). An average probability provides a measure of the effectiveness of a particular type of transfer evidence at distinguishing between two randomly selected individuals (Thompson and Williams, 1991). In the context of bloodstains, it is so called because it is the average of the probabilities that an innocent person will be found to have an allele which matches that of a crime stain. These probabilities are of the type considered in Example 1.1. For example, for locus *TPOX*, if the crime stain were of allele 8, the probability an innocent suspect matches this is just the probability he has allele 8, namely 0.554. A similar argument gives probabilities of 0.093, 0.054, 0.259 and 0.040 for matches between crime stain alleles and that of an innocent person of alleles 9, 10, 11 and 12, respectively. The average probability is the average of these four probabilities, weighted by their relative frequencies in the population, which are 0.554, 0.093, 0.054, 0.259 and 0.040, respectively. The average probability is then just Q, given by $(0.554 \times 0.554) + (0.093 \times 0.093) + (0.054 \times 0.054) + (0.259 \times 0.259) + (0.040 \times 0.040) = 0.3872$.

4.5.3   Finite samples

The relationship between the general result for a population, which is conceptually infinite in size, and a sample of finite size is explained by Jones (1972). Consider a test to distinguish between k classes $C_1, \ldots, C_k$. A sample of n individuals is taken from the relevant population. The numbers of individuals in each class are $c_1, c_2, \ldots, c_k$ with $\sum_{j=1}^{k} c_j = n$. An estimate of the probability that a randomly selected individual will be in class C_j is $\hat{p}_j = c_j/n$, $j = 1, 2, \ldots, k$.

There are $n(n-1)/2$ possible pairings of individuals. For any particular class j, the number of pairings of individuals within the class is $c_j(c_j - 1)/2$, $j = 1, 2, \ldots, k$. Thus, the overall proportion of pairings which result in a match is

$$\hat{Q} = \left\{ \sum_{j=1}^{k} c_j(c_j - 1) \right\} \Bigg/ \{n(n-1)\}.$$

Then

$$\hat{Q} = \left(\sum_{j=1}^{k} c_j^2 - \sum_{j=1}^{k} c_j \right) \Bigg/ \{n(n-1)\}$$

$$= \left(\sum_{j=1}^{k} c_j^2 - n \right) \Bigg/ (n^2 - n) \text{ since } \sum_{j=1}^{k} c_j = n,$$

$$= \left\{ \sum_{j=1}^{k} (c_j^2/n^2) - 1/n \right\} \Bigg/ (1 - 1/n)$$

$$= \left(\sum_{j=1}^{k} \hat{p}_j^2 - 1/n \right) \Bigg/ (1 - 1/n). \tag{4.3}$$

When the class frequencies are known, the probability of a match *PM* is given by Q (4.2). The result above (4.3) gives an exact expression for the probability of a match for any given sample for all values of n and $\{\hat{p}_j, j = 1, \ldots, k\}$. As n increases towards the population size it is to be expected that the observed sample probability will converge to the population probability. For n large it can be seen that $\hat{Q}$ tends to

$$\hat{Q}_1 = \sum \hat{p}_j^2,$$

since $1/n$ becomes vanishingly small. As $\hat{p}_j$ tends to p_j so $\hat{Q}_1$ will tend to Q. However, as well as n being large, it is necessary for at least one of the $\hat{p}_j^2$ to be much greater than $1/n$ in order that $\sum \hat{p}_j^2 - 1/n \simeq \sum p_j^2$. This should be so for k not too close to n or for n very much larger than k; that is, the number, n, of individuals should be much greater than the number of categories k.

Two examples (Jones, 1972) are given in which the probability of a match estimated by $\hat{Q}_1 = \sum \hat{p}_j^2$ is not a good approximation to the probability of a match estimated by $\hat{Q}$ from (4.3). The true probability of a match, $Q = \sum p_j^2$, will not often be known exactly, except in situations like blood grouping systems where the sample sizes are extremely large and the $\{p_j\}$ are known accurately.

Example 4.2 (Small sample size n) If n is small, then $1/n$ is not very small compared to 1. Consider four playing cards, of which two are red (R_1, R_2: Category 1) and two are black (B_1, B_2: Category 2). Thus, $n = 4$, $c_1 = c_2 = 2$, $p_1 = \hat{p}_1 = p_2 = \hat{p}_2 = 1/2$ and $Q = \hat{Q}_1 = 1/4 + 1/4 = 1/2$. Note that $1/n = 1/4$, which is not very much less than 1. There are six possible pairings of cards ($R_1R_2, R_1B_1, R_1B_2, R_2B_1, R_2B_2, B_1B_2$) of which two result in a match (R_1R_2, B_1B_2). Thus $\hat{Q} = 1/3$, which may be verified from $\hat{Q} = (\sum \hat{p}_j^2 - 1/n)/(1 - 1/n) = (1/2 - 1/4)/(1 - 1/4) = 1/3$. The failure of $1/n$ to be very small has led to a discrepancy between $\hat{Q}$ and $\hat{Q}_1$.

Example 4.3 Very small values of p_j^2 Consider the paper by Tippett *et al.* (1968) in which 1969 sets of paint fragments were considered ($n = 1969$). Two pairs were found to be indistinguishable. Label these pairs as belonging to classes 1 and 2. The other paint fragments may be said to belong to 1965 different classes labelled $3, \ldots, 1967$, each with only one member so that $\hat{p}_1 = 2/1969$, $\hat{p}_2 = 2/1969$, $\hat{p}_3 = \cdots = \hat{p}_{1967} = 1/1969$. Then

$$\hat{Q}_1 = \left(\frac{2}{1969}\right)^2 + \left(\frac{2}{1969}\right)^2 + \left(\frac{1}{1969}\right)^2 + \cdots + \left(\frac{1}{1969}\right)^2$$

$$= \frac{1973}{1969^2}$$

$$\simeq \frac{1}{1965},$$

whereas

$$\hat{Q} = \left(\frac{1973}{1969^2} - \frac{1}{1969}\right) \bigg/ \left(1 - \frac{1}{1969}\right)$$

$$= \frac{4}{1969 \times 1968}$$

$$= \frac{1}{968\,748},$$

agreeing with the earlier result obtained by the authors. Here the approximate result $\hat{Q}_1$ is very inaccurate because no $\hat{p}_j^2$ is very much greater than $1/n$. In fact, the largest $\hat{p}_j^2$ is $(2/1969)^2$ which is smaller than $1/n$.

4.5.4 Combination of independent systems

Consider Q from (4.2). This is the probability of finding a match between two individuals selected at random using a particular classification system. Suppose now that there are p independent systems with corresponding Q values $Q_1, \ldots, Q_p$. The probability of finding a pair which match on all p tests is $PM_p = \prod_{l=1}^{p} Q_l$. The probability of being able to distinguish between two individuals using these p tests is therefore

$$DP_p = 1 - \prod_{l=1}^{p} Q_l.$$

Consider the following example of a comparison of the allelic frequencies between New Zealand (NZ) and Swiss Caucasians, with frequency results for the *TPOX* and *TH01* loci. The calculations for the discriminating power for the combination of the *TPOX* and *TH01* systems are given below. The New Zealand data are given in Harbison *et al.* (2002).

The allelic frequencies for the *TPOX* and *TH01* systems for NZ and Swiss Caucasians are given in Tables 4.2 and 4.3. The probability that two blood samples match on both criteria is

$$PM_2 = Q_{TPOX} \times Q_{TH01}$$

$$= 0.3872 \times 0.2305 = 0.0892 \text{ (Swiss)},$$

$$= 0.3780 \times 0.2251 = 0.0851 \text{ (NZ)}.$$

The discriminating power is

$$DP_2 = 0.9108 \text{ (Swiss)},$$

$$= 0.9149 \text{ (NZ)}.$$

Table 4.2 Allelic frequencies for *TPOX* locus for Swiss and NZ Caucasians and the probability Q_{TPOX} of a match

Allele	Frequency	
	Swiss	NZ
8	0.554	0.529
9	0.093	0.082
10	0.054	0.063
11	0.259	0.294
12	0.040	0.032
Q_{TPOX}	0.3872	0.3780

Table 4.3 Allelic frequencies for *TH01* locus for Swiss and NZ Caucasians and the probability Q_{TH01} of a match

Allele	Frequency	
	Swiss	NZ
5	0.0	0.002
6	0.219	0.180
7	0.194	0.206
8	0.083	0.102
9	0.144	0.155
9.3	0.342	0.340
10	0.018	0.015
Q_{TH01}	0.2305	0.2251

4.5.5   Correlated attributes

Discrete attributes

Consider the hair example of Gaudette and Keeping (1974) in more detail. (A similar argument holds for the paint example of Tippett *et al.*, 1968.) A series of 366 630 pairwise comparisons between hairs from different individuals were made. Nine pairs of hairs were found to be indistinguishable. These results were used to provide an estimate of the probability that a hair taken at random from one individual, *A* say, would be indistinguishable from a hair taken at random from another individual, *B* say, namely 9/366 630 or 1/40 737. It was then argued that if nine dissimilar hairs were independently chosen to represent the hairs on the scalp of individual *B*, the chance that a single hair for *A* is distinguishable from all nine of *B*'s may be taken as $[1 - (1/40\,737)]^9$ which is approximately $1 - (1/4500)$. The complementary probability, the probability that a single hair from *A* is indistinguishable from at least one of *B*'s hairs is

1/4500. This probability provides, in some sense, a measure of the effectiveness of human hair comparison in forensic hair investigations.

There are various criticisms, though, which can be made regarding this approach, some details of which are in Aitken and Robertson (1987). Comments on the Gaudette–Keeping study are also made in Fienberg (1989). Criticisms of the methodology are presented in Barnett and Ogle (1982) and Miller (1987), with rebuttals in Gaudette (1982, 1999).

First, note that the assumption of independence of the nine hairs used in the calculation is not an important one. The use of an inequality known as the *Bonferroni inequality* gives an upper bound for the probability investigated by the authors of 1/4526 (Gaudette, 1982). The Bonferroni inequality states that the probability that at least one of several events occurs is never greater than the sum of the probabilities of the occurrences of the individual events. Denote the events by $R_1, R_2, \ldots, R_n$; then the inequality states that

$$Pr(\text{at least one of } R_1, \ldots, R_n \text{ occurs}) \leq Pr(R_1) + \cdots + Pr(R_n).$$

Gaudette and Keeping (1974) compared each of nine hairs, known to be from one source, with one hair known to be from another source. The events $R_1, \ldots, R_9$, correspond to the inability to distinguish each of the nine hairs from the single hair. The probability of interest is the probability of at least one indistinguishable pair in these nine comparisons. This is the probability that at least one of $R_1, \ldots, R_9$ occurs. Using the Bonferroni inequality, it can be seen that this probability is never greater than the sum of the individual probabilities. From above, these individual probabilities are all equal to 1/40 737. The sum of the nine of them then equals 9/40 737, which is 1/4526. This is very close to the figure of 1/4500 quoted from the original experiment. Even if independence is not assumed there is very little change in the probability figure quoted as a measure of the value of the evidence.

An important criticism, though, is the following. One probability of interest to the court is the probability that a hair found at a crime scene belonged to a suspect. Other probabilities of interest, the relevance of which will be explained in more detail later in Chapter 8, are the probability of the evidence of the similarity of the hairs (crime and suspect) if they came from the same origin and the probability of the evidence of the similarity of the hairs if they came from different origins. Gaudette and Keeping (1974) provided an estimate of the probability that hairs 'selected at random' from two individuals are indistinguishable. This probability is an average probability (Section 4.5.2). It can be used as a broad guideline to indicate the effectiveness of hair identification in general. However, the use of the figure 1/4500 as the value of the evidence in a particular case could be very misleading.

The average probability is the probability that two individuals chosen at random will be indistinguishable with respect to the trait under examination. However, in a particular investigation one sample is of known origin (in this case, the sample of hair from the suspect, which is the source sample), the other (the crime or receptor sample) is not. If the suspect is not the criminal, someone else is and the correct probability of interest is the probability that a person chosen at random (which is how the criminal must be considered) from some population will have hair which is similar to that of the suspect – see Chapter 9 for a more detailed discussion.

Fienberg (1989) also makes the point that 'even if we interpret 1/4500... as the probability of a match given a comparison of a suspected hair with a sample from a different individual we would still need the probability of a match given a sample from the same individual as well as *a priori* probabilities of "same" and "different"'.

Continuous measurements

The problem of interpreting correlated continuous attributes in which the underlying distribution is multivariate Normal (Section 2.4.6) was discussed by Smalldon and Moffat (1973). Further details are given in Chapter 11.

First, consider a set of p uncorrelated continuous attributes with measurements $\{x_l, l = 1, \ldots, p.\}$ and corresponding probability density functions $f(x_1), \ldots, f(x_p)$. An estimate of the probability that two individuals chosen at random from the population will match, in some way which has to be defined, on all p attributes is required.

For attribute l, the probability that the measurement on the first individual will lie in a small interval t of width dx_l about measurement x_l is $f(x_l)dx_l$. This may be understood intuitively by realising that $f(x_l)$ is the height of the probability density curve at the point x_l. For example, see Figure 2.1; the height of the Normal density curve of mean 0.7 and variance 0.005 at the point $x_l = 0.85$ is 0.60 from (2.15). The probability $f(x_l)dx_l$ is the area of a narrow rectangle of height $f(x_l)$ and width dx_l. This is used as an approximation to the probability that the measurement will lie in the interval t. If $f(x_l)$ can be assumed linear over another small interval $\pm e_l$ centred on x_l, $(x_l - e_l, x_l + e_l)$, then the probability that the second individual will match with the first, in the sense that the measurement of attribute l for the second individual lies within $\pm e_l$ of x_l, is $f(x_l)2e_l$. Thus, the probability PM_l that two individuals selected at random will match is

$$PM_l = 2e_l \int \{f(x_l)\}^2 dx_l.$$

This expression is a version of the third law of probability applied to continuous measurements. It is the product of the probability that the first measurement lies within a certain interval and the probability that the second measurement lies within a certain interval conditional on the first lying within that interval. This product is then integrated over all possible values of the measurement. It will be a reasonable approximation to the true probability of a match provided, as stated above, e_l is sufficiently small that $f(x_l)$ can be assumed reasonably linear over the interval $(x_l - e_l, x_l + e_l)$. The discriminating power can then be derived as before as

$$DP_p = 1 - \prod_{l=1}^{p} PM_l.$$

Correlated Normally distributed attributes

The probability density function $f(\mathbf{x})$ (2.25) for a p-dimensional Normally distributed set of correlated attributes $\mathbf{x} = (x_1, \ldots, x_p)$, where, without loss of generality, the mean can be taken as the origin $(0, \ldots, 0)$, may be expressed as

$$f(\mathbf{x}) = \frac{\sqrt{|\,\Omega\,|}}{(2\pi)^{p/2}} \exp\left\{ -\frac{1}{2} \sum_{i,j=1}^{p} \Omega_{ij} x_i x_j \right\}$$

(Smalldon and Moffat, 1973; Anderson, 1984), where the matrix $\Omega = \{\Omega_{ij}; i, j = 1, \ldots, p.\}$ denotes the inverse Σ^{-1} of the variance–covariance matrix Σ and $|\,\Omega\,|$ is the determinant of Ω. The term $\sum_{i,j=1}^{p} \Omega_{ij} x_i x_j$ is the result of the matrix multiplication $\mathbf{x}^T \Omega \mathbf{x}$ where Ω_{ij} is the entry in the ith row and jth column of Ω.

Assume the correlation coefficients are not too close to unity and that the p-dimensional surface is reasonably linear on the volume of dimensions $\pm e_l$ $(l = 1, \ldots, p)$. Then the probability of a match, generalising the result for uncorrelated attributes, can be shown to be

$$PM = \prod_{l=1}^{p} 2e_l \int \cdots \int \{f(x_l)\}^2 \prod_{l=1}^{p} dx_l$$

$$= \frac{\sqrt{|\,\Omega\,|}}{\pi^{p/2}} \left\{ \prod_{l=1}^{p} e_l \right\}.$$

and

$$DP = 1 - PM.$$

The explicit forms of this equation for $p = 1, 2, 3$ are given in Table 4.4.

Table 4.4 The calculation of discriminating power (DP) for Normal distributions of p dimensions

p	DP
1	$1 - e_1/(\pi^{1/2}\sigma_1)$
2	$1 - e_1 e_2/\{\pi\sigma_1\sigma_2\sqrt{(1 - \rho_{12}^2)}\}$
3	$1 - e_1 e_2 e_3/\{\pi^{3/2}\sigma_1\sigma_2\sigma_3\sqrt{(1 - \rho_{12}^2 - \rho_{13}^2 - \rho_{23}^2 + 2\rho_{12}\rho_{13}\rho_{23})}\}$

The results are clearly dependent on the choice of $\{e_l, l = 1, \ldots, p\}$. Obviously the $\{e_l\}$ have to be chosen so that the probabilities in Table 4.4 are less than 1, so as not to break the first law of probability (1.4). The other criterion is that the p-dimensional surface has to be reasonably linear within the volume of dimensions $\pm e_l (l = 1, \ldots, p)$.

Some idea of what this means can be obtained by considering, in one dimension only, the differences, d_1 and d_2 between the values of the probability density function at x and $x + e$ for d_1 and at x and $x - e$ for d_2. If d_1 and d_2 are close in magnitude then this is indicative of linearity.

For the standard Normal distribution, take x equal to 1 (or, more generally, one standard deviation away from the mean). This is a point of inflection in the density function and it could be expected that the curve is roughly linear in this region. This may be verified by considering values of e equal to $0.1x, 0.2x, 0.3x$ and $0.4x$. For $0.1x$, d_1 and d_2 agree to four significant figures, for $0.2x$, d_1 and d_2 agree to three significant figures, for $0.3x$, d_1 and d_2 agree to two significant figures, but for $0.4x$, d_1 and d_2 do not agree to even one significant figure.

The results when x equals 1 can be contrasted to the case when x equals 2 (or two standard deviations from the mean). When e equals $0.05x$, d_1 and d_2 agree to two significant figures. When e equals $0.1x$, d_1 and d_2 do not agree to one significant figure.

Care has to be taken in the consideration of correlated attributes. It is possible for two variables to have a very high positive correlation, to fail to discriminate between two groups on either variable separately but discriminate perfectly when considered together. This may seem counter-intuitive (see, for example, Smalldon and Moffat, 1973). A glance at Figure 4.1, however, should suffice to convince sceptics as to how this may be done.

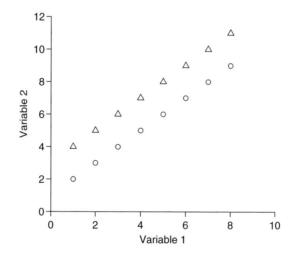

Figure 4.1   Discrimination between two groups, identified by △ and by ○, with two perfectly correlated variables.

Evaluation of evidence for multivariate data is described in Chapter 11, with more details in Aitken and Lucy (2004).

4.6 SIGNIFICANCE PROBABILITIES

4.6.1 Calculation of significance probabilities

During the course of a crime a window has been broken. A suspect is apprehended soon afterwards and a fragment of glass found on his clothing. Denote this fragment by F. It is of interest to assess the uncertainty relating to whether the fragment came from the broken window. The assessment will be discussed for the moment in the context of an approach based on what are known as *significance probabilities*. Later, in Sections 4.7 and 10.4.2, two other approaches to this assessment problem, based on coincidence probabilities and likelihood ratios respectively, will be discussed.

Let θ_0 be the value of the parameter representing the refractive index of the broken window. This is assumed constant. In Sections 4.7 and 10.4.2 this assumption will be dropped and θ_0 will be replaced by a set of sample measurements of the refractive index from the broken window.

Let x be the refractive index for F. This measurement may be considered as an observation of a random variable X, there being variation in the refractive index within a particular window. It is assumed that if F came from a window of refractive index θ then X is such that

$$X \sim N(\theta, \sigma^2)$$

(Section 2.4.2). The question at issue is whether F came from the window at the crime scene (and, by association, that the suspect was present at the crime scene). If this is so then θ will be equal to θ_0.

An argument based on significance probabilities is as follows. Suppose $\theta = \theta_0$. The further inference that F came from the window at the crime scene requires the assumption that the mean refractive index is unique to that window; this is not a particularly statistical assumption and is something that should perhaps be part of I, the background information.

The supposition that $\theta = \theta_0$ will be referred to as the *null hypothesis* and denoted H_0. Other names for such a hypothesis are *working hypothesis* or *status quo*. This nomenclature is not particularly appropriate here. It does not seem reasonable to start the analysis with the hypothesis that the suspect was at the scene of the crime. Nonetheless, this line of reasoning is pursued as the statistical ideas on which it is based are in very common usage.

Under the supposition that $\theta = \theta_0$, the deviation (in absolute terms, independent of sign) of x, the refractive index of F, from θ_0 would be expected to be small − just what is meant by 'small' depends on σ, the standard deviation. The distribution of X has been taken to be Normal. The deviation of an observation x from the mean θ of the distribution is measured in terms of the probability of observing a value for the random variable X as extreme as x. If H_0 is true,

$$X \sim N(\theta_0, \sigma^2).$$

Let $Z = (X - \theta_0)/\sigma$. Then

$$Z \sim N(0, 1).$$

Also $Pr(|X| > x) = Pr(|Z| > z)$.

The probability $Pr(|X| > x)$ is the probability of what has been observed (x) or anything more extreme if H_0 ($\theta = \theta_0$) is true (and hence, as discussed above, that F came from the window at the crime scene). The phrase 'anything more extreme' is taken to mean anything more extreme in relationship to an implicit alternative hypothesis that if H_0 is not true then $\theta \neq \theta_0$. The distance of an observation x from a mean θ is measured in terms of the standard deviation σ. For example, if $\theta = 1.518\,458$ and $\sigma = 4 \times 10^{-5}$, a value of x of $1.518\,538$ is $(1.518\,538 - 1.518\,458)/4 \times 10^{-5}$, or 2.0, standard deviations from the mean. A value of the refractive index x which is more extreme than $1.518\,538$ is one which is more than 2 standard deviations from the mean in *either* direction, that is, a value of x which is greater than $1.518\,538$ or less than $1.518\,378$.

The probability of what is observed or anything more extreme, calculated assuming the null hypothesis is true, is known as the *significance probability*. It may be thought to provide a measure of compatibility of the data with the null hypothesis. It is conventionally denoted P. A small value of P casts doubt on the null hypothesis. In the example discussed here, a small value would cast doubt on the hypothesis that F came from the window at the crime scene and a decision could be taken to act as if the null hypothesis were false. However, it is not clear what value to take for 'small'.

Certain values of P have been used to provide values known as *significance levels* at which a scientist decides to act as if the null hypothesis is false. Typical values are 0.10, 0.05 and 0.01. Thus, for example, a value x of the refractive index for which $P < 0.05$ would be said to be *significant at the 5% level*. If an approach to the evaluation of evidence based on significance levels is taken then the choice of what level of P to choose is obviously of crucial importance. It is helpful when deciding on a level for P to bear in mind the implications of the decision. The significance probability P is the probability of what is observed or anything more extreme if the null hypothesis is true. Suppose a significance level of 0.05 has been chosen. Then, by chance alone, on 5% of occasions on which this test is conducted, a decision will be made to act as if the null hypothesis is false and a wrong decision will have been made.

Consider the glass example again. On 5% of occasions, using a significance level of 5%, in which such a test was conducted and in which F came from the window at the crime scene, it will be decided to act as if F did not come from the window. This decision will be wrong. Obviously there will be many other factors which affect the decision in any particular case. However, the principle remains. The use of this type of analysis gives rise to the probability of an error. It is a well-known error and is given the name *type 1 error* or *error of the first kind*.

The probability of a type 1 error can be reduced by reducing the significance level to, for example, 0.01 or 0.001. However, this can only be done at the expense of increasing the probability of another type of error, known as a *type 2 error* or *error of the second kind*. A type 2 error is the error of failing to reject the null hypothesis when it is false. In the example, it will be the error of deciding that F did come from the window at the crime scene when in fact it did not. In general, if other factors, such as the number of fragments considered, remain constant, it is not possible to choose a significance level so as to decrease the probabilities of both type 1 and type 2 errors simultaneously.

Assume F came from the window at the scene of the crime. Then $\theta = \theta_0$. As a numerical illustration, let $\theta_0 = 1.518458$ and $\sigma = 4 \times 10^{-5}$. Illustrations of the calculations for the significance probability are given in Table 4.5.

Determination of the probability of a type 2 error requires knowledge of the value of θ if F did not come from the window at the crime scene. Normally such knowledge is not available and the probability cannot be determined. There may, occasionally, be circumstances in which the probability of a type 2 error can be determined. For example, this could happen if the defence were to identify another source for F. However, even if a probability cannot be determined, one has to be aware that such an error may be made.

Notice that the philosophy described here fits uncomfortably with the legal philosophy that a person is innocent until proven guilty. The null hypothesis in the example is that F came from the window at the crime scene. Failure to reject this hypothesis is an implicit acceptance that the suspect was at the crime scene. Yet, it is the null hypothesis which is being tested. It is the one which has to be rejected or not. The calculation of the significance probability P is based on the assumption that the null hypothesis is true. Only if P is small (and it is values of 0.05, 0.01 and 0.001 which have been suggested as small)

Table 4.5 Significance probabilities P for refractive index x of glass for mean $\theta_0 = 1.518458$ and standard deviation $\sigma = 4 \times 10^{-5}$ and decisions assuming a significance level of 5%

x	$z = (x - \theta_0)/\sigma$	$P = Pr(\lvert X \rvert > x)$ $= Pr(\lvert Z \rvert > z)$	Decide to act as if F *did* or *did not* come from the crime window
1.518 500	1.05	0.29	Did
1.518 540	2.05	0.04	Did not
1.518 560	2.55	0.01	Did not

is it considered that the null hypothesis may be false. Evidence is required to show that the suspect has *not* been at the crime scene. The principle that one is innocent until proven guilty requires that evidence be produced to show that the suspect has been at the crime scene, not the opposite.

This point is also made by Gaudette (1999) in the context of hair examination. An analogy is made with a fire alarm. A type 1 error corresponds to the alarm not ringing when there is a fire. A type 2 error corresponds to the alarm ringing when there is not a fire. With a fire alarm a type 1 error is more serious error than a type 2 error. In forensic science a type 2 error is more serious than a type 1 error as it results in the false incrimination of an innocent person.

The interpretation of P has to be made carefully. Consider a result from Table 4.5: $\theta_0 = 1.518\,458$, $\sigma = 4 \times 10^{-5}$, $x = 1.518\,560$ and $P = 0.01$. This may be written more explicitly in the notation of conditional probability as

$$Pr(|X| > x \mid \theta = \theta_0, \sigma) = 0.01. \tag{4.4}$$

It is difficult to relate this to the matter at issue: was the suspect at the crime scene? A small value would seem indicative of the falsity of the null hypothesis, but it is not the probability that the null hypothesis is true. It is incorrect to use the value for P as the probability that the suspect was at the crime scene. The transposed probability

$$Pr(\theta = \theta_0 \mid |X| > x, \sigma) \tag{4.5}$$

would be much more useful but this is not what has been calculated. The relationship between (4.4) and (4.5) is similar to that between the probability of the evidence given the suspect is guilty and the probability the suspect is guilty given the evidence. The interpretation of the first of the last two probabilities as the second is the fallacy of the transposed conditional (Section 3.3.1). It is possible, however, to discuss the relationship between the significance probability and the probability that the suspect was at the crime scene, through the use of likelihood ratios.

4.6.2   Relationship to likelihood ratio

The relationship between significance probabilities and the likelihood ratio has been investigated by many authors. Early references include Good (1956), Lindley (1957) and Edwards *et al.* (1963). The discussion here in the context of the refractive index of glass is based on Berger and Sellke (1987).

Consider two competing and complementary propositions:

- H_p, the fragment of glass on the suspect's clothing came from the window at the crime scene;

- H_d, the fragment of glass on the suspect's clothing did not come from the window at the crime scene.

Let p denote the probability that H_p is true, $Pr(H_p)$, and $1 - p$ the probability that H_d is true, $Pr(H_d)$. If H_p is true then θ, the mean refractive index of the source of the fragment F on the suspect's clothing, is θ_0. If H_d is true, then F is assumed to come from a window whose mean refractive index is not equal to θ_0.

Assume H_p is true. Denote the probability density function of X, the refractive index of F, by $f(x \mid \theta_0, H_p)$. In this context, this is a Normal density function.

Assume H_d is true. Denote the probability density function of X by $f(x \mid H_d)$. The mean θ of the refractive index of the source of the fragment on the suspect's clothing may also be thought of as a random variable which varies from window to window over some relevant population of windows. As such it also has a probability density function; denote this by $f(\theta)$. If θ is known, the probability density function of x is given by $f(x \mid \theta)$. An extension of the law of total probability (Section 1.6.7) to continuous data, with integration replacing summation, may be used to give the expression

$$f(x \mid H_d) = \int f(x \mid \theta) f(\theta) d\theta.$$

The probability density function of X, independent of H_p and H_d, is then

$$f(x) = p f(x \mid \theta_0, H_p) + (1 - p) f(x \mid H_d).$$

Thus, the probability that H_p is true, given the measurement x, is

$$Pr(H_p \mid x) = f(x \mid \theta_0, H_p) p / f(x)$$

$$= \left\{ 1 + \frac{(1 - p) f(x)}{p f(x \mid \theta_0, H_p)} \right\}^{-1}, \qquad (4.6)$$

an expression similar to one used in paternity cases (see Section 9.8.1).

The posterior odds in favour of H_p is then, using a version of (3.11),

$$\frac{Pr(H_p \mid x)}{1 - Pr(H_p \mid x)} = \frac{p}{1 - p} \frac{f(x \mid \theta_0, H_p)}{f(x \mid H_d)}$$

where $\{ p / (1 - p) \}$ is the prior odds in favour of H_p and $\{ f(x \mid \theta_0, H_p) / f(x \mid H_d) \}$ is the likelihood ratio V of (3.12).

As an illustration of the calculation of the likelihood ratio, assume that if θ is not equal to θ_0 it is a random variable which has a Normal distribution with mean θ_0 and variance τ^2, where, typically, $\tau^2 \gg \sigma^2$. Then

$$f(x \mid H_d) = \int f(x \mid \theta) f(\theta) d\theta$$

and so

$$(X \mid H_d) \sim N(\theta_0, \sigma^2 + \tau^2).$$

(See Section 10.2.2 for a fuller derivation of this result.) With $\tau^2 \gg \sigma^2$, the distribution of $(X \mid H_d)$ is approximately $N(\theta_0, \tau^2)$. The likelihood ratio is, thus,

$$V = \frac{f(x \mid \theta_0, H_p)}{f(x \mid H_d)} = \frac{(2\pi\sigma^2)^{-1/2} \exp\{-(x-\theta_0)^2/2\sigma^2\}}{(2\pi\tau^2)^{-1/2} \exp\{-(x-\theta_0)^2/2\tau^2\}}.$$

Consider $\tau = 100\sigma$. Let $z^2 = (x - \theta_0)^2/\sigma^2$ be the square of the standardised distance between the observation x and the mean specified by the null hypothesis, θ_0. Then

$$V = 100 \exp\left(-\frac{z^2}{2} + \frac{z^2}{2 \times 10^4}\right)$$

$$\simeq 100 \exp(-z^2/2).$$

For example, consider $x = 1.518\,540$, $\theta_0 = 1.518\,458$, $\sigma = 4 \times 10^{-5}$, $\tau = 4 \times 10^{-3}$. Then $z^2 = 2.05^2$ and $P = 0.04$, as before (see Table 4.5) which, at the 5% level, would lead to rejection of the hypothesis that the fragment of glass came from the window at the scene of the crime. However,

$$V = 100 \exp(-2.05^2/2)$$

$$= 12.2,$$

a value for V which, on the verbal scale in Table 3.10, represents moderate evidence to support H_p against H_d. Such an apparent contradiction between the two approaches is not a new idea and has been graced by the name of Lindley's paradox (see, for example, Good, 1956; Lindley, 1957; Edwards *et al.*, 1963; and Lindley, 1980, for a reference by Lindley to 'Jeffreys' paradox').

Suppose that several, n say, fragments of glass were found on the suspect's clothing rather than just one. Let $\bar{x}$ denote the mean of these fragments. Then,

$$(\bar{X} \mid \theta) \sim N(\theta, \sigma^2/n).$$

If H_d is true

$$\bar{X} \sim N(\theta_0, \tau^2 + \sigma^2/n).$$

The likelihood ratio is

$$V = \frac{(2\pi\sigma^2/n)^{-1/2} \exp\{-n(\bar{x}-\theta_0)^2/2\sigma^2\}}{\{2\pi(\tau^2+\sigma^2/n)\}^{-1/2} \exp[-(\bar{x}-\theta_0)^2/\{2(\tau^2+\sigma^2/n)\}]}$$

$$\simeq \frac{\tau\sqrt{n}}{\sigma} \exp\left[-\frac{(\bar{x}-\theta_0)^2}{2}\left\{\frac{n}{\sigma^2} - \frac{1}{\tau^2}\right\}\right]$$

$$\simeq 100\sqrt{n} \exp\left\{-\frac{n(\bar{x}-\theta_0)^2}{2\sigma^2}\right\}.$$

Table 4.6 Variation in the likelihood ratio V, as given by (4.7), with sample size n, for a standardised distance $z_n = 2$, a result which is significant at the 5% level

n	V
1	14
5	30
10	43
20	61

The square of the standardised distance, z_n, between $\bar{x}$ and θ_0 is

$$z_n^2 = n(\bar{x} - \theta_0)^2/\sigma^2$$

and thus

$$V = 100\sqrt{n}\exp(-z_n^2/2), \tag{4.7}$$

a value which increases in direct proportion to the square root of the sample size. Suppose $z_n = 2$, a value which is significant at the 5% level in a test of the hypothesis $\theta = \theta_0$ against the alternative hypothesis $\theta \neq \theta_0$. The value of V for various values of n is given in Table 4.6. In each case, a result which is significant at the 5% level has a likelihood ratio which lends support to the hypothesis that the fragments of glass found on the suspect came from the window at the crime scene.

4.6.3 Combination of significance probabilities

Significance probabilities also combine in a different way from the probabilities of events. From the third law of probability (1.6), the product of two probabilities, for dependent or independent events, will never be greater than either of the individual components of the product. Let A and B be the two events. Then

$$Pr(AB) = Pr(A)Pr(B \mid A) \leq Pr(A)$$
$$= Pr(B)Pr(A \mid B) \leq Pr(B),$$

with equality in the first case if and only if $Pr(B \mid A) = 1$, $Pr(A) \neq 0$ and in the second case if and only if $Pr(A \mid B) = 1$, $Pr(B) \neq 0$. Note also that $Pr(AB) > Pr(A)Pr(B)$ if $Pr(B \mid A) > Pr(B)$ or $Pr(A \mid B) > Pr(A)$.

However, it is possible, for characteristics which are dependent, that the significance probability of the joint observation may be greater than either of the individual significance probabilities.

Suppose, for example, that as well as the refractive index of the glass in the current example the density has been measured. The suffixes 1 and 2 are introduced to denote refractive index and density, respectively. Then $\mathbf{x} = (x_1, x_2)^T$ is a vector which denotes the measurements of the two characteristics of the glass found on the suspect's clothing and $\boldsymbol{\theta} = (\theta_1, \theta_2)^T$ is a vector which denotes the mean values of refractive index and density of glass from the window at the crime scene.

Let $\theta_1 = 1.518458$ (the original θ_0 above), $x_1 = 1.518540$ and $\sigma_1 = 4 \times 10^{-5}$. Then, the significance probability (P_1, say) for the refractive index of F is 0.04 (see Table 4.5). Suppose $\theta_2 = 2.515\,\mathrm{g\,cm^{-3}}$ with standard deviation $\sigma_2 = 3 \times 10^{-4}\,\mathrm{g\,cm^{-3}}$, and let $x_2 = 2.515615\,\mathrm{g\,cm^{-3}}$ be the density measured on F. The standardised statistic, z_2 say, is then

$$z_2 = (2.515615 - 2.515)/0.0003 = 2.05$$

and the significance probability (P_2, say) for the density measurement is also 0.04.

The product of P_1 and P_2 is 0.0016. However, this is not the overall significance probability. The correlation between the refractive index and the density has to be considered.

Let the correlation coefficient, ρ say, between refractive index and density be 0.93 (Dabbs and Pearson, 1972) and assume that the joint probability density function of the refractive index and density is bivariate Normal (2.26) with mean $\boldsymbol{\theta}$ and covariance matrix $\boldsymbol{\Sigma}$. For bivariate Normal data $\mathbf{x}$ it can be shown that the statistic U, given by

$$U = (\mathbf{x} - \boldsymbol{\theta})^T \boldsymbol{\Sigma}^{-1} (\mathbf{x} - \boldsymbol{\theta}),$$

has a chi-squared distribution with 2 degrees of freedom: $U \sim \chi_2^2$ (Mardia *et al.*, 1979, p. 39). Values of the χ^2 distribution may be found from statistical software or books of tables.

The overall significance probability for the two characteristics can be determined by calculating U and referring the answer to the χ^2 tables. The covariance matrix $\boldsymbol{\Sigma}$ is given by (2.27)

$$\begin{aligned}
\boldsymbol{\Sigma} &= \begin{pmatrix} \sigma_1^2 & \rho\sigma_1\sigma_2 \\ \rho\sigma_1\sigma_2 & \sigma_2^2 \end{pmatrix} \\
&= \begin{pmatrix} (4 \times 10^{-5})^2 & 1.116 \times 10^{-8} \\ 1.116 \times 10^{-8} & (3 \times 10^{-4})^2 \end{pmatrix} \\
&= \begin{pmatrix} 1.6 \times 10^{-9} & 1.116 \times 10^{-8} \\ 1.116 \times 10^{-8} & 9 \times 10^{-8} \end{pmatrix}.
\end{aligned}$$

The deviation of the observation $\mathbf{x}$ from the mean $\boldsymbol{\theta}$ is $(\mathbf{x} - \boldsymbol{\theta})^T = (8.2 \times 10^{-5}, 6.15 \times 10^{-4})$. Some rather tedious arithmetic gives the result that

$(\mathbf{x} - \boldsymbol{\theta})^T \boldsymbol{\Sigma}^{-1} (\mathbf{x} - \boldsymbol{\theta})$ equals 4.204, a result which has significance probability $P = 0.1225$. Each individual characteristic is conventionally significant at the 5% level, yet together the result is not significant at the 10% level.

Considerable care has to be taken in the interpretation of significance probabilities. As a measure of the value of the evidence their interpretation is difficult.

4.7 COINCIDENCE PROBABILITIES

4.7.1 Introduction

One of the criticisms which was levelled at the probabilistic evidence in the Collins case (Section 4.4) was the lack of justification for the relative frequency figures which were quoted in court. The works of Tippett *et al.* (1968) and Gaudette and Keeping (1974) were early attempts to collect data. The lack of data relating to the distribution of measurements on a characteristic of interest is a problem which still exists and is a major one in forensic science. If measurements of certain characteristics (such as glass fragments and their refractive indices) of evidence left at the scene of a crime are found to be similar to evidence (measurements of the same characteristics) found on a suspect, then the forensic scientist wants to know how similar they are and whether this similarity is one that exists between rare characteristics or common characteristics. In certain cases, such data do exist which can help to answer these questions.

The most obvious example of this is DNA profiling. There are various loci and the relative population frequencies of the various categories for each of the loci are well tabulated for different populations. Thus if a bloodstain of a particular profile is found at the scene of a crime there are several possibilities:

(a) it came from the victim;
(b) it came from the criminal;
(c) it came from some third party.

If the first and third possibilities can be eliminated and the profile is only found in $x\%$ of the population, then the implication is that the criminal belongs to that $x\%$ of the population; that is, he belongs to a subset of the population with $Nx/100$ members, where N is the total population size and $Nx/100$ is of necessity no less than 1. At its simplest this implies that, *all other things being equal*, a person with this allele has probability $100/Nx$ of being the criminal. This is a consequence of the defender's fallacy explained in Section 3.3.4.

Another example in which data exist, though not to the extent to which they do for DNA profiling, is the refractive index of glass particles. Data relating to the refractive index have been collected over several years by different laboratories to be able to assess the relative frequency of glass fragments found on different locations (e.g., clothing, footwear, and hair combings). A full list up to 2000 is provided by Curran *et al.* (2000). From a statistical point of view this approach may lead to a biased sample since only glass fragments which have been connected with crimes will be collected. However, as an example of the recognition of this problem, a survey of glass found on footpaths has been conducted (Walsh and Buckleton, 1986) to obtain data unrelated with crimes.

An approach based on probabilities, known as *coincidence probabilities*, with relation to the refractive index of glass was developed by Evett and Lambert in a series of papers (Evett, 1977, 1978; Evett and Lambert, 1982, 1984, 1985), using data from Dabbs and Pearson (1972). Two questions were asked:

1. Are the control and recovered fragments similar in some sense?
2. If they are similar, are the characteristics rare or common?

A very simple – and trivial – example of this is the following. An eyewitness to a crime says that he saw a person running from the scene of the crime and that this person had two arms. A suspect is found who also has two arms. The suspect is certainly similar to the person running away from the crime, but the characteristic – two arms – is so common as not to be worth mentioning. The elementary nature of the example may be extended by noting that if the eyewitness said the criminal was a tall man then any women or short men, despite having two arms, would be eliminated. In contrast, suppose that the eyewitness says that the person running away had only one arm and a suspect is found with only one arm. This similarity is much stronger evidence that the suspect was at the scene of the crime since the characteristic – one arm – is rare.

A more complicated – and considerably less trivial – application relates to the interpretation of evidence on the refractive index of glass. The coincidence method to be explained is capable of development in this context because of the existence of data for the distribution of measurements of the refractive index of glass. Development is not possible in the examples mentioned earlier of paint fragments and hair comparisons because appropriate data do not exist. A crime is committed and fragments of glass from a window are found at the scene of the crime; these are the source fragments. A suspect is found and he has fragments of glass on his person; these are the receptor (or transferred particle) fragments. The interpretation of the data is treated in two stages corresponding to the two questions asked earlier. First, the refractive index measurements are compared by means of a statistical criterion which takes account of between- and within-window variation. Secondly, if the two sets of measurements are found to be similar then the significance of the result is assessed by referring to suitable data.

Various assumptions are needed for these two stages to be applied. Those adopted here are the following:

(a) The refractive index measurements on glass fragments from a broken window have a Normal probability distribution centred on a mean value θ which is characteristic of that window and with a variance σ^2.

(b) The mean θ varies from window to window and in the population of windows θ has its own probability distribution of, as yet, unspecified form.

(c) The variance σ^2 is the same for all windows and it is known.

(d) All the transferred particle fragments are assumed to have come from the same source. A criterion based on the range of the measurements can be devised to check this (e.g., Evett, 1978).

(e) The transferred particle fragments are all window glass.

4.7.2 Comparison stage

The comparison is made by considering the difference, suitably scaled, between the mean of the source fragment measurements and the mean of the receptor fragment measurements. The test statistic is

$$Z = \frac{\bar{X} - \bar{Y}}{\sigma(n^{-1} + m^{-1})^{1/2}},$$

where $\bar{X}$ is the mean of m measurements on source fragments and $\bar{Y}$ is the mean of n measurements on receptor fragments. It is assumed that Z has a standard $N(0, 1)$ distribution. Thus if $|Z| > 1.96$ it is concluded that the receptor fragments are not similar to the source fragments, and if $|Z| < 1.96$ it is concluded that the receptor fragments are similar to the source fragments. The value 1.96 is chosen so that the probability of a type 1 error (deciding that receptor fragments have a different origin from source fragments) is 0.05. Other possible values may be chosen such as 1.64, 2.33, 2.58, which have type 1 error probabilities of 0.10, 0.02 and 0.01, respectively; see Table 2.5.

This statistic and the associated test provide an answer to question (1).

4.7.3 Significance stage

If it is concluded that the receptor and source fragments are not similar, the matter ends there. If it is concluded that they are similar, then there are two possibilities.

1. The receptor and source fragments came from the same source.
2. The receptor and source fragments did not come from the same source.

For the assessment of significance the probability of a coincidence (known as a *coincidence probability*) is estimated. This is defined as the probability that a set of n fragments taken at random from some window selected at random from the population of windows would be found to be similar to a control window with mean refractive index $\bar{X}$. It is denoted $C(\bar{X})$.

Compare this definition of a probability of coincidence with that given in the discussion of discriminating power (Section 4.5). There the receptor and source fragments were both taken to be random samples from some underlying population and the probability estimated was that of two random samples having similar characteristics. Here the mean $\bar{X}$ of a sample of fragments from the source window is taken to be fixed and the concern is with the estimation of the probability that one sample, namely the receptor fragments, will be found to be similar to this source window. Any variability in the value of $\bar{X}$ is ignored; methods of evaluation which account for this variability are discussed in Chapter 10.

Note that there are two levels of variation which have been considered. Further details of how to consider these two levels are given in Chapter 11. First, there is the probability that a window selected at random from a population of windows will have a mean refractive index in a certain interval, $(u, u + du)$, say; denote this probability by $p(u)$. In practice, for this method the data used to estimate $p(u)$ are represented as a histogram and the probability distribution is taken as a discrete distribution over the categories forming the histogram, for example, $\{p(u_1), \ldots, p(u_k)\}$ where there are k categories. Secondly, the probability is required that n fragments selected at random from a window of mean value u would prove to be similar to fragments from a source window of mean $\bar{X}$; denote this probability by $S_{\bar{X}}(u)$. It is then possible to express $C(\bar{X})$ as a function of $p(u)$ and $S_{\bar{X}}(u)$, namely:

$$C(\bar{X}) = \sum_{i=1}^{k} p(u_i) S_{\bar{X}}(u_i).$$

The derivation of this is given in Evett (1977).

Let $\{x_1, \ldots, x_m\}$, with mean $\bar{x}$, denote the source measurements of refractive index from fragments from a window W broken at the scene of a crime; have

$$\bar{X} \sim N(w, \sigma^2/m),$$

where w is the mean refractive index of window W.

Let $\{y_1, \ldots, y_n\}$, with mean $\bar{y}$, denote the transferred particle measurements of refractive index from fragments T found on a suspect's clothing. If the fragments T have come from W, then

$$\bar{Y} \sim N(w, \sigma^2/n).$$

Since the two sets of measurements are independent,

$$(\bar{X} - \bar{Y}) \sim N\left(0, \sigma^2(1/m + 1/n)\right),$$

and, hence,

$$\frac{\bar{X} - \bar{Y}}{\sigma(1/m + 1/m)^{1/2}} \sim N(0, 1).$$

Denote this statistic by Z. Thus, it can be said that if the fragments T came from W and if the distributional assumptions are correct, then the probability that Z has a value greater than 1.96 in absolute terms,

$$|Z| > 1.96,$$

is 0.05. Such a result may be thought unlikely under the original assumption that the fragments T have come from W. Hence the original assumption may be questioned. It may be decided to *act as if* the fragments T did not come from W. Evett (1977) showed that with this decision rule at the comparison stage, the probability $S_{\bar{X}}(u)$ at the significance stage is given by

$$S_{\bar{X}}(u) = \Phi\left\{(\bar{X} - u)m^{1/2}/\sigma + 1.96(1 + m/n)^{1/2}\right\}$$
$$- \Phi\left\{(\bar{X} - u)m^{1/2}/\sigma - 1.96(1 + m/n)^{1/2}\right\}, \tag{4.8}$$

from which the coincidence probability $C(\bar{X})$ may be evaluated in any particular case. Some results of the application of this method in comparison with others are given in Section 10.4.2 and Table 10.6.

4.8 LIKELIHOOD RATIO

This is discussed at considerable length in later chapters. The likelihood ratio has been introduced in Section 3.5.1 as the ratio $Pr(E \mid H_p)/Pr(E \mid H_d)$ which converts prior odds in favour of the prosecution proposition (H_p) into posterior odds in favour of the prosecution proposition, given evidence E.

From a historical point of view, it is useful to remember that, in the Dreyfus case discussed in Section 4.2, Poincaré and his colleagues supported such a Bayesian approach and proposed the use of the likelihood ratio. In fact, as Poincaré wrote and as has already been discussed in earlier sections,

> an effect may be the product of either cause A or a cause B. The effect has already been observed. One wants to know the probability that it is the result of cause A; this is the *a posteriori* probability. But I am not able to calculate in advance the *a priori* probability for the cause producing the effect. I want to speak of the probability of this eventuality, for one who has never before observed the result. (Poincaré, 1992, at p. 229.)

However,

> since it is absolutely impossible for us [the experts] to know the *a priori* probability, we cannot say: this coincidence proves that the ratio of the forgery's probability to the inverse probability is a real value. We can only say: following the observation of this coincidence, this ratio becomes *X* times greater than before the observation. (Darboux *et al.*, 1908, p. 504)

For more information on this statistical argument and an example of application to shoeprints, see Taroni *et al.* (1998).

Similarly, in 1930, Bruno de Finetti expressed the same view:

> Il calcolo delle probabilità è la logica del probabile. Come la logica formale insegna a dedurre la verità o la falsità di certe conseguenze della verità o falsità di certe premesse, così il calcolo delle probabilità insegna a dedurre la maggiore o minore verosimiglianza o probabilità di certe conseguenze dalla maggiore o minore verosimiglianza o probabilità di certe premesse. (de Finetti, 1930, p. 259)

which can be translated as follows:

> Probability calculus is the logic of the probable. As logic teaches the deduction of the truth or falseness of certain consequences from the truth or falseness of certain assumptions, so probability calculus teaches the deduction of the major or minor likelihood, or probability, of certain consequences from the major or minor likelihood, or probability, of certain assumptions.

Olkin (1958) proposed an evaluation of the identification problem in terms of the likelihood ratio statistic, which – he said – is the ratio of the probability of the characteristics under the assumption of the identity to the probability of the characteristics under the assumption of non-identity.

Kingston and Kirk (1964) make the following comment on the likelihood ratio:

> Now consider a problem of evaluating the significance of the coincidence of several properties in two pieces of glass. Suppose that the probability of two fragments from different sources having this coincidence of properties is 0.005, and that the probability of such a coincidence when they are from the same source is 0.999. What do these figures mean? They are simply guides for making a decision about the origin of the fragments.

An interesting and somewhat prophetic comment was made by Parker and Holford (1968), in which the comparison problem was treated as in two stages (similarity and discrimination) rather than in one, as an analysis using the likelihood ratio does. The two-stage approach follows, they said, the traditions of forensic scientists. Interestingly, they then went on to make the following remark:

> We could (therefore) set up an index *R* whose numerator is the likelihood that the crime hair comes from the suspect and whose denominator is the likelihood that it comes from the population at large. In ordinary language one would then

assert that it was R times more likely for the hair to have come from the suspect than from someone drawn at random from the population. But a statement of this nature is of rather limited use to forensic scientists, by-passing as it does the similarity question altogether. (Parker and Holford, 1968)

It will be explained in later chapters how the use of the likelihood ratio, which is all that Parker and Holford's index R is, has rather more than 'limited use' and how its use considers similarity in a natural way.

The next comment was made in 1975:

> Bayesianism does not take the task of scientific methodology to be that of establishing the truth of scientific hypotheses, but to be that of confirming or disconfirming them to degrees which reflect the overall effect of the available evidence – positive, negative, or neutral, as the case may be. (Jeffrey, 1975)

Finally:

> Bayes' rule tells us that we then take those prior odds and multiply them by the likelihood ratio of the blood/DNA evidence in order to arrive at the posterior odds in favour of the defendant's paternity. The Court then has to consider whether those odds meet the required standard of proof. Thus the expert should say 'however likely you think it is that the defendant is the father on the basis of the other evidence, my evidence multiplies the odds X times'. (Robertson and Vignaux, 1992)

The importance of the likelihood ratio for evidence evaluation is emphasised by consideration of the various disciplines of the people who have made the above statements. These disciplines include mathematics (de Finetti, Olkin), philosophy (Jeffrey), forensic science (Kingston, Kirk, Parker, Holford), law (Robertson) and operations research (Vignaux).

5

Bayesian Inference

5.1 INTRODUCTION

Bayesian inference can incorporate subjective information about a problem into the analysis. Objections are raised to the loss of objectivity that results from using the analyst's subjective information. However, either the data are strong enough for reasonable people to agree on their interpretation, regardless of the prior information, or the analysts should be using their subjective prior information in order to make appropriate decisions related to the data. It is also possible to choose a so-called ignorance prior in which the subjective prior information is minimal. The discussion of Bayesian inference in Chapter 3 is concerned with discrete events. This chapter extends these ideas to continuous variables, discusses prior and posterior distributions and likelihoods and includes a comparison of frequentist confidence intervals with Bayesian probability intervals. Later we refer to Bayesian ideas in sampling (Chapter 6) and paternity (Section 9.8). General introductions to Bayesian ideas are given in Berry (1996), Antelman (1997) and Lee (2004).

Throughout this chapter and, indeed, in the book in general it should be borne in mind that 'Bayesian methodology does not pretend to get the "true" probabilities: it is an effective method to analyse, criticise, check the "coherence" of people's opinions and help people revise their opinions in a coherent way. No more, but not less, than that' (Taroni *et al.*, 2001).

There is an interpretation of probability, described in Chapter 1, as a measure of belief, which is the property of an individual. This interpretation says that probability is subjective. In contrast, the frequency definition of probability is a property of a sequence; all who observe the sequence will agree its value. It is objective.

An excellent argument in support of subjective probability is given by Lindley (1991). He states that objectivity is the hallmark of science and subjectivity is thought to be undesirable. In science, if a hypothesis is thought worthy of study and yet individual scientists disagree as to whether it is true or not, experiments are performed and repeated (analogous to the presentation and corroboration of evidence in a court of law) until there is general agreement. Like scientists, the

Statistics and the Evaluation of Evidence for Forensic Scientists: Second Edition
C.G.G. Aitken and F. Taroni © 2004 John Wiley & Sons, Ltd   ISBN: 0-470-84367-5

members of a jury may all have different prior beliefs (i.e., prior to conducting the 'experiment' of hearing the evidence).

Consider the following. Given initial information I, two people have different beliefs in the truth of an event G: their probabilities $Pr(G \mid I)$ are not the same. Evidence, E, is then produced and posterior probabilities $Pr(G \mid E, I)$ are determined for each person. Then it can be shown that E brings these probabilities closer together. For sufficiently large amounts of evidence, the two posterior probabilities will agree for all practical purposes. Lindley (1991) argues that this is exactly what happens in a courtroom. The jurors are brought together by what they hear. There is nothing to force agreement, but experience shows that agreement is usually reached.

Another champion of the cause of subjectivity has written:

> We therefore accept the subjective definition of probability as the degree of belief of a given individual in the occurrence of a certain event . . . It is worth considering the precise, technical, above-defined meaning of the difference between objective and subjective. In fact, I feel that many misunderstandings and many oppositions derive from the more or less unconscious and vague interpretation of objective, as synonymous with 'founded, reasonable, serious', while one would call subjective a judgement that is 'hurried, without foundation, improvised or pulled out of a hat'. Nothing could be further from the true intentions of the subjective theory: this aims at studying and promoting probability evaluations carried out with all the attention of those who consider them as objective and, if necessary, with a greater sense of responsibility deriving from not having illusions regarding their pretended objective nature. Those who are not satisfied and disdain subjective probability and believe that they remedy it by declaring it objective, do not in fact reach a better result than those who, unhappy about the imperfect level of rigidity of the available materials, would decide to use them by labelling them as 'perfectly rigid materials'. (de Finetti, 1952)

Much of what will be discussed here concerns estimation of the values of parameters. Variation in a parameter is modelled with a *prior distribution*, which is a probability distribution. Uncertainty about the value of a parameter is represented by probability. It is possible to make statements about the value of a parameter in the form of a probability distribution. This distribution is itself characterised by one or more parameters (known as prior parameters).

One of the criticisms of the Bayesian approach to evidence evaluation (and also to other areas of statistical analysis) is the use of subjective probabilities. However, subjectivity should not be confused with arbitrariness. A commonly held view is that if probability represents a degree of belief then it has to be arbitrary because one person's belief would be different from another person's. However, just because a degree of belief is personal it does not mean it is arbitrary. Probability may represent how much a person believes something to be true but this belief is based on all relevant information available to the person. Such information will almost certainly be different from that available to another person and thus the other person will have a different degree of belief from the first person. This difference is not an arbitrary difference. The implication is that the degree of belief, or probability, is conditional on what a person knows. However, all probabilities are conditional and the conditions should be stated explicitly. The need for this explicit statement of conditions is one of the benefits of a Bayesian approach. Eliciting opinions (or prior measures of belief or prior probabilities), updating the opinion after the observation of data (to produce a posterior measure of belief or posterior probability) and quantifying the uncertainty through the use of probability distributions are all part of Bayesian statistics.

People with different prior probabilities will observe the same data (evidence), for which they will have the same *likelihood*. 'Likelihood' is the word used to describe the probability of observing the data, conditional on the values of the parameters, but taking the expression as a function of the parameters. The posterior probabilities will come closer together as evidence accumulates and the information in the likelihood increases. The influence of prior personal degrees of belief will decrease as the evidence accumulates. The likelihood may be said to dominate the prior. Posterior probabilities are less subjective and more objective than prior probabilities.

Subjective probabilities may be measured by considering betting analogies. An important concept here is one of coherence. Betting odds are said to be coherent if an individual who bets on all possible outcomes breaks even. If the odds (probabilities) are arranged otherwise so that, whatever happens, either the bookmaker wins or the individual wins then the odds (probabilities) are said to be incoherent. As an example, consider a three-horse race in which the probabilities as set by the bookmaker are $1/4, 1/3, 1/2$, for each of the horses to win. These probabilities add up to more than 1 and are thus incoherent. An individual who puts money on each horse will lose money. Suppose he bets £13 000 in total with £3000 on the first horse, £4000 on the second and £6000 on the third. The choice of the size of the bet is made by the individual to ensure he breaks even in the long run for each horse. However, whichever horse wins, the individual receives £12 000 and the bookmaker retains £1000. (Of course, in practice, on individual races it is possible for bookmakers to lose money, but in the long run, with incoherent odds, they make money, which is how they earn a living.) When a scientist is considering probabilities, the consideration has to be such that the probabilities are coherent, otherwise an incoherent (and possibly nonsensical) outcome will occur.

Subjectivity enters in the choice of the model (probability distribution) and in the choice of the values of the prior parameters. Thus, given initial information I, two people may agree on the form of the model, as this is usually sufficiently flexible and general to satisfy most beliefs in a particular context. They may disagree on the values of the parameters, however. This may not be too important if the data are sufficiently informative that the likelihood is able to dominate the prior, as will be shown in examples. In other cases, it may be important because the data are not very informative. However, this situation only reflects the reality that uninformative data do not provide much information.

Denote the prior parameter by θ, which may be vector-valued, and the prior density by $f(\theta)$. The data x are modelled by the likelihood. The likelihood involves x and θ. It is proportional to a probability density function (for continuous x) and to a probability function (for discrete x). Denote the likelihood by $L(\theta \mid x)$ to emphasise that it is of interest as a function of θ, conditional on the value of x. An example in the context of binomial sampling is given in Section 5.6.4. The likelihood and the prior combine to form a posterior density function $f(\theta \mid x)$ using Bayes' theorem,

$$f(\theta \mid x) \propto L(\theta \mid x) f(\theta)$$

or

$$f(\theta \mid x) = \frac{L(\theta \mid x) f(\theta)}{f(x)},$$

where $f(x)$ is the probability density function for x, not conditional on θ. It can be determined as

$$f(x) = \int f(x \mid \theta) f(\theta) d\theta.$$

In the Bayesian paradigm, the data are taken as fixed and known. Uncertainty resides in the parameters. The data modify the prior distribution via the likelihood to provide the posterior distribution.

5.2   BAYESIAN INFERENCE FOR A BERNOULLI PROBABILITY

Inference is to be made of a binomial probability θ, that is, one which is associated with a Bernoulli trial (one with only two possible outcomes; see Section 2.3.2). For example, θ may be the true proportion of illicit drugs in a consignment. Within the Bayesian paradigm, θ has a probability distribution and it is desired to determine this in order to make inferences about θ. The most common distribution for θ is the beta distribution, introduced in Section 2.4.4. An example of its use for drug sampling (Chapter 6) is given later. The beta distribution is a so-called *conjugate prior* distribution for the binomial distribution. The beta prior distribution and binomial distribution combine together to give a posterior distribution which is also a beta distribution. See Berry (1991c, 1993) for further details. Another example of conjugacy, not discussed further here, is that of the gamma distribution, which has two parameters α and β and for which further details may be found in Evans *et al.* (2000). This is a conjugate prior for the Poisson distribution with probability density function

$$f(d \mid \alpha, \beta) = c d^{\alpha-1} e^{-\beta d}, \qquad d > 0, \ \alpha > 0, \ \beta > 0, \qquad (5.1)$$

where

$$c = \beta^\alpha / \Gamma(\alpha)$$

and Γ is the gamma function (2.8).

The beta distribution for θ, parameterised by α and β, is

$$f(\theta \mid \alpha, \beta) = \theta^{\alpha-1}(1-\theta)^{\beta-1}/B(\alpha, \beta), \qquad 0 < \theta < 1 \qquad (5.2)$$

(from Section 2.4.4). Consider a random variable X which has a binomial distribution (from Section 2.3.3) with parameters n and θ, so that

$$Pr(X = x \mid n, \theta) = \binom{n}{x}\theta^x(1-\theta)^{n-x}, \qquad x = 0, 1, \ldots, n.$$

This is the likelihood function and, in this context, is denoted $L(\theta \mid n, x)$. Then, the posterior distribution of θ is given by

$$f(\theta \mid x+\alpha, n-x+\beta) = \theta^{x+\alpha-1}(1-\theta)^{(n-x+\beta-1)}/B(x+\alpha, n-x+\beta), \quad 0 < \theta < 1, \tag{5.3}$$

denoted $Be(x+\alpha, n-x+\beta)$. In the particular case where $x = n$, the density function is given by

$$f(\theta \mid n+\alpha, \beta) = \theta^{n+\alpha-1}(1-\theta)^{\beta-1}/B(n+\alpha, \beta), \qquad 0 < \theta < 1. \tag{5.4}$$

Similarly, when $x = 0$,

$$f(\theta \mid \alpha, n+\beta) = \theta^{\alpha-1}(1-\theta)^{n+\beta-1}/B(\alpha, n+\beta), \qquad 0 < \theta < 1. \tag{5.5}$$

A special case of the beta distribution occurs when $\alpha = \beta = 1$. This is a *uniform* prior for which $f(\theta \mid 1, 1) = 1$ for $0 < \theta < 1$ from (5.2). This is taken to represent prior ignorance of the value of θ as the function takes a constant value, 1, over the range of possible values of θ.

Note from (5.3) that if α and β are small relative to n, x and $n-x$ then the choice of the parameters for the prior distribution is not very important. This is exemplified in Chapter 6 for the choice of sample size when sampling from consignments of discrete units such as tablets, compact discs or video tapes. An application to the detection of nandrolone metabolites in urine is described in Robinson *et al.* (2001).

An application to blood group frequencies is given in Weir (1996a), with reference to data from Gunel and Wearden (1995). The parameter, θ, of interest is the frequency of allele M in the MN blood group system. From previous

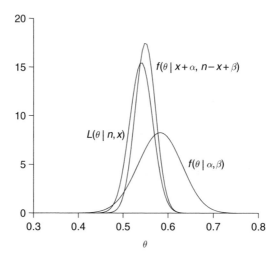

Figure 5.1 Prior density function $f(\theta \mid \alpha, \beta)$ with $\alpha = 61$, $\beta = 44$, likelihood function $L(\theta \mid n, x)$ with $n = 372$, $x = 201$, and posterior density function $f(\theta \mid x + \alpha, n - x + \beta)$ for a Bernoulli parameter. (Reproduced from Weir, 1996a, by permission of Sinauer Associates, Inc.)

samples, Gunel and Wearden (1995) have a prior $Be(61, 44)$ distribution ($\alpha = 61, \beta = 44$) for θ. The frequency $x = 201$ of the M allele in a sample of size $n = 372$ is taken as the data for the likelihood, from Race *et al.* (1949). Assuming Hardy–Weinberg equilibrium, the posterior distribution $f(\theta \mid 372, 201, 61, 44)$ is $Be(61 + 201, 44 + 171) = Be(262, 215)$. The prior and posterior density functions and the likelihood function are plotted in Figure 5.1. The posterior density function $f(\theta \mid 262, 215)$ is more narrow, less dispersed, than the prior density function $f(\theta \mid 61, 44)$, showing that the posterior density function has more precise information about θ. This is to be expected since the likelihood function $L(\theta \mid 372, 201)$ contains information about θ.

Inference about θ, given the prior distribution and the data (n, x), may be made with reference to the posterior beta distribution (5.3). Thus, for example, the shortest 95% probability interval may be determined by determining the shortest interval in $(0, 1)$ within which 95% of the distribution lies.

5.3   ESTIMATION WITH ZERO OCCURRENCES IN A SAMPLE

Consider the following scenario:

The scientist found a match in several physical properties between glass from a broken crime scene window and glass fragments found in connection with a suspect. He then surveyed 64 glass objects and found that none of these other 64 samples agreed with the glass found in his case. (Stoney, 1992, p. 383)

There is no match with the case sample in a sample of 64 objects. Estimation of the probability of a match by the sample relative frequency alone would give a value of zero and a likelihood ratio of infinity. However, the sample is only of size 64. Intuitively, this is insufficient to say that the glass in the crime scene window was unique.

An upper bound on the probability for the proportion of positive outcomes (e.g., characteristics of glass, illicit tablets, pirated CDs) in a population when there has been no observation of a positive outcome in a sample of n members of the population may be obtained using a Bayesian analysis with a beta prior and a binomial likelihood. Let θ denote the true, but unknown, proportion of positive outcomes in the population from which the sample has been taken. The prior distribution for θ is taken to be a beta distribution with parameters $\alpha = \beta = 1$ corresponding to a uniform prior. The likelihood is a binomial likelihood with n trials and the probability of a positive outcome is θ. In this example, there are no positive outcomes. The posterior probability distribution for θ is then a beta distribution with parameters $x = 0$, n and $\alpha = \beta = 1$ in (5.5). The probability density function is

$$f(\theta \mid 0+1, n-0+1) = \theta^0 (1-\theta)^n / B(1, n+1)$$
$$= (n+1)(1-\theta)^n.$$

The probability that $\theta > \theta_0$ is

$$Pr(\theta > \theta_0) = 1 - Pr(\theta < \theta_0)$$
$$= 1 - (n+1) \int_0^{\theta_0} (1-\theta)^n d\theta$$
$$= (1-\theta_0)^n.$$

For Stoney's (1992) example, $n = 64$ and $f(\theta \mid 1, 65) = 65(1-\theta)^{64}$.

A probabilistic upper bound θ_0 for θ may then be determined by choosing a probability level ϵ, say, and solving the equation

$$\epsilon = Pr(\theta > \theta_0)$$
$$= 1 - 65 \int_0^{\theta_0} (1-\theta)^{64} d\theta$$
$$= (1-\theta_0)^{65}.$$

The solution is

$$\theta_0 = 1 - \epsilon^{1/65}.$$

Table 5.1 Probabilistic upper bounds θ_0 for the proportion θ of occurrences of an event in a population when there have been no occurrences in a sample of size n, for given values of $\epsilon = Pr(\theta > \theta_0)$, with a beta prior for which $\alpha = \beta = 1$

ϵ	θ_0
0.05	0.0450
0.01	0.0684

Note that the solution, not surprisingly, depends on the sample size in the current survey. Solutions for two values of ϵ are given in Table 5.1. Thus, in the context of Stoney's (1992) example, it can be said that:

- there is a probability of 0.95 that the true proportion of matches in the population from which the crime scene window, the suspect's fragments and the 64 surveyed objects were taken is less than 0.0450;

- there is a probability of 0.99 that the true proportion of matches in the population from which the crime scene window, the suspect's fragments and the 64 surveyed objects were taken is less than 0.0684.

Zero occurrences of an event are often observed with sequence analyses of human mitochondrial DNA (mtDNA); mtDNA is widely used to characterise forensic biological specimens, particularly when there is insufficient nuclear DNA in samples for typing.

The sequences of a mtDNA region from a recovered and control sample are compared. If the sequences are unequivocally different, then the sample can be excluded as originating from the same source. If the sequences are the same, then the recovered and control samples cannot be excluded as potentially being from the same source (note that no problems of nucleotide differences, mutations or heteroplasmy are considered in what follows).

When there is no difference between the two samples, it is desirable to convey some information about the weight of the evidence. Presently, the practice is to count the number of times a particular sequence (or haplotype) is observed in a relevant database and apply a correction for sampling error (such as a confidence interval, a bootstrap value or a Balding and Nichols correction; see, for example, Balding and Nichols, 1994, Curran *et al.*, 2002). A measure of sampling error is provided for these estimates, which is especially necessary when estimates are based on samples of just a few hundred profiles. This approach allows the scientist to communicate the value of the mtDNA evidence using the reciprocal of the relative frequency as a likelihood ratio (Carracedo *et al.*, 2000).

Analogously, the use of Y-chromosome sequence-tagged site polymorphisms has become commonplace in forensic laboratories (Sinha, 2003). Their applications include paternity cases which are deficient in information (e.g., where the father is not available for analysis and inferences are made by reference to relatives) and, especially, the discrimination of stains in forensic investigations when a male suspect is involved (e.g., a male–female mixture in sexual assault cases). The counting method where the number of observations may be declared in a relevant database plays an important role in the assessment of the evidence (Gill *et al.*, 2001). If there are no observations of a particular profile, either in the literature or the current survey, then a probabilistic upper bound on the profile may be obtained as described above for the glass example of Stoney (1992).

Other relevant references for the estimation of Bernoulli parameters include Crow *et al.* (1960), Louis (1981), Kaye (1987), Balding and Nichols (1994), Taroni and Mangin (1999) and Curran *et al.* (2002).

5.4 ESTIMATION OF PRODUCTS IN FORENSIC IDENTIFICATION

Another example of the use of Bayesian inference for discrete data and a continuous parameter is described by Balding (1995) for estimating a likelihood ratio from DNA profiles. This approach includes what has been called a *size-biased correction* (Curran *et al.*, 2002) in that the result represents the effect of adding the profile in question to the database exactly twice.

Consider two bands at a locus to be mutually independent. Let the two bands be denoted $X = (X_1, X_2)$ with frequencies $\theta = (\theta_1, \theta_2)$ where $\theta_1 + \theta_2 \leq 1$. The observation X may be considered a trinomial random variable (Section 2.3.4) with sample size n, where n is the total number of observations in the sample.

The derivation of the size-biased correction will be given for the genotype relative frequency $\phi = 2\theta_1\theta_2$. The corresponding result for homozygotes (with $\phi = \theta_1^2$) will be quoted at the end of the section. Further details are available in Balding (1995), Evett and Weir (1998) and Curran *et al.* (2002). The usual estimator for ϕ is $2X_1X_2/n^2$, but this takes no account of any measure of uncertainty in the estimation procedure.

Uncertainty may be accounted for through the use of a prior distribution for θ. The appropriate prior distribution for a trinomial distribution is a Dirichlet distribution (Section 2.4.5), with probability density function $f(\theta \mid \alpha_1, \alpha_2, \alpha_3)$ proportional to

$$\prod_{i=1}^{3} \theta_i^{\alpha_i - 1}$$

where $\theta_3 = 1 - \theta_1 - \theta_2$. The posterior expectation of ϕ^s, $s \geq 0$, is given by

$$E(\phi^s) = 2^s \frac{\Gamma(\alpha_. + n)}{\Gamma(\alpha_. + n + 2s)} \prod_{i=1}^{2} \frac{\Gamma(\alpha_i + x_i + s)}{\Gamma(\alpha_i + x_i)}$$

and

$$E(\phi^{s+1}) = 2E(\phi^s)\frac{\Gamma(\alpha_1 + x_1 + s)\Gamma(\alpha_2 + x_2 + s)}{\Gamma(\alpha_. + n + 2s)\Gamma(\alpha_. + n + 2s + 1)},$$

where $\alpha_. = \alpha_1 + \alpha_2 + \alpha_3$. For a uniform prior, $\alpha_1 = \alpha_2 = \alpha_3 = 1$. Assuming independence to justify the product rule, ϕ can be interpreted as the ratio of the likelihood of the data under the assumption of innocence (ϕ^2) to its likelihood under the assumption of guilt (ϕ). Thus, consideration of the ratio of $E(\phi^2)$ to $E(\phi)$, with $\alpha_i = 1(i = 1, 2, 3)$, gives

$$\frac{E(\phi^2)}{E(\phi)} = 2\frac{(x_1 + 2)(x_2 + 2)}{(2n + 5)(2n + 6)}.$$

For the case when $\phi = \theta_1^2$,

$$\frac{E(\phi^2)}{E(\phi)} = \frac{(x_1 + 3)(x_1 + 4)}{(2n + 5)(2n + 6)}$$

(Balding, 1995).

5.5   BAYESIAN INFERENCE FOR A NORMAL MEAN

Consider the mean θ of a Normal distribution. For example, θ may be the true level of alcohol in the blood of someone suspected of driving while under the influence of alcohol. Within the Bayesian paradigm, θ has a probability distribution and it is desired to determine this in order to make inferences about θ. The most common distribution for θ, the mean of a Normal distribution, is itself a Normal distribution. The Normal distribution is a conjugate prior distribution for the Normal distribution. In the situation where the variances of the two Normal distributions are known, they combine together to give a posterior distribution which is also a Normal distribution.

The prior distribution for θ, parameterised by ν, the mean, and τ^2, the variance, is represented by

$$(\theta \mid \nu, \tau^2) \sim N(\nu, \tau^2)$$

(from Section 2.4.2). Consider a random variable X which has a Normal distribution with mean θ and variance σ^2 (assumed known), so that

$$(X \mid \theta, \sigma^2) \sim N(\theta, \sigma^2).$$

Then it can be shown (e.g., Lee, 2004) that the posterior distribution of θ, given a value x for X, and given σ^2, ν and τ^2, is

$$(\theta \mid x, \sigma^2, \nu, \tau^2) \sim N(\theta_1, \tau_1^2), \tag{5.6}$$

where

$$\theta_1 = \frac{\sigma^2 \nu + \tau^2 x}{\sigma^2 + \tau^2} \tag{5.7}$$

and

$$(\tau_1^2)^{-1} = (\sigma^2)^{-1} + (\tau^2)^{-1} \tag{5.8}$$

or, equivalently,

$$\tau_1^2 = \frac{\sigma^2 \tau^2}{\sigma^2 + \tau^2}.$$

The posterior mean, θ_1, is a weighted average of the prior mean ν and the observation x, where the weights are the variance of the observation x and the variance of the prior mean ν, respectively, such that the component (observation or prior mean) which has the smaller variance has the greater contribution to the posterior mean.

The reciprocal of the variance is known as the *precision*. Thus the precision of the posterior distribution of μ is the sum of the precisions of the prior distribution and the observation.

This result can be generalised to consider the distribution of the mean θ of a set of n independent, identically Normally distributed observations $x_1, \ldots, x_n$ with mean θ and variance σ^2. The generalisation follows from the result that $\bar{X}$, the random variable corresponding to the sample mean $\bar{x}$ of a sample of size n, has a distribution

$$(\bar{X} \mid \theta, \sigma^2) \sim N(\theta, \sigma^2/n).$$

The posterior distribution of θ is Normally distributed with mean

$$\theta_1 = \frac{\dfrac{\sigma^2}{n} \nu + \tau^2 \bar{x}}{\dfrac{\sigma^2}{n} + \tau^2} \tag{5.9}$$

and precision

$$(\tau_1^2)^{-1} = n(\sigma^2)^{-1} + (\tau^2)^{-1}. \tag{5.10}$$

Consider again the example in Section 2.4.2 of a person whose blood alcohol level has been measured as $0.85\,\text{g/kg}$. It is of obvious interest to determine the probability that the true level θ of blood alcohol is greater than $0.8\,\text{g/kg}$. From (5.6), a posterior distribution for θ may be obtained and the posterior probability that θ is greater than $0.8\,\text{g/kg}$ may be obtained using the methods described in

Section 2.4.2. Simple application of these methods requires the prior distribution for θ, including values of ν and τ^2. The choice of these values may well be subjective and would have to be made carefully given the legal context in which the measurement is being made. As illustrated at the beginning of the chapter, there is considerable debate about the role of prior distributions in the law and forensic science; further comments are made in Chapter 6. It may be that one day databases of the distributions of the measurements of blood alcohol levels in various groups of individuals, of different sexes, weights, health, etc., will be constructed and be accepted in courts of law, but this is a long way in the future at present. Of course, prior information here can include the evidence of the investigating officer who may form an opinion about the blood alcohol level of the suspect before taking the measurement.

One approach which has been suggested to overcome the subjectivity of the choice of prior is the use of a uniform prior, with the distribution taken to be constant over the range of the variable of interest. This idea was introduced in Section 5.2 in the context of the binomial distribution where the range of interest was $(0, 1)$. In the context of the mean of a Normal distribution the range of interest is from $-\infty$ to $+\infty$. For a Normal distribution it is not possible to take a constant value over the range of interest and retain the properties of a probability distribution, as the probability density function will not integrate to 1. The prior distribution is then known as an *improper* or *vague* prior distribution. However, such a choice of prior may be acceptable if it combines with a likelihood to give a proper posterior distribution. In this example, the prior is uniform as it takes a constant value over the whole real line.

This is so for the Normal distribution. The uniform prior is defined as the limiting distribution in which $\nu \to 0$ and $\tau^2 \to \infty$. Inspection of (5.7) and (5.8) shows that the limiting values for the posterior mean and variance are simply $\theta_1 = x$ and $\tau^2 = \sigma^2$; that is, for a uniform prior, the posterior distribution of θ is

$$\theta \sim N(x, \sigma^2). \tag{5.11}$$

A similar result is used in Section 10.2.3 to illustrate an approximate value of evidence from measurements on the refractive index of glass. In the blood alcohol example, given a measurement $x = 0.85$, a known variance for measurements from this procedure of 0.005, and a uniform (improper) prior, the distribution of the true blood alcohol level, θ, is $N(0.85, 0.005)$. The standard deviation is $\sigma = 0.0707$. The probability that the true blood alcohol level is greater than 0.8 is then

$$Pr\left(\theta > 0.8 \mid \theta \sim N(0.85, 0.005)\right)$$
$$= 1 - Pr\left(\theta < 0.8 \mid \theta \sim N(0.85, 0.005)\right)$$
$$= 1 - \Phi\left(\frac{0.80 - 0.85}{0.0707}\right)$$

$$= 1 - \Phi(-0.7071)$$

$$= \Phi(0.7071)$$

$$= 0.76.$$

This result rather invites the question as to whether this reading of 0.85, combined with a uniform prior, is sufficient to find the suspect guilty beyond reasonable doubt of having a blood alcohol content greater than 0.8 g/kg. The assumption of a uniform prior is very supportive of the defence and implies the procedures are very imprecise. In practice, it is expected that the variability of the procedure will be very small and the posterior estimate of τ will be small. This will lead to a higher value for the probability that the true blood alcohol level is greater than 0.8 g/kg. Apart from some additional discussion in Chapter 6, further detailed discussion of the role of prior distributions is beyond the scope of this book. However, consideration of them draws attention to the factors which have to be considered when assessing the role of probability in the law.

So far, the variance of the distribution of the observations X has been assumed known. If this is not the case then the posterior distribution for θ is related to the t-distribution (Section 2.4.3). Suppose there are n observations of X. The sample mean is $\bar{x}$ and the sample variance is

$$s^2 = \sum_{i=1}^{n}(x_i - \bar{x})^2/(n-1).$$

Uniform priors can be taken for θ and $\log(\tau^2)$ (the logarithmic transformation in the latter case ensures a range from $-\infty$ to $+\infty$). The probability density function of

$$t = \frac{\theta - \bar{x}}{s/\sqrt{n}}$$

is a t-density with $n-1$ degrees of freedom. Note that $\bar{x}$ and s are taken to be fixed, as functions of the observations. The unknown quantity is θ and probabilistic inferences about θ can be made with reference to the t-distribution.

As an illustration of this approach, consider a prior

$$\theta \sim N(1.30, 0.0004)$$

with data from $n = 31$ observations such that

$$\bar{x} = 1.35, \quad s^2 = 0.0003.$$

Then

$$t = \frac{\theta - \bar{x}}{s/\sqrt{n}} = \frac{\theta - 1.35}{\sqrt{0.0003/31}} \qquad (5.12)$$

has a t-distribution with 30 degrees of freedom. The 95% lower probability bound for θ is obtained from the 95% lower bound for a t-distribution with 30 degrees of freedom. This is -1.697. Then, solution of

$$\frac{\theta - 1.35}{\sqrt{0.0003/31}} = -1.697$$

gives $\theta = 1.3447$. There is a probability of 0.95 that the true value of θ is greater than 1.345.

5.6   INTERVAL ESTIMATION

Once it is accepted that uncertainty about a parameter may be represented by a probability distribution, then it is a straightforward matter to determine a probability interval for the parameter. Various approaches to the estimation of intervals within which the true value of a parameter may lie are described. These include confidence intervals, with which many people will be familiar, and other less familiar intervals such as highest posterior density intervals, bootstrap intervals and likelihood intervals.

5.6.1   Confidence intervals

A confidence interval is defined by data (e.g., mean plus or minus a multiple of a standard deviation) and is fixed by the data. There is no probability distribution associated with it and thus no probability can be attached to it. For a given sample size, the width of the confidence interval increases with the level of confidence. Thus, a 95% confidence interval is wider than a 90% confidence interval but narrower than a 99% confidence interval. A common application is in the estimation of the mean μ of a Normal distribution of unknown variance σ^2. A random sample $\{x_1, \ldots, x_n\}$ of size n is taken from the distribution. The sample mean, $\bar{x}$, and sample standard deviation, s, are calculated. As the standard deviation σ is unknown it is estimated by the sample standard deviation s and a t-distribution is the appropriate one for use in determining the confidence interval. Denote the $100(1 - \alpha/2)\%$ point of the t-distribution with $n - 1$ degrees of freedom by $t_{(n-1)}(\alpha/2)$ as in Section 2.4.3. Then the $100(1 - \alpha)\%$ confidence interval for μ is

$$\left(\bar{x} - t_{(n-1)}(\alpha/2)\frac{s}{\sqrt{n}}, \bar{x} + t_{(n-1)}(\alpha/2)\frac{s}{\sqrt{n}} \right). \tag{5.13}$$

It is also possible to consider just one end of an interval, a so-called confidence limit. For example, consider the problem of estimating the quantity of drugs in a consignment from the measurement of the quantity of drugs in a sample from the consignment. A lower limit on the quantity of drugs is desired as it enables the courts to determine a limit at the appropriate level of proof above which the true quantity lies. For example, Frank *et al.* (1991) suggest a statement of the form 'at the 95% confidence level, 90% or more of the packages in an exhibit contain the substance' as being sufficient proof in cases of drug handling that 90% or more of the packages contain the substance. The nature of the construction of the confidence limit is that, in 95% of cases studied for which these results are obtained, 90% of the packages will contain the substance. However, the probability with which a particular interval in a particular case contains the true proportion is not known.

A confidence interval derives its validity as a method of inference on a long-run frequency interpretation of probability. For example, consider specifically the 95% confidence interval for a proportion. The probability with which a particular 95% confidence interval contains the true proportion is not known. However, suppose the experiment which generated the 95% confidence interval is repeated many times (under identical conditions, a theoretical stipulation which is impossible to fulfil in practice) and on each of these occasions a 95% confidence interval for the true proportion is calculated. Then, it can be said that 95% of these (95%) confidence intervals will contain the true proportion. This does not provide information concerning the one 95% confidence interval which has been calculated. It is not known whether it does or does not contain the true proportion, and it is not even possible to determine the probability with which it contains the true proportion (Kaye, 1987).

It is also possible to obtain confidence intervals for the true proportion in a binomial model. Consider a 95% confidence interval and the binomial model. The 97.5% point of the Normal distribution is 1.96, thus one can be 95% confident that the true proportion lies within 1.96 standard deviations of the sample proportion. Denote the sample proportion x/n by $\hat{p}$. The sample standard deviation is then estimated by $[\hat{p}(1-\hat{p})]^{1/2}/n$ (replacing θ by $\hat{p}$). The 95% confidence interval for the true proportion θ is then

$$\hat{p} \pm 1.96[\hat{p}(1-\hat{p})]^{1/2}/n. \tag{5.14}$$

An example of the use of the Normal approximation in the estimation of a binomial proportion is given by McDermott *et al.* (1999). They chose to survey at least 1000 vehicles in keeping with the estimate of the confidence interval by Ryland *et al.* (1981) for this sample number. If a specified colour has a true probability of occurrence θ then an approximate 95% confidence interval estimate for θ is given by (5.14), where $\hat{p}$ is the sample estimate of θ and n is the sample size.

Let $y_1, \ldots, y_n$ be n observations from a Poisson distribution with mean λ. Then, from (2.20) the 95% confidence interval for the true mean λ is

$$\bar{y} \pm 1.96(\bar{y}/n)^{1/2}, \tag{5.15}$$

replacing λ by $\bar{y}/n$ as the estimate of the standard deviation.

A confidence interval from (5.13) and the corresponding probability interval derived from expressions similar to (5.12) are formally identical. However, the philosophies underlying their respective constructions are very different.

5.6.2 Highest posterior density intervals

This approach provides a probability statement about the true proportion. This is in contrast to the inference obtainable from an approach which provides a confidence interval for the true proportion. It is no accident that the word *probability* is not used to describe a confidence interval. The term 'highest posterior density' is used to reflect the fact that the interval chosen for a particular probability, say 0.95, is the one for which the posterior probability density function takes its highest values whilst ensuring that the total probability within the interval is 0.95.

Consider (5.7) and (5.8). These are the posterior mean and variance for a posterior Normal distribution. Because of the symmetry of the Normal distribution, the highest posterior density interval for a given probability value is symmetric about the posterior mean. Thus, the 95% highest posterior density interval is

$$(\theta_1 - 1.96\tau_1, \ \theta_1 + 1.96\tau_1).$$

The use of the method is illustrated in the context of the estimation of the quantity, θ, of alcohol in blood under certain circumstances. For these, the prior distribution of θ is taken to be $N(1.30, 0.0004)$. This distribution can be determined from past experiments, from the experience of the expert or from the literature. A sample is analysed and a measurement X obtained. The distribution of $(X \mid \theta)$ is taken to be $N(1.35, 0.0003)$. The posterior distribution for $(\theta \mid X)$, from (5.7) and (5.8), is $N(1.34, 0.0003)$ and a 95% highest posterior density interval (by symmetry) is

$$(1.34 \pm 1.96 \times \sqrt{0.0003}) = 1.34 \pm 0.034 = (1.306, 1.374).$$

5.6.3 Bootstrap intervals

Sometimes (Dujourdy *et al.*, 2003) it is of interest to determine an interval estimate for the likelihood ratio. A comparison of likelihood ratios from different samples from the same consignment of drugs, for example, may produce different values. The distribution of the likelihood ratio, V, may be difficult to determine and direct construction of a probability interval is also difficult. An alternative approach is to consider resampling techniques (Davison and Hinkley, 1997) such as the so-called *bootstrap*. In a bootstrap operation, new samples, usually of the same size as the original data set, are drawn with replacement from the observed data. This sampling procedure is repeated many times, N say. Dujourdy *et al.* (2003) use 1000 repeats, though N can be considerably larger. The statistic of interest, in this case V, is calculated for each new set of data. Thus, N observations of the likelihood ratio are observed. The histogram of the N values of V provides an estimate of the distribution of V. From this distribution, measures of location (e.g., mean and median) and dispersion (e.g., standard deviation) and probability intervals may be estimated. For example, a 95% probability interval for V may be obtained by observing the 2.5% and 97.5% sample quantiles from the bootstrap sample. Thus, for $N = 1000$, the 2.5% quantile is the value below which 25 of the observations of V lie; denote this value as $V_{0.025}$. The 97.5% quantile is the value below which 975 of the observations of V lie or above which 25 of the observations lie; denote this value as $V_{0.975}$. The 95% bootstrap probability interval is then $(V_{0.025}, V_{0.975})$.

5.6.4   Likelihood intervals

Consider a large population of unrelated individuals in which it is desired to determine the proportion γ of people of blood group Γ. A sample of size n is taken. The sample is sufficiently small with respect to the size of the population that the proportion γ is effectively unchanged by the removal of the sample. The number X of people of blood group Γ in a sample of size n is a random variable with a binomial distribution (Section 2.3.3) such that

$$Pr(X = x \mid n, \gamma) = \binom{n}{x} \gamma^x (1 - \gamma)^{n-x}. \tag{5.16}$$

However, γ is unknown. It is estimated by determining the value of $\gamma, \hat{\gamma}$ say, for which (5.16) is maximised. Suppose that the number of people of group Γ in this particular sample is x_0. It can be shown that the value of γ which maximises (5.16) is the intuitively desirable $\hat{\gamma} = x_0/n$, the sample proportion.

Another approach may be considered. A sample is taken until it contains x_0 people with blood group Γ. It is then noticed that the total sample size is n. Then it can be shown that

$$Pr(X = x_0 \mid n, \gamma) = \binom{n-1}{x_0 - 1} \gamma^{x_0} (1 - \gamma)^{n-x_0}. \tag{5.17}$$

$((x_0 - 1)$ of the first $(n - 1)$ are γ; the nth is also γ.) Again, the value of γ for which this probability is maximised is $\hat{\gamma} = x_0/n$.

Thus, the estimate of γ does not depend on the sampling scheme used. The part of the expressions (5.16) and (5.17) which contains γ is the same in both (5.16) and (5.17). These are likelihood functions for γ, to which reference has been made in Section 5.1. Both (5.16) and (5.17) are proportional to $\gamma^x(1-\gamma)^{(n-x)}$, which can be written as $L(\gamma \mid n,x)$. The value $\hat{\gamma}$ is known as the *maximum likelihood estimate* of γ; $\hat{\gamma}$ is the value of γ for which the likelihood function is maximised. The exact value of $L(\gamma \mid n,x)$ is not itself of interest and can be standardised so that its maximum value is 1. An example of a likelihood function for $x = 6, n = 30$ (with $x/n = 0.2$) is shown in Figure 5.2, standardised so that $L(0.2 \mid 30,6) = 1$.

A frequentist approach to determining an interval estimate for γ could use a Normal approximation to the binomial distribution (Section 2.4.2), and the 95% confidence interval would then be

$$\frac{x}{n} \pm 1.96\sqrt{\frac{x(n-x)}{n^3}}, \tag{5.18}$$

but it would be subject to the usual criticisms levied against confidence intervals.

The likelihood approach considers as an interval estimate of γ, all values of γ for which $L(\gamma)$ is greater than some fraction of $L(\hat{\gamma})$. Examples quoted in Royall (1997) are $1/8$ and $1/32$; the interval estimate of γ would correspond to those values of γ for which $L(\gamma) > L(\hat{\gamma})/8$ and for which $L(\gamma) > L(\hat{\gamma})/32$, respectively. The values 8 and 32 are suggested by Royall (1997) as values to define interval estimates of parameters such that for values of γ lying outside the intervals the data provide 'fairly strong' or 'strong' evidence in support of $\hat{\gamma}$ over γ.

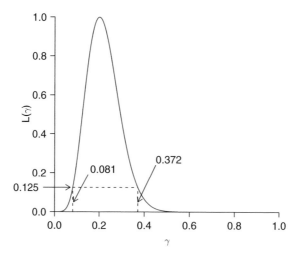

Figure 5.2 Standardised likelihood function for the proportion of people of blood group Γ, from a sample of size 30 in which 6 were of group Γ. A likelihood interval, derived from the observed data, of fairly strong support for the values of γ included within it is indicated with the dotted lines: (0.081, 0.372).

The use of such adjectives for the support for $\hat{\gamma}$ is not to be confused with similar adjectives given in Table 3.10 for likelihood ratio values. Consider the value 8, and values of γ for which $L(\gamma) > L(\hat{\gamma})/8$; these lie in an interval (γ_1, γ_2) (see Figure 5.2). If γ lies in the interval (γ_1, γ_2) then there is no alternative value of γ, $\hat{\gamma}$ say, for which the observations (x, n) (such that $\hat{\gamma} = x/n$) represent 'fairly strong' evidence in favour of $\hat{\gamma}$ rather than γ. For a value of γ outside (γ_1, γ_2) there is at least one alternative value, namely $\hat{\gamma}$, that is better supported by a factor greater than 8. The end-points of the horizontal line in Figure 5.2 are (γ_1, γ_2) for the example $x = 6, n = 30$. For this example, the maximum value of $L(\gamma)$ is

$$L(0.2) = \binom{30}{6} 0.2^6 0.8^{24} = 0.179.$$

The end-points of the interval for 'fairly strong' evidence are the values (γ_1, γ_2) for which $L(\gamma_1) = L(\gamma_2) = L(0.2)/8 = 0.022$. These points may be verified to be $\gamma_1 = 0.081$ and $\gamma_2 = 0.372$. There is, thus, 'fairly strong' evidence from a sample of size 30 in which there are 6 'successes' that the true proportion of successes in the population from which the binomial sample was drawn is in the interval $(0.081, 0.372)$.

The corresponding 95% interval is, from (5.18),

$$\frac{6}{30} \pm 1.96\sqrt{\frac{6(30-6)}{30^3}} = 0.2 \pm 1.96\sqrt{\frac{0.2 \times 0.8}{30}}$$

$$= 0.2 \pm 0.143$$

$$= (0.057, 0.343).$$

An interval determined by 32 (instead of 8) has a similar interpretation but with 'strong' replacing 'fairly strong'.

5.7 ODDS RATIOS

The use of odds ratios to compare populations to aid in the determination of the source of a particular sample (Royall, 1997; Taroni and Aitken, 1999a) is illustrated with an example concerning the frequency of contamination of bank-notes.

The numbers of banknotes contaminated with cocaine for two groups are shown in Table 5.2. The first group is banknotes seized during drug trafficking investigations, and the second group is notes in general circulation.

Let X ($= 382$) and Y ($= 562$) denote the numbers of notes contaminated with cocaine from drug dealing and from general circulation respectively, and m ($= 462$) and n ($= 992$) denote the total numbers of notes inspected from drug trafficking investigations and from general circulation, respectively.

Table 5.2 Numbers of banknotes contaminated

	Number contaminated	Number not contaminated	Total
Notes seized in drug dealing	382	80	462
Notes in general circulation	562	430	992

Let p_x and p_y denote the probability that a note is contaminated with cocaine from the drug dealing and from general circulation respectively. From Table 5.2, p_x is estimated by $\hat{p}_x = 382/462 = 0.83$ and p_y is estimated by $\hat{p}_y = 562/992 = 0.57$.

Assume that the two groups of banknotes are independent. Then X and Y have binomial distributions (Section 2.3.3) such that $X \sim Bin(m, p_x)$ and $Y \sim Bin(n, p_y)$. The likelihood function for p_x and p_y combined is given by

$$L(p_x, p_y) \propto p_x^x (1 - p_x)^{m-x} \times p_y^y (1 - p_y)^{n-y}.$$

Substitution of the observed values for m, n, x and y gives

$$L(p_x, p_y) \propto p_x^{382} (1 - p_x)^{462-382} \times p_y^{562} (1 - p_y)^{992-562}.$$

The expression $p_x/(1 - p_x)$ is the odds (Section 3.1) in favour of a banknote from the seized group being contaminated with cocaine. The expression $p_y/(1 - p_y)$ is the odds in favour of a banknote from general circulation being contaminated with cocaine. The expression

$$\theta = \frac{p_x(1 - p_y)}{p_y(1 - p_x)} \tag{5.19}$$

formed from the ratio of these odds is known as the *odds ratio*. Note that the odds ratio is not defined for p_x and p_y equal to 0 or 1. The odds ratio is always positive and takes values between 0 and infinity. The value θ of an odds ratio in this context may be interpreted as meaning that the odds in favour of a banknote from the seized group being contaminated with cocaine are θ times greater than the odds in favour of a banknote from the general circulation being contaminated with cocaine.

The maximum values of the binomial probabilities are attained when $p_x = 0.83$ for seized notes and when $p_y = 0.57$ for notes in general circulation. Given these values for p_x and p_y, it is possible to determine the odds ratio θ as $\theta = p_x(1 - p_y)/p_y(1 - p_x) = 0.83(1 - 0.57)/0.57(1 - 0.83) = 3.7$.

In this example, the odds ratio is 3.7: the odds in favour of contamination are 3.7 times greater for seized banknotes than for notes in general circulation.

A likelihood function for the odds ratio is given by the conditional distribution of $(X \mid X + Y = x + y)$, which is

$$L(\theta) \propto Pr(X = x \mid X + Y = x + y, p_x, p_y)$$

$$\propto \left\{ \sum_j \binom{m}{j} \binom{n}{x+y-j} \theta^{j-x} \right\}^{-1} \tag{5.20}$$

$$\propto \left\{ \sum_0^{462} \binom{462}{j} \binom{992}{944-j} \theta^{j-x} \right\}^{-1} ;$$

the lower limit of summation is $\max(0, x + y - n)$ and the upper limit is $\min(m, x + y)$ (Royall, 1997; Taroni and Aitken, 1999a).

The maximum value of the odds ratio may be standardised to 1. The neutral value of the (unstandardised) odds ratio is 1; this corresponds to an absence of association between the proportion of notes contaminated and the source (general circulation or seizure) of the notes. If there is an association between the source and the proportion contaminated then the odds ratio is different from 1.

The observed probabilities $p_x = 382/462$ and $p_y = 562/992$ give an odds ratio of 3.7 and a likelihood ratio of $L(3.7)/L(1)$ of 7×10^{21}, where $L(3.7)$ and $L(1)$ are obtained from (5.20). This value provides extremely strong evidence of an association between the contamination of the banknotes and the source of the banknotes.

The likelihood approach described here for the comparison of two groups by means of the proportion of notes with cocaine on them avoids the problems of interpretation that are associated with confidence intervals. It measures the size of the difference between the two populations. It shows whether the proportion of contamination in one population is different from the proportion of contamination in another population. It does not provide a measure of the value of the evidence in a particular case. The example discussed here concerned banknotes gathered over several different cases. Naturally, there will be debate over how relevant such collections of banknotes are; it is an important debate and one worthy of resolution.

The value of the evidence in a particular case can be determined by considering the ratio of two binomial probabilities. Some banknotes are seized. The number seized equals n and, after inspection, z are found to be contaminated with cocaine. These numbers are the evidence E. Consider two propositions:

- H_p, the banknotes have been involved in drug dealing;
- H_d, the banknotes are part of the general circulation.

If H_p is true then the probability an individual banknote is contaminated with cocaine is estimated to be 0.83, from the data in Table 5.2; similarly, if H_d is true then the probability an individual banknote is contaminated with cocaine

Table 5.3   The value of evidence in comparing two binomial proportions for different values of sample size n and number of 'successes' z, where the samples are banknotes and a 'success' is a contaminated banknote. The binomial proportion in the numerator is 0.83, estimated from banknotes known to be used for drug dealing, and the binomial proportion in the denominator is 0.57, estimated from banknotes known to be from general circulation

Sample size n	Number contaminated z	Proportion z/n	Evidential value V
100	66	0.66	1/854
100	72	0.72	2.9
100	76	0.76	538
50	33	0.66	1/29
50	36	0.72	1.7
50	38	0.76	23

is estimated to be 0.57(to two decimal places). The value V of the evidence E in favour of the proposition that the banknotes have been involved in drug dealing is then $Pr(E \mid H_p)/Pr(E \mid H_d)$, which is

$$V = \frac{\binom{n}{z} 0.83^z(1-0.83)^{(n-z)}}{\binom{n}{z} 0.57^z(1-0.57)^{(n-z)}} = \left(\frac{0.83}{0.57}\right)^z \left(\frac{0.17}{0.43}\right)^{(n-z)}.$$

The use of the binomial distribution may be questioned here. One of the modelling assumptions for a binomial distribution is that all members of the sample have a constant probability of 'success', independent of other members of the sample. Models which allow for dependence amongst members of the sample are beyond the scope of this book.

The value of the evidence for certain values of n and z is given in Table 5.3. The evidence ($n = 100, z = 66$) and ($n = 50, z = 33$) is such that it supports H_d. As the ratio z/n increases for fixed n so the strength of the evidence in support of H_p increases. For the same z/n, the larger z and n the stronger the support in favour of H_p or H_d.

6

Sampling

6.1 INTRODUCTION

A general introduction to sampling techniques is given in Cochran (1977). Ideas on sample size determination are discussed in Smeeton and Adcock (1997). A very good discussion of various types of samples, such as random samples, representative samples and convenience samples, is given in Evett and Weir (1998). A review of statistical and legal aspects of the forensic study of illicit drugs is given by Izenman (2001). This includes a discussion of various sampling procedures, various methods of choosing the sample size, a strategy for assessing homogeneity and the relationship between quantity and the possible standards of proof. Further comments on sampling issues are given in Bring and Aitken (1997), in various chapters of Gastwirth (2000), such as Aitken (2000), Gastwirth *et al.* (2000) and Izenman (2000a, b, c), and in Izenman (2003). A relevant case is that of *U.S. v. Shonubi.*

Only inferences from simple random samples are discussed here. It may be that it is not possible to take a simple random sample. If so, the following comments are of relevance. The comments are made in the context of sampling for the estimation of allelic frequencies in DNA profiles but are applicable to other areas of forensic science, including drug sampling, which is the main example in this chapter.

> Of course, a real crime laboratory would not attempt . . . to take a random, representative, stratified sample of individuals to address the question of issue. In the vast majority of cases the laboratory will have one or more *convenience* samples. Such a sample may be of laboratory staff members, or from blood donor samples with the cooperation of a local blood bank, or from samples from victims and suspects examined in the course of casework.
>
> . . . [In] the forensic context, we will generally be dealing, not with random but, with convenience samples. Does this matter? The first response to that question is that every case must be treated according to the circumstances within which it has occurred, and the next response is that it is always a matter of judgement. . . . In the last analysis, the scientist must also convince a court of the reasonableness of his or her inference within the circumstances as they are presented as evidence. (Evett and Weir, 1998, pp. 44–45)

Statistics and the Evaluation of Evidence for Forensic Scientists: Second Edition
C.G.G. Aitken and F. Taroni © 2004 John Wiley & Sons, Ltd   ISBN: 0-470-84367-5

Several sampling procedures, including random sampling, are discussed in Izenman (2001). First, for single containers, examination by a chemist of a random sample of a substance seized within a single bag or container has been accepted by the courts to prove the identity of the remainder of the substance in the container. For multiple containers, without homogeneity, a rule is that at least one sample from each container should be conclusively tested for the presence of an illicit drug. Another procedure is that of composite sampling. In this procedure, a sample is taken from each source, the samples are then thoroughly mixed and a subsample is taken from the mixture. The mixture is the composite sample.

This chapter is concerned with the choice of sample size and the interpretation of data from samples (in particular, for quantity estimation). Two questions in particular are addressed and answered:

- What proportion of a consignment of discrete, homogeneous items is illicit?

- Given a sample from a consignment of homogeneous material, what is the quantity of illicit material?

Both questions have probabilistic answers, determined using a Bayesian framework. There will be considerable reference to Chapter 2 for definitions of probability distributions and to Chapter 5 for philosophical considerations.

The main results of the chapter are phrased in the context of inspecting a consignment of drugs. However, the ideas expressed are just as applicable to other forensic contexts, such as the inspection of computer disks for pornographic images. In such a situation sampling may be beneficial as it exposes the investigators to as little stress as possible.

Consider a population or consignment which consists of discrete units, such as individual tablets in a consignment of tablets or individual computer disks in a consignment of disks. Each unit may, or may not, contain something illegal, such as illicit drugs or pornographic images. It is of interest to an investigating scientist to determine the proportion of the consignment which contains something illegal. This may be done exactly (assuming no mistakes are made) by examination of every unit in the consignment. Such an examination can be extremely costly. Considerable resources can be saved if information, sufficient to satisfy the needs of the investigators, may be gained from examination of a sample from the consignment. Uncertainty is introduced when inference is made from the sample to the population, because the whole population is not inspected. However, this uncertainty may be quantified probabilistically.

With reasonable assumptions, a probability distribution for the true proportion of units in the consignment may be derived, based on the scientist's prior beliefs (i.e., prior to the inspection of individual units) and the outcome of the inspection of the sample. The strength of the scientist's prior beliefs may be expressed by a probability density function as described in Chapter 5. It is possible to choose the function in such a way that the effect of the scientist's prior beliefs is very small (or very large). The choice of the model which is used to represent the uncertainty introduced by the sampling process is a subjective choice influenced by the scientist's prior beliefs. The choice of the binomial model used here requires assumptions about independence of the probability for each unit being illegal and the choice of a constant value for this probability.

A comparison is made here of the results obtained from the Bayesian and frequentist approaches to assessing uncertainty, to contrast the clarity of the inferences obtainable from the Bayesian approach with the lack of clarity associated with the frequentist approach and to illustrate the greater flexibility of the Bayesian approach with the inflexibility of the frequentist approach.

The methods are illustrated with reference to sampling from consignments of drugs. However, they apply equally well to sampling in other forensic contexts, for example, glass (Curran *et al.*, 1998b) and pornographic images.

Frequentist procedures are described in Tzidony and Ravreboy (1992) for choosing a sample size from a consignment. A distinction is drawn between an approach based on the binomial distribution (Section 2.3.3) and one based on the hypergeometric distribution (Section 2.3.5). It is argued in Tzidony and Ravreboy (1992) that the former can be used for large consignments in which the sampling of units may be taken to be sampling with replacement. For small samples, the sampling units may be taken to be sampling without replacement and the hypergeometric approach is used (Frank *et al.*, 1991). The Bayesian approach also has different methods for analysing large and small samples.

Various methods for selecting the size of a random sample from a consignment have been accepted by the US courts (Frank *et al.*, 1991; Izenman, 2001). A summary of different procedures used in 27 laboratories around the world is given in Colón *et al.* (1993). Various procedures suggested for the choice of sample size include methods based on the square root of the consignment size, a percentage of the consignment size and a fixed number of units regardless of the consignment size, as well as the hypergeometric distribution. The formula

$$m = 20 + 10\%(N - 20) \quad \text{(for } N > 20) \quad (6.1)$$

where m is the sample size, the number of items inspected, and N is the consignment size is proposed by Colón *et al.* (1993). As well as being simple to implement, this approach, as the authors rightly claim, provides the opportunity to discover heterogeneous populations before the analysis is completed. It has been suggested that 'an inference made at the 95% confidence level, that 90% or more of the packages in an exhibit contain the controlled substance, should be accepted as sufficient proof in such cases' (Frank *et al.*, 1991). These summaries are given as confidence limits using a frequentist approach and not in probabilistic terms. A Bayesian approach provides summaries in probabilistic terms such as 'How big a sample should be taken for it to be said that there is a $100p\%$ probability that the proportion of units in the consignment which contain drugs is greater than $100\theta\%$?' or, for a particular case, with $p = 0.95$ and $\theta = 0.50$, 'How big a sample should be taken for it to be said that there is a

95% probability that the proportion of units in the consignment which contain drugs is greater than 50%?'.

6.2   CHOICE OF SAMPLE SIZE

6.2.1   Large consignments

A large consignment is taken to be one which is sufficiently large that sampling is effectively with replacement (see Section 2.3.5 for a discussion of this). This can be as small as 50, though in many cases it will be of the order of many thousands.

A consignment of drugs containing N units will be considered as a random sample from some super-population (Section 3.3.5) of units which contain drugs. Let θ $(0 < \theta < 1)$ be the proportion of units in the super-population which contain drugs. For consignment sizes of the order of several thousand all realistic values of θ will represent an exact number of units. For small sample sizes less than 50, θ can be considered as a nuisance parameter (i.e., one which is not of primary interest) and integrated out of the calculation, leaving a probability distribution for the unknown number of units in the consignment which contain drugs as a function of known values. For intermediate calculations, θ can be treated as a continuous value in the interval $(0 < \theta < 1)$, without any detriment to the inference. As before in (6.1), let m be the number of units sampled. The ratio m/N is known as the *sampling fraction*. Denote the number which are found to contain drugs by z.

Frequentist approaches to the estimation of θ

The sample proportion $p = z/m$ is an unbiased estimate of θ (Section 2.3.3). The variance of p is given by Cochran (1977) as

$$\frac{\theta(1 - \theta)}{m} \left(\frac{N - m}{N - 1} \right). \tag{6.2}$$

The factor $(N - m)/(N - 1)$ is known as the *finite population correction* (fpc). Interestingly, provided that the sampling fraction m/N is low, the size of the population has no direct effect on the precision of the estimate of θ. For example, if θ is the same in the two populations, a sample of 500 from a population of 200 000 gives almost as precise an estimate of the population proportion as a sample of 500 from a population of 10 000. The estimated standard deviation of θ in the second case is 0.98 times the estimated standard deviation in the first case. Little is to be gained by increasing the sample size in proportion to the population size.

Consider the following example. To simplify matters, the fpc is ignored and the sample size and proportion are such that the sample proportion p may be assumed to be Normally distributed. It is necessary to have some prior belief about the value of θ before determining the sample size. The expression $\theta(1 - \theta)$ takes its maximum value at $\theta = 1/2$ and its minimum value of 0 when $\theta = 0$ or 1. A conservative choice of sample size is to take $\theta = 1/2$. Assume that θ is thought to be about 75%. It is stipulated that a sample size m is to be taken to estimate θ to within 25%, that is, in the interval (0.50, 1.00), with 95% confidence. (This may be thought a very wide interval but it is consistent with the form of problem to be discussed later in this section.) It is desired to determine a sample size such that it can be said that, if all of the sample are found to contain drugs, then there is a 95% probability that θ is greater than 50%. The criterion for the sample size is that there should be a confidence of 0.95 that the sample proportion p lies in the interval 0.75 ± 0.25. From Section 2.4.2, this implies that two standard deviations equal 0.25. The standard deviation of p, ignoring the fpc, is

$$\sqrt{\frac{\theta(1 - \theta)}{m}}.$$

Setting two standard deviations equal to 0.25, and solving for m, gives the following expression for m:

$$m = \frac{4\theta(1 - \theta)}{0.25^2}.$$

When $\theta = 0.75$, $m = 12$. Thus a sample of size 12 is sufficient to estimate θ to be greater than 0.5 with confidence 0.95.

A Bayesian approach to the estimation of θ

As already discussed, a criterion has to be specified in order that the sample size may be determined. Consider the criterion from Section 6.1 that the scientist wishes to be 95% certain that 50% or more of the consignment contains drugs when all units sampled contain drugs. Then the criterion may be written mathematically in the notation of (5.3) as

$$Pr(\theta > 0.5 \mid m + \alpha, m - m + \beta) = 0.95$$

or

$$\int_{0.5}^{1} \theta^{m+\alpha-1}(1 - \theta)^{\beta-1} d\theta / B(m + \alpha, \beta) = 0.95, \tag{6.3}$$

using a beta conjugate prior distribution and binomial distribution to give a beta posterior distribution (5.3), with the special case where the number of 'successes' equals the number of trials (5.4). The general question in which p

and θ are specified at the end of Section 6.1 may be answered by finding the value of m which solves the equation

$$\int_{\theta}^{1} \theta^{m+\alpha-1}(1-\theta)^{\beta-1}d\theta/B(m+\alpha,\beta) = p. \qquad (6.4)$$

Such integrals are easy to evaluate using standard statistical packages given values for m, α and β. Given specified values for θ and p and values for α and β chosen from prior beliefs, the appropriate value of m to solve (6.4) may be found by trial and error. Note from Table 6.1 that differing values of α and β, so long as both are small, have little effect on m.

For large consignments, of whatever size, the scientist need only examine 4 units, in the first instance. If all are found to contain drugs, there is a 95% probability that 50% of the consignment contains drugs. Compare this with the result derived from a frequentist approach using a Normal approximation to the binomial distribution which gave a value of 12 for the sample size. These sample sizes are not large. However, there is not very much information gained about the exact value of θ. It is only determined that there is probability of 0.95 that $\theta > 0.5$. This is a wide interval (from 0.5 to 1) within which the true proportion may lie.

There may be concerns that it is very difficult for a scientist to formalise his prior beliefs. However, if α and β are small, large differences in the probabilities associated with the prior beliefs will not lead to large differences in the conclusions.

The methodology can be extended to allow for units which do not contain drugs. For example, if one of the original four units inspected is found not to contain drugs then three more should be inspected. If they all contain drugs, then it can be shown that the probability that $\theta > 0.5$, given that six out of seven contain drugs, is 0.96.

The dependency of the sample size on the values of p and θ is illustrated in Table 6.2. The prior parameters α and β are set equal to 1. Consider $p = 0.90$, 0.95 and 0.99 and consider values of $\theta = 0.5$, 0.6, 0.7, 0.8, 0.9, 0.95, 0.99. The sample size m required to be $100p\%$ certain that θ is

Table 6.1 The probability that the proportion of drugs in a large consignment is greater than 50% for various sample sizes m and prior parameters α and β. (Reprinted with permission from ASTM International)

α	β	m			
		2	3	4	5
1	1		0.94	0.97	
0.5	0.5	0.92	0.97	0.985	0.993
0.065	0.935		0.90	0.95	0.97

Table 6.2   The sample size required to be $100p\%$ certain that the proportion of units in the consignment which contain drugs is greater than θ, when all the units inspected are found to contain drugs. The prior parameters $\alpha = \beta = 1$. (Reprinted with permission from ASTM International)

θ	p		
	0.90	0.95	0.99
0.5	3	4	6
0.6	4	5	9
0.7	6	8	12
0.8	10	13	20
0.9	21	28	43
0.95	44	58	89
0.99	229	298	458

greater than the specified value is then given by the value of m which satisfies the equation

$$Pr(\theta > \theta_0 \mid m+1, m-m+1) = 1 - \theta_0^{m+1} = p,$$

a special case of (6.4). The value of m is thus given by the smallest integer greater than

$$[\log(1-p)/\log(\theta_0)] - 1.$$

Obviously, when considering the results in Table 6.2, the consignment size has to be taken into account in order that the sample size may be thought small with respect to the consignment size. Thus, for the last row in particular to be useful, the size of the consignment from which the sample is to be taken will have to be of the order of several tens of thousands.

There may be situations in which different choices of α and β may be wanted. It may be the scientist has some substantial prior beliefs about the proportion of the consignment which may contain drugs to which he is prepared to testify in court. These beliefs may arise from previous experiences of similar consignments, for example. In such cases, use can be made of various properties of the beta distribution (2.21) to assist the scientist in choosing values for α and β. For example, a prior belief concerning the proportion of the consignment which contains drugs would set that proportion equal to the mean of the distribution, and a belief about the precision of belief would provide a value for the variance (Section 2.4.4). Alternatively, if it was felt that β could be set equal to 1, so that the probability density function is monotonic increasing with respect to θ,

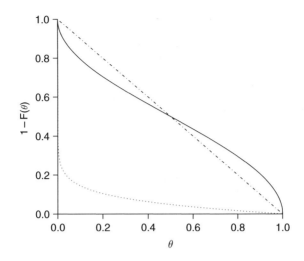

Figure 6.1   The prior probability $1 - F(\theta)$ that the proportion of units in a consignment is greater than θ, for various choices of α and β: $\alpha = \beta = 1$(dot-dashed curve), $\alpha = \beta = 0.5$(solid), $\alpha = 0.065$, $\beta = 0.935$(dotted). (Reprinted with permission from ASTM International.)

and that there was a prior belief about a lower bound for the proportion, say that

$$Pr(\text{Proportion} > \theta \mid \alpha, \beta) = p,$$

then

$$\alpha = \log(1 - p)/\log(\theta).$$

Variation in the prior beliefs, expressed through variation in the values of α and β, may have little influence on the conclusions, once some data have been observed. Figure 6.1 illustrates the prior probability that the true proportion of illegal units in a consignment is greater than a value θ, for $0 < \theta < 1$ for three choices of α and β. Figure 6.2 illustrates the posterior probability that the true proportion of illegal units in a consignment is greater than θ, for these choices of α and β, once four units have been examined and all found to be illegal.

6.2.2   Small consignments

Suppose now that the consignment size N is small. A sample of m units from the consignment is examined and $z(\leq m)$ units are found to contain drugs.

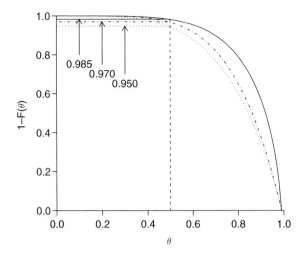

Figure 6.2   The posterior probability $1 - F(\theta)$ that the proportion of units in a consignment is greater than θ, for various choices of α and β: $\alpha = \beta = 1$(dot-dashed curve), $\alpha = \beta = 0.5$(solid), $\alpha = 0.065, \beta = 0.935$(dotted), after observation of four units, all found to be illegal. The corresponding probabilities that at least 50% of the consignment contains illegal units are marked as 0.985 ($\alpha = \beta = 0.5$), 0.970 ($\alpha = \beta = 1$), 0.950 ($\alpha = 0.065, \beta = 0.935$). (Reprinted with permission from ASTM International.)

A frequentist approach

Consider a frequentist approach based on the hypergeometric distribution (2.3). Examples are given in Bates and Lambert (1991) and Faber *et al.* (1999). Let $R = Z + Y$ be the total number of units in the consignment which contain illicit drugs, where Z is the number of units in the sample of size m and Y is the number of units in the remainder which contain drugs. Then the distribution of Z is hypergeometric (see Section 2.3.5) with

$$Pr(Z = z) = \frac{\binom{R}{z}\binom{N-R}{m-z}}{\binom{N}{m}}, \qquad z = 0, 1, \ldots, \min(R, m).$$

When $m = z$ this is

$$\frac{R!(N-m)!}{N!(R-m)!}. \tag{6.5}$$

Denote this probability as P_m. Consider an example where $N = 10$ and $\theta = 0.7$. Then $R = N\theta = 7$. For $m = 5$, substitution of $N = 10, R = 7$ and $m = 5$ into (6.5) gives $P_m = P_5 = 0.08$. For $m = 6, P_6 = 0.03$ and for $m = 4, P_4 = 0.17$ (with $N = 10, R = 7$). With $N = 10$ still, and $\theta = 0.6$ so that $R = 6$, and $m = 5$, P_5, from (6.5), equals $1/42 = 0.02$. Thus, for $N = 10$ and $m = 5$, if all five tablets sampled contain illicit drugs then one can be 92% confident that the proportion of illicit drugs in the consignment is at least 0.7 and 98% confident that the proportion of illicit drugs in the consignment is at least 0.6. Similarly, for $N = 10$ and $m = 4$, if all four tablets sampled contain illicit drugs then one can be 83% confident that the proportion of illicit drugs in the consignment is at least 0.7. For $N = 10$ and $m = 6$, if all six tablets sampled contain illicit drugs then one can be 97% confident that the proportion of illicit drugs in the consignment is at least 0.7.

A Bayesian approach for small consignments, using the hypergeometric distribution, is described in Coulson *et al.* (2001b). A discrete prior distribution is chosen for the $N + 1$ possible divisions of the consignment into licit and illicit items. The likelihood function is based on the hypergeometric distribution sampling m from N and a posterior distribution obtained.

The beta-binomial distribution

The above interpretation is not so clear as that obtained from the Bayesian method which uses the beta-binomial distribution (2.7). The beta-binomial distribution provides a probability statement about the number of units in the consignment which contain drugs.

As before, let θ ($0 < \theta < 1$) be the proportion of units in the super-population which contain drugs. The probability distribution of z, given m and θ, may be taken to be binomial. For each unit, independently of the others in the consignment, the probability it contains drugs is taken to be equal to θ. The posterior distribution of θ is another beta distribution with parameters $\alpha + z$ and $\beta + m - z$.

Since the consignment size is small, a better representation of the variability of the number of units in the consignment which contain drugs is obtained by considering a probability distribution for this number, Y say, explicitly. There are n units in the remainder of the consignment ($m + n = N$) which have not been inspected. Then Y (unknown and no greater than n) is the number of units in this remainder which contain drugs. Given θ, the distribution of ($Y \mid n, \theta$), like that of ($Z \mid m, \theta$), is binomial. However, θ has a beta distribution, and the distribution of ($Y \mid n, \theta$) and of ($\theta \mid m, z, \alpha, \beta$) can be combined to give a Bayesian predictive distribution for ($Y \mid m, n, z, \alpha, \beta$), a beta-binomial distribution (Section 2.3.7):

$$Pr(Y = y \mid m, n, z, \alpha, \beta)$$

$$= \frac{\Gamma(m + \alpha + \beta)\binom{n}{y}\Gamma(y + z + \alpha)\Gamma(m + n - z - y + \beta)}{\Gamma(z + \alpha)\Gamma(m - z + \beta)\Gamma(m + n + \alpha + \beta)} \quad (y = 0, 1, \ldots, n). \quad (6.6)$$

From this distribution, inferences can be made about Y, such as probability intervals or lower bounds for Y.

The beta distribution can be generalised to the Dirichlet distribution (Section 2.4.5). The binomial distribution can be generalised to the multinomial distribution (Section 2.3.4). A generalisation may be made, analogously, from the beta-binomial distribution to a so-called Dirichlet-multinomial distribution.

Comparison of Bayesian and frequentist approaches

It can be shown that in a limiting case, the beta distribution and the binomial distribution give the same numerical answers for large consignments and the beta-binomial and hypergeometric distributions give the same numerical answers for small consignments, though the Bayesian and frequentist approaches give different interpretations to the results (Aitken, 1999).

Use of the beta-binomial distribution

As an example, consider a consignment of size $N = 10$, where five units are inspected and all five are found to contain drugs ($m = z = 5$). For the proportion of units in the consignment which contain drugs to be at least 0.7 ($\theta \geq 0.7$), it is necessary for the number of units Y in the five units not inspected to be at least 2 ($Y \geq 2$). The beta-binomial probability (2.7) with a uniform prior $\alpha = \beta = 1$, is given by

$$Pr(Y \geq 2 \mid 5, 5, 5, 1, 1,) = \sum_{y=2}^{5} \frac{6\binom{5}{5}\binom{5}{y}}{11\binom{10}{5+y}} = 0.985.$$

The hypergeometric distribution has the interpretation that if $m = z = 5$, one is 92% confident that $\theta \geq 0.7$. The beta-binomial approach enables one to assign a probability of 0.985 to the event that $\theta \geq 0.7$.

As with large consignments, values for α and β may be chosen subjectively to represent the scientist's prior beliefs before inspection about the proportion of the units in the consignment (as a random sample from the super-population) which contain drugs.

General results can be obtained. The problem is to choose m such that, given n, α, and β (and possible values for z, consequential on the choice of m and the outcome of the inspection), a value for y can be determined to satisfy some probabilistic criterion – for example, the value y_0 such that $Pr(Y \geq y_0 \mid m, n, z, \alpha, \beta) = p$. Results are given in Aitken (1999) for $p = 0.9$, where the consignment size N is taken to be 30.

If six units are inspected and one or two do not contain drugs, then the number of units in the remainder of the consignment which can be said, with probability 0.9, to contain drugs drops from 17 to 12 to 9. Even if 16 units (out of 30) are inspected and all are found to contain drugs, then it can only be said, with probability 0.9, that 12 of the remaining 14 contain drugs (and this is so even with $\alpha = 4, \beta = 1$).

These approaches to sample size estimation assume that the classification as licit or illicit is free of error. It is obviously desirable that this assumption be true. An additional benefit of this assumption is that the posterior distribution of the proportion of the consignment which is illicit is robust to the choice of the prior parameters. When there is a possibility of misclassification the posterior distribution is no longer robust to the choice of the prior parameters. Such a situation is not discussed here, but details are available in Rahne *et al.* (2000). A frequentist approach using the hypergeometric distribution, with an adaptation to allow for false positives and false negatives, is described in Faber *et al.* (1999).

Application of these ideas to the sampling of glass fragments is described in Curran *et al.* (1998b).

6.3 QUANTITY ESTIMATION

The estimation of the quantity of a drug will be treated in two stages. First, the proportion of the units in the consignment that contain illicit drugs will be modelled. Secondly, the total weight of the illicit material in those packets that do contain anything illicit is estimated. Uncertainty in the prior belief in the proportion of packets that are 'illicit' may be represented by a beta distribution. It is assumed there is no prior information for the mean and variance of the distribution of the quantity of drugs in the packages. Details of how such prior information may be considered are given in Aitken *et al.* (1997).

Given the sample size, and thus an estimate of the proportion of a consignment which contains drugs and an estimate of the mean and standard deviation of the weight in the consignment, a confidence interval for the true quantity of drugs may be calculated (Tzidony and Ravreboy, 1992). A probability interval is appropriate in a Bayesian context. In this context, a probability distribution is associated with a parameter (Q, say) denoting the total quantity of illicit material in the consignment, and probability statements of any desired kind may be made. For example, these could include the probability that Q is greater than a certain value, q say, which will be of importance in sentencing hearings.

6.3.1 Frequentist approach

It is only possible to make a statement about the consignment as a whole with certainty if the whole consignment is analysed. Once it is accepted that a sample has to be considered, then it is necessary to consider what level of proof is adequate. This is strictly a matter for the court to decide.

The method described by Tzidony and Ravreboy (1992) considers the consignment as a population and the packages (or units) examined as a sample. The quantities (weights) of drugs in the units are assumed to be random variables which are Normally distributed, with population mean μ and population variance σ^2, say. The mean quantity in a unit in the consignment is estimated by the mean, denoted $\bar{x}$, of the quantities found in the sample. A confidence interval is determined for μ based on the sample size m, the sample mean $\bar{x}$, the sample standard deviation s of the quantities of drugs in the units examined, and an associated t-distribution. An estimate of the total quantity of drugs in the consignment is then determined by considering the size N of the consignment and the proportion θ of packages in the consignment thought to contain drugs.

For example, the inequalities in expression (7) of Tzidony and Ravreboy (1992) are, as a generalisation of (5.13),

$$\bar{x} - t_{(m-1)}(\alpha/2) \frac{s}{\sqrt{m}} \sqrt{\frac{(N-m)}{N}} \leq \mu \leq \bar{x} + t_{(m-1)}(\alpha/2) \frac{s}{\sqrt{m}} \sqrt{\frac{(N-m)}{N}},$$

where $\sqrt{\frac{(N-m)}{N}}$ is the fpc factor and the interval is the $100(1-\alpha)\%$ confidence interval for the mean quantity in a package.

The corresponding confidence interval for Q, the total quantity of drugs in the consignment, is obtained by multiplying all entries in the inequalities by $N\hat{\theta}$, where $\hat{\theta}$ is an estimate for θ based on the sample of size m. This gives, as a $100(1-\alpha)\%$ confidence interval for Q (expression (9) of Tzidony and Ravreboy, 1992),

$$N\hat{\theta} \left\{ \bar{x} - t_{(m-1)}(\alpha/2) \frac{s}{\sqrt{m}} \sqrt{\frac{(N-m)}{N}} \right\} \leq Q \leq N\hat{\theta} \left\{ \bar{x} + t_{(m-1)}(\alpha/2) \frac{s}{\sqrt{m}} \sqrt{\frac{(N-m)}{N}} \right\}.$$

$$(6.7)$$

However, no account is taken of the uncertainty in the estimation of θ, only a point estimate of θ is used.

A corresponding $100(1-\alpha)\%$ lower bound for Q is given by the left-hand side of the inequality

$$N\hat{\theta} \left\{ \bar{x} - t_{(m-1)}(\alpha) \frac{s}{\sqrt{m}} \sqrt{\frac{(N-m)}{N}} \right\} \leq Q. \qquad (6.8)$$

6.3.2 Bayesian approach

Procedures for the estimation of the quantity of drugs using informative prior information are described in Aitken *et al.* (1997) for statistical considerations and in Bring and Aitken (1997) for legal considerations.

Consider the consignment as itself a random sample from a larger super-population of units or packages, some or all of which contain illegal material. Then θ $(0 < \theta < 1)$ is the proportion of units in the super-population which contain drugs. The variability in θ may be modelled by a beta distribution.

Let n be the number of packages in the consignment which are not examined. Then N equals $m + n$. As before, let z $(\leq m)$ be the number of units in those examined which contain drugs and let y ($\leq n$) be the number of units which contain drugs amongst those units which are not examined. Let $(x_1, \ldots, x_z)$ be measurements of the quantities of drugs in those units examined which contain drugs. Let $(w_1, \ldots, w_y)$ be measurements of the quantities of drugs in those units not examined which contain drugs. Let $\bar{x} = \sum_{i=1}^{z} x_i/z$ be the sample mean quantity of drugs in units containing drugs amongst those examined, and let s be the sample standard deviation, where the sample variance $s^2 = \sum_{i=1}^{z}(x_i - \bar{x})^2/(z-1)$. Let $\bar{w} = \sum_{j=1}^{y} w_j/y$ be the mean quantity of drugs in units containing drugs amongst those not examined. The total quantity q of drugs in the exhibit is then $z\bar{x} + y\bar{w}$, and the problem is one of first estimating $\bar{w}$, given $\bar{x}, s$ and z, whilst not knowing y, and then of finding $y\bar{w}$ by finding the posterior distribution of $f(y \mid \bar{x})$. An estimative approach is one in which the parameters (μ, σ^2) of the Normal distribution representing the quantity of drugs in an individual unit are *estimated* by the corresponding sample mean $\bar{x}$ and sample variance s^2 (Tzidony and Ravreboy, 1992). A *predictive* approach is one in which the values of the unknown measurements $(w_1, \ldots, w_y)$ are *predicted* by values of known measurements $(x_1, \ldots, x_z)$ (Aitchison and Dunsmore, 1975; Aitchison *et al.*, 1977; Evett *et al.*, 1987; Geisser, 1993).

The predictive approach *predicts* the values of $\bar{w}$ (and hence q) from $\bar{x}$ and s through the probability density function

$$f(\bar{w} \mid \bar{x}, s) = \int f(\bar{w} \mid \mu, \sigma^2) f(\mu, \sigma^2 \mid \bar{x}, s) d\mu \, d\sigma^2,$$

where $f(\mu, \sigma^2 \mid \bar{x}, s)$ is a Bayesian posterior density function for (μ, σ^2) based on a prior density function $f(\mu, \sigma^2)$ and the summary statistics $\bar{x}$ and s. When prior information for μ and σ^2 is not available a uniform prior for μ and $\log(\sigma^2)$ may be used as in Section 5.5. The predictive density function for $f(y \mid \bar{x})$ is then a generalized t-distribution as described below.

There are two advantages of the predictive approach relative to the estimative approach. First, any prior knowledge of the variability in the parameters (μ, σ^2) of the Normal distribution can be modelled explicitly. Suggestions as to how this may be done are given by Aitken *et al.* (1997) with reference to *U.S. v. Pirre*.

A consignment of $m + n$ ($= N$) units is seized. A number (m) of the units are examined; the choice of m may be made following the procedures described in Section 6.2. On examination it is found that z ($\leq m$) units contain drugs and that $m - z$ do not. The contents of the z units which contain drugs are weighed and their weights $(x_1, \ldots, x_z)$ recorded. The remainder (n) are not examined. All of m, z and n are known.

First, consider a small consignment. Let $Y(\leq n)$ denote the unknown number of units not examined which contain drugs. The estimation of quantity is able to take account of the lack of knowledge of Y. A probability function for Y may be determined using the methods described in Section 6.2. A weighted average of the quantities obtained for each value of Y is taken with weights the probabilities of Y obtained from an appropriate beta-binomial distribution (2.7).

Let $(X_1, \ldots, X_z)$ and $(W_1, \ldots, W_y)$ be the weights of the contents of the units examined and not examined, respectively, which contain drugs. It is assumed that these weights are Normally distributed. Let $\bar{X} = \sum_{i=1}^{z} X_i / z$ and $\bar{W} = \sum_{j=1}^{y} W_j / y$. The total weight, Q, of the contents of the units in the consignment is then given by

$$Q = z\bar{x} + Y\bar{W}.$$

Let $(x_1, \ldots, x_z)$ be the observed values of $(X_1, \ldots, X_z)$. The distribution of $(Q \mid x_1, \ldots, x_z)$, which is a predictive distribution, is of interest. Once known, it is possible to make probabilistic statements, as distinct from confidence statements, about Q.

Let $\bar{x} = \sum_{i=1}^{z} x_i / z$, $\bar{w} = \sum_{j=1}^{y} w_j / y$. Also, let $s^2 = \sum_{i=1}^{z} (x_i - \bar{x})^2 / (z - 1)$ be the variance of the measurements on the units which were examined and found to contain drugs. In the absence of prior information about the mean or variance of the distribution of the weights of drugs in the packages, a uniform prior distribution is used. The probability density function of $(\bar{w} \mid z, y, \bar{x}, s^2)$ is such that

$$\frac{\bar{w} - \bar{x}}{s\sqrt{\dfrac{1}{z} + \dfrac{1}{y}}}$$

is a t-distribution with $z - 1$ degrees of freedom. Quantiles of this distribution and hence lower bounds for the quantity $q = z\bar{x} + y\bar{w}$, according to appropriate burdens of proof, may be determined.

For given values of m, z, n, y, $\bar{x}$ and s, lower bounds for $\bar{w}$, and hence q, can be determined from the formula

$$\bar{w} = \bar{x} + s\, t_{(z-1)}(\alpha)\sqrt{\frac{1}{z} + \frac{1}{y}}. \tag{6.9}$$

For a small consignment, the value of y is a realisation of a random variable which has a beta-binomial distribution (2.7). The distribution of $(\bar{w} \mid z, y, \bar{x}, s^2)$ can be combined with (2.7) to give a distribution of $(\bar{w} \mid s^2, \bar{x}, z)$. The distribution and corresponding probability density function of Q may then be determined from the relationship $Q = z\bar{x} + y\bar{W}$. Let $f_{t,z-1}(.)$ denote the probability density function of the t-distribution with $z - 1$ degrees of freedom. The probability density function $f(q)$ of Q is then given by

$$f(q) = \sum_{y=0}^{n} f_{t,z-1}\left\{ \frac{q - (z+y)\bar{x}}{sy\sqrt{\dfrac{1}{z} + \dfrac{1}{y}}} \right\} \left\{ sy\sqrt{\frac{1}{z} + \frac{1}{y}} \right\}^{-1} Pr(Y = y) \tag{6.10}$$

(Aitken and Lucy, 2002).

An example in which a seized drug exhibit contained 26 street doses is given in Tzidony and Ravreboy (1992). A sample of six ($m = 6$) units was taken and each was analysed and weighed. Twenty ($n = 20$) units were not examined. All six of the units examined contained drugs. The average net weight $\bar{x}$ of the powder in the six units was 0.0425 g, with a standard deviation s of 0.0073 g. A 95% confidence interval for the total quantity Q in the 26 doses is 1.105 $\pm$ 0.175 g. Note that this interval incorporates the fpc factor from (6.2) to allow for the relatively large sample size ($m = 6$) compared with the consignment size ($N = 26$). The Bayesian approach described here does not require such a correction.

The values for Q corresponding to appropriate percentage points of the distribution may be determined from (6.10). Some results are given in Table 6.3, together with corresponding results with the method of Tzidony and Ravreboy, and in Figure 6.3.

The lower end 0.930 g of the 95% confidence interval (1.105 ± 0.175)g for the quantity Q of drugs in the 26 packages may be thought of as a 97.5% lower confidence limit for Q. This can be compared with the value 0.689 g in the corresponding cell of Table 6.3 which is the amount such that $Pr(Q > 0.689) = 0.975$ obtained from the predictive approach. The predictive approach produces a lower value because of the uncertainty associated with the values determined for the number of unexamined units which contain drugs. This difference is repeated for different probabilities. In general, the Bayesian approach gives smaller values for the quantities than the frequentist approach.

Further details are available in Aitken *et al.* (1997), Izenman (2001) and Aitken and Lucy (2002), where it is shown that as the burden of proof, concerning the amount of drugs in the packages, increases, the quantity for which charges may be brought decreases, thus lowering the length of any

Table 6.3 Estimates of quantities q of drugs (in grams), in a consignment of $m + n$ units, according to various possible burdens of proof, expressed as percentages $P = 100 \times Pr(Q > q \mid m, z, n, \bar{x}, s)$ in 26 packages when 6 packages are examined ($m = 6, n = 20$) and $z = 6, 5,$ or 4 are found to contain drugs. The mean ($\bar{x}$) and standard deviation (s) of the quantities found in the packages examined which contain drugs are 0.0425 g and 0.0073 g. The parameters for the beta prior are $\alpha = \beta = 1$. Numbers in brackets are the corresponding frequentist lower bounds using the fpc factor (6.2). (Reprinted with permission from ASTM International)

Percentage P	Number of units examined which contain drugs			Possible burden of proof (Illustrative)
	6	5	4	
97.5	0.689 (0.930)	0.501 (0.744)	0.345 (0.575)	
95	0.750 (0.968)	0.559 (0.785)	0.397 (0.613)	Beyond reasonable doubt
70	0.944 (1.067)	0.770 (0.885)	0.603 (0.704)	Clear and convincing
50	1.015 (1.105)	0.862 (0.921)	0.704 (0.737)	Balance of probabilities

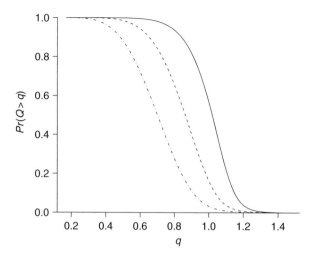

Figure 6.3 The probability that the total quantity Q of drugs (in grams) in a consignment of 26 packages is greater than q when 6 packages are examined and 6 (solid curve), 5 (dashed), or 4 (dot-dashed) are found to contain drugs. The mean and standard deviation of the quantities found in the packages examined which contain drugs are 0.0425 g and 0.0073 g. The parameters for the beta prior are $\alpha = \beta = 1$. (Reprinted with permission from ASTM International.)

sentence which may be related to quantity. For example, if proof is required *beyond reasonable doubt* and a probability of 0.95 is thought to meet this burden, then the quantity associated with this is 0.750 g (assuming all six units examined contain drugs) since, from Table 6.3, $Pr(Q > 0.750) = 0.95$. Alternatively, if proof is required on *the balance of probabilities* and a probability of 0.50 is thought to satisfy this, then the quantity associated with this is 1.015 g since, again from Table 6.3, $Pr(Q > 1.015) = 0.50$. If less than six of the units examined are found to contain drugs then the estimates for q decrease considerably, as can be seen from the second and third columns of Table 6.3.

Second, for a large consignment the data are used to provide a beta posterior for the proportion of illicit drugs in the whole consignment. It is assumed that the consignment size is known. The total weight, Q, of the contents of the units in the consignment is given as before, by

$$Q = z\bar{x} + y\bar{W}.$$

The distribution of Q is then given by the t-density, conditional on y, with $Pr(Y = y)$ replaced by an appropriate part of a beta distribution over the interval $(0, n)$ (2.22). Results for a large consignment, obtained by scaling up by a factor of 100 from the results in Table 6.3, are shown in Table 6.4 and Figure 6.4, with a similar pattern of results to those for small consignments. Note that in the t-density component of the expression y is treated as a discrete variable in

Table 6.4   Estimates of quantities q of drugs (in grams), in a consignment of $m + n$ units, according to various possible burdens of proof, expressed as percentages $P = 100 \times \Pr(Q > q \mid m, z, n, \bar{x}, s)$ in 2600 packages when 6 packages are examined ($m = 6, n = 2594$) and $z = 6, 5$, or 4 are found to contain drugs. The mean ($\bar{x}$) and standard deviation (s) of the quantities found in the packages examined which contain drugs are 0.0425 g and 0.0073 g. The parameters for the beta prior are $\alpha = \beta = 1$. Numbers in brackets are the corresponding frequentist lower bounds without using the fpc factor (6.8). (Reprinted with permission from ASTM International).

Percentage P	Number of units examined which contain drugs		
	6	5	4
97.5	63 (95)	44 (78)	30 (61)
95	69 (98)	51 (80)	36 (63)
70	91 (106)	74 (88)	58 (70)
50	98 (110)	84 (92)	69 (74)

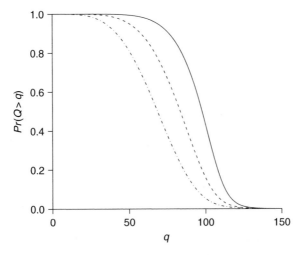

Figure 6.4   The probability that the total quantity Q of drugs (in grams) in a consignment of 2600 packages is greater than q when 6 packages are examined and 6 (solid curve), 5 (dashed), or 4 (dot-dashed) are found to contain drugs. The mean and standard deviation of the quantities found in the packages examined which contain drugs are 0.0425 g and 0.0073 g. The parameters for the beta prior are $\alpha = \beta = 1$. (Reprinted with permission from ASTM International.)

the interval $\{0, \ldots, n\}$ and in the beta component of the expression it is treated as a continuous variable. The treatment of y as a continuous variable for the beta integral enables the calculation of the probability that y takes a particular integer value for use with the t-density.

6.4 MISLEADING EVIDENCE

Consider the following quote:

Statistical thinking concerns the relation of qualitative data to a real-world problem, often in the presence of variability and uncertainty. It attempts to make *precise and explicit* what the data has [*sic*] to say about the problem of interest. (Mallows, 1998; emphasis added.)

Two of the requirements of the courts in the United States for scientific evidence are that it be relevant and reliable. Likelihood ratios consider the matter of relevance. Evidence can be said to be relevant if the likelihood ratio is different from one; that is, the posterior odds after presentation of the evidence is different (larger or smaller) than the prior odds before presentation of the evidence. In a series of papers (Royall, 1997, 2000; Mellen, 2000; Mellen and Royall, 1997) Mellen and Royall discuss the issue of reliability through the concept of weak evidence and strong misleading evidence. Weak evidence is evidence with a low likelihood ratio. Strong misleading evidence is evidence with a high likelihood ratio in favour of the wrong proposition – for example, evidence which has a high likelihood ratio in favour of the prosecution proposition when the defence proposition is true.

Relevance is addressed in the United States through Rule 401 of the Federal Rules of Evidence, where it is defined as 'having any tendency to make the existence of any fact that is of consequence to the determination of the action more probable or less probable than it would be without the evidence'.

A change in the odds in favour of the prosecution's proposition, through a value for the evidence different from 1, is a change in the probability of the prosecution's proposition. Thus, there is a connection between Rule 401 and the likelihood ratio. Note that this concept of relevance is different from that discussed in Chapter 9, where it is taken to be the probability that trace evidence which is recovered from the victim/suspect and which matches (in some sense) trace evidence associated with the suspect/victim is connected with the crime (Stoney, 1991a, 1994; Evett *et al.*, 1998a). The relevance which is a probability is used as a term in the expression for the likelihood ratio, as illustrated in Chapter 9. The relevance as defined in Rule 401 of the Federal Rules of Evidence is a statement about the value of the likelihood ratio.

Rule 702 of the Federal Rules of Evidence lays out when expert witnesses may be allowed to testify:

If scientific, technical, or other specialized knowledge will assist the trier of fact to understand the evidence or to determine a fact in issue, a witness qualified as an expert by knowledge, skill, experience, training, or education may testify thereto in the form of an opinion or otherwise.

In 1993, the US Supreme Court ruled that scientific knowledge will assist the trier of fact only if it is also *reliable*, or trustworthy:

The requirement that an expert's testimony pertain to 'scientific knowledge' establishes a standard of . . . evidentiary reliability – that is, trustworthiness.

> In a case involving scientific evidence, evidentiary reliability will be based on scientific validity. (*Daubert v. Merrell Dow Pharmaceuticals*)

In 1999, the US Supreme Court stated that

> *Daubert's* general principles apply to the expert matters described in Rule 702. The Rule, in respect to all such matters 'establishes a standard of evidentiary reliability' (509 U.S. at 590).
>
> ... the trial judge must determine whether the testimony has 'a reliable basis in the knowledge and experience of [the relevant] discipline' (509 U.S. at 592). (*Kumho Tire Co., Ltd. v. Carmichael*.)

Reliability is the probability of observing strong misleading evidence. This is related to the amount of evidence one has. If one wishes to improve the reliability of one's evidence then the amount collected has to be increased. This is intuitively reasonable.

Consider two competing propositions, A and B, for evidence E. The likelihood ratio is $Pr(E \mid A)/Pr(E \mid B)$. Denote this as V_{AB}. Strong evidence in favour of A will be taken to be evidence for which the likelihood ratio is greater than a specified value k, say. Thus the probability of strong misleading evidence will be evidence for which $Pr(V_{AB} > k)$ when B is the correct proposition. The subscript B can be used in the notation to clarify under which proposition the probability is being determined, so that the probability of strong misleading evidence can be denoted $Pr_B(V_{AB} > k)$. Consider E as a set of measurements x (with corresponding random variable X), such as DNA profiles (suspect, victim and background data) or refractive indices of glass (suspect, victim and background data). Then, it can be shown (Royall, 1997) that it is unlikely there will be strong evidence favouring A, when B is true. In particular,

$$Pr_B(V_{AB} > k) = Pr_B(Pr(X = x \mid A)/Pr(X = x \mid B) > k) < 1/k,$$

where $X = x$ has been substituted for E.

Consider the set S of all possible values of X which produce a value V_{AB} greater than k. For each of these values x of X,

$$Pr(X = x \mid B) < Pr(X = x \mid A)/k$$

by rearrangement of the probabilistic inequality above. It is then possible to sum over all values x of X in S to obtain

$$Pr(S) = \sum_{x \in S} Pr(X = x \mid B) < \sum_{x \in S} Pr(X = x \mid A)/k.$$

The right-hand sum $\sum_{x \in S} Pr(X = x \mid A)$ will not be greater than 1, as it is a sum of mutually exclusive probabilities. Thus, $\sum_{x \in S} Pr(X = x \mid A)/k < 1/k$ and, hence, $Pr(S) < 1/k$. Further details and a stronger result that 'if an unscrupulous researcher sets out deliberately to find evidence supporting his favourite but erroneous hypothesis over his rival's, which happens to be correct, by a factor of k, then the chances are good that he will be eternally frustrated' are given in Royall (1997).

Values of k of 8 and 32 are proposed by Royall (1997, p. 25) to represent 'fairly strong' and 'strong' evidence respectively and have already been used in Section 5.6.4. These are justified with reference to an urn example (see Section 1.6.2). Consider an urn which may contain all white balls or half white balls and half black balls. If three balls are drawn without replacement and all are found to be white, this may be thought of as 'fairly strong' evidence that the urn contains only white balls. The probability of this event if the urn contains half white and half black balls is $(1/2)^3 = 1/8$. If five balls are drawn without replacement and all are found to be white, this may be thought of as 'strong' evidence that the urn contains only white balls. The probability of this event if the urn contains half white and half black balls is $(1/2)^5 = 1/32$. Similar benchmarks have been proposed by Edwards (1992), Jeffreys (1983) and Kass and Raftery (1995). Comparison should also be made with Section 3.5.3.

An example is given in Mellen (2000) of the application of these ideas to DNA evidence. Let s denote the source of the DNA and d denote the defendant. Consider two propositions:

- the defendant is the source of the crime scene DNA ($s = d$);
- another possible suspect is the source ($s \neq d$).

Suppose a match in DNA profiles is observed between the DNA from the source and from the defendant. Let z denote the genotype that is observed. Let Z_i denote the random variable corresponding to the genotype from person i; Z_s is the genotype from the source of the DNA at the crime scene and Z_d is the genotype from the defendant. Then $Z_d = Z_s = z$. The probability of misleading evidence (evidence whose value V is greater than k) is evaluated assuming $s \neq d$. Thus

$$Pr(V > k \mid Z_s = z) = Pr(V > k, Z_d = z \mid Z_s = z)$$

$$= Pr(V > k \mid Z_s = z, Z_d = z)Pr(Z_d = z \mid Z_s = z)$$

$$< Pr(Z_d = z \mid Z_s = z).$$

Assuming that $s \neq d$, this final probability is equal to the genotype frequency (after allowance for co-ancestry and relatives). The probability of strong misleading evidence is not greater than the genotype frequency. As stated in Mellen (2000),

> as might be expected, if the genotype z tends to be rare among individuals in the same genetic subset of the population as the defendant, then the probability of observing genotypes in the defendant and the reference sample that constitute strong misleading evidence is not great. If, on the other hand, the genotype z tends to be quite common in this subpopulation, then the probability might be larger.

In the discussion at the end of Mellen and Royall (1997) comment is made on several useful features of the analysis. These include:

- *separation* between measures of evidence and reliability of the process which produces the evidence;
- *distinction* between the strength of implicating evidence and the improbability of its occurrence – there is a low probability of misleading strong implicating evidence;
- *explicit conditioning* on the circumstances of a case – condition on the non-DNA evidence to delimit the suspect population and condition on the DNA type whose source is known in probabilities of strong implicating evidence;
- *generality* of the methods, the importance of conditional probabilities (Balding and Donnelly, 1995b) and the extension of the methods to identification evidence other than DNA evidence.

Royall (2000) extends these ideas to continuous data. Consider two propositions for evidence in the form of measurements X, such that for the first, denoted f_1,

$$X \sim N(\theta_1, \sigma^2)$$

and for the second, denoted f_2,

$$X \sim N(\theta_2, \sigma^2).$$

Let there be data $x_1, \ldots, x_n$. Then the likelihood functions, in the two propositions, are

$$f_{2n} = \prod_{i=1}^{n} (2\pi\sigma^2)^{-\frac{1}{2}} \exp\left\{-\frac{1}{2\sigma^2}(x_i - \theta_2)^2\right\}$$

$$= (2\pi\sigma^2)^{-\frac{n}{2}} \exp\left\{-\frac{1}{2\sigma^2}\sum(x_i - \theta_2)^2\right\}$$

$$f_{1n} = (2\pi\sigma^2)^{-\frac{n}{2}} \exp\left\{-\frac{1}{2\sigma^2}\sum(x_i - \theta_1)^2\right\}$$

$$\frac{f_{2n}}{f_{1n}} = \exp\left\{-\frac{1}{2\sigma^2}\left[\sum(x_i - \theta_2)^2 - \sum(x_i - \theta_1)^2\right]\right\}$$

$$= \exp\left\{\frac{n(\theta_2 - \theta_1)}{\sigma^2}\left(\bar{x} - \frac{\theta_1 + \theta_2}{2}\right)\right\}.$$

If the first proposition is true then

$$\bar{X} \sim N(\theta_1, \sigma^2/n)$$

and it can be shown (Royall, 2000) that

$$Pr_1\left(\frac{f_{2n}}{f_{1n}} > k\right) = \Phi\left(-\frac{\Delta\sqrt{n}}{2\sigma} - \frac{\sigma\log_e(k)}{\Delta\sqrt{n}}\right)$$

where $\Delta = |\theta_2 - \theta_1|$ and the subscript 1 associated with the Pr indicates that the first proposition is taken to be true. In analogous notation, Pr_2 will indicate that the probability is to be determined assuming the second proposition to be true. If Δ, expressed as a multiple c of the standard error of $\bar{X}$, is such that $\Delta = |\theta_2 - \theta_1| = c\sigma/\sqrt{n}$, then

$$Pr_1\left(\frac{f_{2n}}{f_{1n}} > k\right) = \Phi\left(-\frac{c}{2} - \frac{\log_e(k)}{c}\right),$$

assuming θ_1 to be the true mean. This function is a so-called *bump* function. See Figure 6.5. If θ_1 is true then there is very little chance of observing strong evidence supporting θ_2 over θ_1 when the difference Δ between the two parameter values is a small fraction of the standard error $\sigma/\sqrt{n}$.

These ideas may be used to determine a sample size based on the criteria of controlling for the probability of strong misleading evidence and for the probability of weak evidence. Consider a likelihood ratio f_1/f_2 of density functions where the subscripts denote the two propositions being compared.

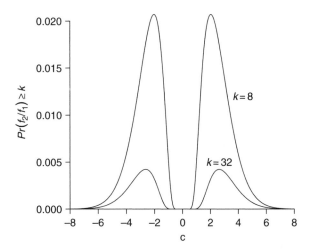

Figure 6.5 Bump function for the probability of misleading evidence $Pr_1\left(\frac{f_{2n}}{f_{1n}} > k\right)$ for $k = 8$ and $k = 32$ as a function of c, the distance from the true mean to the alternative, in standard errors. (Royall, 2000; reprinted with permission from the *Journal of the American Statistical Association*. Copyright 2000 by the American Statistical Association. All rights reserved.)

- *Strong* evidence is defined as evidence for which f_1/f_2 is greater than a pre-specified value k, or, conversely, less than $1/k$.
- *Strong misleading* evidence is defined as evidence for which f_1/f_2 is greater than the pre-specified value k when the second proposition is assumed true, or, conversely, less than $1/k$ when the first proposition is assumed true.
- *Weak* evidence is evidence which is not strong, that is, evidence for which $1/k < f_1/f_2 < k$.

The probability, $M(n)$, of observing strong misleading evidence, as a function of the sample size n, is given by

$$
M(n) = Pr_1\left(\frac{f_{2n}}{f_{1n}} > k\right)
$$

$$
= Pr_2\left(\frac{f_{1n}}{f_{2n}} > k\right)
$$

$$
= \Phi\left(-\frac{\Delta\sqrt{n}}{2\sigma} - \frac{\sigma\log_e(k)}{\Delta\sqrt{n}}\right) \tag{6.11}
$$

and the probability, $W(n)$, of observing weak evidence, as a function of the sample size n, is given by

$$
W(n) = Pr_1\left(\frac{1}{k} < \frac{f_{2n}}{f_{1n}} < k\right)
$$

$$
= Pr_2\left(\frac{1}{k} < \frac{f_{2n}}{f_{1n}} < k\right)
$$

$$
= \Phi\left(-\frac{\Delta\sqrt{n}}{2\sigma} + \frac{\sigma\log_e(k)}{\Delta\sqrt{n}}\right) - M(n). \tag{6.12}
$$

Consider an example where the characteristic of interest is the refractive index of glass. A window is broken at the crime scene. A suspect is apprehended soon afterwards, and he has fragments of glass on his clothing. He explains the presence of the fragments by saying that he had just broken a glass. The two propositions of interest are:

- H_p, the glass fragments on the suspect's clothing came from the crime scene window;
- H_d, the glass fragments on the suspect's clothing came from the broken glass.

The window at the crime scene is of a very common type. There is a population database with a refractive index of known mean (θ_1) and standard deviation σ for variation within windows. The glass from which the broken glass was manufactured is also of a common type with a refractive index of known mean (θ_2) and standard deviation σ for variation within glasses (the same as for the crime scene window glass).

A pre-assessment question (see Section 7.2 for further discussion) is the determination of the number of fragments of glass from the suspect's clothing to be examined. Once this number has been determined then refractive indices of fragments of glass from the suspect's clothing and fragments from the crime scene window can be measured and compared using the likelihood ratio expressions of Chapter 10.

From (6.11) and (6.12), with $\Delta = |\theta_1 - \theta_2|$ and σ known, the two unknowns are the sample size n and the criterion k for strong evidence. Both n and k can be varied and the corresponding values of $M(n)$ and $W(n)$ investigated. For the determination of the sample size in the pre-assessment stage it is necessary to consider three criteria:

- the meaning of 'strong' (the value for k), and, as a consequence,
- the probability of strong misleading evidence;
- the probability of weak evidence.

This procedure is illustrated using the following values for the parameters: $\theta_1 : 1.519\,507\,3$, $\theta_2 : 1.519\,573\,0$, $\sigma : 0.000\,049\,2$. Then $\Delta = 0.000\,065\,7$ and, from (6.11) and (6.12), Table 6.5 can be obtained.

Suppose it is decided that strong evidence (either in support of H_p or of H_d) is evidence with a value greater than 8, and that it is tolerable to have a probability of strong misleading evidence no greater than 0.005 and to have a probability of weak evidence no greater than 0.05. These criteria will be satisfied with a sample of size 10. This follows from inspection of Table 6.4, in the row for $k = 8$ and the columns for $M(10)$ and $W(10)$ where the corresponding cell values are $0.0046\,(< 0.005)$ for $M(10)$ and $0.0481\,(< 0.05)$ for $W(10)$.

A probability of 0.005 for strong misleading evidence is the probability that strong evidence will be obtained that the glass fragments on the suspect's clothing came from the crime scene window when in fact they came from the broken glass, or that the glass fragments on the suspect's clothing came from the broken glass when in fact they came from the crime scene window.

Other applications can be considered. For example, consider the sampling of a consignment of drugs as described in Section 6.2. The sample size n is determined by the criterion of satisfying a pre-specified probability that the true proportion of illicit drugs in the consignment is greater than a pre-specified value.

Table 6.5 Probabilities of strong misleading evidence $M(n)$ and weak evidence $W(n)$ for boundary values k of 8 and 32 for strong evidence and sample sizes n of 5, 10 and 20

k	$M(5)$	$W(5)$	$M(10)$	$W(10)$	$M(20)$	$W(20)$
8	0.0143	0.1985	0.0046	0.0481	0.0004	0.0038
32	0.0040	0.3658	0.0017	0.0967	0.0002	0.0079

In contrast to that criterion, consider two propositions about the possible source of the consignment:

- H_p, the drugs are from a source with mean quantity of drug per tablet of θ_1;
- H_d, the drugs are from a source with mean quantity of drug per tablet of θ_2.

The criterion k for strong evidence can be chosen. Then the probabilities for strong misleading evidence and for weak evidence may be determined.

A procedure for estimating the quantity of drug in a consignment, based on a sample from the consignment, was described in Section 6.3.2. This procedure could be adapted to estimate the mean amount of drug in one tablet in the consignment.

There are, thus, two procedures. In the first, probabilities for strong misleading evidence and for weak evidence are determined in a pre-assessment stage before sampling to compare two propositions about the source of the consignment. In the second, an estimate of the quantity of drug in the consignment is obtained, without reference to any possible source of the consignment.

The results from the sampling can be used in the determination of a likelihood ratio to evaluate the evidence from the consignment in support of either H_p or H_d.

It is also possible to determine a sample size from the criteria based on (6.11) and (6.12). This may give a different sample size from that obtained using the criterion described in Section 6.2. However, the two sets of criteria are designed to answer two different questions and thus may give different answers. The criteria based on (6.11) and (6.12) are designed to compare two propositions about a mean value. The criterion described in Section 6.2 is designed to satisfy a pre-specified probability that the true proportion of illicit drugs in the consignment is greater than a pre-specified value.

7

Interpretation

In this chapter various court cases are introduced and discussed in which issues of evaluation and interpretation of a probabilistic nature were raised. The nature of the propositions which have to be addressed is of particular interest. Principles of interpretation and a methodology to choose relevant propositions are presented. Later in the book, in Chapters 12 and 13, the evaluation of evidence for two evidential types, fibres and DNA, is described in detail. The evaluation of glass evidence is the forerunner of many of the ideas applied to other evidential types. Glass evidence is discussed in various different chapters (rather than in one of its own) throughout the book as it is one of the best evidential types for the discussion of the different ideas. There are many other evidential types for which the techniques of evidence evaluation are not so well developed, and some of these are described briefly here. These include earprints, firearms and toolmarks, fingerprints, speaker recognition, hair, documents and handwriting, and paint. Various general principles are presented which it is hoped will help the development of statistical methods for evidence evaluation in these areas. Reference is made to these types (as well as glass, fibres and DNA) elsewhere as appropriate.

7.1 CONCEPTS AND COURT CASES

The use of probabilistic reasoning in the legal process is not without controversy. Various possible sources of controversy, with reference to specific cases, are given here. A Bayesian approach to the evaluation of evidence is discussed with reference to these cases.

7.1.1 Relevant population

The evaluation of evidence requires a population from which the probabilities may be estimated (see also Section 8.5). At its simplest, this could be taken to be the population of the whole world. However, this should quickly be reduced

---

Statistics and the Evaluation of Evidence for Forensic Scientists: Second Edition
C.G.G. Aitken and F. Taroni © 2004 John Wiley & Sons, Ltd   ISBN: 0-470-84367-5

to more manageable proportions through the use of Bayes' theorem as the prior odds for members of certain sub-populations of the world will be small as will the likelihood ratio.

An interesting idea has arisen from a ruling in *R. v. Doheny and Adams*.

> Members of the jury, if you accept the scientific evidence called by the Crown, this indicates that there are probably only four or five white males in the United Kingdom from whom that semen could have come. The defendant is one of them. If that is the position, the decision you have to reach, on all the evidence, is whether you are sure that it was the defendant who left that stain or whether it is possible that it was one of the other small group of men who share the same DNA characteristics.

The choice of the United Kingdom as the relevant population here is merely an illustration. The Court acknowledged that it might be appropriate to choose a more limited 'suspect population' such as 'the Caucasian, sexually active males in the Manchester area', the area where the crime was committed. Note that the Court is using the word 'suspect', whereas here the word 'relevant' is preferred. The population is defined initially as white males in the UK and then refined to 'Caucasian, sexually active males in the Manchester area'. The definition of the population is important because it restricts the number of potential criminals but also because it defines the population from which the relative frequency of the characteristics of the trace evidence under consideration is derived. This would often include DNA profiles but could include other evidential types such as characteristics of hair or fibres of clothing.

7.1.2   Consideration of odds

The forensic scientist, in consideration of the evidence, is concerned with the evaluation of the likelihood ratio. The judge and the jury are concerned with the odds in favour of the propositions (prosecution and defence). The combination of the likelihood ratio and the prior odds gives the posterior odds. The likelihood ratio is a numerical quantity. Thus, the prior odds has also to be a numerical quantity if it is to be combined with the likelihood ratio to give a posterior odds (Berry, 1990). The prior odds could be provided as a range of values which would then give a range of values for the posterior odds. Some thoughts on appropriate values for the odds are given here. The figures given are for illustration and the stimulation of discussion. They are used for the same purpose in Thompson *et al.* (2003) in a discussion of error rates in DNA analysis, of which further details are given in Chapters 13 and 14. The context here is that of the assessment of trace evidence for which the prosecution proposition is that the suspect is the source.

Some suggestions for prior odds in favour of the prosection proposition are the following:

- 2:1 – other evidence prior to that under consideration is fairly strong but probably not sufficient by itself to suggest the suspect is the source.
- 1:10, 1:100 – other evidence suggests a possibility the suspect is the source without creating a strong reason to believe he is.
- 1:1000 – almost no evidence apart from the trace evidence.

There have also been various ideas which help consideration of the posterior odds in favour of the prosection proposition. For example:

- What is meant by the phrase 'beyond reasonable doubt'?
- As discussed earlier in Section 3.5.6, in an analysis of questionnaire returns the phrase 'beyond reasonable doubt' is thought to vary between 0.8 and 0.99 (Simon and Mahan, 1971).
- It may depend on the severity of the possible sentence in that a juror may wish to be more convinced of guilt in a capital case than in a minor assault case before returning a verdict of guilty.
- The phrase 'it is better that ten guilty people escape than that one innocent suffer' (Blackstone, quoted by Ceci and Friedman, 2000) has been used, and this is equivalent to odds of 10 to 1.
- The phrase 'it is better that ninety nine offenders should suffer than that one innocent man should be condemned' (Starkie, quoted by the US Supreme Court, 1995) has also been used, and this is equivalent to odds of 99 to 1.

The quote from Jaynes (2003) in Section 3.5.6 is particularly apposite here, and one part in particular bears repetition: 'if you were a judge, would you rather face one man whom you had convicted falsely or 100 victims of crime that could have been prevented?'

It may be that some further interpretation of the prior and posterior odds is required. An approach based on a community scale is described in Aitken and Taroni (1998) and has been described earlier (Section 3.5.3). The scale is based on a logarithmic scale for the prior and posterior odds, thus enabling one to consider very large probabilities. The logarithms (to base 10) are related to communities to which juries will be able to relate with greater ease than the original probabilities. The communities range in size from the individual, with log(odds) of 0 (odds of 1 or evens), to the world, with log(odds) of 10 (odds of 10 billion). For example, consider a city of 1 million people (a little bigger than the UK city of Glasgow). The criminal is one person in this city. All others are innocent. A person is selected at random from the city. The odds are a million to 1 against that person being the criminal. Now consider a village of 2000 people. There is one innocent person in the village. All others are guilty. A person is selected at random from the township. The odds are 2000 to 1 in favour of that person being a guilty person. These examples are illustrations of the island problem described in Section 3.5.6.

7.1.3   Combination of evidence

See Sections 8.1.1 and 8.1.3 for a technical description of the problems here.

The combination of evidence was relevant in the case of Splatt (Shannon, 1984; Davis, 1986). A statistician, John Darroch, provided advice to a Royal Commission investigating the case. In his presidential address to the 7th Australian Statistical Conference in August 1984 (Darroch, 1985, 1987), he provided the following quote from *The Belhaven and Stenton Peerage Case* of 1875:

> Lord Cairns (1875): It is sometimes said that in dealing with circumstantial evidence you should consider the weight which is to be given to the united force of all the circumstances put together. You may have a ray of light so feeble that by itself it will do little to elucidate a dark corner but, on the other hand, you may have a number of such rays of light, each of them insufficient in itself to light up that dark corner but all converging and brought to bear upon the same point and, when united, producing a body of illumination which will clear away the darkness you are endeavouring to dispel.

As Darroch commented, the rays of light only combine to produce a greater body of illumination if they relate to independent pieces of evidence. In the Splatt case, much of the trace evidence was not independent as the different items all came from the same source. The discovery of one set of fragments was not independent of the discovery of another set.

7.1.4   Specific cases

Applications of the above arguments can be illustrated from several cases.

R. v. Adams, D.J. (Dawid, 2002)

Adams was arrested for rape. The evidence E linking him to the crime was, first, a match between his DNA and that of semen obtained from the victim, and second, the fact that he lived locally. A match probability of 1 in 200 million for the DNA was reported. The defence challenged this and suggested a figure of 1 in 20 million or even 1 in 2 million could be more appropriate. There was other (prior) information:

- Identification (I_1) – the victim gave a description of her attacker which was hard to reconcile with the defendant and did not pick out the defendant in an identity parade.

- Alibi (I_2) – a former girlfriend of Adams gave an alibi which was not challenged.

At the trial, with the consent of the prosecution, the defence and the Court, the jury was given instruction in the correct way to combine all the evidence. A prior probability of guilt was introduced, followed by what were claimed to be plausible likelihood ratios for I_1 and I_2. The DNA evidence (E) was then introduced to provide a final posterior probability of guilt. Of course, the figures which follow could be challenged or the jury could substitute their own figures.

Consider two propositions:

- H_p, Adams is guilty;
- H_d, Adams is not guilty.

The prior probability of guilt was assessed as follows: there were thought to be approximately 150 000 males between 18 and 60 in the area who, in the absence of other evidence, may have committed the crime. Another 50 000 were added to this to allow for the possibility that the attacker may have come from outside the area (i.e., a probability of 0.25 for the possibility that the attacker came from outside the area was thought appropriate). Thus $Pr(H_p) = 1/200\,000$, $Pr(H_p)/Pr(H_d) = 1/199\,999 \simeq 1/200\,000$. The other two pieces of evidence, that of identification (I_1) and alibi (I_2), were assessed as follows:

- Identification evidence: I_1 was assigned a probability $Pr(I_1 \mid H_p) = 0.1$ if he were guilty, and a probability $Pr(I_1 \mid H_d) = 0.9$ if he were innocent. (Note that these two probabilities sum to 1 but they need not necessarily do so; see the alibi evidence below.) These assignations provided a likelihood ratio $Pr(I_1 \mid H_p)/Pr(I_1 \mid H_d) = 1/9$ (or a likelihood ratio $Pr(I_1 \mid H_d)/Pr(I_1 \mid H_p) = 9$ in favour of the defence).

- Alibi evidence: I_2 was assigned a probability $Pr(I_2 \mid H_p) = 0.25$ if he were guilty, and a probability $Pr(I_2 \mid H_d) = 0.50$, if he were innocent. These assignations provided a likelihood ratio $Pr(I_2 \mid H_p)/Pr(I_2 \mid H_d) = 1/2$ (or a likelihood ratio $Pr(I_2 \mid H_d)/Pr(I_2 \mid H_p) = 2$ in favour of the defence).

These two items of evidence were assumed to be independent. Thus, placing H_d in the numerator and H_p in the denominator,

$$\frac{Pr(I_1, I_2 \mid H_d)}{Pr(I_1, I_2 \mid H_p)} = \frac{Pr(I_1 \mid H_d)}{Pr(I_1 \mid H_p)} \times \frac{Pr(I_2 \mid H_d)}{Pr(I_2 \mid H_p)}$$
$$= 18,$$

an overall likelihood ratio of 18 in favour of the defence. This likelihood ratio can then be combined with the prior odds $Pr(H_d)/Pr(H_p)$ of 200 000, to give odds before consideration of the DNA evidence of

$$\frac{Pr(H_d \mid I_1, I_2)}{Pr(H_p \mid I_1, I_2)} = \frac{Pr(I_1 \mid H_d)}{Pr(I_1 \mid H_p)} \times \frac{Pr(I_2 \mid H_d)}{Pr(I_2 \mid H_p)} \times \frac{Pr(H_d)}{Pr(H_p)}$$

which equals 3.6 million in favour of the defence or against the prosecution.

Now the DNA evidence E (with a likelihood ratio of 200 million to 2 million in favour of the prosecution) is included through multiplication with the odds of 1 in 3.6 million, from consideration of I_1 and I_2. The DNA evidence E is assumed independent of I_1 and I_2:

$$
\begin{aligned}
V &= \frac{Pr(H_p \mid E, I_1, I_2)}{Pr(H_d \mid E, I_1, I_2)} \\
&= \frac{Pr(E \mid H_p)}{Pr(E \mid H_d)} \times \frac{Pr(I_1 \mid H_p)}{Pr(I_1 \mid H_d)} \times \frac{Pr(I_2 \mid H_p)}{Pr(I_2 \mid H_d)} \times \frac{Pr(H_p)}{Pr(H_d)} \\
&= \frac{Pr(E \mid H_p)}{Pr(E \mid H_d)} \times \frac{Pr(I_1 \mid H_p)}{Pr(I_1 \mid H_d)} \times \frac{Pr(I_2 \mid H_p)}{Pr(I_2 \mid H_d)} \times \frac{Pr(H_p)}{Pr(H_d)},
\end{aligned}
$$

where $Pr(E \mid H_p)/Pr(E \mid H_d)$ may be taken to 200 million (the prosecution's suggestion) or 2 million (the lower of the suggestions of the defence). These results give posterior odds from 56 to 1 (200/3.6) in favour of guilt to 1.8 to 1 in favour of innocence. These odds in turn give a posterior probability of guilt in the interval from 0.98 (56/57) to 0.36 ($1 - 1.8/2.8$). The defence argued from these figures that guilt was not proved 'beyond reasonable doubt'.

The jury returned a verdict of guilty. The Appeal Court rejected the attempt to introduce probabilistic reasoning, saying that 'it trespasses on an area peculiarly and exclusively within the province of the jury', and 'to introduce Bayes' theorem, or any similar method, into a criminal trial plunges the jury into inappropriate and unnecessary realms of theory and complexity deflecting them from their proper task'. (This is reminiscent of comments made in the appeal for the Collins case; see Section 4.4.) The task of the jury was said to be to 'evaluate evidence and reach a conclusion not by means of a formula, mathematical or otherwise, but by the joint application of their individual common sense and knowledge of the world to the evidence before them'. This fails to recognise that so-called common sense usually fares very badly when it comes to manipulating probabilities.

The appeal was granted on the basis that the trial judge had not adequately dealt with the question of what the jury should do if they did not want to use Bayes' theorem. A retrial was ordered. Attempts were made to describe the Bayesian approach to the integration of all the evidence. Once again the jury convicted, once again the case went to appeal and once again the Bayesian approach was rejected as inappropriate to the courtroom, and the appeal was dismissed.

Note that Dennis Adams had a full brother whose DNA was not investigated. The probability that the brother had the same DNA profile as the defendant was calculated as 1 in 220. This was claimed to weaken the impact of the DNA evidence against Dennis Adams. This point was dismissed on the grounds that there was no evidence as to the brother's actual DNA profile, nor any suggestion that he might have committed the offence. However, there was also no other evidence against Dennis Adams except the DNA evidence.

The Lashley and Smith cases(Redmayne, 2002)

In the context of the ruling in *Doheny*, consider the case of *R. v. Lashley*. Lashley was convicted of the robbery of a Liverpool post office. The only evidence against him was a DNA match which left him as a suspect along with seven to ten other males in the UK (in the argument used in *Doheny*). There was no evidence linking him to the Liverpool area. His conviction was quashed on appeal.

However, the case of *R. v. Smith*, with the same judges on the same day as *Lashley*, was treated differently. He was convicted of the robbery of a post office in Barley, Hertfordshire. The principal evidence against him was a DNA match (with probability 1 in 1.25 million), which (as in *Doheny*) left him as a suspect along with 43 other males in the UK. His appeal was quashed because the DNA evidence 'did not stand alone because there was also quite clearly evidence of this man having been arrested some shortish distance away'.

Smith was arrested at a place called Potton, which is 13 miles from Barley. A circle centred on Barley with Potton on its border encloses several cities and, at a rough estimate, some 80 000 men of appropriate age who live at least as close to Barley as Smith. This figure could be used to provide prior odds of 1 in 80 000 against Smith's guilt. This may be combined with the DNA evidence to produce a probability of guilt of 0.94. If it is thought at least 16 times as bad to convict an innocent person as to acquit a guilty person then Smith would not be convicted. This would be the situation with the Blackstone dictum but not with the Starkie dictum, both quoted in Section 7.1.2.

Also, Smith came from a large family – his father had 13 brothers and sisters. The Court of Appeal did not pursue this. As in *R. v. Adams, D.J.*, there was no evidence to implicate the relatives, but, apart from the DNA (and geography), neither was there evidence to implicate Smith. If it is assumed that all members of the population of potential suspects are equally likely to have committed the crime then any relatives among the population are, prior to the DNA test, as likely to be guilty as anyone else. Relatives are far more likely to match than other members of the population and so should not be ignored (Lempert, 1991, 1993; Balding, 2000; Redmayne, 2002; see also Section 13.6).

R. v. Clark

Sally Clark's first child Christopher died unexpectedly at the age of about 3 months when she was the only other person in the house. The death was initially treated as a case of sudden infant death syndrome (SIDS). Her second child, Harry, was born the following year. He died in similar circumstances. She was arrested and charged with murdering both her children. At trial, a professor of paediatrics testified that, in a family of background similar to Sally Clark's, the probability that two babies would both die of SIDS was around 1 in 73 million. This was based on a study which estimated the probability of a single SIDS death in such a family as 1 in 8500, and then squared to obtain the probability of two deaths. Squaring of the figure of 1 in 8500 makes an assumption of independence of the two deaths. The probability that the second child suffers a SIDS death, given that the first child has died from SIDS, is assumed to be the same as the probability that the first child dies from SIDS. (Again, this is reminiscent of the Collins case (Section 4.4), this time because an inappropriate assumption of independence has been made.) The case and the following analysis are described in Dawid (2002). Two propositions may be considered:

- H_p, the mother did indeed murder two of her babies;
- H_d, the babies died of SIDS.

The Bayesian approach considers the relative likelihoods of the evidence E of the two deaths under the two propositions. This is the ratio of Pr(two dead babies | murdered) / Pr(two dead babies | SIDS deaths). Both probabilities in this ratio are 1: the probability of two dead babies if they have been murdered is 1 and the probability of two dead babies if they have died of SIDS is also 1. For these two propositions, the value of the evidence of the deaths is not relevant. However, the determination of the posterior odds $Pr(H_p \mid E)/Pr(H_d \mid E)$ requires consideration of the prior probabilities for H_p and H_d as well as the likelihood ratio. First, consider H_d: the probability of the two deaths by SIDS is taken to be 1 in 73 million (assuming the figure given in the trial to be accurate). For the probability of death by murder (H_p), Dawid (2002) cites Office of National Statistics (ONS) data for 1997 in which there were 642 093 live births and seven babies were murdered in the first year of life. The probability of being murdered is then estimated by Dawid (2002) to be 7/642 093 or approximately 1 in 90 000. Assuming independence, as for the analysis assuming SIDS deaths (not a particularly reasonable assumption here either), and squaring the probability for two murders, a probability is obtained of about 1 in 8.4 billion for the probability of two murders of babies in the same family. The ratio $Pr(H_p)/Pr(H_d)$ of 1 in 8.4 billion to 1 in 73 million is $1/115$. This figure is the prior odds for H_p relative to H_d. As the likelihood ratio has a value of 1, the posterior odds for H_p relative to H_d is also 1 in 115. Thus, the posterior probability that the babies died of SIDS rather than being murdered is $115/116$ or 0.99.

It is important that the propositions are specified carefully. Dawid (2002) provides an alternative analysis in which the propositions considered are not so restricted as those above (see Section 7.2.1 and the distinction between *proposition* and *explanation*). In the alternative analysis, let the propositions be:

- H_p, the babies were murdered;
- H_d, the babies were not murdered.

The evidence E is, as in the previous analysis, that the babies died. Here it is further assumed that if the babies were murdered then it was Sally Clark that did the murders and if the babies were not murdered then they died of SIDS. Thus $Pr(E \mid H_p) = 1$, as before. However, H_d does not include the implication that the babies died. Thus, $Pr(E \mid H_d)$ is the probability that the babies died, assuming they were not murdered. This is taken to be the probability that they died of natural causes, and more specifically SIDS. This is then 1 in 73 million (still using the figure provided in the original trial). The likelihood ratio is then 1 divided by 1 in 73 million, or 73 million, a figure which provides very strong evidence in support of the proposition that the babies were murdered by Sally Clark. However, the change in propositions from the initial analysis means that the prior odds also change. In this (second) analysis, $Pr(H_p) = 1/8.4$ billion using the ONS figures provided above. Thus $Pr(H_p)/Pr(H_d) \simeq 1 / 8.4$ billion. The combination of the likelihood ratio and the prior odds gives the same posterior odds as before: 73 million divided by 8.4 billion or 1 in 115. These give a posterior probability that the babies were not murdered of 115/116, as before.

The probability of other evidence has also to be assessed under each of the two propositions. The choice of probabilities for the other evidence is subjective. However, the Bayesian approach makes it very clear what features of the evidence should be taken into account and what the effect of the choices of probabilities is.

Note that the Court of Appeal heard new medical evidence in January 2003 and Sally Clark's conviction was overturned. A commentary on this case is also given in Aitken (2003).

Summary

The Bayesian approach provides an intellectually rigorous approach to the analysis of uncertainty in the evaluation of evidence. It enables the implications of assumptions to be considered thoroughly. It enables questions of uncertainty to be answered in a coherent probabilistic manner. The cases discussed here have illustrated how various aspects of the legal process may be assisted with the Bayesian approach:

- The likelihood ratio – illustrated here as to how the combination of more than one piece of evidence (*R. v. Adams, D.J.*) may be evaluated and how the interpretation of certain cases (*R. v. Clark*) may be aided. See also Kaye and Koehler (2003) for further justification of the use of the likelihood ratio to measure the probative value of evidence.
- Consideration of all the evidence – illustrated by the case of *R. v. Adams, D.J.*
- Definition of propositions – the evaluation of evidence depends on the propositions put forward by both the prosecution and the defence (*R. v. Clark*).
- Relevant population – this needs to be defined carefully in order that the prior odds are assessed properly. Also, for DNA profiles, the frequency of the profiles depends on the ethnic origins of the donor of the profile, and consideration has also to be given to the possible involvement of relatives.

7.2 PRE-ASSESSMENT AND RELEVANT PROPOSITIONS

7.2.1 Levels of proposition

It is widely accepted for the assessment of scientific evidence that the forensic scientist should consider different propositions which commonly represent alternatives proposed by the prosecution and the defence to illustrate their description of the facts under examination. The importance of these has been illustrated in *R. v. Clark*. These alternatives are formalized representations of the framework of circumstances. The forensic scientist evaluates the evidence under these propositions. The formulation of the propositions is a crucial basis for a logical and scientific approach to the evaluation of evidence (Cook *et al.*, 1998b). The framing of the propositions is an important and difficult stage of the evaluation process, which can be specified by three key principles (Evett and Weir, 1998):

- Evaluation is only meaningful when at least one alternative proposition (two or more competing propositions) is addressed, conventionally denoted throughout this book as H_p and H_d.
- Evaluation of scientific evidence (E) considers the probability of the evidence given the propositions that are addressed, $Pr(E \mid H_p)$ and $Pr(E \mid H_d)$.
- Evaluation of scientific evidence is carried out within a framework of circumstances, denoted I. The evaluation is conditioned not only by the competing propositions but also by the structure and content of the framework.

Therefore, propositions play a key role in this process.

Generally, propositions are considered in pairs. There will be situations where there will be three or more, and comments on these situations are given in Section 8.1.3. This happens quite often with DNA mixtures, for example, where the number of contributors to the mixture is in dispute (Buckleton *et al.*, 1998; Lauritzen and Mortera, 2002). It is generally possible to reduce the number of propositions to two, which will be identified with the respective prosecution and defence positions. Clearly the two propositions must be mutually exclusive. It is tempting to specify that they are exhaustive, but this is not necessary. The simplest way to achieve this is with the addition of the word 'not' into the first proposition, saying, for example: 'Mr C is the man who kicked Mr Z', and 'Mr C is not the man who kicked Mr Z'. However, this gives the court no idea of the way in which the scientist has assessed the evidence with regard to the second proposition. Mr C may not have kicked the victim, but he may have been present at the incident. Analogously, consider the propositions 'Mr B had sexual intercourse with Miss Y' and 'Mr B did not have sexual intercourse with Miss Y'. In fact, if semen has been found on the vaginal swab then it may be inferred that someone has had sexual intercourse with Miss Y and, indeed, the typing results from the semen would be evaluated by considering their probability given that it came from some other man. It will help the court if this is made plain in the alternative proposition that is specified. So the alternative could be 'Some unknown man, unrelated to Mr B, had sexual intercourse with Miss Y' (with no consideration of relatives). In summary, the simple use of 'not' to frame the alternative proposition is unlikely to be particularly helpful to the court (Cook *et al.*, 1998b). In the same sense, it is useful to avoid the use of misleading words like 'contact' to describe the type of action in the propositions. In fact, there is a danger in using such a vague word. As stated by Evett *et al.* (2000a), the statement that a suspect has been in recent contact with broken glass could mean many things. There is a clear need to specify correctly the proposition in a framework of circumstances. Moreover, the scientist may confuse *propositions* with *explanations*. For example, propositions like 'the crime stain originated from the suspect', and 'the crime stain originated from some unknown person who happened to have the same genotype as the suspect' represent explanations. The probability of the evidence that the DNA profile of the crime stain genotype matches the profile of the suspect given the first proposition is one; the probability of the evidence given the alternative proposition is also one. Thus, the likelihood ratio is simply 1. The interpretation is uninformative because the alternative proposition explains the observation but does not enable the weight of the evidence to be determined (Evett *et al.*, 2000b). The discussion above of Dawid's (2002) analysis of *R. v. Clark* explains how the prior odds and the likelihood ratio interact so that even though the likelihood ratio is one, the posterior odds are intuitively reasonable. Explanations can be useful as a kind of exploratory tool and they play an important role in reconstruction (which normally contributes to the investigative phase). Details and examples of the distinction are presented by Evett *et al.* (2000a). Therefore, identification of the propositions to be considered is not an easy task for the scientist. A fruitful approach to assist the scientist has been proposed by Cook *et al.* (1998b). Practically speaking, the propositions that are addressed in a judicial case depend on (a) the circumstances of the case, (b) the observations that have been made and (c) the available background data. A classification (called *hierarchy*) of these propositions into three main categories or levels has been proposed, notably the source level (level I), the activity level (level II) and the crime (or offence) level(level III).

The assessment of the level I category (the source) depends on analyses and measurements on the recovered and control samples. The value of a trace (or a stain) under source level propositions (such as 'Mr X's pullover is the source of the recovered fibres' and 'Mr X's pullover is not the source of the recovered fibres', so that another item of clothing is the source of the trace) does not need to take account of any more than the analytical information obtaining during examination. The probability of the evidence under the first proposition (numerator) is considered from a careful comparison between two samples (the recovered and the control). The probability of the evidence under the second proposition (denominator) is considered by comparison of the characteristics of the control sample and some kind of population of alternative sources.

The next level (level II) is related to an activity. This implies that the definitions of the propositions of interest have to include an action. Such propositions could be, for example, 'Mr X assaulted the victim' and 'Mr X did not assault the victim' (some other man assaulted her, and Mr X is not involved in the offence), or 'Mr X sat on the car driver's seat' and 'Mr X never sat on the car driver's seat'. The consequence of this activity (the assault or the sitting on a driver's seat) is the contact (between the two people involved in the assault, or the contact between the driver and the seat of the car) and consequently a transfer of material (fibres in this example). So the scientist needs to consider more detailed information about the case under examination relative to the transfer and persistence of the fibres on the receptor (e.g., the victim's pullover). Circumstances of the case (e.g., the distance between the victim and the criminal, the strength of the contact and the *modus operandi*) are needed to be able to answer relevant questions like 'is this the sort of trace that would be seen if Mr X were the man who assaulted the victim?' or 'is this the sort of trace that would be seen if Mr X were not the man who assaulted the victim?'. The assessment of evidence under level I propositions requires little in the way of circumstantial information. Only *I*, the background information, is needed. This could be useful in order to define the relevant population for use in the assessment of the rarity of the characteristic of interest. Activity level propositions cannot be addressed without a framework of circumstances. The importance of this will clearly appear in the pre-assessment approach (see Sections 7.2.2 and 12.3) when the expert is obliged to examine the scenarios of the case and to verify that all relevant information for the proper assessment of the evidence is available. The main advantage of level II over level I propositions is that the evaluation of evidence under level II propositions does not strictly depend on the recovered material; for example, it is possible to assess the fact that no fibres have been recovered. It is clearly important to assess the importance of the absence of material (such absence of material is evidence of interest).

Level III, the so-called 'crime level' or 'offence level', is close to the activity level. At level III, the propositions are really those of interest to the jury. Non-scientific information, such as whether or not a crime occurred, or whether or not an eyewitness is reliable, plays an important role in the decision. In routine work, forensic scientists generally use the source level to assess scientific evidence, notably for DNA evidence. Evidence under the activity level propositions requires that an important body of circumstantial information is available to the scientist (see Section 7.2.3). Unfortunately, this is often not the case because of a lack of interaction between the scientists and investigators. There are limitations in the use of a source level evaluation in a criminal investigation compared with an activity level evaluation. The lower the level at which the evidence is assessed, the lower is the relevance of the results in the context of the case discussed in the courts. For ease of simplicity, note that even if the value,

Table 7.1   Examples of the hierarchy of propositions. (Reproduced by permission of The Forensic Science Society)

Level	Generic		Examples
III	Offence	A	Mr A committed the burglary
			Another person committed the burglary
		B	Mr B raped Ms Y
			Some other man raped Ms Y
		C	Mr C assaulted Mr Z
			Mr C had nothing to do with the assault of Mr Z
II	Activity	A	Mr A is the man who smashed window X
			Mr A was not present when window X was smashed
		B	Mr B had sexual intercourse with Ms Y
			Some other man had sexual intercourse with Ms Y
		C	Mr C is the man who kicked Mr Z in the head
			Mr C was not present at the kicking of Mr Z
I	Source	A	The glass fragments came from window X
			They came from some other broken glass object
		B	The semen came from Mr B
			The semen came from some other man
		C	The blood on Mr C's clothing came from Mr Z
			The blood on Mr C's clothing came from an unknown person

V, of the evidence is such as to add considerable support to the proposition that the stain comes from the suspect, this does not help determine whether the stain had been transferred during the criminal action or for an innocent reason. Consequently, there is often dissatisfaction if the scientist is restricted to level I propositions.

There can be uncertainty around the relevance of the evidence for other reasons also. Because of the sensitivity of DNA profiling technology, it is now possible to envisage situations where it is not necessarily the case that a particular profile actually came from what was observed as a discernible region of staining. In such cases, it may be necessary to address what are termed 'sub-level I' propositions. In a DNA context, level I propositions such as 'The semen came from Mr Smith', and 'The semen came from some other man' have to be replaced by 'DNA came from Mr Smith', and 'DNA came from some other person' (Evett *et al.*, 2000a). The information available and the context of the case influence the choice of proposition. Examples of the hierarchy of propositions are given in Table 7.1.

7.2.2   Pre-assessment of the case

The evaluation process should start when the scientist first meets the case. It is at this stage that the scientist thinks about the questions that are to be addressed and the outcomes that may be expected. The scientist should attempt to frame propositions and think about the weight of evidence that is expected (Evett *et al.*, 2000a). However, there is a tendency to consider evaluation of evidence as a final step of a casework examination, notably at the time of preparing the formal report. This is so even if an earlier interest in the process enables the scientist to make better decisions about the allocation of resources. For example, consider a case of assault involving the possible cross-transfer of textile fibres between a victim and an assailant. The scientist has to decide whether to look first for potential transferred fibres on the victim's pullover or for fibres on the suspect's pullover. If traces compatible with the suspect's pullover are found on the victim's pullover, then the expectation of the detection of traces from the victim's pullover on the suspect's pullover has to be assessed. This includes the possibility of reciprocal transfer. Should we have expectations? How can we quantify them? If so, what is the interpretative consequence when those expectations are or are not met (presence or absence of evidence)? Matters to be considered include:

- the appropriate nature of the expectations;
- the quantification of the expectations;
- the interpretation of the presence or absence of evidence, through the success or failure of the expectations.

The scientist requires an adequate appreciation of the circumstances of the case so that a framework may be set up for consideration of the kind of examinations that may be carried out and what may be expected from them (Cook *et al.*, 1998a), in order for a logical decision to be made.

Such a procedure of pre-assessment can be justified on a variety of grounds. Essentially it is justified because the choice of level for the propositions for the evaluation of scientific evidence is carried out within a framework of circumstances, and these circumstances have to be known before any examination is made in order that relevant propositions may be proposed (e.g., at the activity instead of source level). This procedure may inform a discussion with the customer before any substantial decision is reached (e.g., about expense). Moreover, this process provides a basis for consistency of approach by all scientists who are thereby encouraged to consider carefully factors such as circumstantial information and data that are to be used for the evaluation of evidence and to declare them in the final report.

The scientist should proceed by considering estimation of the probability of whatever evidence will be found given each proposition. Consider, for example, a case where a window is smashed, and assume that the prosecution and defence propose the following alternatives: 'The suspect is the man who smashed the window' and 'The suspect was not present when the window was smashed'. The examination of the suspect's pullover will reveal a quantity Q of glass fragments, where Q can be, for example, one of the states {no, few, many}.

- To assess the numerator of the likelihood ratio, the first question asked is 'what is the probability of finding a quantity Q of matching glass fragments if the suspect is the man who smashed the window?'

- To assess the denominator of the likelihood ratio, the second question asked is 'what is the probability of finding a quantity Q of matching glass fragments if the suspect was not present when the window was smashed?'

The scientist is asked initially to assess six different probabilities using data coming from surveys, relevant publications on the matter, or subjective assessments:

1. The probability of finding *no* matching glass fragments if the suspect is the man who smashed the window.
2. The probability of finding *few* matching glass fragments if the suspect is the man who smashed the window.
3. The probability of finding *many* matching glass fragments if the suspect is the man who smashed the window.
4. The probability of finding *no* matching glass fragments if the suspect was not present when the window was smashed.
5. The probability of finding *few* matching glass fragments if the suspect was not present when the window was smashed.
6. The probability of finding *many* matching glass fragments if the suspect was not present when the window was smashed.

These probabilities may not be easy to derive because of a lack of information available to the scientist (Cook *et al.*, 1998b). For example, it will be very difficult to assess transfer probabilities (see Sections 8.3.3 and 10.5.4) if the scientist has no answer to questions like the following.

- Was the window smashed by a person or by a vehicle? The fact that a window was smashed by a person or by a vehicle changes the amount of glass fragments the scientist expects to be transferred.

- How (*modus operandi*) was the window smashed? If it was smashed by a person, then was that person standing close to it? Was a brick thrown through it? Information about the way a window is smashed is important because it provides information on the amount of glass potentially projected. Information of the distance between the person who smashed the window and the window offers relevant information on the amount of glass fragments the scientist will expect to recover.

Where there is little information about the time of the alleged offence and the time of the investigators taking clothing, the lapse of time between the offence and the collection of evidence cannot precisely be estimated. It is also difficult to assess the probability of persistence of any transferred glass fragment (see Section 10.5.5). Therefore, if the scientist has a small amount of information about the case under examination, then assessment has to be restricted to level I (or sub-level I) propositions.

The case pre-assessment process can be summarised by the following steps.

- Collection of information that the scientist may need about the case.
- Consideration of the questions that the scientist can reasonably address, and, consequently, the level of propositions for which the scientist can reasonably choose to assess evidence.
- Identification of the relevant parameters which will appear in the likelihood ratio.
- Assessment of the strength of the likelihood ratio expected given the background information.
- Determination of the examination strategy.
- Conduct of tests and observation of outcomes.
- Evaluation of the likelihood ratio and report of the value.

A practical procedure is presented in Section 12.3.

7.2.3   Pre-assessment of the evidence

Examples considering the pre-assessment of various types of scientific evidence are presented in the literature. Cook *et al.* (1998b) present pre-assessment through the example of a hypothetical burglary involving potential glass fragments (an unknown quantity, Q, of recovered fragments). Stockton and Day (2001) consider an example involving signatures in questioned documents. Champod and Jackson (2000) consider a burglary case involving fibres evidence. Booth *et al.* (2002) discuss a drugs case. A cross-transfer (also called 'two-way transfer') case involving textiles is presented by Cook *et al.* (1999) where it is shown how the pre-assessment can be updated when a staged approach is taken. The results of the examination of one of the garments are used to inform the decision about whether the second garment should be examined.

Puch and Smith (2002) describe a pre-assessment procedure within a training package, Forensic Inference Networks for Decision Support (FINDS), used within the Forensic Science Service of England and Wales to assess fibre evidence. The purpose is to provide support to the forensic scientist to determine whether an analysis of the collected fibres is cost-effective. Possible values for the likelihood ratios, determined under the prosecution and defence propositions, are assigned to seven categories: (1): strong support for the defence, (2) support for the defence, (3) weak support for the defence, (4) no support, (5) weak support for the prosecution, (6) support for the prosecution, (7) strong support for the prosecution. The scientist considers the probabilities for the likelihood ratios, given the proposition of the client (prosecution or defence). If there is high probability for support for the client's proposition then advice can be given to proceed with the analysis.

7.3 ASSESSMENT OF VALUE OF VARIOUS EVIDENTIAL TYPES

This is not meant to be a comprehensive survey of all ideas relating to all evidential types. DNA and fibres, which are two of the most common types of trace evidence, are given chapters of their own (Chapters 13 and 12). Glass has several mentions elsewhere (Sections 10.4.2, 10.5.4, 10.6 and Chapter 11). Paint has a brief mention in this section but has several references elsewhere (Section 8.3.4). For the evidential types presented here, discussion is based primarily on recent ideas in order to illustrate how the concepts of evidential value advocated in this book relate to these types of evidence.

7.3.1 Earprints

First, a distinction has to be drawn between an earprint and an earmark. The distinction made here is that of Champod *et al.* (2001). An *earprint* is a control impression taken from the ear or ears of a person. An *earmark* is the impression of a human ear which has been recovered from a crime scene. A critical review of the role of earprints as evidence is given by Champod *et al.* (2001). They comment that the current identification process for earprints is described mainly as a matching process, which is an assessment of the adequacy of the superposition between the mark and the print. The value of the match is not considered. In order to assess the value of a match a database of prints is needed. However, attempts to build a database have not considered the classification of earprints and earmarks. All that is provided is a 'gallery of images without a forensic structure' (Champod *et al.*, 2001).

Champod *et al.* (2001) describe a procedure which could be used to assist in the determination of the value of earprint evidence. The procedure is of sufficient generality that it could be used for other evidence types (e.g., document examination – see Section 7.3.6). Any such procedure requires at the outset a match algorithm. This is an algorithm which may be used to quantify the match which is apparent when an earprint is compared with an earmark (or when text written by a known person or from a known source is compared with text written by an unknown person or from an unknown source). The match is represented by a so-called *distance*. Thus, the quantification of the match between an earprint and an earmark may be represented by a statistic d, which is the distance between the print and the mark. The value of the evidence is then the likelihood ratio $f(d \mid H_p)/f(d \mid H_d)$ where $f(\cdot \mid \cdot)$ is the probability density of d under each of the two propositions, H_p and H_d, and

- H_p is the proposition that the earmark has been left by the person who provided the earprint and
- H_d is the proposition that the earmark has been left by the ear of some other unknown person.

Consider an earprint P provided by a person X and an earmark M which may or may not originated from X. The distance between P and M is denoted d.

The numerator $f(d \mid H_p)$ provides a measure of within-source variability. In order to estimate this, a set of earmarks provided by X are needed; denote these by $M_1, \ldots, M_m$. The distances from P to $M_1, \ldots, M_m$ are denoted $a_1, \ldots, a_m$. These provide data from which the density function $f(\cdot \mid H_p)$ for the numerator can be determined.

The denominator $f(d \mid H_d)$ provides a measure of between-source variability. In order to estimate this, a set of earprints provided by people other than X are needed; denote these by $P_1, \ldots, P_n$. The distances from M to $P_1, \ldots, P_n$ are denoted $b_1, \ldots, b_n$. These provide data from which the density function $f(\cdot \mid H_d)$ for the denominator can be determined. Whilst straightforward to describe, there are many difficulties to be overcome before this procedure can be used in practice. These difficulties include:

- the characteristics to measure;
- the matching algorithm to use;
- the sample sizes needed to ensure suitable precision in the density estimations.

More generally than in the examples described by Champod *et al.* (2001), there are three levels of variation, to consider:

- *Measurement error.* If the distance between two points on the print and the mark is measured more than once, the same answer may not always be obtained because of this.
- *Within-source variation.* If the distances between a point on the mark and the same point on different prints made from the same ear are compared, the same answer may not always be obtained because of this.
- *Between-source variation.* If the distances between a point on the mark and the same point on prints made from the ears of different people are compared, the same answer may not always be obtained because of this.

The above difficulties are common to many evidential types. For earprints it is also necessary to allow for differing pressure and orientation. Champod *et al.* (2001) comment that 'the variability for a given individual of the prints obtained with various pressures is important and has been shown to be larger than the variability of the marks given by an individual in a series of burglaries'.

The principles enunciated by Champod *et al.* (2001) and repeated above are very good and provide a template for the collection of data for the provision of databases for many different evidential types. Two cases in which evidence of earprints has been an issue are *R. v. Dallagher* and *R. v. Kempster*.

7.3.2   Firearms and toolmarks

A statistic which has been suggested (Biasotti, 1959; Biasotti and Murdock, 1997) for the assessment of identification of firearms and of toolmarks is the number of *consecutive matching striations* (CMS). The method has been reviewed by Bunch (2000). General criteria for firearm and toolmark identification are reviewed in Nichols (1997, 2003).

The determination of striations is partly subjective. Ideas are described here for firearms only. An examiner studies bullets under a macroscope and decides what is a striation and, for comparison, what matches there are between two bullets. The recovered bullet and a bullet fired from the gun which is suspected of being used in the crime (suspect gun) are compared. Observation is made of the matching striations and of the differences. The criterion for the (subjective) opinion that the bullet was fired from the suspect gun is based on a subjective assessment that the number of matching striations and their quality are sufficient for the formation of such an opinion.

For the analysis of CMS, the general procedure set out in Bunch (2000) describing Biasotti's approach is to fire numerous bullets through many firearms of the same make and model. A macroscopic comparison is then made of specimens known to have been fired from the same barrel and specimens known to have been fired from different barrels, and counts are made of the numbers of matching striations. Bunch (2000) describes a model in which the only CMS run on a bullet which matters is the one (or more) which features the maximum CMS count. Two data sets can be compiled, one for bullets fired from the same gun (*SG*) and one for bullets fired from different guns (*DG*). Hypothetical data are given by Bunch (2000) and reproduced in Table 7.2 where the data are the (hypothesised) relative frequencies for the maximum number of CMS obtained from the hypothetical experiment.

It would be unsatisfactory if results were not available for higher numbers of CMS simply because there were no empirical data available. Bunch (2000) explains how the results may be extrapolated to these higher numbers by postulating a probability distribution for the maximum CMS count. The distribution suggested is the Poisson distribution (Section 2.3.6) where the parameter is the weighted average maximum CMS count. Two Poisson distributions are required, one for bullets fired from the same gun and one for bullets fired from different guns. For bullets fired from the same gun, the weighted average maximum CMS count, denoted λ_S, is given by

$$\lambda_S = (0 \times 0.030) + (1 \times 0.070) + \cdots + (8 \times 0.020) = 3.91.$$

For bullets fired from different guns, the weighted average maximum *CMS* count denoted λ_D is given by

$$\lambda_D = (0 \times 0.220) + (1 \times 0.379) + \cdots + (6 \times 0.001) = 1.325.$$

Table 7.2   Hypothetical relative frequencies for the maximum CMS count y for bullets fired from the same gun $f(y \mid SG)$ and for bullets fired from different guns $f(y \mid DG)$, and likelihood ratios $V = f(y \mid SG)/f(y \mid DG)$. (Reprinted with permission from ASTM International)

Maximum CMS count (y)	$f(y \mid SG)$	$f(y \mid DG)$	V
0	0.030	0.220	0.136
1	0.070	0.379	0.185
2	0.110	0.300	0.367
3	0.190	0.070	2.71
4	0.220	0.020	11.0
5	0.200	0.010	20.0
6	0.110	0.001	110
7	0.050	–	*
8	0.020	–	*
9	–	–	*
Total	1.000	1.000	–

* The likelihood ratio is undefined because in the first two cases no cases exist of bullets fired from two different guns having more than six CMS and in the third case no cases exist of bullets fired from the same gun having more than eight CMS.

Let Y be the maximum CMS count for a particular bullet found at a crime scene when compared with a bullet fired from a gun, known as the suspect gun. If the suspect gun is the same gun as the one that fired the bullet found at the crime scene then

$$Pr(Y = y \mid SG) = \frac{\lambda_S^y}{y!} e^{-\lambda_S}, \qquad y = 0, 1, \ldots . \qquad (7.1)$$

If the gun is a different gun than the one that fired the bullet found at the crime scene then

$$Pr(Y = y \mid DG) = \frac{\lambda_D^y}{y!} e^{-\lambda_D}, \qquad y = 0, 1, \ldots . \qquad (7.2)$$

The propositions to be considered are:

- H_p, the bullet found at the crime scene was fired from the suspect gun;
- H_d, the bullet found at the crime scene was fired from a gun other than the suspect gun.

The evidence E to be considered is that the observed number of CMS is y_0. The ratio of $Pr(Y = y_0 \mid SG)$ to $Pr(Y = y_0 \mid DG)$ is then the value $Pr(E \mid H_p)/Pr(E \mid H_d)$ of the evidence, $E = y_0$, of the maximum CMS count. Bunch (2000) gives the results in Table 7.3 for his hypothetical data.

Table 7.3 Poisson probabilities for the maximum CMS count y for bullets fired from the same gun $Pr(Y = y \mid SG)$ (7.1) with mean $\lambda_S = 3.91$ and for bullets fired from different guns $Pr(Y = y \mid DG)$ (7.2) with mean $\lambda_D = 1.325$ and likelihood ratio $V = Pr(Y = y \mid SG)/Pr(Y = y \mid DG)$. (Reprinted with permission from ASTM International)

Maximum CMS count (y)	$Pr(Y = y \mid SG)$	$Pr(Y = y \mid DG)$	V
0	0.020	0.267	0.075
1	0.078	0.353	0.221
2	0.153	0.233	0.657
3	0.200	0.102	1.96
4	0.195	0.034	5.74
5	0.153	0.0089	17.2
6	0.099	0.00196	50.5
7	0.056	0.00037	151
8	0.027	0.000061	443
9	0.0118	0.000009	1311
10	0.0046	0.0000012	3833

The sums of the probabilities are not quite equal to 1 because of rounding error.

A Bayesian approach to the analysis of CMS data with hypothetical data is illustrated in Bunch (2000). As Nichols (2003) points out, CMS is just one of many factors to be considered in the study of firearms and toolmarks. The criterion of CMS, according to Nichols, addresses only the 'fruit of the tree': 'at the root of the entire set of difficulties is the lack of appropriate articulation and communication of what one means'. However, the use of CMS through the likelihood ratio enables a summary of the evidence of CMS to be in a phrase of the form 'the evidence is so many times more likely if H_p is true than if H_d is true'. This provides a very good summary of what the CMS statistic means in the context of determining the origin of the bullet found at a crime scene.

The determination of these means by Bunch (2000) using hypothetical data glosses over considerable problems which will arise if data are to be collected. Studies generally present data collected from 'perfect' test marks. For example, the cuts are made in lead, the bullets captured in water. Moreover, the trials are not blind. As stated earlier in the context of earprints (Section 7.3.1), there will be two levels of variation, in this case for variation in the maximum CMS for bullets fired from the same gun and for variation in the maximum CMS for bullets fired from different guns. However, the likelihood ratio as described by Bunch (2000) provides a good starting point for the resolution of these problems (Taroni *et al.*, 1996). Champod *et al.* (2003) mention the logical differences between the CMS threshold and likelihood ratio approaches, supporting the latter, while Moran (2001, 2002, 2003) favours the former.

7.3.3   Fingerprints

Bunch (2000) compared the counting of consecutive matching striations with the counting of similarities in matching fingerprints. He cited a study conducted by Evett and Williams (1996) in which no fingerprint examiner misidentified the most difficult comparison. This was a set of latent and ink prints from different people but with the prints modified to show many points of similarity. However, the examiners gave many different answers to the number of similarities. The same principle may apply to the counting of the maximum number of consecutive matching striations. Examiners may sometimes report different likelihood ratios for the same bullet. As Bunch (2000) explains, this is the analogue of examiners using traditional methods and drawing different conclusions about the same bullet.

Evaluation of fingerprint evidence is a very complex issue, beyond the scope of this book, and there have been a few attempts to construct a suitable statistical model. Overviews are given by Stoney (2001) and by Pankanti *et al.* (2002); the latter take advantage of an automatic fingerprint identification system. In forensic science, the latest work has been carried out by Champod and Margot (Champod, 1996; Champod and Margot, 1996, 1997).

Champod and Margot (1997) aimed at exploring the validity of a statistical model to compute match probabilities associated with level II features reduced to minutiae. The postulated model suggests that the probability of a configuration of minutiae, $Pr(C)$, can be computed from consideration of:

- $Pr(N)$, the probability of a given number of minutiae on the surface considered;

- $Pr(T)$, the probability of the observed types of minutiae (obtained by the multiplication of the probability of each type);

- $Pr(S)$, the probability of the orientations of the minutiae (obtained by the multiplication of the probability of the orientation for each minutia);

- $Pr(D)$, the probability of the length of minutiae (when applicable, obtained by the multiplication of the probability of each minutia length);

- $Pr(A)$, the probability of the arrangement of the minutiae.

These probabilities have been studied.

Statistical studies on minutiae provide extremely valuable knowledge, but they cannot yet be deployed for large-scale, case-specific calculations. As emphasised by Stoney (2001), none of the proposed models has been subjected to extended empirical validation studies. In the meantime, as proposed by Kingston (1970), such studies provide valuable data to verify the subjective judgements of experts concerning the rarity of given fingerprint features. The discrepancies that may be observed between examiners in the assessment of the relative frequencies of various types of minutiae have been explored (Osterburg and Bloomington, 1964). Such variations have to be minimised, and statistical surveys provide an excellent baseline. These data can be found in Champod *et al.* (2004).

Some of the debate concerning the definition of a suitable population for the assessment of fingerprint evidence is of interest here. Suppose a suspect has been chosen on the basis of fingerprint evidence alone. The underlying population from which the suspect has to be considered as being selected has been argued as being that of the whole world (Kingston, 1964). However, Stoney and Thornton (1986) argued that it is rarely the case that a suspect would be chosen purely on the basis of fingerprint evidence. Normally, there would have been a small group of suspects who would have been isolated from the world population on the basis of other evidence, though Kingston (1988) disagreed with this. The fingerprint evidence has then to be considered relative to this small group only. Further discussion of this point is given in Aitken (1991).

There has been considerable debate about the role of a standard for identity based on a fixed number of matching points between two prints (see Champod, 1995, for a historical account and critique). Thus, in a particular case, if the number of matching points was at least the number specified in the standard then the examiner would testify that the prints came from the person. If the number of matching points was not at least the number specified in the standard then the examiner would testify that the prints did not come from the person. This led to a dichotomy in that the interpretation of prints which just failed to meet the standard would be very different from the interpretation of prints which just did meet the standard.

A non-numeric fingerprint evidence standard was introduced in the UK in 2001 following a report by Evett and Williams (1996). A review of this standard is given by Knowles (2000), where it is commented that it was an overwhelming view of heads of fingerprint bureaux, meeting to discuss the removal of the 16-point standard, that 'positivity of evidence should remain'. This view is criticised by Taroni and Margot (2000) who advocate that fingerprint experts should accept that 'identification is purely a probabilistic assessment of the value of any type of evidence'. They cite Locard (1914): 'there are few characteristics: in that case (the) print(s) show(s) no certainty but a presumption proportional to the number of points and their sharpness'. In conclusion, Taroni and Margot (2000, p. 248) provide three criteria which are applicable to many forms of scientific evidence. These are the need to:

(a) emphasise that an element of scientific evidence provided by the expert is one element (as strong as it may be) among others which helps the judge in deciding on an identification, or more generally, in deciding about the link between the mark discovered and a potential perpetrator;

(b) consider the objective part of this type of evidence as probabilistic, in the sense that, from the characteristics observed on the mark, an expert will exclude a certain population ([from having] caused it), an argument that has to be integrated in the court process of decision;

(c) require that efforts are made to collect data and establish a model to describe the decision process.

The possibility of a probabilistic approach to fingerprint evidence is discussed by Champod and Evett (2001). They conclude that, with proper training, the adoption of a probabilistic approach to the presentation of evidence of finger-prints will be no different than at present and that evidence may well be given in a greater number of cases than at present. Note that such a discussion can be easily translated into odontology, another forensic area where a numerical standard is applied.

In an interesting case in the USA (*U.S. v Llera Plaza*), the judge ruled that expert witnesses will not be permitted to present 'evaluation testimony as to their opinion that a particular latent print is in fact the print of a particular person'. This ruling was later vacated by the judge who said that, 'In short, I have changed my mind'.

7.3.4   Speaker recognition

An approach to evaluation of the evidence of speech using a likelihood ratio has been described by Rose (2002, 2003). He uses formula (10.9) given in Section 10.2.6. Part of the evidence in forensic speaker identification is the result of acoustic analyses. Speech sounds are rapid fluctuations in air pressure. They are made when air is made to move by the vocal organs. The acoustic properties of the radiated speech wave constitute the basis for the phonetician's acoustic analysis and auditory transcription. Data are provided by two structures which are basic to the production of speech. These are the vocal chords and the supralaryngeal vocal tract. They are two independently functioning and controlled modules in speech production. The supralaryngeal vocal tract includes the tongue, lips and teeth.

The continuous data used include the fundamental frequency. Speakers produce different pitches by controlling the rate of vibration of their vocal chords, and these different vibration rates are signalled acoustically by an easily measurable acoustic parameter known as the fundamental frequency, denoted F0. It is measured in Hertz (Hz); average F0 values lie within a range of 180 to 300 Hz for females, and 90 to 140 Hz for males (Rose, 2003). However, there are factors such as race, age, smoking and intoxication which are associated with F0. A distribution of F0 can be built up over a period of time, and Rose (2002) comments that there is experimental support for at least 60 seconds' worth of speech for meaningful long-term F0 measurements. Given a distribution for F0 it is then possible to determine a mean and standard deviation. Rose (2002) fits a Normal distribution to F0.

As well as the fundamental frequency, other frequencies are of importance forensically. These are produced by the supralaryngeal vocal tract. Exciting the air in a supralaryngeal vocal tract, for example with a uniform cross-sectional area and an overall length of 17.5 cm, gives rise to an acoustic response whereby the air vibrates with maximum amplitude at the frequencies of 500 Hz, 1500 Hz, 2500 Hz, etc. In acoustic phonetics, vocal tract resonances are usually called *formants.* The formant with the lowest frequency, 500 Hz, is the first formant or F1. Other formants are labelled F2 (1500 Hz), F3 (2500 Hz), etc. Some, but not all, formants have Normal distributions about their central values.

An artificial example is given by Rose (2002) to illustrate how prior odds may be determined. An incriminating phone call could have been intercepted from a house in which five men, including the suspect, were known to be at the time of the call. Prior to consideration of any voice evidence, the probability that the suspect made the call is 1/5. The prior odds are thus 4 to 1 against the suspect making the call.

Other examples of the use of the likelihood ratio for speech/speaker recognition are given in Champod and Meuwly (2000), Meuwly (2001) and Meuwly and Drygajlo (2001).

7.3.5 Hair

The impracticality of the development of relevant databases, as described by Gaudette (1999), ensures that it is not currently possible to consider the use of a likelihood ratio for the evaluation of hair evidence.

As discussed in Section 4.6, the relative role of type 1 and type 2 errors is of interest. A level of discrimination should be set which minimises the number of type 2 errors without incurring too many type 1 errors. However, when considering what Gaudette (1999) calls *screening* evidence prior to DNA analysis, it is the type 1 errors which are more serious. Rather than 'if in doubt, throw it out', the rule of thumb is 'if in doubt, include it' (Gaudette, 1999).

Average probabilities (see Section 4.5.2) were developed for the use in hair comparisons because of the impracticality of the development of relevant databases. Gaudette and Keeping (1974) determined that if a single scalp hair selected at random from an individual *A* was found to be consistent with a representative known sample (consisting of about nine mutually dissimilar hairs) from *B* then, on average, the chance of a coincidental match was about 1 in 4500. In a similar study with pubic hairs, Gaudette (1986) showed that if a single pubic hair selected at random from an individual *X* was found to be consistent with a single pubic hair selected at random from an individual *Y*, an estimate of the average probability of a coincidental match was about 1/6336. Gaudette (1999) notes that the finding of two or more questioned hairs to be consistent with the known sample will greatly reduce the probability of a coincidental match. It is not possible, however, simply to square the probability of a coincidental match for one hair (e.g., $1/4500 \times 1/4500$ for human head hairs) since independence cannot be assumed.

Gaudette (1999) places average probabilities in the scale of support for the prosecution or defence propositions, as described in Section 4.5.2. The average probability for the hair evidence can be put together with various factors which strengthen or weaken one or other of the propositions. Gaudette cautions that great care must be taken in the interpretation of negative hair comparison results. For example, 'the finding of a few hairs on a suspect's shirt which matched neither his own nor the victim's known sample would have no probative value. On the other hand, it would be highly significant if a clump of hairs found clutched in a murder victim's hand matched neither the victim's nor a suspect's known sample' (Gaudette, 1999)

Likelihood ratios for various situations are presented by Hoffmann (1991). The relative frequencies $(p_1, \ldots, p_k)$ of categories $(x_1, \ldots x_k)$ of a feature X from within a suspected source are distinguished from the relative frequencies $(q_1, \ldots, q_k)$ of the same categories of the same feature determined from within a relevant population by means of an existing database. For the calculation of $(p_1, \ldots, p_k)$, it is necessary to have what Hoffmann (1991) calls 'a representative and large sample of hairs from the suspected source'. There is considerable scope for discussion of what is meant by 'representative' and by 'large'.

Suppose one hair is found at the crime scene and that it is of category x_i. A suspect is found whose hair matches that of the crime scene hair. The propositions to be considered are:

- H_p, the hair in question did come from the suspected source;
- H_d, the hair in question did not come from the suspected source.

Then $V = p_i/q_i$.

Suppose now that n hairs are found, all from one source. This is the evidence E. The propositions under consideration are

- H_p, the n hairs in question did come from the suspected source;
- H_d, the n hairs in question came from one source, different from the suspected source.

Let n_i be the number of hairs in category i found at the crime scene and $n_1 + \cdots + n_k = n$. Often, some of the n_i will be zero. When H_p is true,

$$Pr(E \mid H_p) = \frac{n!}{n_1! \cdots n_k!} p_1^{n_1} \cdots p_k^{n_k},$$

an example of the multinomial distribution (Section 2.3.4).

When H_d is true, the probabilities q_i cannot be used. These probabilities are only applicable if the hairs all come from different sources. They represent the variability of the feature X within the population approximated by the database. However, it is an intra-individual variability that is required. The database must contain records in which each record corresponds to one source and contains the relative frequencies of the categories $\{x_i, i = 1, \ldots, k\}$ within this source. Let N be the number of records within the database and denote the relative frequency of x_i within the tth record by q_{ti}. Then the probability of the evidence, given H_d is true, is estimated by the average over the N records in the database of the individual multinomial probabilities, that is,

$$Pr(E \mid H_d) = \frac{1}{N} \frac{n!}{n_1! \cdots n_k!} \sum_{t=1}^{N} q_{t1}^{n_1} \cdots q_{tk}^{n_k}.$$

The value of the evidence is then

$$V = \left[\frac{1}{N} \sum_{t=1}^{N} \left(\frac{q_{t1}}{p_1} \right)^{n_1} \cdots \left(\frac{q_{tk}}{p_k} \right)^{n_k} \right]^{-1}.$$

A further extension of the situation to one in which n hairs are found from r sources is discussed by Hoffmann (1991). However, as Hoffmann acknowledges, the discussion assumes a rather unrealistic assumption that the number of sources r is known and that the assignment of the hairs to the sources is known. Also, Hoffmann's (1991) procedure requires that the prosecution proposition H_p be divided into r components H_{pj}, where H_{pj} is the proposition that the jth source of hairs is the jth suspect. The defence proposition, H_d, is that all r sources are different from the suspect. It is possible, as Hoffmann (1991) shows, to obtain an overall value for the likelihood ratio. Care has to be taken over the statement of the propositions. Hoffmann (1991) is concerned with r sources but only one suspect.

However, there is still the problem that a database of characteristics and variability between and within individuals is required. In addition, there is a bias in the methodology from the psychological point of view. The comparison is made of hairs known to have come from different people. The judgement on the match or otherwise between two hairs coming from two different persons will be influenced by this knowledge. There is a subjective judgement made by the expert. This point is discussed in Barnett and Ogle (1982), Miller (1987) and the Scientific Sleuthing Newsletter (1988).

Further discussion is to be found in Hoffman (1991), Wickenheiser and Hepworth (1990, 1991) and Ogle (1991).

7.3.6 Documents

The Dreyfus case (see also Section 4.2) is an early use of the likelihood ratio in document examination. Another early example by Souder (1934/1935) is cited by Champod *et al.* (1999), who also cite Hilton (1995). Champod *et al.* (1999) discuss the Dreyfus case with particular reference to a report by Darboux *et al.* (1908). The two propositions under consideration are:

- H_p, the *bordereau* handwriting is a forgery;
- H_d, the *bordereau* handwriting is natural handwriting.

The other components of the Bayesian approach are:

- the evidence, E, which is the alleged coincidences observed on the *bordereau*,
- the background information, I, related to the case (police enquiry, eyewitness testimony, etc.).

Bertillon (1899, 1905) was one of those who argued that Dreyfus was the author of the *bordereau*. Part of his report provided an early example of the prosecutor's fallacy. Since $Pr(E \mid H_d, I)$ was a small probability p, then $Pr(H_d \mid E, I)$ equalled p and hence $Pr(H_p \mid E, I)$ was a large probability $(1 - p)$.

Bertillon also made a mistake in the probability calculations. He traced 4 mm interval vertical lines on the *bordereau* and showed that four pairs of polysyllabic words (amongst 26) had the same relative position with respect to the grid of vertical lines he had drawn. Bertillon then stated that the coincidences of the same relative positions could not be attributed to normal handwriting. He claimed that it would be necessary to repeat the composition of the document many hundreds of thousands of times in order to observe a chance occurrence of such an arrangement of coincidences. Bertillon concluded that the *bordereau* was a forged document.

The following is an example of the probability calculation. Let the probability of an individual coincidence be 0.2. The probability of observing four coincidences out of four events is then $0.2^4 = 0.0016 = 1/625$. For N coincidences, the probability is 0.2^N. For Bertillon, the four coincidences had such a small probability $(1/625)$ if the *bordereau* were the result of normal handwriting that he accepted the alternative proposition that the *bordereau* was a forgery, an example of the prosecutor's fallacy. Note that $1/625$ is rather much larger than the probability implied by the 'many hundred thousand times' initially suggested by Bertillon.

Bertillon, however, has not only made the mistake of the prosecutor's fallacy. He has also not calculated the probability of the four coincidences correctly. He looked at 26 pairs of polysyllabic words and identified four coincidences. Thus there were 22 pairs which were not coincidences. The probability of four coincidences out of 26, when the probability of a coincidence is 0.2, and assuming the conditions for a binomial distribution (see Section 2.3.3), is

$$\binom{26}{4} 0.2^4 \, 0.8^{22} = 14\,950 \times 0.0016 \times 0.0074 = 0.176.$$

Another famous case of disputed authorship, not discussed here, is that of the *Federalist* papers, a set of 77 essays published anonymously in 1787–1788 to persuade the citizens of the State of New York to ratify the Constitution. The authorship of 12 of the papers is in dispute, it being uncertain in each case whether Alexander Hamilton or James Madison was the author. There are three other papers where the dispute concerns the extent of each man's contribution. Details may be found in Mosteller and Wallace (1984). In an earlier publication,

Mosteller and Wallace (1963), on the same topic, comment in their conclusions that '[S]tudy of the variation of results with different priors is recommended. Bracketing the prior is often easy. When founded in data, the choice of the prior has a status like that of the data distribution – subjectivity tempered with empiricism.' The comment that the choice of the data distribution is based on 'subjectivity tempered with empiricism' is pertinent for those who argue in favour of the objectivity of the frequentist approach to statistics over the perceived inadequacy of the subjectivity of the Bayesian approach.

7.3.7 Envelopes

A simple example from a real case is given of the assessment of the probability that an envelope S recovered from a suspect's home has the same source as questioned envelopes Q. There can be one or more questioned envelopes. The case described here has two. The aim of this 'naive' approach is essentially to show that a Bayesian framework can also be easily used in a forensic field traditionally not related to the routine use of probabilities.

The structure of the analysis requires specification of:

- two propositions, that of the prosecution (H_p) and that of the defence (H_d);
- evidence (E) to be evaluated;
- background information (I).

Here the background information I includes evidence that is accepted prior to the evaluation of E and assumptions that are made following investigations of the background of the evidential type. For the example described here, these investigations concern the printing of the envelopes and the occurrence of identifying characteristics on them. Envelopes are sold in packets. It is not possible to sell individual envelopes. Packets are of particular types, identified by a specific code. All three envelopes in the case are of the same type. Because of the manufacturing process, only n packets of that type have been sold. In order to evaluate the evidence, it is necessary to have a background population of packets. Denote the size of this population by N; that is, there are N packets, which may be subdivided into different types. Only n of the N are of the type seen in Q and S. The background information is then:

- the questioned envelopes are from the same packet and are thus of the same type;
- there are exactly n packets of this type in a population of size N packets.

The evidence E to be evaluated is that the envelope S in the suspect's house is of the same type as the questioned envelopes.

The propositions are

- H_p, S is from the same packet as the questioned envelopes Q;
- H_d, S is from a different packet of envelopes to Q.

The numerator, $Pr(E \mid H_p, I)$, of the likelihood ratio is 1. If S is from the same packet as Q then it will be of the same type. The denominator, $Pr(E \mid H_d, I)$, is n/N. There are N packets, of which n are of the same type as Q but not from the same packet. Thus, the probability that S is of the same type as Q but not from the same packet is the ratio n/N. The value of the evidence is then N/n.

It is possible to evaluate a posterior probability $Pr(H_p \mid E, I)$ for H_p, as a function of n, N and a prior probability $Pr(H_p \mid I)$.

The above analysis is a simple approach to the evaluation. There are several levels of evidence which can be evaluated. Denote these as E_1, E_2, E_3. There are initial prior odds, $Pr(H_p)/Pr(H_d)$, which are then updated to $Pr(H_p \mid E_1)/Pr(H_d \mid E_1)$, and these become the prior odds for the evaluation of E_2. The posterior odds for the evaluation of E_2 become the prior odds for the evaluation of E_3. Consider the following descriptions of the evidence:

- E_1, the brand and model of Q and S,
- E_2, the fabrication code of Q and S,
- E_3, the defects in the printing process found in Q and S.

The initial prior odds relates to the total number of packets of envelopes produced in Switzerland, the country in which the crime was committed. Packets are considered because, in the case of concern, envelopes were only sold in packets. The packets contain 50 envelopes, and it is assumed (perhaps unrealistically) that this is constant over all manufacturers of envelopes. The analysis could be extended to include envelopes produced in other countries, but these could either be eliminated altogether or could be assigned a prior probability which would then be built into the analysis. Consideration of E_1 provides odds for the brand and model, consideration of E_2 reduces this to odds for the fabrication code and, finally, E_3 enables evaluation of the defects in the printing process. The size, N, of the population of envelopes in the case under discussion is then the number of envelopes with the fabrication code of Q and S.

Let N_0 be the total number of packets of envelopes in Switzerland. Let N_1 be the total number of packets of envelopes of the brand and model specified by Q and S. This enables evaluation of E_1. The total number of packets of envelopes of the fabrication code specified by Q and S is N. This enables evaluation of E_2. The number of packets of envelopes of the defects specified by Q and S is n. This enables evaluation of E_3.

The initial prior odds are $1/N_0$ (strictly speaking $1/(N_0 - 1)$). After E_1 is considered the odds are $1/N_1$. After E_2 is considered the odds are $1/N$. The value of E_3 is then n/N and the posterior odds are $1/n$.

7.3.8   Handwriting

One typical problem which requires examination of handwriting is the comparison of two pieces of handwriting to determine if they were written by the same person. The two pieces of writing could be signatures, for example. It could be that the writer of one was known and the writer of the second was unknown but was claimed by one party to a dispute to be the writer of the first one. Alternatively, the writer of neither signature would be known.

Another problem would be to identify the writer of a document from a set of possible writers to which it is known the writer belongs.

Unlike DNA or glass, for example, there are no general databases from which underlying variation in handwriting may be modelled. Thus, a direct determination of a likelihood ratio from measurements on control and recovered items (e.g., documents or signatures) is not possible.

One possible approach, in the absence of a database, is that used in the case of the Howland will, from the 1860s, and described by Meier and Zabell (1980), who describe an analysis by Benjamin Peirce, Professor of Mathematics at Harvard, and his son Charles Sanders Peirce (already cited in Sections 4.1 and 3.4.2) of an alleged forged signature on a will. There was a collection of n signatures ($n = 42$ in the Howland will case) $S_1, \ldots, S_n$ known to be written by a particular person, A say. There was a signature, S_0, accepted to have been written by A on a will, and a disputed signature, S_d, alleged to have been written by A. The question of interest was whether the disputed signature was written by A. A set of m features were considered. A comparison between signatures recorded the number of coincidences in the m features. A perfect match would be m coincidences. The approach described by Meier and Zabell (1980) was to develop a statistic to assess the similarity between two signatures. All possible pairwise comparisons amongst the n signatures may be made. The number of comparisons of features is then $\binom{n}{2} \times m$. For the Howland will case $m = 30$, and the number of comparisons is thus

$$\binom{42}{2} \times 30 = 25\,830.$$

The number of agreements, m_0, is recorded. Thus, the probability P_0 of an agreement in a feature in a comparison of two signatures, known to have been written by the same person, may be estimated as

$$m_0 \bigg/ \left\{ \binom{n}{2} \times m \right\}.$$

In the Howland will case, m_0 was $5\,325$ so $P_0 = 0.206$, or approximately 1 in 5. The number m_1 ($\leq m$) of coincidences in features between the disputed signature S_d and S_0 may then be noted (for the Howland will, $m_0 = m$). The probability of a coincidence between features in signatures known to have been written by the same person is then

$$\left[m_0 \middle/ \left\{ \binom{n}{2} \times m \right\} \right]^{m_1}.$$

For the Howland will case, this is approximately $(1/5)^{30}$, a very small number. Meier and Zabell (1980) discuss the merits of this procedure with particular reference to the independence assumption. The inference is that the disputed signature is a copy, and hence a forgery, of the known signature. The number of coincidences in features may be considered as a measure of distance between two signatures. The greater the number of coincidences, the more similar the two signatures are.

Another approach, exemplified by a study conducted by Srihari *et al.* (2002) on handwritten document examination, uses a distance measure where the smaller the distance, the more similar the handwriting in the two documents is. The objective of the study was to obtain a set of handwriting samples that captured variations in handwriting between and within writers. This required the collection of handwriting samples from multiple writers as well as multiple samples from each writer.

The population to which the study was to refer was to be as representative of the population of the USA as possible. The database had about 1500 people in it. Various features were measured. The data derived from these features are multivariate. Further details of the analysis of multivariate data are given in Chapter 11. Note, in particular, that there is a change of notation for multivariate data. For univariate data, control data are denoted x and recovered data y, with background data denoted z. For multivariate data, the control data are denoted $\mathbf{y}_1$ and the recovered data $\mathbf{y}_2$. Background data are denoted $\mathbf{x}$. The example here of multivariate data in handwriting is purely illustrative. The features measured included pen pressure, writing movement, stroke formation, slant and height. From these features, vectors associated with each writer in the sample may be derived. Consider two writers with corresponding vectors $(y_{11}, \ldots, y_{1p})^T$ and $(y_{21}, \ldots, y_{2p})^T$, where p is the number of features under consideration. The distance between the writers can then be defined by the Euclidean distance $\sqrt{\{\sum_{i=1}^{p} (y_{1i} - y_{2i})^2\}}$, though other measures could be used. See Section 10.1 and Chapter 11 for further details of vector notation.

Accuracy of identification was evaluated by comparing writing within and between writers. The accuracy of the methods investigated was assessed through a series of experiments using type 1 and type 2 errors. A type 1 error was defined as the probability of misclassifying two handwriting samples as written by two different writers when they were actually written by the same writer. A type 2 error was defined as the probability of misclassifying two handwriting samples as written by the same writer when they were actually written by two different writers. The discriminatory power of handwriting was assessed by considering two different approaches: the identification of a writer from a set of possible writers and the determination of whether two documents were written by the same writer. The use of type 1 and type 2 errors may be criticised on the same grounds as those used for significance probabilities (see Section 4.6). Evaluation of the evidence using a likelihood ratio requires probability distributions for the variation in the statistic between and within writers.

A sample of writers is needed, for example the 1500 used by Srihari *et al.* (2002). Measurements are made of the variables for each writer. Thus, assume n samples for each of m writers and p variables for each sample of writing. Then, for each sample, the variable of characteristics is denoted

$$\mathbf{x}_{ij} = (x_{ij1}, \ldots, x_{ijp})^T, \; i = 1, \ldots, m; \; j = 1, \ldots, n.$$

From these data, it is possible to estimate measurements of covariance (variances and correlations) between and within writers and derive distributions for variation between and within writers. Let the data on the control and recovered samples of writing be denoted $\mathbf{y}_1$ and $\mathbf{y}_2$, respectively. The likelihood ratio is then

$$\frac{f_p(\mathbf{y}_1, \mathbf{y}_2 \mid H_p)}{f_d(\mathbf{y}_1, \mathbf{y}_2 \mid H_d)}, \tag{7.3}$$

where f_p, f_d denote the probability density functions under the prosecution and defence propositions, respectively, and I has been omitted. This ratio would aid determination as to whether two samples were written by the same person. A more general discussion of a likelihood ratio of this form for univariate data is given in Chapter 10, with (10.1) corresponding to (7.3). More details of the likelihood ratio for multivariate data are given in Section 11.6 for an application and Section 11.9.2 for the components of the formula.

The following approach may be used to aid the identification of a writer from a set of possible writers (see also Chapter 8). Let $W_1, \ldots, W_s$ be the set of possible writers (assumed known). Let $\mathbf{y}$ be the vector of measurements on the questioned document. The probability density functions for $\mathbf{y}$ for each of $W_1, \ldots W_s$ are $f_l(\mathbf{y} \mid W_l)$, $l = 1, \ldots, s$. If the prior probabilities $Pr(W_l)$ that writers W_l wrote the questioned document are all equal then the probability that W_{l^*} wrote the document is

$$Pr(W_{l^*} \mid \mathbf{y}) = f_{l^*}(\mathbf{y} \mid W_{l^*}) / \sum_{l=1}^{s} f_l(\mathbf{y} \mid W_l),$$

where l^* is one of $1, \ldots, s$. If the probabilities $Pr(W_l)$ are not all equal then

$$Pr(W_{l^*} \mid \mathbf{y}) = f_{l^*}(\mathbf{y} \mid W_{l^*}) \times Pr(W_{l^*}) / \sum_{l=1}^{s} f_l(\mathbf{y} \mid W_l) \times Pr(W_l).$$

The two approaches discussed here (Srihari *et al.*, 2002; and Meier and Zabell, 1980) illustrate a standard method of assessing evidence when there is no population database and no underlying model with which to develop a likelihood ratio. The two approaches both

- develop a statistic to assess the similarity between two items of evidential significance,
- develop a database of measurements $(x_1, \ldots x_N$, say) of that statistic for items of known origin,
- from the database, construct an empirical distribution for the statistic,
- determine the measurement (x_0) of the statistic for the two items of evidential significance, and
- consider x_0 in the context of the empirical distribution.

An average probability, as described in Gaudette and Keeping (1974) and in Section 4.5.2, provides a measure of how good a particular evidential type is at discriminating between items from different sources. It is determined by making many pairwise comparisons and counting the proportion of these in which it is not possible to distinguish between items from different sources. The average probability does not make an assessment of the value of the evidence in a particular case.

The approaches of Meier and Zabell (1980) and of Srihari *et al.* (2002) take the idea of pairwise comparisons a stage further. An average probability is determined from a dichotomous matching process: the items are distinguishable or not. The other approach quantifies the difference between two items. It is thus able to consider a particular case. In the particular case, the difference between the two items, control and recovered, is measured and can then be compared with the set of differences in the database.

A likelihood ratio could be developed from this. First, a database of distances in pairwise comparisons from items known to have come from the same source could be constructed. From this, a probability density function f_p could be determined. For a particular case, the distance d_0 between the control and recovered items could be measured. The value of f_p could be calculated at d_0. Similarly, a database of distances of pairwise comparisons from items known to have come from different sources could be constructed. From this, a probability density function f_d could be determined. The value of f_d could be calculated at d_0. The values of $f_p(d_0)$ and $f_d(d_0)$ could then be compared using a likelihood ratio.

This is rather different from the use of the likelihood ratio advocated elsewhere in the book in which the databases are of measurements or characteristics of individual items. The likelihood ratio is then determined from the ratio of the joint distributions of the measurements or characteristics on the control and recovered items under the propositions H_p and H_d of the prosecution and defence, respectively, where the information in the database is used to aid development of the probability distributions for the measurements or characteristics on the control and recovered items.

7.3.9 Paint

The evidential value of paint is discussed in McDermott and Willis (1997) and McDermott *et al.* (1999). Note the two propositions:

- H_p, the paint on the injured party's vehicle originated from the suspect vehicle;
- H_d, the paint on the injured party's vehicle originated from a random vehicle.

There is an implication in the choice of defence proposition that the defence accept that the paint originated from a vehicle. If that premiss was not accepted by the defence, an alternative version of H_d could be that the paint on the injured party's vehicle did not originate from the suspect vehicle. Further discussion of the interpretation of paint evidence is given in Chapters 8 and 9.

7.4 PRE-DATA AND POST-DATA QUESTIONS

The worth of some evidential types may be assessed in a general way. For example, discriminating power (Section 4.5) provides a measure of the worth of hair examinations. Such a measure is provided independently of any particular crime. As such, the question of the worth of evidence assessed in this way has been termed a *pre-data* question. The worth of evidence garnered from a particular crime and used as a measure of support for one or other of the competing propositions about the crime is termed a *post-data* question. See Koons and Buscaglia (1999b) and Curran *et al.* (1999a) for a discussion of these ideas in the context of glass evidence.

A statistic known as *information content* has been proposed (Koons and Buscaglia, 1999a) as a measure of the worth of trace evidence. The use of the statistic is illustrated with reference to the elemental composition of glass. Ten elements are considered, and measurements of the composition of these ten elements are taken to be independent. Histograms of the distributions of the measurements on each of these ten elements are constructed from data collected from crime scene analysis.

For ease of reference and of generalisation, consider k variables (which may or may not be elements) and denote these as $x_1, \ldots, x_k$. Denote the number of cells used for each of the k variables as $n_1, \ldots, n_k$. The choice of the number of cells is subjective. As always, a balance has to be made between too few cells, which obscures the underlying variation in the data through too much smoothing, and too many cells, which obscures the underlying variation in the data with too much random variation (or *noise*). The information content in the data is then defined as the product of the number of cells in each of the histograms and is considered as 'the total number of possibly analytically distinguishable results'. Thus,

$$\text{information content} = n_1 \times \cdots \times n_k \qquad (7.4)$$

(Koons and Buscaglia, 1999a).

The larger this statistic, the more information is in the data. Thus the better the evidential type is at determining the worth of the finding of a match (in some sense) between evidence of the same type found at a crime scene and on a suspect in some future case.

Care has to be taken with such an interpretation for three reasons. First, the assumption of independence amongst the variables has to be considered. It is not sufficient simply to consider correlations and to discard one of two variables which are perfectly correlated. Figure 4.1 shows two measurements (denoted variable 1 and variable 2) of eight items in each of two groups of data, indicated by a $\triangle$ and by a $\bigcirc$. Separately, they are perfectly positively correlated. Together, ignoring the grouping, the correlation is 0.9. Measurements of variable 1 alone do not discriminate between the two groups. Measurements of variable 2 alone do not discriminate between the two groups. However, measurements of variable 1 and variable 2 provide perfect discrimination.

The second reason for care is the subjective choice of the number of cells chosen for the histograms for each of the variables. The more cells are chosen for each variable, the greater the information content. However, it is difficult to know the optimal number of cells to choose. Too many cells and there may be a good fit to the current data but poor predictive ability for future data. Too few cells and the fit to the data may be poor, also leading to poor predictive ability.

The third reason for care is that not all $n_1 \times \cdots \times n_k$ cells in the combined data set will be equally likely to occur. As an extreme example, some combinations of cells may be empty, because of correlations amongst the variables, and so provide no information but add to the value of the information content statistic.

Information content answers a 'pre-data question: "what is the probability I would make a mistake if I carried out this matching procedure?' (see the correspondence in Curran *et al*, 1999a, and Koons and Buscaglia, 1999b). A post-data question is one which evaluates the evidence in a particular case (Curran *et al.*, 1997a, b). A few further comments are made in Section 11.7.

Consider now a general post-data question. A database exists of some evidential type but only of general qualitative (and possibly subjective) assessments of the variability in the characteristics of this evidential type. A crime has been committed. A trace has been found at the crime scene; this is the crime trace. A trace of the same evidential type has been found in association with a suspect; this is the recovered trace. The two traces are indistinguishable in some sense. The two traces are compared with all traces in the database. The crime and recovered traces are distinguishable from all traces in the database.

Fingerprint evidence is an obvious illustration of this scenario. However, there may be other evidential types for which the database is not large, and may even be quite small. Consider earprints (Section 7.3.1). We have two propositions and some evidence.

- the prosecution's proposition (P), that the earprint from the suspect and the earmark at crime scene are from the same person;
- the defence proposition (D), that the earprint from suspect and the earmark at crime scene are not from the same person;
- the evidence (E), that the earprint from suspect and the earmark at the crime scene are indistinguishable, according to the opinion of an expert witness.

Assume that if P is true, $Pr(E|P) = 1$: there are no false negatives. If D is true, let $Pr(E|D) = \phi$. This is the probability of a match by chance between the suspect's earprint and the earmark left by another person in some population; it is hoped that this is small.

There is a database of n earprints, all known to be from different people. The prints are all identifiably different according to an expert witness. The database is assumed to be representative of the population from which the criminal is assumed to have come. It is assumed, also, that the suspect is a member of this population (otherwise he would not be a suspect).

All earprints in the database can be distinguished from the earprint of suspect. The earprint is known to come from the suspect. If the earmark at the crime scene came from someone other than the suspect the probability that the earmark would be determined to be indistinguishable from the earprint of the suspect is ϕ.

The database is assumed by the expert to be representative of the population. Thus, the information in the database and the suspect's earprint can be used to provide a meaningful estimate of ϕ.

The number of matches, X, found when earprints in the database are compared with the suspect's earprint is zero. Thus, a point estimate of ϕ based on this result would be zero, which takes no account of the variability implied from consideration of a sample rather than the entire population. A more relevant estimate would be that provided by a probabilistic upper bound for ϕ. This would enable it to be said that one was $100(1 - \gamma)\%$ certain that the true value of ϕ was less than ϕ_0, where γ is taken to be small. For example, γ could be taken to be 0.01 or 0.001, and the percentage certainty would be 99% or 99.9%.

Certain modelling assumptions require to be made in order to consider answers to the above comments.

1. There are only two possible outcomes (is or is not distinguishable from) when comparing earprints of the suspect and a member of the database.
2. Characteristics of members of the database and characteristics of the suspect are independent.
3. The probability ϕ that characteristics of the suspect and a member of the database are indistinguishable is constant for all members of the database.

If this is thought untrue, then future analysis requires data on individual characteristics.

4. The number of matches, X, by chance alone, in n comparisons, is modelled by a binomial distribution such that $Pr(X = x|n, \phi) \propto \phi^x(1 - \phi)^{n-x}$.

5. The prior distribution for ϕ is a beta distribution such that $f(\phi|\alpha, \beta) \propto \phi^{\alpha-1}(1 - \phi)^{\beta-1}$.

These assumptions imply a beta posterior distribution for ϕ, namely

$$f(\phi|x + \alpha, n - x + \beta) \propto \phi^{x+\alpha-1}(1 - \phi)^{n-x+\beta-1},$$

from which $Pr(\phi < \phi_0)$ may be determined. The constant of proportionality (the term to be inserted in the above expression to ensure that $f(\phi|x + \alpha, n - x + \beta)$ is a probability density function) may be determined with reference to (2.21), and in this example is equal to $1/B(x + \alpha, n - x + \beta)$; that is, the first term of the beta function is one more than the power of ϕ and the second term of the beta function is one more than the power of $(1 - \phi)$.

Some illustrative values for ϕ_0 are given in Tables 7.4 and 7.5 for different values of n and γ. It is assumed that $\alpha = \beta = 1$ and that $X = 0$.

The use of these tables is best illustrated by an example. Suppose 900 earprints, all known to be from different people, have been examined and found

Table 7.4 Values of a probabilistic γ upper bound for the probability ϕ_0 of a false match for different values of n and γ

n	γ	
	0.01	0.001
250	0.0182	0.0271
600	0.0076	0.0114
900	0.0051	0.0076

Table 7.5 Values of a probabilistic γ lower bound for the likelihood ratio $1/\phi_0$ for different values of n and γ, with values rounded down

n	γ	
	0.01	0.001
250	54	36
600	131	87
900	196	131

to be distinguishable, one from another. They are also all distinguishable from that of the suspect. The earprint from the suspect is indistinguishable from the earmark found at the crime scene. From these sample data, which are assumed to be representative of the population to which the criminal belongs, it can be said that, with reference to the last row of Table 7.5, there is a probability of

- 0.99 that the likelihood ratio for the evidence that the earprint of the suspect is indistinguishable from the earmark at the crime scene is greater than 196, and

- 0.999 that the likelihood ratio for the evidence that the earprint of the suspect is indistinguishable from the earmark at the crime scene is greater than 131.

Note that this analysis is not specific to a particular case. The only evidence which has been used is that of the size of the database (900) and the fact that the suspect's earprint is distinguishable from all members of the database. If examination of any future database provides examples in which there are earprints from different people which are found to be indistinguishable, then the analysis can be altered to allow for this. Also, all 900 earprints have been found to be indistinguishable from each other. This is $900 \times 899/2$ or $404\,550$ pairwise comparisons. This is the discriminating power of the evidential type, as discussed in Section 4.5, and is a measure of the pre-data worth of the evidence.

Finally, the matter of what is, or is not, distinguishable, is a matter of expert judgement and relies considerably on the expertise of the person making the judgement.

8

Transfer Evidence

8.1 THE LIKELIHOOD RATIO

8.1.1 Probability of guilt

Uncertainty may be measured by probability. However, attempts to measure uncertainty in forensic evidence by probability alone lead to difficulties in interpretation. The two-stage approach described in Chapter 4 illustrates this. Also, as discussed in Section 3.3.1, there is considerable potential for confusion between the following two conditional probabilities: the probability of the evidence given the guilt of the suspect and the probability of the guilt of the suspect given the evidence. Attempts to evaluate combinations of different pieces of evidence are also fraught with difficulty, requiring consideration of the dependence of these different pieces; see Cohen (1977, 1988) and Dawid (1987) for a debate about this, a debate which is also summarised below.

The scientist is concerned with the uncertainty regarding his evidence. He is not concerned with the guilt or otherwise of the suspect. This is the concern of the court. The court has to take account of much other information and evidence, of which the scientist is not aware, presented to it.

Consider the following example of the confusion that may arise if probabilities of guilt are determined for two pieces of evidence E_1 and E_2. Let H_p denote the proposition that the suspect is guilty. For both E_1 and E_2, suppose the court determines $Pr(H_p \mid E_1)$ and $Pr(H_p \mid E_2)$ to be 0.7. On a balance of probabilities, E_1 and E_2, separately, imply the guilt of the suspect. The court then multiplies these together, a multiplication which assumes some sort of independence, to produce a probability of $0.7^2 = 0.49$. This last probability is less than 0.5. Superficially, this seems to imply that two pieces of evidence, which separately imply the guilt of the suspect, when combined imply the innocence of the suspect. This apparent contradiction is part of the basis of the criticism by Cohen (1977) of the relevance of the calculations of standard probability (Cohen's term is *Pascalian*). However, Dawid (1987) explained how a rigorous Bayesian analysis is able to counter this criticism.

Statistics and the Evaluation of Evidence for Forensic Scientists: Second Edition
C.G.G. Aitken and F. Taroni © 2004 John Wiley & Sons, Ltd   ISBN: 0-470-84367-5

The posterior probability of guilt depends on the prior probability of guilt. Consider the two pieces of evidence above such that $Pr(H_p \mid E_1) = Pr(H_p \mid E_2) = 0.7$. The temptation, succumbed to above, which suggests that these two separate probability statements when combined imply that $Pr(H_p \mid E_1, E_2) = 0.49$ should be resisted. The odds form of Bayes' theorem (Section 3.4) clarifies the position. The version of the theorem which is applicable here states that

$$\frac{Pr(H_p \mid E_1, E_2)}{Pr(H_d \mid E_1, E_2)} = \frac{Pr(E_1, E_2 \mid H_p)}{Pr(E_1, E_2 \mid H_d)} \times \frac{Pr(H_p)}{Pr(H_d)}.$$

A clear understanding is needed of the meaning of 'independence' when applied to the two pieces of evidence, E_1 and E_2. In this context independence means that the joint probability of the two pieces of evidence, given guilt (H_p) or innocence (H_d), is equal to the product of the individual probabilities. It does not mean, for example, that $Pr(H_p \mid E_1, E_2) = Pr(H_p \mid E_1) \times Pr(H_p \mid E_2)$. Thus

$$Pr(E_1, E_2 \mid H_p) = Pr(E_1 \mid H_p) \times Pr(E_2 \mid H_p),$$

$$Pr(E_1, E_2 \mid H_d) = Pr(E_1 \mid H_d) \times Pr(E_2 \mid H_d), \text{ and}$$

$$\frac{Pr(E_1, E_2 \mid H_p)Pr(H_p)}{Pr(E_1, E_2 \mid H_d)Pr(H_d)} = \frac{Pr(E_1 \mid H_p)Pr(E_2 \mid H_p)Pr(H_p)}{Pr(E_1 \mid H_d)Pr(E_2 \mid H_d)Pr(H_d)}. \qquad (8.1)$$

If $Pr(H_p) = Pr(H_d)$, this equals

$$\frac{Pr(E_1 \mid H_p)Pr(H_p)}{Pr(E_1 \mid H_d)Pr(H_d)} \times \frac{Pr(E_2 \mid H_p)Pr(H_p)}{Pr(E_2 \mid H_d)Pr(H_d)}$$

$$= \frac{Pr(H_p \mid E_1)}{Pr(H_d \mid E_1)} \times \frac{Pr(H_p \mid E_2)}{Pr(H_d \mid E_2)}$$

$$= \frac{0.7}{0.3} \times \frac{0.7}{0.3}$$

$$= \frac{0.49}{0.09}.$$

Thus

$$Pr(H_p \mid E_1, E_2) = \frac{0.49/0.09}{1 + (0.49/0.09)} = \frac{0.49}{0.58} = 0.84 > 0.7.$$

Hence, if the prior odds are equal and if the two pieces of evidence are independent, conditional on the hypotheses of guilt or innocence, then the conjunction of E_1 and E_2 is such as to strengthen the probability of guilt.

8.1.2   Justification

The odds form of Bayes' theorem presents a compelling intuitive argument for the use of the likelihood ratio as a measure of the value of the evidence.

A mathematical argument exists also to justify its use. A simple proof is given in Good (1991) and repeated here for convenience.

It is desired to measure the value V of evidence E in favour of guilt H_p. There will be dependence on background information I, but this will not be stated explicitly. It is assumed that this value V is a function only of the probability of E given that the suspect is guilty, H_p, and of the probability of E given that the suspect is innocent, H_d.

Let $x = Pr(E \mid H_p)$ and $y = Pr(E \mid H_d)$. The assumption above states that

$$V = f(x, y)$$

for some function f.

Consider another piece of evidence T which is irrelevant to (independent of) E and H_p (and hence H_d) and which is such that $Pr(T) = \theta$. Then

$$Pr(E, T \mid H_p) = Pr(E \mid H_p)Pr(T \mid H_p)$$

by the independence of E and T

$$= Pr(E \mid H_p)Pr(T)$$

by the independence of T and H_p

$$= \theta x.$$

Similarly,

$$Pr(E, T \mid H_d) = \theta y.$$

The value of the combined evidence (E, T) is equal to the value of E, since T has been assumed irrelevant. The value of (E, T) is $f(\theta x, \theta y)$ and the value of $E = V = f(x, y)$. Thus

$$f(\theta x, \theta y) = f(x, y)$$

for all θ in the interval $[0, 1]$ of possible values of $Pr(T)$. It follows that f is a function of x/y alone and hence that V is a function of

$$Pr(E \mid H_p)/Pr(E \mid H_d),$$

namely, the likelihood ratio.

8.1.3   Combination of evidence and comparison of more than two propositions

Notice that the representation of the value of evidence as a likelihood ratio enables successive pieces of evidence to be evaluated sequentially in a much more intuitive and simpler way than can be done with significance probabilities, for example as in Section 4.6. The posterior odds from one piece of evidence, E_1 say, become the prior odds for the following piece of evidence, E_2 say. Thus

$$\frac{Pr(H_p \mid E_1)}{Pr(H_d \mid E_1)} = \frac{Pr(E_1 \mid H_p)}{Pr(E_1 \mid H_d)} \times \frac{Pr(H_p)}{Pr(H_d)}$$

and

$$\frac{Pr(H_p \mid E_1, E_2)}{Pr(H_d \mid E_1, E_2)} = \frac{Pr(E_2 \mid H_p, E_1)}{Pr(E_2 \mid H_d, E_1)} \times \frac{Pr(H_p \mid E_1)}{Pr(H_d \mid E_1)}$$

$$= \frac{Pr(E_2 \mid H_p, E_1)}{Pr(E_2 \mid H_d, E_1)} \times \frac{Pr(E_1 \mid H_p)}{Pr(E_1 \mid H_d)} \times \frac{Pr(H_p)}{Pr(H_d)}, \tag{8.2}$$

using a generalisation of (8.1). The possible dependence of E_2 on E_1 is recognised in the form of the probability statements within the likelihood ratio.

If the two pieces of evidence are independent (as may be the case with genetic marker systems) this leads to the likelihood ratios combining by simple multiplication,

$$\frac{Pr(E_1, E_2 \mid H_p)}{Pr(E_1, E_2 \mid H_p)} = \frac{Pr(E_1 \mid H_p)}{Pr(E_1 \mid H_p)} \times \frac{Pr(E_2 \mid H_p)}{Pr(E_2 \mid H_d)},$$

as in (8.1). Thus, if V_{12} is the likelihood ratio for the combination of evidence (E_1, E_2), and if V_1 and V_2 are the likelihood ratios for E_1 and E_2, respectively, then

$$V_{12} = V_1 \times V_2.$$

If the weight of evidence is used (Section 3.4.2), different pieces of evidence may be combined by addition. This is a procedure which has an intuitive analogy with the scales of justice.

Example 8.1   Consider the case of *State v. Klindt* discussed in Lenth (1986). This case has two aspects of interest. It illustrates a method for combining evidence and a method for weighing the evidence for more than two propositions. The example uses blood grouping, although DNA profiling would now be used. However, the principles remain the same and this example provides a good example of them.

The case involved the identity of a portion of a woman's body. The portion was analysed and it was determined that the woman was white, aged between 27 and 40, had given birth to at least one child and had not been surgically sterilised. Also, seven genetic markers were identified. All but four missing persons in the area of the four states around where the body was found were eliminated as possible identities for the woman. Label these four persons as P, Q, R and S. Women Q, R and S had been missing for 6 months, 6 years and 7 years, respectively, and their last known locations were at least 200 miles from where the body was found. Woman P had been missing for a month at the time the body was discovered and had last been seen in the same area.

The blood type of P was known to be A. The blood types of Q, R and S were not known. The other markers were unknown for all four women. For the remaining six phenotypes, samples of tissue from the parents of P enabled a value of 0.5 to be calculated for the probability that the woman whose body was found had the phenotypes it had if it were that of P; that is, Pr(given phenotypes| P) equals 0.5. (See also Section 9.8 for another example of this so-called parentage testing.) For some general population the incidence of these phenotypes was 0.007 64. No familial testing was done for Q, R or S. The ages of all four women were known and they were all mothers. However, in order to illustrate the procedure for combining evidence, Lenth (1986) made the following alterations to the actual data: it was not known whether Q was a mother, R's age was taken to be unknown, and S was known to have type A blood.

The above information is summarised in Table 8.1 using frequency information given by Lenth (1986). If a particular characteristic is known to hold for a particular woman then a probability of 1 (corresponding to certainty) is entered. If the presence or absence of an attribute is not known the incidence probability of the attribute amongst the general population is given. There are four propositions to be compared, one for each of the four women whose body may have been found. The evidence to be assessed is $E = \{E_1, \ldots, E_4\}$, the four separate pieces of evidence listed in Table 8.1. The probabilities to be determined

Table 8.1 Probability evidence in *State v. Klindt* (altered for illustrative purposes) from Lenth (1986)

Indicator	Attribute	Woman, X			
		P	Q	R	S
	Age (years)	33	27	unknown	37
E_1	Mother|age	1.0	0.583	1.0	1.0
E_2	Sterilised | mother, age	1.0	0.839	0.662	0.542
E_3	Type A blood	1.0	0.362	0.362	1.0
E_4	Other six phenotypes	0.500	0.007 64	0.007 64	0.007 64
$Pr(E_1, E_2, E_3,$ $E_4 \mid$ age, $X)$		0.5000	0.001 35	0.001 83	0.004 14
Fraction of total, $Pr(X \mid E,$ age$)$		0.9856	0.0027	0.0036	0.0082

are of the form $Pr(E_1 \ldots E_4 \mid age, X)$ where X is one of P, Q, R or S. These pieces of evidence are not all mutually independent. Denote the evidence as follows:

- E_1: mother, yes or no;
- E_2: not sterilised, yes or no;
- E_3: type A blood, yes or no;
- E_4: other six phenotypes.

Using the third law of probability for dependent events (1.7),

$$Pr(E \mid age, X) = Pr(E_1, E_2, E_3, E_4 \mid age, X)$$
$$= Pr(E_1 \mid age, X) \times Pr(E_2 \mid E_1, age, X) \times Pr(E_3 \mid E_2, E_1, age, X)$$
$$\times Pr(E_4 \mid E_3, E_2, E_1, age, X)$$
$$= Pr(E_1 \mid age, X) \times Pr(E_2 \mid E_1, age, X) \times Pr(E_3 \mid X)Pr(E_4 \mid X),$$

where X is one of P, Q, R or S, and the final expression depends on certain independence relationships amongst the four pieces of evidence. For example, the probability of sterilisation is dependent on age and being a mother or not, whilst the probability an individual has type A blood (E_3) is independent of age, whether a mother or not (E_1), and whether or not sterilised (E_2). The probabilities for the combined evidence E are given in the penultimate row of Table 8.1, namely $Pr(E \mid age, P) = 0.5000$, $Pr(E \mid age, Q) = 0.00135$, $Pr(E \mid age, R) = 0.00183$ and $Pr(E \mid age, S) = 0.00414$. Explicit dependence on 'age' is now omitted for ease of notation. The posterior probability that the body is that of P, $Pr(P \mid E)$, may be determined so long as information concerning the identity of the body prior to the discovery of the evidence contributing to E is available and is represented by the prior probabilities $Pr(P), Pr(Q), Pr(R)$ and $Pr(S)$. From Bayes' theorem (3.3),

$$Pr(P \mid E) = \frac{Pr(E \mid P)Pr(P)}{Pr(E)}$$

with similar results for Q, R and S. From the law of total probability (1.9), since the events P, Q, R and S are mutually exclusive and exhaustive,

$$Pr(E) = Pr(E \mid P)Pr(P) + Pr(E \mid Q)Pr(Q) + Pr(E \mid R)Pr(R) + Pr(E \mid S)Pr(S).$$

If the four prior probabilities are assumed mutually exclusive, exhaustive and equal,

$$Pr(P) = Pr(Q) = Pr(R) = Pr(S) = 1/4.$$

Hence,

$$Pr(P \mid E) = \frac{Pr(E \mid P)}{Pr(E \mid P) + Pr(E \mid Q) + Pr(E \mid R) + Pr(E \mid S)}.$$

From Table 8.1,

$$Pr(P \mid E) = \frac{0.5000}{0.5000 + 0.001\,35 + 0.001\,83 + 0.004\,14} = 0.9856,$$

the value given in the final row of the table. The posterior odds in favour of P versus Q, R or S then equal $0.5000/(0.001\,35 + 0.001\,83 + 0.004\,14) \simeq 68$. This can be verified by determining $Pr(P \mid E)/Pr(\bar{P} \mid E) = 0.9856/0.0144 \simeq 68$, where $\bar{P}$ is the complement of P, which, for the moment is taken to be Q, R or S.

The value V of the evidence, which equals $Pr(E \mid P)/Pr(E \mid \bar{P})$, may be determined, again if the prior probabilities are available. From Bayes' theorem (3.3),

$$
\begin{aligned}
Pr(E \mid \bar{P}) &= \frac{Pr(\bar{P} \mid E)Pr(E)}{Pr(\bar{P})} \\
&= \frac{\{Pr(Q \mid E) + Pr(R \mid E) + Pr(S \mid E)\}Pr(E)}{Pr(Q) + Pr(R) + Pr(S)} \\
&= \frac{Pr(E \mid Q)Pr(Q) + Pr(E \mid R)Pr(R) + Pr(E \mid S)Pr(S)}{Pr(Q) + Pr(R) + Pr(S)}.
\end{aligned}
\tag{8.3}
$$

If $Pr(Q) = Pr(R) = Pr(S)$ then

$$
\begin{aligned}
Pr(E \mid \bar{P}) &= \frac{Pr(E \mid Q) + Pr(E \mid R) + Pr(E \mid S)}{3} \\
&= \frac{0.007\,32}{3} \\
&= 0.002\,44.
\end{aligned}
$$

Thus

$$\frac{Pr(E \mid P)}{Pr(E \mid \bar{P})} = \frac{0.5000}{0.002\,44} = 204.92.$$

The evidence is about 205 times more likely if the body is that of P than that of one of the other three women.

If $Pr(P) = Pr(Q) = Pr(R) = Pr(S) = 1/4$ then $Pr(P)/Pr(\bar{P}) = 1/3$ and the posterior odds in favour of P equals

$$\frac{0.5000}{0.002\,44} \times \frac{1}{3} = \frac{0.5000}{0.007\,32} \simeq 68,$$

as determined previously.

It is also possible to determine the value of the evidence in favour of P relative to one other woman, Q say, in the usual way of comparing two propositions:

$$\frac{Pr(E \mid P)}{Pr(E \mid Q)} = \frac{0.5000}{0.001\,35} \simeq 370.$$

The evidence is about 370 times more likely if the body is that of P rather than that of Q.

As well as age, the conditioning on background information I has, as usual, been omitted for clarity of exposition. However, there is information available regarding the length of time for which the women have been missing and their last known locations. This may be considered as background information I. As such it may be incorporated into the prior probabilities which may then be written as $Pr(P \mid I)$, $Pr(Q \mid I)$, $Pr(R \mid I)$ and $Pr(S \mid I)$. Suppose $(P \mid I)$ is thought the most likely event and that the other three events are all equally unlikely. Represent this as $Pr(P \mid I) = 0.7, Pr(Q \mid I) = Pr(R \mid I) = Pr(S \mid I) = 0.1$ (though this is not the only possible combination of probabilities which satisfy this criterion). The likelihood ratio $Pr(E \mid P)/Pr(E \mid \bar{P})$ is the same as before (204.92) since it was determined assuming only that $Pr(Q) = Pr(R) = Pr(S)$, without specifying a particular value. The posterior odds alter, however. The prior odds are

$$\frac{Pr(P \mid I)}{Pr(\bar{P} \mid I)} = \frac{0.7}{0.3}.$$

The posterior odds then become

$$204.92 \times \frac{0.7}{0.3} \simeq 478$$

and $Pr(P \mid E) = 0.998$.

The posterior probabilities in Table 8.1 have been calculated assuming $Pr(P) = Pr(Q) = Pr(R) = Pr(S) = 1/4$. If these probabilities are not equal, the posterior probabilities have to be calculated taking account of the relative values of the four individual probabilities. For example,

$$Pr(P \mid E) = \frac{Pr(E \mid P)Pr(P)}{Pr(E \mid P)Pr(P) + Pr(E \mid Q)Pr(Q) + Pr(E \mid R)Pr(R) + Pr(E \mid S)Pr(S)}.$$

The likelihood ratio also has to be calculated again if $Pr(P)$, $Pr(Q)$, $Pr(R)$ and $Pr(S)$ are not equal. From (8.3),

$$\frac{Pr(E \mid P)}{Pr(E \mid \bar{P})} = \frac{Pr(E \mid P)\{Pr(Q) + Pr(R) + Pr(S)\}}{Pr(E \mid Q)Pr(Q) + Pr(E \mid R)Pr(R) + Pr(E \mid S)Pr(S)}. \tag{8.4}$$

These results for comparing four propositions may be generalised to any number, n say, of competing exclusive propositions. Let $H_1, \dots, H_n$ be n exclusive propositions and let E be the evidence to be evaluated. Denote the probability of E under each of the n propositions by $Pr(E \mid H_i), i = 1, \dots, n$. Let $p_i = Pr(H_i)$, $i = 1, \dots, n$, be the prior probabilities of the propositions, such that $\sum_{i=1}^{n} p_i = 1$. Then consider the value E for comparing H_1 with $(H_2, \dots, H_n) = \bar{H}_1$, say:

$$\frac{Pr(E \mid H_1)}{Pr(E \mid \bar{H}_1)} = \frac{Pr(E \mid H_1)(\sum_{i=2}^{n} p_i)}{\sum_{i=2}^{n} Pr(E \mid H_i)p_i}$$

from a straightforward extension of (8.4). Thus

$$\frac{Pr(E \mid H_1)}{Pr(E \mid \bar{H}_1)} = \frac{Pr(E \mid H_1)(1 - p_1)}{\sum_{i=2}^{n} Pr(E \mid H_i)p_i},$$

since $\sum_{i=2}^{n} p_i = 1 - p_1$, and

$$Pr(H_1 \mid E) = \frac{Pr(E \mid H_1)p_1}{Pr(E)}$$

$$= \frac{Pr(E \mid H_1)p_1}{\sum_{i=1}^{n} Pr(E \mid H_i)p_i}.$$

The posterior odds are best evaluated by writing $Pr(\bar{H}_1 \mid E)$ as $1 - Pr(H_1 \mid E)$ and then

$$\frac{Pr(H_1 \mid E)}{Pr(\bar{H}_1 \mid E)} = \frac{Pr(E \mid H_1)p_1 / \sum_{i=1}^{n} Pr(E \mid H_i)p_i}{1 - \{Pr(E \mid H_1)p_1 / \sum_{i=1}^{n} Pr(E \mid H_i)p_i\}}$$

$$= \frac{Pr(E \mid H_1)p_1}{\sum_{i=2}^{n} Pr(E \mid H_i)p_i}.$$

Example 8.1 (Continued): Population considerations The probability figures for non-sterilisation of women, and for women who are mothers and of a certain age, are based on information about white women in general. It is unlikely that the probabilities are the same among missing white women (Lenth, 1986). However, these results are the best available and serve to illustrate the methodology.

There is a possibility, not so far accounted for, that the body may be that of a woman other than P, Q, R or S. If such a possibility is to be considered then it can be done by adding an extra proposition to the four under consideration and using the general results with $n = 5$. Information is then needed concerning the probability p_5 to be assigned to this proposition, remembering to adjust $p_1, \ldots, p_4$ appropriately so that $p_1 + \cdots + p_5 = 1$. Information is also needed concerning $Pr(E \mid H_5)$. Some is available from Table 8.1 by taking relevant information from that pertaining to the other women. Thus $Pr(\text{mother} \mid \text{age unknown}) = 0.583$, from Q. However, Pr (not sterilised | mother unknown, age unknown) is unavailable. Probabilities are given in Table 8.1 for mothers of unknown age and for women of a certain age but about whom it is not known if they are mothers or not. A probability is not given for the probability that a woman would be sterilised when her age is unknown and it is not known whether she is a mother or not. The probabilities for E_3 (type A blood) and E_4 (other six phenotypes) would remain the same since the unknown woman has been identified from her remains as being white, and the appropriate probabilities have been given.

Further discussion of these ideas in the context of DNA profiles is given in Section 13.7.

8.2   CORRESPONDENCE PROBABILITIES

A *correspondence probability* is the probability that evidence related to the suspect corresponds to evidence found at the scene of the crime. It is discussed by Stoney (1991a), where it is called a *correspondence frequency*. It provides the investigator with an answer to the question, given the evidence found at the scene of the crime, what is the probability of encountering corresponding evidence related to the suspect. The evidence found at the scene of the crime is accepted. The rarity of a correspondence with this evidence is of interest. Consider Example 1.2, in which the evidence at the scene of the crime is the bulk (source) form.

Given the properties of the broken crime scene window, what is the probability of the random occurrence of corresponding glass fragments? The estimation of this probability requires information on a population of glass samples. From this population, the probability of a correspondence with the properties of the window at the scene of the crime can be determined. The frequency of occurrence of this set of properties among glass fragments (the transferred particle form) is the frequency of interest. It is not necessarily the frequency of occurrence amongst windows (the bulk (source) form).

Extensive data may exist on the properties of window glass but it cannot be assumed that the same frequencies of properties necessarily exist amongst glass fragments found on people. It is not necessarily the case that all glass fragments found on people come from broken windows. They may come from containers or headlamps, for example. The frequencies of properties in the different populations may differ. Rare properties in window glass may be common properties in glass fragments.

The results of a search of footpaths in metropolitan Auckland, New Zealand, for pieces of broken, colourless glass were reported by Walsh and Buckleton (1986). The predominant type of broken glass was that of container glass. The assessment of the probability of a coincidental match of refractive indices can be affected by the type of glass in a survey. In 1984, a refractive index of 1.5202 had a frequency of 3.1% in container glass in the United Kingdom but a frequency of 0.23% amongst building glass (Lambert and Evett, 1984). These results are used here for illustrative purposes only. Consider a crime in which the refractive index of glass from a broken window at the scene has a refractive index of 1.5202. A suspect may have fragments of container glass of refractive index 1.5202 embedded in his footwear. Reference of this to a database constructed from a survey which is predominantly of building glass will provide a probability of a coincidental match which is too low by a factor greater than 10 (see Section 8.3.3). The value of the evidence against the suspect will be exaggerated. As Walsh and Buckleton (1986) comment, a survey used to assess the significance of glass evidence 'should realistically reflect the type of broken glass likely to be encountered at random in the community'.

The appropriate survey results may depend on where on the suspect the fragments were found. Fragments found on footwear may be expected to have come from a street and have greater proportions of automobile or bottle glass. Fragments found on clothing may be expected to have come from contact with loose fragments on a surface or as a result of being close to a breaking object. In general, population data relevant to the type of data and the environment in which the material associated with the suspect has been found should be used. The survey reported by Walsh and Buckleton (1986) has direct relevance only for the consideration of glass fragments found in footwear. It is not applicable to the consideration of glass fragments found on the clothing of a suspect. A survey carried out by Harrison *et al.* (1985) of refractive indices of glass recovered from persons suspected of a crime and unrelated to the crime reference samples reported, unsurprisingly, a predominance of fragments of container glass in footwear but suggested a significant proportion of building glass on clothing. The results of more recent surveys are reported in Curran *et al.* (2000).

8.3   DIRECTION OF TRANSFER

It is a feature of likelihood ratios that it is not necessary to distinguish between the scene-anchored and the suspect-anchored perspective; see below and Stoney (1991a). However, despite this feature, it is still necessary to recognize the possible importance of the direction of the transfer of the evidence. General results are given in Evett (1984).

8.3.1   Transfer of evidence from the criminal to the scene

Consider Example 1.1 again. A bloodstain is found at the scene of the crime. It is of genotype Γ for locus *LDLR*. All innocent sources of the stain have been eliminated, the knowledge of which may be recorded as relevant background information I. (Note here that if it is thought unreasonable to be able to eliminate all innocent sources of the stain then consider, analogously, an example of a rape case in which a semen stain replaces the bloodstain.) There will normally be other information which a jury, for example, will have to consider. However, in this context, I is restricted to information considered relevant in the sense that the evidence for which probabilities are of interest is dependent on I. A suspect has been identified. He is also of genotype Γ. Blood of genotype Γ is not common amongst the general population to which the criminal is thought to belong, being found in only 4% of this population. However, the suspect is discovered to be of Ruritanian ethnicity and blood of genotype Γ at locus *LDLR* is found in 70% of Ruritanians. How, if at all, should knowledge of the suspect's ethnicity be taken into account? There is assumed at present to be no other evidence, such as eyewitness evidence, to provide information about the ethnic group of the criminal.

Consider Table 8.2. Notice that 800/20 000 (4%) of the general population are of genotype Γ whereas 700/1000 (70%) of Ruritanians are of genotype Γ, satisfying the description above.

The relevance of the Ruritanian ethnicity of the suspect can be determined by evaluating the likelihood ratio. The evidence E may be partitioned into three parts:

- E_r, the racial grouping (Ruritanian) of the suspect;
- E_s, the genotype (Γ) of the suspect;
- E_c, the genotype (Γ) of the crime stain.

The likelihood ratio is then

$$V = \frac{Pr(E \mid H_p)}{Pr(E \mid H_d)}$$

$$= \frac{Pr(E_r, E_s, E_c \mid H_p)}{Pr(E_r, E_s, E_c \mid H_d)}$$

where the two propositions to be compared are:

- H_p, there was contact between the suspect and the crime scene;
- H_d, there was not contact between the suspect and the crime scene.

Note that explicit mention is not made of the background information I that there was no innocent explanation for the crime stain. However, it should be remembered that all the probabilities under discussion here are conditional on I. It is also assumed implicitly that if contact has taken place (H_p) then the evidence (bloodstain) was left by the suspect. This may be thought to imply guilt, but the inference of guilt from contact is not one which the forensic scientist should make. Rather it is for the jury to make this inference, bearing in mind all other evidence presented at the trial.

Table 8.2 Frequencies of Ruritanians and those of genotype Γ for locus *LDLR* in a hypothetical population

	Ruritanians	Others	Total
Genotype Γ	700	100	800
Others	300	18 900	19 200
Total	1000	19 000	20 000

Scene-anchored perspective

The scene-anchored perspective is one in which the suspect evidence (E_1, E_2) is conditioned on the scene evidence E_3. Using the odds form of Bayes' theorem (3.6),

$$V = \frac{Pr(E_r, E_s, E_c \mid H_p)}{Pr(E_r, E_s, E_c \mid H_d)}$$

$$= \frac{Pr(E_r, E_s \mid H_p, E_c)Pr(E_c \mid H_p)}{Pr(E_r, E_s \mid H_d, E_c)Pr(E_c \mid H_d)}.$$

Consider, first, the ratio $Pr(E_c \mid H_p)/Pr(E_c \mid H_d)$. If no more is assumed of the suspect than that he was at the crime scene (H_p, the numerator) or that he was not at the crime scene (H_d, the denominator) then the frequency of Γ is the same whether he was present or not. Thus, $Pr(E_c \mid H_p) = Pr(E_c \mid H_d)$ and

$$V = \frac{Pr(E_r, E_s \mid H_p, E_c)}{Pr(E_r, E_s \mid H_d, E_c)}.$$

This may be written as

$$\frac{Pr(E_r \mid H_p, E_c)Pr(E_s \mid E_r, H_p, E_c)}{Pr(E_r \mid H_d, E_c)Pr(E_s \mid E_r, H_d, E_c)}.$$

If the suspect was present at the crime scene (H_p) and if the crime stain is of genotype $\Gamma(E_c)$ then the probability he is Ruritanian is $7/8$, the proportion of Ruritanians amongst those of genotype Γ. Thus $Pr(E_r \mid H_p, E_c) = 7/8$.

If the suspect was present at the crime scene (H_p) and if the crime stain is of genotype $\Gamma(E_c)$ then the probability the suspect's genotype is $\Gamma(E_s)$ is 1, independent of his ethnicity (E_r). Thus $Pr(E_s \mid E_r, H_p, E_c) = 1$.

If the suspect was not at the scene of the crime (H_d) the blood group (E_c) of the crime stain gives no information about his ethnicity (E_r). Thus $Pr(E_r \mid H_d, E_c) = Pr(E_r \mid H_d) = 1/20$, the proportion of Ruritanians in the general population.

Similarly, if the suspect was not at the scene of the crime, the blood group of the crime stain gives no information about the blood group of the suspect. Thus $Pr(E_s \mid E_r, H_d, E_c) = Pr(E_s \mid E_r, H_d)$ and this is the proportion of Ruritanians which are of genotype Γ or, alternatively, the probability that a Ruritanian, selected at random from the population of Ruritanians, is of blood group Γ. This probability is $7/10$. Then

$$V = \frac{\dfrac{7}{8} \times 1}{\dfrac{1}{20} \times \dfrac{7}{10}}$$

$$= \frac{7/8}{7/200}$$

$$= \frac{200}{8}.$$

This is the reciprocal of the proportion of people of genotype Γ in the general population. The ethnicity of the suspect is not relevant. The general proof of this result is given in Evett (1984).

Suspect-anchored perspective

The suspect-anchored perspective is one in which the scene evidence (E_c) is conditioned on the suspect evidence (E_r, E_s). From the odds form of Bayes' theorem (3.6),

$$V = \frac{Pr(E_r, E_s, E_c \mid H_p)}{Pr(E_r, E_s, E_c \mid H_d)}$$

$$= \frac{Pr(E_c \mid H_p, E_r, E_s)Pr(E_r, E_s \mid H_p)}{Pr(E_c \mid H_d, E_r, E_s)Pr(E_r, E_s \mid H_d)}.$$

Consider the ratio $Pr(E_r, E_s \mid H_p)/Pr(E_r, E_s \mid H_d)$. Assume there is no particular predisposition towards (or away from) criminality amongst Ruritanians (E_r) or those of genotype Γ (E_s). Then $Pr(E_r, E_s \mid H_p) = Pr(E_r, E_s \mid H_d)$ and

$$V = \frac{Pr(E_c \mid H_p, E_r, E_s)}{Pr(E_c \mid H_d, E_r, E_s)}.$$

The numerator $Pr(E_c \mid H_p, E_r, E_s)$ of this ratio equals 1. The suspect is assumed to have been present at the scene of the crime (H_p), to be Ruritanian (E_r) and of genotype Γ (E_s). The background information I, assumed implicitly, is that the criminal left the bloodstain. Thus, given $E_s, Pr(E_c \mid H_p, E_r, E_s) = 1$. (Remember it has also been assumed that if contact did take place then it was the suspect who left the crime stain.)

If the suspect is assumed innocent (H_d), the information that he is Ruritanian and of genotype Γ is not relevant for determining the probability that the crime stain is of group Γ. Thus, the estimate to be used for its probability is just $800/20\,000$ $(8/200)$, its proportion in the general population. Then

$$Pr(E_c \mid H_d, E_r, E_s) = 8/200,$$

$$V = 200/8.$$

As before, the ethnicity of the suspect is not relevant. Also, the scene-anchored and suspect-anchored perspective provide the same result.

Suppose now that all innocent sources of the stain have not been eliminated. Consider the scene-anchored perspective. Assume there was contact between the suspect and the crime scene (H_p) and that the stain at the crime scene (though not necessarily the 'crime' stain, in that it may not have been left by the criminal) is of group $\Gamma(E_c)$. No information is contained in this evidence about the ethnicity of the suspect. Thus

$$Pr(E_r \mid H_p, E_c) = 1000/20\,000 = 1/20.$$

The stain at the crime scene may not have come from the criminal. (See Section 9.5 for a more detailed discussion of this idea, known as *relevance*; see also Stoney, 1991a, 1994.) Thus, the probability that the genotype of the suspect is Γ, given he is Ruritanian (E_r) and that there was contact (H_p), is just the relative frequency of Γ amongst Ruritanians, which is 700/1000 (7/10). Thus

$$Pr(E_s \mid E_r, H_p, E_c) = 7/10.$$

The probabilities in the denominator have the same values as before, namely, $Pr(E_r \mid H_d, E_c) = 1/20$ and $Pr(E_s \mid E_r, H_d, E_c) = 7/10$. Hence $V = 1$. In this case the evidence has no probative value since the stain may have come from somewhere else. A similar line of argument derives the same result for the suspect-anchored perspective.

Eyewitness evidence

Consider now eyewitness evidence which states that the criminal is Ruritanian. This eyewitness evidence is assumed to be completely reliable; how to account for evidence which is less than completely reliable is not discussed. Note, however, that in such a discussion there are two conditional probabilities to be considered. Let T be an event and let W_T be an eyewitness report of T. Then it is necessary to consider both $Pr(T \mid W_T)$, the probability the event happened given the eyewitness said it did, and $Pr(W_T \mid T)$, the probability the eyewitness said the event happened given it did. The purpose of including eyewitness evidence here is to illustrate the effect of restricting the population of potential offenders to a particular subgroup of a more general population.

Suppose now, once more, that the crime stain has been identified as coming from the criminal. Thus, the relevant background information, I, is in two parts: the ethnicity of the criminal and the identification of the crime stain as coming from the criminal. The evidence E is now only of two parts:

- E_s, the genotype Γ of the suspect;
- E_c, the genotype Γ of the crime stain.

Evidence E_r of the ethnicity of the suspect has been subsumed into I and is now evidence of the ethnicity of the criminal.

For the scene-anchored perspective, with I assumed implicitly,

$$V = \frac{Pr(E_s \mid H_p, E_c)}{Pr(E_s \mid H_d, E_c)}.$$

The numerator $Pr(E_s \mid H_p, E_c) = 1$ since, if contact is assumed and the blood group of the crime stain is Γ, the blood group of the suspect is certain to be Γ also. The denominator $Pr(E_s \mid H_d, E_c) = 7/10$, the frequency of Γ amongst Ruritanians. Hence, $V = 10/7$. The eyewitness evidence is such as to ensure that the ethnicity of the suspect (Ruritanian) is relevant.

For the suspect-anchored perspective, again with I assumed implicitly,

$$V = \frac{Pr(E_c \mid H_p, E_s)}{Pr(E_c \mid H_d, E_s)}.$$

If the suspect was in contact with the crime scene and is of genotype Γ then the crime stain is certain to be of genotype Γ. Thus, the numerator, $Pr(E_c \mid H_p, E_s) = 1$. The denominator $Pr(E_c \mid H_d, E_s) = 7/10$ since I includes the information that the criminal is Ruritanian. The frequency of genotype Γ amongst Ruritanians is $7/10$.

The scene-anchored and suspect-anchored perspectives give the same result.

8.3.2   Transfer of evidence from the scene to the criminal

This situation is analogous to the earlier situation where the stain at the crime scene could not be identified as coming from the criminal. It is discussed in detail in Evett (1984). The complication introduced when it cannot be assumed that the transferred particle form of the evidence is associated with the crime is made more explicit when transfer is in the direction from the scene to the criminal.

A crime has been committed during which the blood of a victim has been shed. A suspect has been identified. A single bloodstain of genotype Γ has been found on an item of the suspect's clothing. The suspect's genotype is *not* Γ. The victim's genotype is Γ. There are two possibilities:

- A_1, the bloodstain came from some background source;
- A_2, the bloodstain was transferred during the commission of the crime.

As before, there are two propositions to consider:

- H_p, the suspect assaulted the victim;
- H_d, the suspect did not assault the victim.

The evidence E to be considered is that a single bloodstain has been found on the suspect's clothing and that it is of genotype Γ. The information that the victim's genotype is Γ is to be considered as part of the relevant background information I. This is a scene-anchored perspective. The value of the evidence is then

$$V = \frac{Pr(E \mid H_p, I)}{Pr(E \mid H_d, I)}.$$

Consider the numerator first and event A_1 initially. Then the suspect assaulted the victim (H_p) and no blood has been transferred to the suspect. This is an event with probability $Pr(A_1 \mid H_p)$. Also, a stain of genotype Γ has been transferred by some other means, an event with probability $Pr(B, \Gamma)$ where B refers to the event of a transfer of a stain from a source (the background source) other than the crime scene.

Consider A_2. Blood has been transferred to the suspect, an event with probability $Pr(A_2 \mid H_p)$; given A_2, H_p and the genotype Γ of the victim, it is certain that the group of the transferred stain is Γ. This implies also that no blood has been transferred from a background source.

8.3.3 Transfer probabilities

Before continuing with consideration of the transfer of bloodstains, some thoughts on how transfer probabilities may be estimated are pertinent. An example is described in Cook *et al.* (1993) of a case study involving fibres evidence conducted as part of the training programme of the Forensic Science Service of England and Wales.

An assault was described in which a suspect answering a description given by the victim was arrested shortly after the assault was committed. Six fibres were found on the sweatshirt worn by the suspect which were indistinguishable from those of a jumper worn by the victim. The transfer probability of interest is the probability of more than one fibre of the relevant type being transferred, persisting and being recovered from the suspect's clothing if the suspect committed the crime. Cook *et al.* (1993) note that, whilst it would be satisfying to consider the probability of exactly six fibres being found on the suspect's clothing, the imponderables of a particular case would not enable a probability distribution to be established with any degree of precision. Factors listed by Cook *et al.* (1993) which have to be considered include pressure and duration of contact, nature of the donor and recipient fibre surfaces, types of fibres involved and elapsed time. These factors can be contributors to a debate amongst investigators and probabilities for transfer of 0, 1 or more than one fibre can be determined by a consensus. These probabilities can be thought of as three positive numbers which add up to 1. Cook *et al.* (1993) suggest a graphical approach using a pie chart. Thus, a circle is to be split into three segments corresponding to the three probabilities for 0, 1 and more than one fibre being transferred. The investigators can agree on the relative areas of the segments of the pie chart and the corresponding probabilities can be obtained. Such an approach will commend itself to those who find it easier to think visually rather than numerically!

Another example of the modelling of transfer probabilities is given by Evett *et al.* (1995). This is a possible model for the probability distribution for the number of glass fragments remaining at time t after the breaking of a window. This is modelled by a Poisson distribution (Section 2.3.6). Let X be the number of fragments remaining at time t, and let λ_t be the mean number of fragments remaining at time t, determined from experimental data on the persistence of glass fragments. Then

$$Pr(X = x \mid \lambda_t) = \frac{e^{-\lambda_t} \lambda_t^x}{x!}, \qquad x = 0, 1, \ldots .$$

Thus, if λ_t is known, the probabilities are known. It is a matter of judgement by the expert, perhaps informed by experimental data, as to what value to choose for λ_t. However, as Evett *et al.* (1995) note, the assumption of a Poisson distribution means that the variance of the distribution is also λ_t. This may give a value for the precision (thought of as the reciprocal of variance) which may disagree with the expert's view. If this is the case, then a different model will need to be assumed for the number of fragments transferred in a time interval of length t.

The probability that no blood has been transferred by other means has also to be included. Let $t_0 = Pr(A_1 \mid H_p)$ and $t_1 = Pr(A_2 \mid H_p)$ denote the probabilities of no stain or one stain being transferred during the course of a crime. Let b_0 and $b_{1,1}$, respectively, denote the probabilities that a person from the relevant population will have no bloodstain or one bloodstain on his clothing. The probabilities are taken with respect to an object, here a person but not necessarily always so, which has received the evidence. Such a body is a *receptor* (Evett, 1984; see also Section 1.4). Let γ denote the probability that a stain acquired innocently on the clothing of a person from the relevant population will be of genotype Γ. This probability may be different from the relative frequency of Γ amongst the general population (Gettinby, 1984). Then $Pr(B, \Gamma) = \gamma b_{1,1}$. The numerator can be written as (Evett, 1984)

$$t_0 \gamma b_{1,1} + t_1 b_0.$$

The first term accounts for A_1, the second for A_2.

Now consider the denominator. The suspect and the victim were not in contact. The denominator then takes the value $Pr(B, \Gamma)$, which equals

$$\gamma b_{1,1} .$$

The value of the evidence is thus

$$V = \frac{t_0 \gamma b_{1,1} + t_1 b_0}{\gamma b_{1,1}}$$

$$= t_0 + \frac{t_1 b_0}{\gamma b_{1,1}}. \tag{8.5}$$

Consider the probabilities which have to be estimated:

- t_0, that no stain was transferred during the commission of a crime;
- t_1, that one stain was transferred during the commission of a crime;
- b_0, that no stain was transferred innocently;
- $b_{1,1}$, that (one group of) one stain was transferred innocently;
- γ, the frequency of genotype Γ amongst stains of body fluids on clothing.

The first four of these probabilities relate to what has been called *extrinsic evidence*, the fifth to *intrinsic evidence* (Kind, 1994). The estimation of probabilities for extrinsic evidence is subjective and values for these are matters for a forensic scientist's personal judgement. The estimation of probabilities for intrinsic evidence may be determined by observation and measurement. Extrinsic evidence in this example is the evidence of transfer of a stain. Intrinsic evidence is the result of the frequency of the profile. More generally, extrinsic evidence can be physical attributes (the number, position and location of stains) and intrinsic evidence can be descriptors of the stains (profiles).

Note that, in general, t_0 is small in relation to $t_1 b_0/\{b_{1,1} \gamma\}$ and may be considered negligible.

Let Γ be a DNA profile which has a match probability (Section 13.5) of approximately 0.01 amongst Caucasians in England. Assume that the distribution of DNA profiles among stains on clothing is approximately the distribution among the relevant population. This assumption is not necessarily correct (Gettinby, 1984) and there is further discussion of it below. Then $\gamma = 0.01$. A survey of men's clothing was conducted by Briggs (1978) from which it appears reasonable that $b_0 > 0.95, b_{1,1} < 0.05$. The transfer probabilities t_0 and t_1 require to be estimated from a study of the circumstances of the crime and, possibly, experimentation. Suppose $t_1 > 0.5$ (Evett, 1984). Then, irrespective of the value of t_0 (except that it has to be less than $1 - t_1$ that is, less than 0.5 in this instance),

$$V > \frac{0.5 \times 0.95}{0.05 \times 0.01} = 950,$$

a value which indicates very strong evidence to support the hypothesis that the suspect and the victim were in contact. The evidence is at least 950 times more likely if the suspect and victim were in contact than if they were not.

Notice that (8.5) is considerably different from $1/\gamma$ ($= 100$ in the numerical example). This latter result would hold if $t_1 b_0 / b_{1,1}$ were approximately 1, which may mean unrealistic assumptions about the relative values of the transfer probabilities would have to be made.

Various data have been published regarding studies carried out to investigate numerous aspects of *transfer* and *persistence* and offer estimates for probabilities. For glass particles, see Allen and Scranage (1998), Allen *et al.* (1998 a, b, c, d) and Coulson *et al.* (2001a). A summary concerning transfer and persistence studies can be found in Curran *et al.* (2000, Chapter 5).

Another example (McDermott *et al.*, 1999) concerns evidence of the transfer of paint fragments between a suspect vehicle and another (injured party's) vehicle. Consider the two propositions:

- H_p, the suspect vehicle and the injured party's vehicle were in contact;
- H_d, the suspect vehicle and the injured party's vehicle were not in contact.

The value of the evidence is approximately

$$V = \frac{t_n(1 - b_{i,m})}{b_{i,m}\gamma},$$

where

- t_n is the likelihood of paint transferring in the course of an automobile accident and consisting of one top coat layer;
- $b_{i,m}$ is the probability that a random vehicle will have foreign paint on it;
- γ is the probability that the foreign paint would match that from the injured party's vehicle (when paint transfer from the injured party's vehicle to the suspect vehicle is considered).

McDermott *et al.* (1999) provide the following values. The value 0.8 is suggested for t_n on the basis of experience; paint consisting of at least a top layer is transferred in 80% of collisions investigated. A value for $b_{i,m}$ of 0.094 is used, having been obtained from a survey of damage to vehicles in which 9.4% had foreign paint on the surface. A value of 0.127 is used for γ, this being the proportion of vehicles with white solid paint. The value of the evidence is then

$$V = \frac{0.8 \times 0.906}{0.094 \times 0.127} = 61.$$

A similar argument applies when paint is transferred in the opposite direction to the injured party's vehicle from the suspect vehicle. Note this is in disagreement with the argument proposed by McDermott *et al.* (1999) who simply use the reciprocal of the relative frequency for the value of transfer evidence in this direction. The statement of the propositions is important. Consider

- H_p, the paint on the injured party's vehicle originated from the suspect vehicle, and

- H_d, the paint on the injured party's vehicle did not originate from the suspect vehicle.

Then the value of the evidence is the reciprocal of the relative frequency of the paint colour in some population of vehicle paints.

However, consider

- H_p, the injured party's vehicle and the suspect vehicle were in contact, and

- H_d, the injured party's vehicle and the suspect vehicle were not in contact.

Now the evaluation of the evidence has to take into account the possibility of transfer of paint from a source other than the suspect or injured party's vehicle. Transfer and background probabilities are relevant.

Bloodstains on clothing

The assumption that the distribution of blood groups among stains on clothing is approximately the distribution among the relevant population has been questioned (Gettinby, 1984). This is because the blood on a piece of clothing may have come from the wearer of the clothing and thus there is a bias in favour of the genotype of the wearer. The following argument is based on that in Gettinby (1984).

Consider a population of size N in which a proportion p have innocently acquired bloodstains on their clothing. Let p_o, p_a, p_b and p_{ab} be the proportions with which blood groups O, A, B and AB occur in the population, such that $p_o + p_a + p_b + p_{ab} = 1$.

Consider people of group O. Bloodstains detected on clothing may arise from several sources:

- by self-transfer (O type stains), with probability α, say,

- O stains from somewhere else, with probability β_0, say,

- stains of type A, B or AB, necessarily from somewhere else, with probability γ_0, say,

such that

$$\alpha + \beta_0 + \gamma_0 = 1,$$
$$\beta_0 = (1 - \alpha)p_o. \tag{8.6}$$

The proportion α is independent of the blood grouping of the individuals under consideration, unlike β_0 and γ_0.

With an intuitively obvious notation, the following results for individuals of types A, B and AB can be stated:

$$\alpha + \beta_a + \gamma_a = 1,$$
$$\beta_a = (1-\alpha)p_a; \tag{8.7}$$
$$\alpha + \beta_b + \gamma_b = 1,$$
$$\beta_b = (1-\alpha)p_b; \tag{8.8}$$
$$\alpha + \beta_{ab} + \gamma_{ab} = 1,$$
$$\beta_{ab} = (1-\alpha)p_{ab}. \tag{8.9}$$

Of those individuals who have bloodstains which have arisen from a source other than themselves (*non-self stains*) only a proportion γ will be distinguishable as such, where

$$\gamma = p_o\gamma_o + p_a\gamma_a + p_b\gamma_b + p_{ab}\gamma_{ab}.$$

For example, $p_o\gamma_o = Pr$(type A, B or AB stain found on clothing | person is of type O) $\times Pr$(person is of type O). Multiplication of pairs of equations (8.6) to (8.9) by p_o, p_a, p_b and p_{ab}, respectively, gives:

$$p_o\alpha + (1-\alpha)p_o^2 + p_o\gamma_o = p_o,$$
$$p_a\alpha + (1-\alpha)p_a^2 + p_a\gamma_a = p_a,$$
$$p_b\alpha + (1-\alpha)p_b^2 + p_b\gamma_b = p_b,$$
$$p_{ab}\alpha + (1-\alpha)p_{ab}^2 + p_{ab}\gamma_{ab} = p_{ab},$$

and summing gives

$$\alpha + (1-\alpha)(1-\delta) + \gamma = 1, \tag{8.10}$$

where

$$\delta = 1 - p_o^2 - p_a^2 - p_b^2 - p_{ab}^2,$$

the discriminating power (Section 4.5) of the *ABO* system. From (8.10),

$$\alpha = 1 - \frac{\gamma}{\delta}.$$

Values of $\gamma = 0.182$ and $\delta = 0.602$ are used by Gettinby (1984) who cites Briggs (1978). From these a value of $\alpha \simeq 0.7$ is obtained for the estimate of the probability of a bloodstain being acquired from oneself, given that a bloodstain has been found on the clothing; that is, approximately 70% of bloodstains on clothing are acquired by self-transfer.

Consider a person of blood group O. Denote the probability that he innocently bears a bloodstain and the bloodstain is of type O by $C_O(O)$. Then

$$C_O(O) = Pr(\text{suspect has stain from self})$$

$$+ Pr(\text{suspect has stain not from self but of type } O)$$

$$= p\alpha + p(1 - \alpha)p_o.$$

With a similar notation, for a person of blood group O to bear bloodstains of type A, B or AB, the probabilities are

$$C_A(O) = p(1 - \alpha)p_a,$$
$$C_B(O) = p(1 - \alpha)p_b,$$
$$C_{AB}(O) = p(1 - \alpha)p_{ab}.$$

The sum

$$C_O(O) + C_A(O) + C_B(O) + C_{AB}(O) = p$$

gives the probability of innocently acquiring a bloodstain. A value of $p = 0.369$ is given by Briggs (1978) and used by Gettinby (1984). Also, the distribution of blood groups amongst innocently acquired bloodstains on clothing of people of type O may be determined. For example, the probability a person of type O has a bloodstain of type O on his clothing, given it was acquired innocently, is $C_O(O)/p = \alpha + (1 - \alpha)p_o$. The distribution of bloodgroups, for people of type O, is thus:

$$Pr(\text{type} O \mid \text{innocent acquisition of a bloodstain}) = \alpha + (1 - \alpha)p_o,$$
$$Pr(\text{type} A \mid \text{innocent acquisition of a bloodstain}) = (1 - \alpha)p_a,$$
$$Pr(\text{type} B \mid \text{innocent acquisition of a bloodstain}) = (1 - \alpha)p_b,$$
$$Pr(\text{type} AB \mid \text{innocent acquisition of a bloodstain}) = (1 - \alpha)p_{ab},$$

with similar results for people of type A, B and AB. The comparison of this distribution with the general distribution is made in Table 8.3.

Table 8.3   Distribution of blood groups of innocently acquired bloodstains on clothing of people of type O, compared with the distribution in the general population

Bloodgroup	O	A	B	AB	Total
Clothing of people of type O	$\alpha + (1-\alpha)p_o$	$(1-\alpha)p_a$	$(1-\alpha)p_b$	$(1-\alpha)p_{ab}$	1
General population	p_o	p_a	p_b	p_{ab}	1

Populations

The importance of the population from which transfer evidence may have come has long been recognized. In 1935, it was remarked that:

> One need only consider the frequency with which evidence regarding blood and semen stains is produced in court to realise the need for data relating to the relative frequency of occurrence of such stains on garments in no wise related to crimes; for example, on one hundred garments chosen at random from miscellaneous sources, how many would show blood stains, how many semen stains? Questions such as these must arise in court, and answers based on experimental investigation would prove of considerable value in assessing evidence of this type. (Tryhorn, 1935, quoted by Owen and Smalldon, 1975)

Various data have been published regarding what has been called *environment-specific population data* concerned with the incidence of the transferred particle form of materials on clothing (Stoney, 1991a). These include glass and paint fragments on clothing (Pearson *et al.*, 1971; Dabbs and Pearson, 1970, 1972; Pounds and Smalldon, 1978; Harrison *et al.*, 1985; McQuillan and Edgar, 1992; Lau and Beveridge, 1997; Petterd *et al.*, 2001), glass in footwear (Davis and DeHaan, 1977; Walsh and Buckleton, 1986; Coulson *et al.*, 2001a; Roux *et al.*, 2001), loose fibres on clothing (Owen and Smalldon, 1975; Briggs, 1978; Fong and Inami, 1986; and references quoted in Chabli, 2001). These surveys are relevant when the material associated with the suspect is in particle form on clothing. If the material is in source form then population data for the source form are needed. An example of the evidential value of paint on vehicles has already been discussed (McDermott *et al.*, 1999, Section 7.3.9).

An example of the importance of the choice of the correct population is given by Walsh and Buckleton (1986), already discussed in Section 8.2. A crime is committed in which a window is broken. The refractive index of glass from the scene of the crime is 1.5202. A suspect is identified and he has fragments of glass embedded in his footwear whose refractive index is also 1.5202. However, using the survey of Lambert and Evett (1984), the glass in the footwear is container glass of which 3.1% has a refractive index of 1.5202, whilst only 0.23% of building glass in the window has a refractive index of 1.5202. The frequency which is relevant for assessing the value of the evidence E is 3.1%, giving a probability of 0.031 of finding E on an innocent (H_d) person, $Pr(E \mid H_d)$.

Notice that the relevant population is that which is applicable to innocent people. This is so because it is the population from which the transferred particle may have come. An example of how the background information I and H_d influence the definition of the relevant population is given for fibres evidence in Chapter 12.

Consider, however, transfer to the scene of a bloodstain. The relevant population is that from which the criminal has come, and has nothing to do with a population which may be defined by the suspect, as often suggested (see, for example, Decorte and Cassiman, 1993, and comments thereon by Taroni and Champod, 1994, or the discussion on criminal cases presented in Wooley, 1991, Weir and Evett, 1992, 1993, and Lewontin, 1993). This argument is one of the most persistent fallacies in the DNA debate (Weir, 1992). A full description of the fallacy is presented in Evett and Weir (1992).

The choice of population is not so clear, however, in surveys conducted into bloodstains on clothing. The survey by Owen and Smalldon (1975) was conducted by visiting a dry cleaning establishment. Briggs (1978) used data from two large-scale murder investigations during the course of which large numbers of articles of clothing from numerous male suspects were examined in the laboratory for the presence of bloodstaining. In these investigations, Briggs (1978) argued, very reasonably, that the sampling was biased in that it was all derived from people involved in a crime investigation. This may be expected to lead to a higher incidence of bloodstains which could be grouped on clothing than a survey of clothing from the general population would give. Briggs (1978) reported that, for the *ABO* blood group, in the second murder investigation, a total of 966 items of clothing from 122 suspects were examined. Forty-five of these had blood on their clothing $(45/122 = 0.369$, the value for p used by Gettinby, 1984). There were 22 suspects for which a grouping result was obtained from their clothing and of these only four provided clothing with stains where the grouping result differed from the donor's blood group; this gives a value of $4/22$ (0.182) for γ as used by Gettinby (1984). From the original 122 suspects it was possible for only four (3.2%), with nine articles, to show the presence of blood which did not originate from the wearer. Eight of these nine articles originated from three of the four suspects, all three of whom had a history of violence. The conclusion drawn by Briggs was that the proportion of people who have blood on their clothing which is of a different group from their own is 3.2%. Gettinby's parameter γ is the proportion of those people who have bloodstains on their clothing which is not of their own blood where the bloodstain is distinguishable as such: if n is the number of people who have bloodstains on their clothing which is not of their own blood and n_0 is the number of people who have bloodstains on their clothing which is not of their own blood and is of a different blood group (and so is distinguishable as coming from someone else), then $\gamma = n_0/n$.

This example refers to blood groups, which are an outdated type of evidence. However, the underlying idea that the relative frequencies of characteristics of stains on clothing is not necessarily the same as those of the characteristics of the stains in the relevant population is still important.

Owen and Smalldon (1975) reported that out of 100 pairs of trousers examined at the dry cleaning establishment 16 had bloodstains on them. However, it was not possible to determine if the blood group was different from that of the wearer. Of 100 jackets, five had bloodstains on them. As to trousers, 44% had semen stains. The results of this survey may be of little relevance nearly 30 years later, but the principles of the survey itself point to an important area of future research.

8.3.4   Two-way transfer

Consider again the example on the transfer of paint between vehicles. Consider

- H_p, paint on the injured party's vehicle originated from the suspect vehicle, and

- H_d, paint on the injured party's vehicle did not originate from the suspect vehicle.

Then the value of the evidence is the reciprocal of the relative frequency of the paint colour in some population of vehicle paints. This is the value when evidence transfer in one direction only is considered. It may be thought that when transfer in both directions is to be considered, the value of the evidence is simply the product of the reciprocals of the relative frequencies of the paint colours of the two vehicles concerned. However, such an evaluation does not account for the association of the two pieces of evidence. Note that in such a context, propositions change from source to activity level.

If there is evidence of transfer in one direction, say from a suspect vehicle to an injured party's vehicle, then the probability of transfer in the other direction, to the suspect vehicle from the injured party's vehicle, is considerably increased. Thus, the factor t_n in the expression

$$t_n(1 - b_{i,m})/(b_{i,m}\gamma) \tag{8.11}$$

is increased.

The background probabilities for the two directions of transfer may be different because of the nature of the vehicles involved. It may be that a car from an impoverished part of a large city will have a different probability of background paint on it than a car from a prosperous part of the same city.

Issues of two-way transfer in the context of fibres (Champod and Taroni, 1999; Cook *et al.*, 1999) and body fluids (Aitken *et al.*, 2003) are discussed in Chapters 12 and 14, respectively.

8.3.5 Presence of non-matching evidence

In transfer evidence it may be that there is evidence present on the suspect which does not match that found at the scene of the crime, or evidence present at the scene of the crime which does not match that found on the suspect. For example, consider a case involving the transfer of fibres from the scene of a crime to the criminal. A suspect is found with fibres on his clothing which match, in some sense, fibres from the scene of the crime. However, he also has fibres of many different types on his clothing which do not match those found at the scene (so-called *foreign* fibres).

A likelihood ratio for such a situation has been derived by Grieve and Dunlop (1992). It includes factors to allow for transfer probabilities, probabilities of foreign fibre types being found on the person, relative frequencies of occurrence of the matching fibre types and a factor to account for the number of matching fibres amongst the total number of fibres. There is considerable subjectivity in the determination of these figures. The importance of the work lies in the recognition that the number of items found which do *not* match has to be considered when assessing the evidence as well as the number of items found which do match (for details, see also Section 12.3).

8.4 GROUPING

A *group* is defined as a set of material (such as individual items of trace evidence) which share the same forensic characteristics (Champod and Taroni, 1999). A group is not necessarily defined by any spatial separation concerning where the items were found.

When items of trace evidence are recovered from a suspect and are to be considered as evidence, the possibility that the items may have come from more than one source has to be considered. The sample which forms the entirety of the evidence may be divisible into groups. For items of trace evidence, a group is declared only if there is sufficient specificity in the shared features to link these items reasonably with a unique source. A group is defined by the similarity of the characteristic within a group and the dissimilarity of the characteristic between groups. The number of groups in the sample is not known. In some cases, such as for fibres, decisions about the compositions of groups can only be made with difficulty through complete numerical demonstration, though guided by logical, well-qualified opinions.

Statistical grouping methods, illustrated with refractive index (r.i.) measurements on fragments of glass, are available (Evett and Lambert, 1982; Triggs *et al.*, 1997; Curran *et al.*, 1998c, 2000). The methods are illustrated with an example of r.i.s on glass fragments but could be used for other characteristics of trace evidence which is of the form of continuous data, such as elemental compositions, or for discrete data in which a suitable distance measure may be used.

Consider a sample of n fragments of glass whose r.i.s are $x_1, \ldots, x_n$. The fragments may have come from one or more groups. A group of fragments is defined by the similarity of the r.i.s within a group and the dissimilarity of r.i.s between groups. The number of groups in the sample is not known. Two methods of grouping are considered. The first is an agglomerative approach. This starts with assuming the n fragments are in n separate groups. Consideration is then given to agglomerating the fragments into smaller numbers of groups. The second is a divisive approach. This starts with assuming the fragments all belong to one group. Consideration is then given to dividing the fragments into larger numbers of groups.

An agglomerative method was first described by Evett and Lambert (1982). A modification by Triggs *et al.* (1997) is outlined here. The basic principle is the comparison of distances between measurements of fragments with a table of critical values determined from an assumption of the underlying distribution of the distances. The procedure described by Triggs *et al.* (1997), called by them ELM2 and applied to the r.i.s $x_1, \ldots, x_n$, is as follows. Assume the standard deviation, σ, of the r.i. measurements is known and that the measurements are Normally distributed. The standardised range, $(x_{\max} - x_{\min})/\sigma$, denoted R say, is compared with critical values of the sample range from a standard Normal distribution. Tables of the values of R for given sample sizes n and critical values α have been published in which the value $r(\alpha, n)$ is tabulated such that $Pr(R < r(\alpha, n)) < (1 - \alpha)$ for given values of n and α; see, for example, Owen (1962) or Pearson and Hartley (1966).

1. Sort $x_1, \ldots, x_n$ into ascending order and label the sorted measurements $x_{(1)}, \ldots, x_{(n)}$ such that $x_{(1)}$ is the minimum and $x_{(n)}$ is the maximum.
2. Check that the standardised range of the measurements, $(x_{(n)} - x_{(1)})/\sigma = r$, say, is less than $r(\alpha, n)$. If $r \leq r(\alpha, n)$ then the fragments are considered to have come from only one source and there is no need to proceed further. If $r > r(\alpha, n)$ then proceed to step 3.
3. Find the smallest gap $| x_{(i)} - x_{(j)} |$, for $1 \leq j < i \leq n$, between the measurements.
4. If the corresponding standardised range $| x_{(i)} - x_{(j)} | / \sigma \leq r(\alpha, 2)$, find the nearest neighbour to $x_{(i)}$ or $x_{(j)}$ and compare the standardised range of these three measurements with $r(\alpha, 3)$, etc.
5. Repeat step 4 until all the measurements are either in one group, or go to step 2 for any subgroups of size greater than 1.

Some values of $r(\alpha, n)$ are given in Table 8.4.

As an example of the use of Table 8.4, consider r.i. measurements of 1.515 57 and 1.515 59 on two fragments of glass. The standardised range is $0.000\,02/\sigma = 0.000\,02/(4 \times 10^{-5}) = 0.5$, which is less than the 90% critical value for samples of size 2 of 2.326. The two fragments may then be deemed to belong to the same group. Alternatively, consider r.i. measurements of 1.515 57

Table 8.4 Critical values $r(\alpha, n)$ for the range of a sample of size n from a standardised Normal distribution, from Owen(1962)

n	Percentile		
	90% $r(0.10, n)$	95% $r(0.05, n)$	99% $r(0.01, n)$
2	2.326	2.772	3.643
5	3.478	3.858	4.603
10	4.129	4.474	5.157
20	4.694	5.012	5.645

and 1.519 14 on another two fragments of glass. The standardised range is $0.003\,57/\sigma = 0.003\,57/(4 \times 10^{-5}) = 89.25$ which is considerably in excess of the 99% critical value of 3.643. The two fragments may then be deemed to belong to different groups.

The divisive approach described by Triggs *et al.* (1997) is based on an approach by Scott and Knott (1974). A statistic λ is determined and critical values $\lambda(\alpha, n)$, dependent on a critical measure α and the sample size n, are given in a look-up table (Triggs *et al.*, 1997).

1. Sort $x_1, \ldots, x_n$ into ascending order as before, and label the sorted measurements $x_{(1)}, \ldots, x_{(n)}$.
2. For $j = 1, \ldots, n-1$, calculate

$$B_j = j(\bar{x}_{1j} - \bar{x})^2 + (n-j)(\bar{x}_{2j} - \bar{x})^2$$

where $\bar{x}_{1j}$ is the mean of the fragments $x_{(1)}, \ldots, x_{(j)}, \bar{x}_{2j}$ is the mean of the fragments $x_{(j+1)}, \ldots, x_{(n)}$ and $\bar{x}$ is the mean of all the fragments. This is the between-group sum of squares for each of the $n-1$ ordered partitions of the data into two subgroups. Find the maximum value of $B_j, j = 1, \ldots, n$, denote this by B_0 and denote the corresponding value of j as j_0.
3. Calculate the statistic

$$\lambda = \frac{\pi}{2(\pi - 2)} \frac{B_0}{s^2}$$

(Scott and Knott, 1974).
4. Refer λ to the look-up table of critical values for the chosen value of α and the appropriate value of n. If $\lambda > \lambda(\alpha, n)$, then split the corresponding fragments into two groups, $x_{(1)}, \ldots, x_{(j_0)}$ and $x_{(j_0+1)}, \ldots, x_{(n)}$.
5. If there was a split in step 4, repeat steps 2–5 for each new subgroup until no more splits can be made.

Some critical values of $\lambda(\alpha, n)$ are given in Table 8.5.

The divisive method has lower failure rates than the other methods (Triggs *et al.*, 1997).

Table 8.5   Critical values $\lambda(\alpha, n)$ for samples of size n for the divisive grouping algorithm from Triggs *et al.* (1997)

n	Percentile		
	90% $\lambda(0.10, n)$	95% $\lambda(0.05, n)$	99% $\lambda(0.01, n)$
2	3.73	5.33	9.29
5	8.88	10.91	15.21
10	15.18	17.54	23.09
20	26.45	29.20	35.45

8.5   RELEVANT POPULATIONS

A similarity is found between transfer evidence (E_c) of a trait (e.g., DNA profile) found at the crime scene and evidence of the same trait found on the suspect (E_s). The value of the similarity between E_c and E_s is assessed partly by comparison with respect to some population. The similarity may be purely coincidental. A more general derivation of the results in Sections 5.3.1 and 5.3.2 is now given.

Consider transfer from the criminal to the scene of the crime. A suspect-anchored perspective is taken. The value of the evidence is given by

$$V = \frac{Pr(E_c, E_s \mid H_p)}{Pr(E_c, E_s \mid H_d)}$$

$$= \frac{Pr(E_c \mid E_s, H_p)}{Pr(E_c \mid E_s, H_d)} \times \frac{Pr(E_s \mid H_p)}{Pr(E_s \mid H_d)}.$$

When there is transfer from the criminal to the scene, the evidence found on the suspect, E_s, is independent of whether the suspect was present (H_p) at the scene of the crime or not (H_d). Thus,

$$Pr(E_s \mid H_p) = Pr(E_s \mid H_d)$$

and

$$V = \frac{Pr(E_c \mid E_s, H_p)}{Pr(E_c \mid E_s, H_d)}.$$

If H_d is true then E_c is independent of E_s and thus

$$Pr(E_c \mid E_s, H_d) = Pr(E_c \mid H_d).$$

Assume for the present that $Pr(E_c \mid E_s, H_p) = 1$. This is a reasonable assumption for blood group frequency data, for example. Then

$$V = \frac{1}{Pr(E_c \mid H_d)}.$$

Now consider transfer from the scene of the crime to the criminal. A scene-anchored perspective is taken. The value of the evidence is given by

$$V = \frac{Pr(E_c, E_s \mid H_p)}{Pr(E_c, E_s \mid H_d)}$$
$$= \frac{Pr(E_s \mid E_c, H_p)}{Pr(E_s \mid E_c, H_d)} \times \frac{Pr(E_c \mid H_p)}{Pr(E_c \mid H_d)}.$$

The evidence at the scene of the crime, E_c, is independent of whether the suspect was present (H_p) at the scene or not (H_d). Thus,

$$Pr(E_c \mid H_p) = Pr(E_c \mid H_d)$$

and

$$V = \frac{Pr(E_s \mid E_c, H_p)}{Pr(E_s \mid E_c, H_d)}.$$

If H_d is true, then E_s is independent of E_c and

$$Pr(E_s \mid E_c, H_d) = Pr(E_s \mid H_d).$$

Assume for the present that $Pr(E_s \mid E_c, H_p) = 1$. Then

$$V = \frac{1}{Pr(E_s \mid H_d)}.$$

The results from the two directions may be represented in one result by noticing that in both cases the evidence of interest is the transferred particle form. Thus, if E_{tp} denotes transferred particle form of the evidence then

$$V = \frac{1}{Pr(E_{tp} \mid H_d)}. \tag{8.12}$$

With reference to which population is $Pr(E_{tp} \mid H_d)$ to be evaluated? In the case of DNA evidence found at the scene of the crime the population could be based on the ethnic group of the suspect, but this is incorrect, as earlier discussion (Sections 8.3.1 and 8.3.3) has shown. The accused's ethnic group is not relevant to the probability of obtaining the evidence under the alternative proposition that someone else left the mark. The relevant question is 'what sort of someone else?' and the answer depends on what is known about the *perpetrator*, not the accused (Robertson and Vignaux, 1995b).

The concept of a *relevant population* was introduced by Coleman and Walls (1974) who said that

> The relevant population are those persons who could have been involved; sometimes it can be established that the crime must have been committed by a particular class of persons on the basis of age, sex, occupation or other sub-grouping, and it is then not necessary to consider the remainder of, say, the United Kingdom.

The concept of a *suspect population*, defined as 'the smallest population known to possess the culprit as a member' was introduced by Smith and Charrow (1975). The term *suspect population* was also used by Lempert (1991), referring to the population of possible perpetrators of the crime. These populations should not be confused with any population which may be defined by reference to the suspect only, despite their name. It could be argued that the smallest 'suspect population' is just the population of the world, but this is not helpful.

The *total population involved*, as an ideal of a population suitable for study in order to evaluate significance, was suggested by Kirk and Kingston (1964). An *appropriate population* from which a sample can be taken for the study was suggested by Kingston (1965a) without further elaboration of the meaning of 'appropriate'. The estimation of the expected number of people in a population who would be expected to have a characteristic in question was considered by Kingston (1965b), who concluded it was best to base calculations on the *maximum possible population*.

Evaluation of fingerprint evidence is not discussed here. However, some of the debate (presented in Section 7.3.3) concerning the definition of a suitable population for the assessment of fingerprint evidence is of interest here.

More recent discussion about populations in the context of the island problem (Section 3.5.6; Eggleston, 1983; Yellin, 1979; Lindley, 1987), including the effect of search procedures on estimations of the probability of guilt through reduction of the population size, may be found in Balding and Donnelly (1995b), Dawid (1994) and Section 13.9.

Many important aspects of evidence evaluation are illustrated by the island problem. A crime has been committed on an island. Consider a single stain left at the crime scene. The island has a population of $N + 1$ individuals. In connection with this crime, a suspect has been identified by other evidence. The genotype G_s of the suspect and the genotype G_c of the crime stain are the same. The probability that a person selected at random from the island population has this genotype is γ. The two propositions are:

- H_p, the suspect has left the crime stain;
- H_d, some other person left the crime stain.

There is considerable debate as to how prior odds may be determined (see Section 3.5.6).

An extension to this analysis is to allow for varying match probabilities amongst the inhabitants of the island. Let ϕ_0 be the prior probability that the suspect left the crime stain and identify inhabitant number $N+1$ with the suspect. The inhabitants $i = 1, \ldots, N$ are innocent. Let ϕ_i denote the probability that inhabitant i left the crime stain ($i = 1, \ldots, N$), such that $\sum_{i=1}^{N} \phi_i + \phi_0 = 1$. Assume as before that $Pr(G_c \mid G_s, H_p, I) = 1$. Now generalise the match probability π to allow it to be different for each individual, such that $Pr(G_c \mid G_s, H_d, I) = \pi_i$. Then, it can be shown that

$$Pr(H_p \mid G_c, G_s, I) = \phi_0 \left/ \left(\phi_0 + \sum_{i=1}^{N} \phi_i \pi_i \right) \right. .$$

From this expression it is possible to allow for different match probabilities and for different prior probabilities for different individuals. It is possible to write the expression in terms of $w_i = \phi_i / \phi_0$, where the w_i can be regarded as a weighting function for how much more or less probable than the suspect the ith person is to have left the crime stain, based on the other evidence as recorded in I (Balding and Nichols, 1995; Evett and Weir, 1998; see also Chapter 13 below).

$$Pr(H_p \mid G_c, G_s) = 1 \left/ \left(1 + \sum_{i=1}^{N} w_i \pi_i \right) \right. . \tag{8.13}$$

Notice that a population need not necessarily be described by a geographical boundary. Further details are given in Section 13.9.

A suitable name for the population in question is required. One possible though rather lengthy name is the *potential perpetrator population*. When there is information about the source of the transfer evidence, $Pr(E_{tp} \mid H_d)$ can be calculated with respect to a *potential source population*. In many cases these two populations are approximately the same. However, Koehler (1993a) cites the novel *Presumed Innocent* by Scott Turow in which a woman commits a murder and plants her husband's semen in the victim in an effort to incriminate him. The woman may be a member of the potential perpetrator population but not of the potential source population. Alternatively, consider a case in which a woman is murdered in her bed one week after her husband died (Koehler, 1993a). Hairs recovered at the scene of the crime may belong to the woman's deceased husband. This would place him in the potential source population but he would not be a member of the potential perpetrator population.

For cases involving the analysis of DNA evidence, the consideration of the close relatives of a suspect as part of the potential source population will lead to values of $Pr(E_{tp} \mid H_d)$ larger than values based on a general population, and hence to smaller values for V (Lempert, 1991; Robertson and Vignaux, 1993a). A suspect's relatives are more likely to be genetically similar to the suspect than are random members of a general population. The difficulty of defining a potential source population may cause problems with the evaluation of

$Pr(E_{tp} \mid H_d)$. If close relatives are included then there could be dramatic alterations in $Pr(E_{tp} \mid H_d)$.

A case in which there is evidence of a DNA profile against a suspect s and there is a brother, b, and 100 other unrelated men, all of whom are also suspects, is discussed in Sections 13.6 and 13.7. Consideration of the presence of the brother increases the probability of the defence proposition. A corollary of this is that although the value of the evidence is decreased if the alternative perpetrator is a brother, so is the pool of possible suspects. In fact, specifying a brother as the alternative may reduce the prior odds from 1 in several million to 1 in 3 or 4. The combined effect of this and the blood evidence may even be to strengthen the case against the accused (Robertson and Vignaux, 1995a).

In this case evidence about subgroups is still misleading if it fails to take into account that the suspect pool has been sharply reduced. The overall effect, when properly considered, may even be counter-productive to the defence but the emphasis in cross-examination is frequently on producing a number of different values for the likelihood ratio alone. However, if the defence change their proposition then this will change the value of the prior odds as well as that of the likelihood ratio (Robertson and Vignaux, 1995b).

A practical example is the following:

> In some cases it will not be in the interest of the defendant to make such a claim. If, for example, the defendant claims to be a Pitcairn Islander and argues that it will be possible that the perpetrator is also a Pitcairn Islander, the use of a specific Pitcairn Island database might be justified. However, if there are only 50 Pitcairn Islanders in the country where the crime was committed, then the prior odds are dramatically reduced and the impact of the DNA evidence may even be increased. (Robertson and Vignaux, 1993b, p. 4)

As stated earlier, the concept of a *relevant population* was introduced by Coleman and Walls (1974). A definition is given below in the context of a defendant. However, in the context of a police investigation, a relevant population may be considered to be what Lempert (1993) called the *suspect population*. This is the group of people who plausibly may be suspected of having committed the crime. This group or population may contain members of a particular ethnic group. As the proportion of the relevant population that belongs to a particular ethnic group increases, the weight that must be given to allele frequencies within that group also increases. Suppose that the relevant population is such that 50% are Caucasian and 50% are of another ethnic group, and that the allele frequencies within these two groups are meaningfully different. Half of the sample used to generate estimates of the probability that a randomly selected member of the relevant population may have left the DNA evidence should be composed of Caucasians, and half of the other ethnic group. This composition is independent of whether the suspect is a Caucasian or a member of the other ethnic group.

This analysis does not mean, however, that the use of database of allelic frequencies for a black population is appropriate if the victim of a rape committed in a black ghetto states that her assailant was white. So long as the police accept this statement, the relevant population for the purposes of the investigation consists of those men who are white and who are also plausible suspects given any other evidence. If the relevant population should not contain men who do not appear white then the database used to estimate the rarity of a suspects DNA profile should be chosen to reflect that fact.

A relevant population can consist largely of one particular ethnic group when, for example, either living patterns or some other information about the criminal limits potential suspects to those of a particular ethnicity. Thus allele frequencies found in a mixed Caucasian reference sample may provide an under- or overestimate of the rarity of the DNA of a Caucasian suspect who is a member of an ethnically homogeneous isolated community within which a rape occurred. Similarly, allelic frequencies found within an American black reference sample may under- or overestimate the rarity (within the relevant population) of the DNA of a West Indian black defendant arrested in a black ghetto, if the victim's description of the assailant meant that men with West Indian accents determined the relevant population. Whether such under- or overestimates will in fact occur, and their likely magnitudes if they do occur, are empirical questions (Lempert, 1993).

The situation is different, however, if people within the relevant population have allele configurations across the loci tested with a relatively high probability of matching that of the suspect. Often the relevant population will include such people; for example, they could be close relatives of the suspect. Not only are such people likely to live in the same vicinity as the suspect, but they are also likely to share some of the characteristics (e.g., general appearance, accent, mannerisms) that brought the suspect to the attention of the police. If a suspect is innocent, it may be that by concentrating on him, the police have concentrated either on an associate or a relative of the actual criminal (Nichols and Balding, 1991; Thompson, 1993).

As previously mentioned, although the probability may be quite low that the DNA of a randomly selected member of the relevant population would match the suspect's DNA at the tested loci, the probability will be substantially higher that at least one member of the relevant population has matching DNA, because the alleles of relatives are not randomly distributed with respect to those of the suspect.

An interesting contribution to the discussion is made by Walsh and Buckleton (1994). They report the results of a study which attempts to estimate the general place of residence of a criminal, given a crime was committed in Auckland, New Zealand. These are given in Table 8.6.

Let the propositions of interest be that the suspect is the criminal (H_p) and that some other person in the world is the criminal (H_d). In the absence of any information about the crime, including that of its location, the prior

Table 8.6 Probability that a criminal lived in a particular area, given the crime was committed in Auckland, New Zealand

Residence of offender	Auckland	Remainder of North Island	Remainder of New Zealand	Remainder of the world
Approximate population	809 000	1 603 000	1 660 000	$\simeq 4 \times 10^9$
Probability	0.92	0.05	0.03	0.00

odds in favour of contact between the suspect and the crime scene are then $Pr(H_p)/Pr(H_d) = 1/N$, where N is the population of the world, as argued by Kingston (1964). Consider information I which is in two parts:

- I_1, the offence was committed in Auckland;
- I_2, the suspect lives in Auckland.

Thus I, the other information, reference to which has usually been suppressed, may now be written as $I = (I_1, I_2)$ and the prior odds in favour of H_p, given I, may be written as

$$\frac{Pr(H_p \mid I_1, I_2)}{Pr(H_d \mid I_1, I_2)}.$$

Then it can be shown (Walsh and Buckleton, 1994) that

$$\frac{Pr(H_p \mid I_1, I_2)}{Pr(H_d \mid I_1, I_2)} \times \frac{Pr(I_1, I_2)}{Pr(I_1, I_2)} = \frac{Pr(I_1, I_2, H_p)}{Pr(I_1, I_2, H_d)}$$

$$= \frac{Pr(I_1, I_2 \mid H_p)}{Pr(I_1, I_2 \mid H_d)} \times \frac{Pr(H_p)}{Pr(H_d)}$$

$$= \frac{Pr(I_2 \mid I_1, H_p)Pr(I_1 \mid H_p)}{Pr(I_1 \mid I_2, H_d)Pr(I_2 \mid H_d)} \times \frac{Pr(H_p)}{Pr(H_d)}.$$

Assume that $Pr(I_1 \mid H_p) = Pr(I_1 \mid I_2, H_d) = Pr(I_1)$. In other words, it is assumed, firstly, that the probability that the crime was committed in Auckland is independent of whether the suspect committed the crime or not (since nothing else is assumed about the suspect – in particular, it is not assumed that the suspect lives in Auckland). Similarly, it is assumed that the probability that the crime was committed in Auckland is independent of the information that the suspect lived in Auckland given that the suspect was not in contact with the crime scene (proposition H_d).

It is also assumed that $Pr(I_2 \mid H_d) = Pr(I_2)$, that is, the probability that the suspect lives in Auckland given he was not in contact with the crime scene is equal to the probability that he lives in Auckland. Hence

$$\frac{Pr(H_p \mid I_1, I_2)}{Pr(H_d \mid I_1, I_2)} = \frac{Pr(I_2 \mid I_1, H_p)}{Pr(I_2)} \times \frac{Pr(H_p)}{Pr(H_d)}.$$

With the results from the above table, $Pr(I_2 \mid I_1, H_d) = 0.92$ and $Pr(I_2) \simeq$ $809\,000/4\,000\,000\,000$. Also $Pr(H_p)/Pr(H_d) = 1/4\,000\,000\,000$. Hence

$$\frac{Pr(H_p \mid I_1, I_2)}{Pr(H_d \mid I_1, I_2)} = \frac{0.92}{809\,000} = 1.14 \times 10^{-6}.$$

The prior odds of 1/(population of the world) have been considerably reduced. However, the odds of 1/(population of Auckland) have also been altered by a factor of 0.92 to account for the information that the suspect lived in Auckland. This seems intuitively reasonable.

This discussion illustrates the considerable problems surrounding the definitions of populations and the choice of suitable names. For example, the distinction between a potential perpetrator population and a potential source population needs to be borne in mind.

In general discussion, the term 'relevant population' will be used and this will have the following rather technical definition.

Definition   The *relevant population* is the population defined by the combination of the proposition H_d proposed by the defence and the background information I.

However, there will often be occasions during police investigations when there is not a defendant but there is a suspect. The term 'relevant population' will still be used then. Indeed, it has already been so used in the discussion above about ethnic groups and allelic frequencies, where others (e.g., Lempert, 1993) have used the term 'suspect population'.

For transfer from the scene to the criminal, I may include information about the suspect. For example, with bloodstain evidence found on the suspect's clothing, the lifestyle of the suspect is relevant, as is his ethnic group (Gettinby, 1984; see also Section 8.3.2). For transfer from the criminal to the scene, H_d dissociates the suspect from the scene. The population from which the evidence may have come and which should be used for the determination of V is specified with the help of I.

9

Discrete Data

9.1 NOTATION

As usual, let E be the evidence, the value of which has to be assessed. Let the two propositions to be compared be denoted H_p and H_d. The likelihood ratio, V, is then

$$V = \frac{Pr(E \mid H_p)}{Pr(E \mid H_d)}.$$

The propositions will be stated explicitly for any particular context. In examples concerning DNA profiles and a proposition that a stain did not come from a suspect it will be understood that the origin of the stain was from some person unrelated to, and not sharing the same sub-population as, the suspect. For evaluations in these contexts see Sections 13.5 and 13.6.

Background information I has also to be assessed. The likelihood ratio is then

$$V = \frac{Pr(E \mid H_p, I)}{Pr(E \mid H_d, I)}, \tag{9.1}$$

(see (3.12)).

A discussion of the interpretation of the evidence of the transfer of a single bloodstain from the criminal to the scene of the crime, following that of Section 8.3.1, is given. This is then extended to cases involving several bloodstains and several offenders.

9.2 SINGLE SAMPLE

9.2.1 Introduction

From Section 3.5.1, consider

$$\frac{Pr(H_p \mid E, I)}{Pr(H_d \mid E, I)} = \frac{Pr(E \mid H_p, I)}{Pr(E \mid H_d, I)} \times \frac{Pr(H_p \mid I)}{Pr(H_d \mid I)}.$$

Statistics and the Evaluation of Evidence for Forensic Scientists: Second Edition
C.G.G. Aitken and F. Taroni © 2004 John Wiley & Sons, Ltd   ISBN: 0-470-84367-5

Consider Example 1.1 in which a bloodstain has been left at the scene of the crime by the person who committed the crime. A suspect has been identified and it is desired to establish the strength of the link between the suspect and the crime. A comparison between the profile of the stain and the profile of a sample given by the suspect is made by a forensic scientist. The two propositions to be compared are:

- H_p, the crime stain comes from the suspect;
- H_d, the crime stain comes from some person other than the suspect.

The scientist's results, denoted by E, may be divided into two parts (E_c, E_s) as follows:

- E_s, the DNA profile, Γ, of the suspect;
- E_c, the DNA profile, Γ, of the crime stain.

A particular example was discussed in Section 8.3.1. A general formulation of the problem is given here. The scientist knows, in addition, from data previously collected that profile Γ occurs in $100\gamma\%$ of some population, Ψ say.

The value to be attached to E is given by

$$V = \frac{Pr(E \mid H_p, I)}{Pr(E \mid H_d, I)}$$
$$= \frac{Pr(E_c, E_s \mid H_p, I)}{Pr(E_c, E_s \mid H_d, I)}$$
$$= \frac{Pr(E_c \mid E_s, H_p, I)Pr(E_s \mid H_p, I)}{Pr(E_c \mid E_s, H_d, I)Pr(E_s \mid H_d, I)}.$$

Now, E_s is the evidence that the suspect's profile is Γ. As in Section 8.3.1, it is assumed that a person's profile is independent of whether he was at the scene of the crime (H_p) or not (H_d). Thus

$$Pr(E_s \mid H_p, I) = Pr(E_s \mid H_d, I)$$

and so

$$V = \frac{Pr(E_c \mid E_s, H_p, I)}{Pr(E_c \mid E_s, H_d, I)}.$$

If the suspect was not at the scene of the crime (H_d is true) then the evidence (E_c) about the profile of the crime stain is independent of the evidence (E_s) about the profile of the suspect. Thus

$$Pr(E_c \mid E_s, H_d, I) = Pr(E_c \mid H_d, I)$$

and

$$V = \frac{Pr(E_c \mid E_s, H_p, I)}{Pr(E_c \mid H_d, I)}. \tag{9.2}$$

Notice that the above argument takes a suspect-anchored perspective. It is possible to consider a scene-anchored perspective. A similar argument to that used above shows that

$$V = \frac{Pr(E_s \mid E_c, H_p, I)}{Pr(E_s \mid H_d, I)}.$$

This result assumes that $Pr(E_c \mid H_p, I) = Pr(E_c \mid H_d, I)$, that is, the profile of the crime stain is independent of whether the suspect was present at the crime scene or not (remembering that nothing else is known about the suspect, in particular his profile). This assumption is discussed further in Chapter 13. The assumption that the suspect's characteristics are independent of whether he committed the crime or not should not be made lightly. It is correct with reference to the suspect's DNA profile. However, some crime scenes may be likely to transfer materials to an offender's clothing, for example. If the characteristics of interest relate to such materials and the offender is later identified as a suspect, the presence of such material is not independent of his presence at the crime scene. If the legitimacy of the simplifications is in doubt then the original expression (9.1) is the one which should be used.

The background information, I, may be used to assist in the determination of the relevant population from which the criminal may be supposed to have come. For example, consider an example from New Zealand where I may include an eyewitness description of the criminal as Chinese. This is valuable because the frequency of DNA profiles can vary between ethnic groups and affect the value of the evidence; see Section 8.3.1.

First, consider the numerator $Pr(E_c \mid H_p, E_s, I)$ of the likelihood ratio in the suspect-anchored perspective. This is the probability the crime stain is of profile Γ given the suspect was present at the crime scene *and* the suspect has DNA profile Γ *and* all other information, including, for example, an eyewitness account that the criminal is Chinese. This probability is just 1 since if the suspect was at the crime scene and has profile Γ then the crime stain is of profile Γ, assuming as before that all innocent sources of the crime stain have been eliminated. Thus $Pr(E_c \mid H_p, E_s, I) = 1$.

Now consider the denominator $Pr(E_c \mid H_d, I)$. Here the proposition H_d is assumed to be true; that is, the suspect was not present at the crime scene. I is also assumed known. Together I and H_d define the relevant population (see Section 8.5).

9.2.2 General population

Suppose, initially, I provides no information about the criminal which will affect the probability of his blood profile being of a particular type. For example, I may include eyewitness evidence that the criminal was a tall, young male. However, a DNA profile is independent of all three of these qualities, so I gives no information affecting the probability that the DNA profile is of a particular type.

It is assumed the suspect was not at the crime scene. Thus, the suspect is not the criminal. The relevant population (Section 8.5) is deemed to be Ψ. The criminal is an unknown member of Ψ. Evidence E_c is to the effect that the crime stain is of profile Γ. This is to say that an unknown member of Ψ is Γ. The probability of this is the probability that a person drawn at random from Ψ has profile Γ, which is γ. Thus

$$Pr(E_c \mid H_d, I) = \gamma.$$

The likelihood ratio V is then

$$V = \frac{Pr(E_c \mid H_p, E_s, I)}{Pr(E_c \mid H_d, I)} = \frac{1}{\gamma}. \tag{9.3}$$

This value, $1/\gamma$, is the value of the evidence of the profile of the bloodstain when the criminal is a member of Ψ.

9.2.3 Particular population

Suppose now that I does provide information about the criminal, relevant to the allelic frequencies, and the relevant population is now Ψ_0, a subset of Ψ. For example, as mentioned above, I may include an eyewitness description of the criminal as Chinese. Suppose the allelic frequency of Γ amongst Chinese is $100\beta\%$. Then $Pr(E_c \mid H_d, I) = \beta$ and

$$V = \frac{1}{\beta}.$$

9.2.4 Example

For the purpose of illustration, an example with classical genetical markers (*ABO* system) is used. Buckleton *et al.* (1987) provide data for racial blood group gene frequencies in New Zealand from which numerical examples applicable to Sections 9.2.2 and 9.2.3 may be derived. Consider the *ABO* blood grouping system and that the source (suspect) and receptor (crime) stain are both of group *B*.

General population

From Buckleton *et al.* (1987), the gene frequencies for New Zealand in this system are given in Table 9.1. Thus, for a crime for which the relevant population Ψ was the general New Zealand population,

$$V = 1/0.063 = 15.87 \simeq 16.$$

The evidence is 16 times more likely if the suspect was present at the crime scene than if he was not.

Particular population

Background information I has included an eyewitness description of the criminal as Chinese. From Buckleton *et al.* (1987) the gene frequencies for Chinese in the *ABO* system are as given in Table 9.2. Then, for a crime for which the Chinese population, Ψ_0, a subset of Ψ, was the relevant population,

$$V = 1/0.165 = 6.06 \simeq 6.$$

The evidence is 6 times more likely if the suspect was at the crime scene than if he was not. Thus the value of the evidence has been reduced by a factor of $15.87/6.06 = 2.62$ if there is external evidence that the criminal was Chinese.

Note, however, as discussed in Section 2.1, that in cases regarding transfer to the crime scene by the criminal, evidence regarding blood profile frequencies has to relate to the population from which the criminal has come, not that of the suspect (though they may be the same). It is not relevant to an investigation for a suspect to be detained, for it to be noted that he is Chinese and for this to provide the sole ground for the use of blood group frequencies in the Chinese population. In order to use blood group frequencies in the Chinese population I has to contain information, such as eyewitness evidence, about the criminal.

Table 9.1 Gene frequencies for New Zealand in the *ABO* system

Blood group	A	B	O
Relative frequency	0.254	0.063	0.683

Table 9.2 Gene frequencies for Chinese in New Zealand in the *ABO* system

Blood group	A	B	O
Relative frequency	0.168	0.165	0.667

However, for evidence regarding transfer from the crime scene to the criminal, the lifestyle of the suspect may be relevant.

Note also, as discussed in Buckleton *et al.* (1987), the blood group frequencies in the general population have been derived from a weighted average of the blood group frequencies in each race or sub-population which make up the population from which the criminal may be thought to have come. The weights are taken to be the proportion of each race in the general population.

9.3 TWO SAMPLES

9.3.1 Two stains, two offenders

The single-sample case described in Section 9.2 can be extended to a case in which two bloodstains have been left at the crime scene (Evett, 1987b). A crime has been committed by two men, each of whom left a bloodstain at the crime scene. The stains are grouped; one is found to be of group Γ_1, the other of group Γ_2. Later, as a result of information completely unrelated to the blood evidence, a single suspect is identified. His blood is found to be of group Γ_1. It is assumed there is no evidence in the form of injuries. The scientific evidence is confined solely to the results of the profiling. The two propositions to be considered are:

- H_p, the crime stains came from the suspect and one other man;
- H_d, the crime stains came from two other men.

The scientific evidence E consists of:

- E_c, the two crime stains of profiles Γ_1 and Γ_2;
- E_s, the suspect's blood is of profile Γ_1 (without loss of generality since a similar result follows if the suspect is of profile Γ_2).

The value, V, of the evidence is given by

$$\frac{Pr(E_c \mid H_p, E_s, I)}{Pr(E_c \mid H_d, I)}. \tag{9.4}$$

Notice that if the bloodstain did not come from the suspect, his profile is not relevant.

The scientist knows that profiles Γ_1 and Γ_2 occur with probabilities γ_1 and γ_2, respectively, in some relevant population.

Assume that I contains no information which may restrict the definition of the relevant population to a subset of the general population.

Consider the numerator of (9.4) first. This is the probability that the two stains are of profiles Γ_1 and Γ_2, given that

- the suspect is the source of one of the crime stains,
- the suspect is of profile Γ_1;
- all other information, I (from the assumption above, I implies that profile frequencies in the general population are the relevant ones).

Assume that the two crime stains are independent pieces of evidence in the sense that knowledge of the profile of one of the stains does not influence the probability that the other will have a particular profile. Let the two criminals be denoted A and B.

Also, E_c may be considered in two mutually exclusive partitions,

- E_{c1}: A is of profile Γ_1, B is of profile Γ_2,
- E_{c2}: A is of profile Γ_2, B is of profile Γ_1.

Partitions E_{c1} and E_{c2} may be further subdivided using the assumption of independence. Thus $E_{c1} = (E_{c11}, E_{c12})$, where

- E_{c11}: A is of profile Γ_1,
- E_{c12}: B is of profile Γ_2.

Similarly, $E_{c2} = (E_{c21}, E_{c22})$, where

- E_{c21}: A is of profile Γ_2,
- E_{c22}: B is of profile Γ_1.

Thus, since E_{c1} and E_{c2} are mutually exclusive:

$$Pr(E_c \mid H_p, E_s, I) = Pr(E_{c1} \text{ or } E_{c2} \mid H_p, E_s, I)$$
$$= Pr(E_{c1} \mid H_p, E_s, I) + Pr(E_{c2} \mid H_p, E_s, I),$$

from (1.2), the second law of probability for mutually exclusive events. However, only one of these two probabilities is non-zero. If the suspect is A then the latter probability is zero; if the suspect is B then the former probability is zero. Assume, again without loss of generality, that the suspect is A. Then

$$Pr(E_c \mid H_p, E_s, I) = Pr(E_{c1} \mid H_p, E_s, I)$$
$$= Pr(E_{c11} \mid H_p, E_s, I) \times Pr(E_{c12} \mid H_p, E_s, I)$$

by independence (1.6).

Now, $Pr(E_{c11} \mid H_p, E_s, I) = 1$ since, if the suspect was the source of one of the crime stains and his profile is Γ_1 then it is certain that one of the profiles is Γ_1.

Also, $Pr(E_{c12} \mid H_p, E_s, I) = \gamma_2$ since the second bloodstain was left by the other criminal. At present, in the absence of information from I, B is considered as a member of the relevant population. The probability is thus the relative frequency of profile Γ_2 in the relevant population, and this is γ_2. The numerator then takes the value γ_2.

Now consider the denominator of (9.4),

$$
\begin{aligned}
Pr(E_c \mid H_d, I) &= Pr(E_{c1} \mid H_d, I) + Pr(E_{c2} \mid H_d, I) \\
&= Pr(E_{c11} \mid H_d, I) Pr(E_{c12} \mid H_d, I) \\
&\quad + Pr(E_{c21} \mid H_d, I) Pr(E_{c22} \mid H_d, I) \\
&= \gamma_1 \gamma_2 + \gamma_2 \gamma_1 \\
&= 2\gamma_1 \gamma_2.
\end{aligned}
$$

Thus,

$$
V = \gamma_2 / (2\gamma_1 \gamma_2) = 1/(2\gamma_1). \tag{9.5}
$$

This result should be compared with the result for the single-sample case, $V = 1/\gamma$. The likelihood ratio in the two-sample case is one half of what it is in the corresponding single-sample case. This is intuitively reasonable. If there are two criminals and one suspect, one would not expect the evidence of a matching bloodstain to be as valuable as in the case in which there is one criminal and one suspect. Note that if γ_1 is greater than 0.5 then V is less than 1. The evidence is supportive of the proposition that the crime stain came from two men, not including the suspect. As has been commented, 'this appears counterintuitive at first sight but, rather than demonstrating a flaw in the logic, it demonstrates that intuition can be unreliable!' (Evett, 1990). An illustrative example, using an idea from the tossing of two coins, is given by Evett and Weir (1998).

Other pairs of propositions may appear more appropriate in different situations. For example, suppose the stains are individually identified. It could be that one was found on a carpet and another on a pillowcase. The propositions may be then be:

- H_p, the suspect left the stain on the carpet,
- H_d, two unknown people left the two stains;

  or

- H_p, the suspect left the stain on the carpet,
- H_d, an unknown person left the stain on the carpet.

The context of the case or the strategies of the prosecution and defence can influence the choice of the propositions. Different propositions could lead to different values for the evidence. It is necessary then to look at Bayes' theorem (3.9) in its entirety. Different propositions may lead to different prior odds $Pr(H_p \mid I)/\ Pr(H_d \mid I)$ as well as to different values of the evidence. It follows that the posterior odds $Pr(H_p \mid E, I)/Pr(H_d \mid E, I)$ may also be different for different propositions. Further comments on these ideas may be found in Meester and Sjerps (2003). Further discussion of the two-trace problem at the activity level illustrates that several factors have to be considered: the number of reported perpetrators, the relevance of each stain and the specification of transfer probabilities (Triggs and Buckleton, 2003). A generalisation to cases with n stains, k groups of stains and m offenders is given in Section 9.4, but without consideration of relevance or probabilities of transfer.

9.3.2 DNA profiling

The evidence from a DNA profile may be interpreted simplistically as discrete data. An application of this interpretation to DNA profiles from single locus probes from more than one person is described by Evett *et al.* (1991).

A woman has claimed to have been raped by two men. A suspect A has been arrested. A sample from a vaginal swab provides, using a single locus probe, evidence E of four alleles a_1, a_2, a_3 and a_4. The suspect's sample provides alleles a_1 and a_2, matching two of the bands in the crime sample. Assume the relative frequencies of the alleles a_1, a_2, a_3 and a_4 in the general population are p_1, p_2, p_3 and p_4, respectively. (The notation is chosen to distinguish allelic frequencies, denoted p, from profile frequencies, denoted γ.) It is assumed that none of the alleles in the crime sample can be attributed to the victim.

The two propositions to be considered in the evaluation of the likelihood ratio are

- H_p, the crime sample came from the suspect A and one other man;
- H_d, the crime sample came from two other men.

If H_p were true, the suspect has contributed bands a_1 and a_2 with probability 1. The probability that the other man would contribute bands a_3 and a_4 is $2p_3p_4$. Thus

$$\Pr(E \mid H_p) = 2p_3p_4.$$

If H_d were true there are $\binom{4}{2} = 6$ ways in which the two men could contribute bands to make up the observed crime profile. Each way has probability $4p_1p_2p_3p_4$. For example, if one man is a_1a_2, with probability $2p_1p_2$, and one man is a_3a_4, with probability $2p_3p_4$, the joint probability is $4p_1p_2p_3p_4$. Thus,

$$\Pr(E \mid H_d) = 24p_1p_2p_3p_4.$$

The value of the evidence is then

$$V = \frac{\Pr(E \mid H_p)}{\Pr(E \mid H_d)} = \frac{2p_3p_4}{24p_1p_2p_3p_4} = \frac{1}{12p_1p_2}.$$

If there had been only one assailant and only alleles in positions 1 and 2 present then this likelihood ratio would have the value $1/(2p_1p_2)$. The existence of the other two alleles, and by implication another assailant, has reduced the value of the evidence by a factor of 6.

Other examples are given by Evett *et al.* (1991). Suppose another suspect, B, provides a sample which gives a profile consisting of alleles in positions 3 and 4. If the two propositions are now

- H_p, the crime sample came from suspects A and B,
- H_d, the crime sample came from two unknown men,

then

$$V = \frac{1}{24p_1p_2p_3p_4},$$

since the numerator has the value 1.

Further examples include the situation in which the analysis reveals only three bands, in positions 1, 2 and 3, but not 4. The two propositions considered are:

- H_p, the crime sample came from A and one other man,
- H_d, the crime sample came from two other men.

Assume the band in position 3 is a single band and the conservative estimate $2p_3$ (Weir *et al.*, 1997) is used for its frequency. Then

$$V = \frac{1 + p_1 + p_2}{4p_1p_2(3 + p_1 + p_2 + p_3)}.$$

If the estimate p_3^2 is used for its frequency, then

$$V = \frac{p_3 + 2p_1 + 2p_2}{12p_1p_2(p_1 + p_2 + p_3)}.$$

More details on DNA mixtures are presented in Section 13.10.

9.4   MANY SAMPLES

9.4.1   Many different profiles

Consider a crime in which n bloodstains are left at the crime scene, one from each of n offenders. A single suspect is identified whose DNA profile matches that of one of the bloodstains at the crime scene. Assume throughout that I contains no relevant information. While hypothetical, the example illustrates points which require consideration in the evaluation of evidence. An analogous argument can be developed for a scenario in which there are n groups of fibres from n distinct sources and a suspect has been identified in whose possession is clothing whose fibres are consistent with one of the groups. Assume the n bloodstains, one from each of n offenders, all have different profiles. The two propositions to be considered are:

- H_p, the crime sample came from the suspect and $n-1$ other men;
- H_d, the crime sample came from n unknown men.

The scientific evidence E consists of:

- E_c, the crime stains are of profiles $\Gamma_1, \Gamma_2, \ldots, \Gamma_n$,
- E_s, the suspect's profile is Γ_1 (without loss of generality).

The frequencies of $\Gamma_1, \Gamma_2, \ldots, \Gamma_n$ are, respectively, $\gamma_1, \gamma_2, \ldots, \gamma_n$.

Consider the numerator $Pr(E_c \mid H_p, E_s, I)$. The suspect's profile matches that of the stain of profile Γ_1. There are $n-1$ other criminals who can be allocated to the $n-1$ other stains in $(n-1)!$ ways. Thus:

$$Pr(E_c \mid H_p, E_s, I) = (n-1)! \prod_{i=2}^{n} \gamma_i = (n-1)!(\gamma_2 \cdots \gamma_n).$$

Now, consider the denominator. There are $n!$ ways in which the n criminals, of whom the suspect is not one, can be allocated to the n stains. Thus:

$$Pr(E_c \mid H_d, I) = n! \prod_{i=1}^{n} \gamma_i = n!(\gamma_1 \cdots \gamma_n).$$

Hence

$$V = \frac{(n-1)! \prod_{i=2}^{n} \gamma_i}{n! \prod_{i=1}^{n} \gamma_i} = \frac{1}{n\gamma_1}. \tag{9.6}$$

9.4.2   General cases

n stains, k groups, k offenders

Suppose now that there are k different profiles $\Gamma_1, \ldots, \Gamma_k$, with frequencies $\gamma_1, \ldots, \gamma_k$, and that these correspond to k different people among the n stains ($k < n$) forming the crime sample and that the suspect has one of these profiles. The two propositions to be considered are:

- H_p, the crime sample came from the suspect and $k-1$ other men;
- H_d, the crime sample came from k unknown men.

The scientific evidence consists of:

- E_c, the crime stains are of profiles $\Gamma_1, \ldots, \Gamma_k$ and there are $s_1, \ldots, s_k$ ($\sum_{i=1}^{k} s_i = n$) of each;
- E_s, the suspect's profile is Γ_1 (without loss of generality).

The probabilities given below are in the form of the multinomial distribution (Section 2.3.4).

Consider the numerator $Pr(E_c \mid H_p, E_s, I)$. The suspect's profile matches that of the stains of profile Γ_1. There are $n - s_1$ other bloodstains which can be allocated in $(n-s_1)!/(s_2! \cdots s_k!)$ ways, where $\sum_{i=2}^{k} s_i = n - s_1$, to give

$$Pr(E_c \mid H_p, E_s, I) = \frac{(n-s_1)!}{s_2! \cdots s_k!} \gamma_2^{s_2} \cdots \gamma_k^{s_k}.$$

Now, consider the denominator. There are $n!/(s_1! s_2! \cdots s_k!)$ ways in which the n stains, none of which is associated with the suspect, can be allocated to the profiles. Thus

$$Pr(E_c \mid H_d, I) = \frac{n!}{s_1! \cdots s_k!} \gamma_1^{s_1} \cdots \gamma_k^{s_k}.$$

Hence

$$V = \frac{(n-s_1)! s_1!}{n! \gamma_1^{s_1}} = \frac{1}{\binom{n}{s_1} \gamma_1^{s_1}}. \tag{9.7}$$

Notice that V is independent of k, the number of criminals, and that if $s_1 = 1$ the result reduces to that of (9.6).

n stains, k groups, m offenders

A similar result may be obtained in the following situation. There are n blood-stains with k different profiles, with s_i in the ith group ($\sum_{i=1}^{k} s_i = n$). There are m offenders, with m_i in each profile ($\sum_{i=1}^{k} m_i = m$), such that s_{ij} ($j = 1, \ldots, m_i$) denotes the number of stains belonging to the jth offender in the ith group and $\sum_{j=1}^{m_i} s_{ij} = s_i$ when it is assumed, without loss of generality, that the first set of stains in the first group came from the suspect. The denominator equals

$$\frac{n!}{s_{11}! \cdots s_{km_k}!} \gamma_1^{s_1} \cdots \gamma_k^{s_k}.$$

The numerator equals

$$\frac{(n-s_{11})!}{s_{12}! \cdots s_{km_k}!} \gamma_1^{s_1 - s_{11}} \gamma_2^{s_k} \cdots \gamma_k^{s_k}.$$

Then

$$V = \frac{(n - s_{11})! s_{11}!}{n!} \times \frac{1}{\gamma_1^{s_{11}}} = \frac{1}{\binom{n}{s_{11}} \gamma_1^{s_{11}}},$$

a result similar to (9.7)

9.5 RELEVANCE OF EVIDENCE AND RELEVANT MATERIAL

9.5.1 Introduction

An extension of the results of the previous section to deal with two further issues is considered by Evett (1993a). The first issue concerns material that may not be *relevant* (Stoney, 1991a, 1994). Crime material which came from the offender is said to be relevant in that it is relevant to the consideration of suspects as possible offenders. Relevant material as discussed here should be distinguished from relevant populations as defined in Section 8.3.1. The second issue concerns the recognition that if the material is not relevant to the case then it may have arrived at the scene from the suspect for innocent reasons. In this section reference to I has, in general, been omitted for clarity.

A crime has been committed by k offenders. A single bloodstain is found at the crime scene in a position where it may have been left by one of the offenders. A suspect is found and he gives a blood sample. The suspect's sample and the crime stain are of the same profile Γ with relative frequency γ amongst the relevant population from which the criminals have come. As before, consider two propositions:

- H_p, the suspect is one of the k offenders;
- H_d, the suspect is not one of the k offenders.

Notice the difference between these propositions and those of Section 9.4. There, the propositions referred to the suspect being, or not being, one of the donors of the bloodstains found at the crime scene. Now, the propositions are stronger, namely that the suspect is, or is not, one of the offenders. The value V of the evidence is

$$V = \frac{Pr(E_c \mid H_p, E_s)}{Pr(E_c \mid H_d)}, \tag{9.8}$$

where E_c is the profile Γ of the crime stain and E_s is the profile Γ of the suspect.

9.5.2   Subjective probabilities

A link is needed between what is observed, the stain at the crime scene, and the propositions, that the suspect is or is not one of the offenders. The link is made in two steps.

The first step is consideration of a proposition that the crime stain came from one of the k offenders and the alternative proposition that the crime stain did not come from any of the k offenders. These propositions are known as *association propositions* (or *association hypotheses* by Buckleton, personal communication, cited by Evett, 1993a).

Assume that the crime stain came from one of the k offenders. The second step is then consideration of a proposition that the crime stain came from the suspect and the alternative proposition that the crime stain did not come from the suspect. These propositions are known as *intermediate association propositions*.

Development of these propositions indicates that other factors have to be considered. These are innocent acquisition and relevance. The evaluation of these factors may be done by partitioning the expressions in the numerator and denominator of (9.8). There are two types of subjective probabilities of interest:

- Innocent acquisition, usually denoted p, is a measure of belief that evidence has been acquired in a manner unrelated to the crime (Evett, 1993a).

- The probability of relevance usually denoted r (Stoney, 1991a, 1994; Evett *et al.*, 1998a). In the present context this is the probability that the stain recovered from the crime scene is connected with the crime; it has been left by one of the offenders.

9.5.3   Association propositions

Consider the following:

- B, the crime stain came from one of the k offenders;
- $\bar{B}$, the crime stain did not come from any of the k offenders.

The value, V, of the evidence may now be written using the law of total probability (Section 1.6.7) as

$$V = \frac{Pr(E_c \mid H_p, B, E_s)Pr(B \mid H_p, E_s) + Pr(E_c \mid H_p, \bar{B}, E_s)Pr(\bar{B} \mid H_p, E_s)}{Pr(E_c \mid H_d, B)Pr(B \mid H_d) + Pr(E_c \mid H_d, \bar{B})Pr(\bar{B} \mid H_d)}.$$

In the absence of E_c, the evidence of the profile of the crime stain, knowledge of H_p and of E_s does not effect our belief in the truth or otherwise of B. This is what is meant by *relevance* in this context. Thus

$$Pr(B \mid H_p, E_s) = Pr(B \mid H_p) = Pr(B)$$

and

$$Pr(\bar{B} \mid H_p, E_s) = Pr(\bar{B} \mid H_p) = Pr(\bar{B}).$$

Let $Pr(B) = r$, $Pr(\bar{B}) = 1 - r$ and call r the *relevance term*; that is to say, relevance is equated to the probability the stain had been left by one of the offenders. The higher the value of r, the more relevant the stain becomes. Thus

$$V = \frac{Pr(E_c \mid H_p, B, E_s)r + Pr(E_c \mid H_p, \bar{B}, E_s)(1 - r)}{Pr(E_c \mid H_d, B, E_s)r + Pr(E_c \mid H_d, \bar{B}, E_s)(1 - r)}. \tag{9.9}$$

9.5.4 Intermediate association propositions

In order to determine the component probabilities of (9.9) intermediate association propositions are introduced:

- A, the crime stain came from the suspect;
- $\bar{A}$, the crime stain did not come from the suspect.

Now consider the four conditional probabilities from (9.9). The first term in the numerator, $Pr(E_c \mid H_p, B, E_s)$, is the probability that the crime stain would be of profile Γ if it had been left by one of the offenders (B), the suspect had committed the crime (H_p) and the suspect is of profile Γ. It can be written as

$$Pr(E_c \mid H_p, B, E_s) = Pr(E_c \mid H_p, B, A, E_s)Pr(A \mid H_p, B, E_s)$$
$$+Pr(E_c \mid H_p, B, \bar{A}, E_s)Pr(\bar{A} \mid H_p, B, E_s).$$

Here $E_c = E_s = \Gamma$ and $Pr(E_c \mid H_p, B, A, E_s) = 1$. In the absence of E_c, A is independent of E_s and so

$$Pr(A \mid H_p, B, E_s) = Pr(A \mid H_p, B) = 1/k,$$

where it is assumed that there is nothing in the background information I to distinguish the suspect, given H_p, from the other offenders as far as blood shedding is considered. In a similar manner, $Pr(\bar{A} \mid H_p, B, E_s) = (k-1)/k$. Also,

$$Pr(E_c \mid H_p, B, \bar{A}, E_s) = Pr(E_c \mid H_p, B, \bar{A}) = \gamma,$$

since if $\bar{A}$ is true, E_c and E_s are independent and one of the other offenders left the stain (since B holds). Thus

$$Pr(E_c \mid H_p, B, E_s) = \{1 + (k-1)\gamma\}/k.$$

The second term in the numerator of (9.9), $Pr(E_c \mid H_p, \bar{B}, E_s)$, is the probability that the crime stain would be of profile Γ if it had been left by an unknown person who was unconnected with the crime. (This is the implication of assuming $\bar{B}$ to be true.) The population of people who may have left the stain is not necessarily the same as the population from which the criminals are assumed to have come. Thus, let

$$Pr(E_c \mid H_p, \bar{B}, E_s) = \gamma',$$

where γ' is the probability of profile Γ amongst the population of people who may have left the stain (the prime' indicating it may not be the same value as γ, which relates to the population from which the criminals have come).

Consider now that the suspect is innocent and H_d is true. The first term in the denominator of (9.9) is $Pr(E_c \mid H_d, B, E_s) = Pr(E_c \mid H_d, B) = \gamma$, the frequency of Γ amongst the population from which the criminals have come. There is no need to partition these probabilities to consider A and $\bar{A}$ as the suspect is assumed not to be one of the offenders and B is that the stain was left by one of the offenders.

The second term in the denominator is

$$Pr(E_c \mid H_d, \bar{B}, E_s) = Pr(E_c \mid H_d, \bar{B}, A, E_s)Pr(A \mid H_d, \bar{B}, E_s)$$
$$+ Pr(E_c \mid H_d, \bar{B}, \bar{A}, E_s)Pr(\bar{A} \mid H_d, \bar{B}, E_s).$$

If A is true, $Pr(E_c \mid H_d, \bar{B}, A, E_s) = 1$. Also $Pr(A \mid H_d, \bar{B}, E_s) = Pr(A \mid H_d, \bar{B})$. This is the probability p of innocent acquisition, that the stain would have been left by the suspect even though the suspect was innocent of the offence. Here it is assumed that the propensity to leave a stain is independent of the profile of the person who left the stain. Hence $Pr(A \mid H_d, \bar{B}) = p$ and $Pr(\bar{A} \mid H_d, \bar{B}, E_s) = Pr(\bar{A} \mid H_d, \bar{B}) = 1 - p$. Also $Pr(E_c \mid H_d, \bar{B}, \bar{A}) = \gamma'$. Thus

$$Pr(E_c \mid \bar{C}, \bar{B}, E_s) = p + (1 - p)\gamma'.$$

Substitution of the above expressions into (9.9) gives

$$V = \frac{[r\{1 + (k-1)\gamma\}/k] + \{\gamma'(1-r)\}}{\gamma r + \{p + (1-p)\gamma'\}(1-r)}$$
$$= \frac{r\{1 + (k-1)\gamma\} + k\gamma'(1-r)}{k[\gamma r + \{p + (1-p)\gamma'\}(1-r)]}. \tag{9.10}$$

9.5.5   Examples

Example 9.1   Consider the case where it may be assumed that γ and γ' are approximately equal and that $p = 0$. The latter assumption holds if there is no possibility that the suspect may have left the stain for innocent reasons. Then

$$V = \frac{r\{1 + (k-1)\gamma\} + k\gamma(1-r)}{k\{\gamma r + \gamma(1-r)\}}$$
$$= \frac{r + (k-r)\gamma}{k\gamma}. \tag{9.11}$$

If γ is so small that $r/k\gamma \gg 1$ then $V \simeq r/k\gamma$. If $r = 1, V \simeq 1/k\gamma$ (see(9.6)), the value of the evidence has been reduced by a factor corresponding to the number of offenders.

Example 9.2 Assume $p \neq 0$ but that γ and γ' are approximately equal. Then

$$V = \frac{r + (k-r)\gamma}{k[p(1-r) + \gamma\{r + (1-p)(1-r)\}]}$$
$$= \frac{r + (k-r)\gamma}{k[p(1-r) + \gamma\{1 - p + pr\}]}.$$

Example 9.3 An example of the use of subjective probabilities in the evaluation of footmarks is given by Evett *et al.* (1998a). The evaluation is analogous to that derived in Section 9.5.4 with $k = 1$. The propositions are different. For footmarks, the propositions are:

- H_p, the suspect is the offender;
- H_d, some unknown person is the offender.

The association propositions are:

- B, the footmark was left by the offender;
- $\bar{B}$, the footmark was left by someone other than the offender.

The intermediate association propositions are:

- A, the footmark was made by a particular shoe (X, say), owned by the suspect;
- $\bar{A}$, the footmark was made by some unknown shoe, which may or may not have been owned by the suspect.

Now write $Pr(A \mid H_p, B, E_s) = Pr(A \mid H_p, B) = w$, the probability that the suspect was wearing shoe X given that the suspect was the offender and that the suspect left the footmark. An illustration of the determination of w is given by Evett *et al.* (1998a). The suspect was interviewed the day after the commission of the offence, he had ten pairs of shoes in his possession, and if it is assumed that they are all equally likely to have been worn for the commission of the crime, then w could be estimated to be 0.1.

For footmarks, a variant of (9.10) is derived in which $Pr(E_c \mid H_p, B, A, E_s)$ may be different from unity; denote it by p_{mrk}. The frequency of the characteristics of the footmark is γ. Consider γ and γ' to be equal. A more detailed analysis would treat these as different since the frequency of a particular footmark may be dependent on the wearer. Evett *et al.* (1998a) consider the frequency to be of two parts, one relevant to the manufacturer and one relevant to the acquired features. Also, an assumption that the footmark has not been left for an innocent reason enables p to be set equal to zero. Then

$Pr(E_c \mid H_p, B, E_S) = wp_{mrk} + \gamma(1 - w)$ and $Pr(E_c \mid H_p, \bar{B}, E_s) = Pr(E_c \mid H_d, B, E_s) = Pr(E_c \mid H_d, \bar{B}, E_s) = \gamma$. Then, from (9.10),

$$V = \frac{r\{wp_{mrk} + \gamma(1 - w)\} + \gamma(1 - r)}{\gamma r + \gamma(1 - r)}$$

$$= \frac{rwp_{mrk} - rw\gamma + \gamma}{\gamma}$$

$$= (1 - rw) + \frac{rw}{\gamma}p_{mrk}. \tag{9.12}$$

Numerical examples are given in Evett *et al.* (1998a).

Subjective probabilities

At this stage it is necessary to think more about r and p. The first, r, is the probability that the crime stain came from one of the offenders, and this probability has been defined as the *relevance* of the crime stain. The second, p, is the probability the crime stain came from the suspect, given the suspect did not commit the crime and that the crime stain did not come from any of the offenders; that is, it is the probability the crime stain was left there innocently by someone who is now a suspect. The validity of combining probabilities thought of as measures of belief, and probabilities as relative frequencies has been questioned (Freeling and Sahlin, 1983; Stoney, 1994). However, the discussion in Section 1.6.4 and comments on *duality* by Hacking (1975) explain why such combinations can be valid.

Evett (1993a) suggests that determination of the probabilities such as those above may be the province of the court and that it is necessary to establish the conditions under which the scientific evidence can be of any guidance to the court. Evett suggests an examination of the sensitivity of V to values of p and of r. As an illustration, he takes the number of offenders k to be 4 and the frequencies γ and γ' to be 0.001. Then

$$V = \frac{r + 0.004}{4[p(1 - r) + 0.001(1 - p + pr)]},$$

where $r + (k - r)\gamma$ has been approximated by $r + k\gamma$. The variation of V with r and p is shown in Figure 9.1.

The graph has been drawn with a logarithmic scale for V. This is plotted against p for $r = 1$, 0.75, 0.50, 0.25 and 0. It is useful to consider individually the terms within the expression for V for the case in which there is one blood-stain of profile Γ and frequency γ (which is assumed to be well known).

- The number of offenders, k: this is assumed to be well known.
- Relevance, r: the (subjective) probability that the crime stain came from one of the offenders; factors to be considered in its estimation include location, abundance and apparent freshness of the blood.

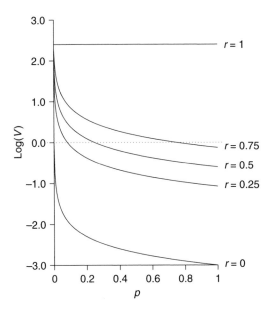

Figure 9.1 Variation in the logarithm to base 10 of the likelihood ratio V of the evidence with p, the probability that the stain would have been left by the suspect even though he was innocent of the offence, for various values of r, the probability that the stain would have been left by one of the offenders. The number of offenders, k, equals 4 and the relative frequency of the profile γ is 0.001. Adapted from Evett (1993a), with the inclusion of a curve for $r = 0$. The dotted line at $\log(V) = 0$ indicates where the evidence is equally likely under both propositions.

- Innocence, p: the (subjective) probability that the crime stain came from the suspect, given the suspect did not commit the crime and the crime stain did not come from any of the offenders.

Values for the probabilities of relevance and innocence may be proposed by the scientist but are matters for the courts to decide. In general, V decreases as r decreases or as p increases.

For $r = 1$, it is certain that the the crime stain came from one of the offenders and

$$V = \frac{1.004}{0.004},$$
$$= 251,$$
$$\log_{10}(V) = 2.40.$$

For $r \neq 1$, V is very sensitive to p. If there is a non-zero probability that the crime stain did not come from one of the offenders then the probability of innocence has a considerable influence on V. For example, if $r = 0.25$, so that there is a small probability the crime stain came from one of the offenders,

V becomes less than 1 for $p > 0.083 \simeq 1/12$. Thus, if $p > 1/12$ (and there is a small probability that the crime stain came from the suspect, conditional on everything else) then the evidence supports the proposition that some other person is the offender rather than the proposition that the suspect is the offender.

9.5.6 Two stains, one offender

The two-trace bloodstain problem of Section 9.3 has been modified by Stoney (1994) to the case where there are two bloodstains, of profile Γ_1 and Γ_2, respectively, as before but only one offender (rather than two as in Section 9.3.1) who left one of the bloodstains (it is not known which bloodstain). A suspect is found who is of group Γ_1. Relevance is applicable in this context since it provides a measure of belief (probability) that the stain at the crime scene which comes from the offender is the one which is of the same group as the suspect. The two competing propositions to be considered are:

- H_p, the suspect is the offender;
- H_d, the suspect is not the offender.

Let r be the probability that the matching stain (Γ_1) was from the offender. As before, this is a subjective probability. It has been assumed that one of the stains is from the offender, so there is a probability $1 - r$ that it is the other stain (Γ_2) that is from the offender.

Suppose H_p is true. Then, through the use of association propositions, the stain may or may not have come from the offender.

$$Pr(\text{suspect's profile and crime profile correspond}$$
$$\mid \text{stain } \Gamma_1 \text{ was from the offender}) = 1.$$
$$Pr(\text{stain } \Gamma_1 \text{ was from the offender}) = r.$$
$$Pr(\text{suspect's profile and crime profile correspond}$$
$$\mid \text{stain } \Gamma_2 \text{ was from the offender}) = 0.$$
$$Pr(\text{stain } \Gamma_2 \text{ was from the offender}) = 1 - r.$$

Thus $Pr(\text{match in } \Gamma_1 \mid H_p) = r$.

Suppose H_d is true. (There is no need to develop the denominator of V using association propositions.) The probability of a correspondence in Γ_1 is the probability that a randomly selected person is of profile Γ_1. This is the profile frequency γ_1. The likelihood ratio is then

$$V = r/\gamma_1.$$

This is a particular case of (9.11) with $k = 1$.

If the two stains have equal probabilities of being left by the offender, $r = 1/2$ and the likelihood ratio equals $1/2\gamma_1$. This is numerically equivalent to the figure $1/2\gamma_1$ derived by Evett (1987b) and quoted above in (9.5) for a problem with two stains, one left by each of the two offenders, and a single suspect whose profile matches that of one of the stains and whose frequency is γ_1. The derivation, however, is very different.

Stoney (1994) continued the development to the case where neither stain may be relevant but there is still a single offender. The suspect has profile Γ_1. Let there be the following probabilities:

$$Pr(\text{The stain of profile }\Gamma_1\text{ is from the offender}) = r_1,$$

$$Pr(\text{The stain of profile }\Gamma_2\text{ is from the offender}) = r_2,$$

$$Pr(\text{Neither stain is from the offender}) = 1 - r_1 - r_2.$$

If H_p is true (i.e., the suspect is the offender), there are three components to the probability:

- The stain of profile Γ_1 is from the offender. There is a match with probability r_1.
- The stain of profile Γ_2 is from the offender. There is no match. This event has probability zero since the suspect is assumed to be the offender and only one offender is assumed.
- Neither stain is from the offender. This event has probability $1 - r_1 - r_2$, and if it is true there is a probability γ_1 of a match between the suspect's profile (Γ_1) and the crime stain of the same profile. The probability of the combination of these events is $(1 - r_1 - r_2)\gamma_1$.

These three components are mutually exclusive and the probability in the numerator of the likelihood ratio is the sum of these three probabilities, namely $r_1 + (1 - r_1 - r_2)\gamma_1$.

If H_d is true (i.e., the suspect is not the offender), the probability of a correspondence is as before, namely γ_1. The likelihood ratio is then

$$V = \frac{r_1 + (1 - r_1 - r_2)\gamma_1}{\gamma_1}.$$

Certain special cases can be distinguished. As r_1 and r_2 tend to zero, which implies that neither stain is relevant, then the likelihood ratio tends to 1. A likelihood ratio of 1 provides no support for either proposition, a result in this case which is entirely consistent with the information that neither stain is relevant. For $r_1 = r_2 = 1/2$, $V = 1/2\gamma_1$. For $r_1 = 1, V = 1/\gamma_1$. As $r_2 \to 1$, then $r_1 \to 0$ and $V \to 0$. All of these are perfectly reasonable results.

9.6 SUMMARY

The results of the previous sections relating to bloodstains may usefully be summarised.

9.6.1 Stain known to have been left by offenders

One stain known to have come from one offender

The profile of the crime stain and of the suspect is Γ with frequency γ. The propositions to be compared are:

- H_p, the stain at the crime scene came from the suspect;
- H_d, the stain at the crime scene did not come from the suspect.

Then

$$V = \frac{1}{\gamma}.$$

Two stains, one from each of two offenders

There are two crime stains, of profiles Γ_1 and Γ_2 with frequencies γ_1 and γ_2. There is one suspect with profile Γ_1 with frequency γ_1. The propositions to be compared are:

- H_p, the crime stain came from the suspect and one other man;
- H_d, the crime stains came from two unknown men.

Then

$$V = \frac{1}{2\gamma_1}.$$

n stains, one from each of n offenders

There is one suspect with profile Γ_1 with frequency γ_1. The propositions to be compared are:

- H_p, the crime sample came from the suspect and $n - 1$ other men;
- H_d, the crime sample came from n unknown men.

Then

$$V = \frac{1}{n\gamma_1}.$$

n stains, k different profiles, k different offenders

There are s_i stains of type i ($i = 1, \ldots, k$; $\sum_{i=1}^{k} s_i = n$). There is one suspect with profile Γ_1 with frequency γ_1. The propositions to be compared are:

- H_p, the crime sample came from the suspect and $k - 1$ other men;
- H_d, the crime sample came from k unknown men.

Then

$$V = \frac{1}{\binom{n}{s_1} \gamma_1^{s_1}}.$$

n stains, k different profiles, m offenders

This situation may arise where there are limited analytical results. There are m_i offenders with profile i ($i = 1, \ldots, k$; $\sum_{i=1}^{k} m_i = m$). There are s_{ij} stains belonging to the jth offender for the ith profile. There is one suspect with blood group Γ_1 with frequency γ_1. Assume this could be the first offender with the first profile. The propositions to be compared are:

- H_p, the crime sample came from the suspect and $m - 1$ other men;
- H_d, the crime sample came from m other men.

Then

$$V = \frac{1}{\binom{n}{s_{11}} \gamma_1^{s_{11}}}.$$

9.6.2   Relevance: stain may not have been left by offenders

One stain, k offenders

The probability that a crime stain came from one of the k offenders is the relevance, r. The propositions to be compared are:

- H_p, the suspect is one of the k offenders;
- H_d, the suspect is not one of the k offenders.

The stain is of profile Γ. It may have been left by an offender. There are k offenders. The suspect is of profile Γ. This profile has frequency γ in the population from which the criminals may be thought to come. It has frequency γ' amongst the population of people who may have left the stain, which may not be the same population as that from which the criminals may be thought to come. For example, there may be eyewitness evidence that the criminals are from one ethnic group, whereas the people normally associated with the crime scene may be from another. The probability that the stain would have been left by the suspect even though he was innocent of the offence is p. Then

$$
\begin{aligned}
V &= \frac{[r\{1+(k-1)\gamma\}/k]+\{\gamma'(1-r)\}}{\gamma r+\{p+(1-p)\gamma'\}(1-r)} \\
&= \frac{r\{1+(k-1)\gamma\}+k\gamma'(1-r)}{k[\gamma r+\{p+(1-p)\gamma'\}(1-r)]}.
\end{aligned}
$$

There are several simplifications.

If $\gamma = \gamma'$ and $p = 0$, then

$$
V = \frac{r+(k-r)\gamma}{k\gamma}.
$$

If, also, $r = 1$,

$$
V = \frac{1+(k-1)\gamma}{k\gamma}.
$$

Compare this with the case in which there are n stains (rather than 1) and the number (k) of offenders equals the number of stains (n). (There is one stain from each of n offenders.) Then $V = 1/n\gamma = 1/k\gamma$.

If there is one stain and k offenders,

$$
\begin{aligned}
V &= \frac{1+(k-1)\gamma}{k\gamma} \\
&= 1+\frac{1}{k\gamma}-\frac{1}{k},
\end{aligned}
$$

an increase from $1/k\gamma$. However, if γ is small, V is approximately equal to $1/k\gamma$. The value of the evidence is the same when there are k offenders and one suspect whether there is one stain, matching the profile of the suspect, or many stains of different profiles.

If $\gamma = \gamma'$ and $p \neq 0$, then

$$
\begin{aligned}
V &= \frac{[r\{1+(k-1)\gamma\}/k]+\{\gamma(1-r)\}}{[\gamma r+\{p+(1-p)\gamma\}(1-r)]} \\
&= \frac{r+(k-r)\gamma}{k[p(1-r)+\gamma\{r+(1-p)(1-r)\}]} \\
&= \frac{r+(k-r)\gamma}{k[p(1-r)+\gamma(1-p+pr)]}.
\end{aligned}
$$

Two stains, one of which is relevant, one offender

The offender left one of the bloodstains but it is not known which one. The propositions to be compared are:

- H_p, the suspect is the offender;
- H_d, the suspect is not the offender.

A suspect is of profile Γ_1, with frequency γ_1. Let r be the probability that the crime stain which matches the group of the suspect is from the suspect. Then $1 - r$ is the probability that the crime stain which does not match the profile of the suspect is from the offender. Then

$$V = \frac{r}{\gamma_1}.$$

Two stains, neither of which may be relevant, one offender

The propositions to be compared are:

- H_p, the suspect is the offender;
- H_d, the suspect is not the offender.

A suspect is of profile $\Gamma_1(\Gamma_2)$ with frequency γ_1. Let $r_1(r_2)$ be the probability that the stain of profile $\Gamma_1(\Gamma_2)$ is from the offender. Then $1 - r_1 - r_2$ is the probability that neither stain is from the offender and

$$V = \frac{r_1 + (1 - r_1 - r_2)\gamma_1}{\gamma_1}.$$

9.6.3   Relevance and the crime level

Relevance is a factor when assessing propositions at the crime level but not at the activity level. Let γ be the relative frequency of a trait in a relevant population. At the crime level, the value of the evidence cannot be greater than $1/\gamma$. It may be reduced below this value because relevance, expressed as a probability, may be less than 1. At the activity level, the value of the evidence can be greater than $1/\gamma$. Here, factors such as transfer and background have to be considered.

However, if the crime level is being considered and it is assumed that the suspect has to be present to commit the crime, then the activity level does not need to be considered. Consider two propositions, H_{p1} that the suspect committed the crime (crime level) and H_{p2} that the suspect was present at the crime scene (activity level). Then

$$Pr(E \mid H_{p1}, H_{p2}) = Pr(E \mid H_{p1}).$$

9.7 MISSING PERSONS

Consider a case of a missing person in which there is evidence of foul play and a suspect has been identified. Evidence of the missing person's blood group and other phenotypic values is not available. Instead, values are available from the parents of the missing person (case couple) from which inferences about the missing person may be made. The discussion refers to cases from twenty years ago and marker systems used at the time. The principles remain true today, however.

Bloodstains found on the suspect's property could have come from one of the parent's offspring. The hypotheses to be compared are:

- H_p, the bloodstains came from a child of the case couple;
- H_d, the bloodstains did not come from a child of the case couple.

The value of the evidence is

$$V = \frac{\text{Probability parents would pass the stain phenotype}}{\text{Probability a random couple would pass the stain phenotype}}.$$

This may be expressed verbally as 'the parents of the missing person are V times more likely than a randomly selected couple to pass this set of genes'. See Section 8.1.3 for an earlier discussion of this problem.

Consider the following two cases of missing persons with phenotypes given in Table 9.3, the first (Case 1) described by Kuo (1982), the second (Case 2) by Ogino and Gregonis (1981), both reviewed by Stoney (1984b).

The stain phenotypic frequencies may be calculated for each case. These are the probabilities that a random couple would pass the corresponding phenotype, and are given in Tables 9.4 and 9.5.

Table 9.3 Phenotypes for two cases of missing persons

Marker system	Case 1			Case 2		
	Father	Mother	Stain	Father	Mother	Stain
ABO	B	O	B	O	O	O
EAP	BA	BA	A	B	BA	B
AK	1	2-1	2-1	1	1	1
ADA	1	1	1	1	1	1
PGM	2-1	1	2-1	1	1	1
Hp	2-1	2	2-1	1	1	1
EsD				1	1	1

Table 9.4 Gene frequencies and phenotypic incidences for Case 1 (Kuo, 1982)

Marker system	Stain phenotype	Gene frequencies	Phenotypic incidences		
ABO	B	B: 0.074	$0.074^2 +$		
		O: 0.664	$2 \times 0.074 \times 0.664$	=	0.1037
EAP	A	A: 0.327	0.327^2	=	0.1069
AK	2-1	2: 0.038			
		1: 0.962	$2 \times 0.038 \times 0.962$	=	0.0731
ADA	1	1: 0.952	0.952^2	=	0.9063
PGM	2-1	2: 0.229			
		1: 0.771	$2 \times 0.229 \times 0.771$	=	0.3531
Hp	2-1	2: 0.578			
		1: 0.422	$2 \times 0.578 \times 0.422$	=	0.4878

Table 9.5 Gene frequencies and phenotypic incidences for Case 2 (Ogino and Gregonis, 1981)

Marker system	Stain phenotype	Gene frequencies	Phenotypic incidences
ABO	O	O: 0.664	$0.664^2 = 0.4409$
EAP	B	B: 0.612	$0.612^2 = 0.3745$
AK	1	1: 0.962	$0.962^2 = 0.9254$
ADA	1	1: 0.952	$0.952^2 = 0.9063$
PGM	1	1: 0.771	$0.771^2 = 0.5944$
Hp	1	1: 0.422	$0.422^2 = 0.1781$
EsD	1	1: 0.884	$0.884^2 = 0.7815$

9.7.1 Case 1 (Kuo, 1982)

A young woman was missing after going on a boat trip with her boyfriend. He became the main suspect. Bloodstains were found on his boat, but there was no known blood of the missing woman to group for comparison with the bloodstains. The blood of the parents of the woman was grouped to try and link the bloodstain on the boat to the missing woman.

The combined phenotypic incidence from Table 9.4 is $0.1037 \times 0.1069 \times 0.0731 \times 0.9063 \times 0.3531 \times 0.4878 = 1.2650 \times 10^{-4}$, the multiplication being justified by the independence of the marker systems. Notice the implicit extension of the approach based on the likelihood ratio to include evidence from more than one marker; see Section 8.1.3.

9.7.2 Case 2 (Ogino and Gregonis, 1981)

A man was reported missing by his family. A suspect, driving the victim's vehicle, was arrested for suspicion of murder. Bloodstains were found in various parts of the vehicle. Blood samples were obtained from the victim's mother and father, as well as other relatives. The combined phenotypic incidence, based on those of the parents (Table 9.3), from Table 9.5 is $0.4409 \times 0.3745 \times 0.9254 \times 0.9063 \times 0.5944 \times 0.1781 \times 0.7815 = 1.146 \times 10^{-2}$.

9.7.3   Calculation of the likelihood ratio

The likelihood ratio V compares two probabilities. The probability parents would pass on a stain phenotype is compared with the probability a random couple would pass a stain phenotype. The latter is simply the product of the gene frequencies and these values have already been calculated for each of the two cases.

The probability that the parents would pass the stain phenotype is calculated for each case as follows.

The possible couple–stain combinations for up to three codominant alleles are given in Table 9.6.

The *ABO* system requires a special treatment because of dominance. The probabilities of passage depend on the frequencies with which homozygous and heterozygous individuals occur; see Table 9.7. Type *A* and type *B* parents may be either homozygous with the dominant allele or heterozygous with the recessive *O* allele. Further details are contained in Stoney (1984b). These frequencies may be calculated directly from the gene frequencies.

The probability that a particular gene will be passed is then determined by combining the likelihoods of heterozygosity and homozygosity with the likelihood of passing the gene in each instance.

Table 9.6   Frequencies for up to three codominant alleles

Couple	Possible stains and frequencies			
PP, PP	*PP* = 1.00			
PP, PQ	*PP* = 0.50	*PQ* = 0.50		
PP, QQ	*PQ* = 1.00			
PQ, PQ	*PP* = 0.25	*PQ* = 0.50	*QQ* = 0.25	
PQ, PR	*PP* = 0.25	*PR* = 0.25	*PQ* = 0.25	*QR* = 0.25
PQ, RR	*PR* = 0.50	*QR* = 0.50		

Table 9.7   Relative frequencies for *ABO* system

Type *P*	Homozygous *PP*	Heterozygous *PO*
A	$(a)/(a+2o)$ $= 0.262/1.590 = 0.1648$	$(2o)/(a+2o)$ $= 1.328/1.590 = 0.8352$
B	$(b)/(b+2o)$ $= 0.074/1.402 = 0.0528$	$(2o)/(b+2o)$ $= 1.328/1.402 = 0.9472$

Case 1

The probability that the couple would pass the stain phenotype equals $0.5264 \times 0.25 \times 0.50 \times 1.00 \times 0.50 \times 0.50 = 0.0164$; see Table 9.8. The probability that the stain phenotype occurs by chance equals 1.2650×10^{-4}. Thus

$$V = \frac{0.0164}{1.2650 \times 10^{-4}} = 130.0.$$

The parents of the missing person in Case 1 may be said to be 130 times more likely to pass the stain phenotype than would be a randomly selected couple. From the qualitative scale of Table 3.10 the evidence provides moderately strong evidence to support the proposition that the bloodstains came from a child of the case couple.

Case 2

The probability that the couple would pass the stain phenotype is 0.50; see Table 9.9. The probability that the stain phenotype occurs by chance is 1.146×10^{-2}, and

$$V = \frac{0.50}{1.146 \times 10^{-2}} = 43.6.$$

Table 9.8 Probabilities that the parents will pass the specified stain phenotype, Case 1

Marker system	Couple phenotypes F	M	Stain phenotype	Probability that couple would pass stain phenotype
ABO	B	O	B	0.5264[a]
EAP	BA	BA	A	0.25
AK	1	2-1	2-1	0.50
ADA	1	1	1	1.00
PGM	2-1	1	2-1	0.50
Hp	2-1	2	2-1	0.50

[a] $(1 \times 0.0528) + (0.5 \times 0.9472)$.

Table 9.9 Probabilities that the parents will pass the specified phenotypes, Case 2

Marker system	Couple phenotypes F	M	Stain phenotype	Probability that couple would pass stain phenotype
ABO	O	O	O	1.00
EAP	B	BA	B	0.50
AK	1	1	1	1.00
ADA	1	1	1	1.00
PGM	1	1	1	1.00
Hp	1	1	1	1.00
EsD	1	1	1	1.00

Table 9.10   Two pieces of evidence on DNA markers

		Profiles		
Evidence	Locus	Child	Mother	Alleged father
E_1	PentaD	13 − 13	9 − 13	11 − 13
E_2	VWA	18 − 19	16 − 19	18 − 18

The parents of the missing person in Case 2 may be said to be about 44 times more likely to pass the stain phenotype than would be a randomly selected couple. Similarly, from Table 3.10, this evidence provides moderate support for the proposition that the bloodstains came from a child of the case couple.

These examples are about 20 years old and use classical marker systems, which have been replaced by DNA markers. However, the principles remain the same. One method of assessment would be to consider the frequency of non-excluded couples. Another method of assessment would be to consider the frequency of non-excluded stains. For the two cases described here, the frequencies of non-excluded couples are 0.0062 (Kuo) and 0.2564 (Ogino and Gregonis) and the frequencies of non-excluded stains are 0.3386 (Kuo) and 0.0237 (Ogino and Gregonis) (Stoney, 1984b). Thus, there are contradictory conclusions. The evidence is rare for Kuo and not rare for Ogino and Gregonis if the frequency of non-excluded couples is considered as the statistic for the assessment of the evidence. The evidence is not rare for Kuo and rare for Ogino and Gregonis if the frequency of non-excluded stains is considered as the statistic for the assessment of the evidence. In contrast to these conflicting assessments, the likelihood ratio considers both the couples and the stains to provide one unified statistic.

Developments using DNA marker systems and for examples in which members of the family additional to the spouse and the child or in which the alleged father is deceased are described in Evett and Weir (1998). A general formula for a likelihood ratio which is appropriate in many potential relationships between two DNA profiles is presented in Brenner and Weir (2003). Fung (2003) proposes an automatic approach for the calculation of complicated pedigrees.

9.8   PATERNITY: COMBINATION OF LIKELIHOOD RATIOS

In paternity testing, the likelihood ratio is used to compare two probabilities, as in the criminal context. In this context the propositions to be compared are:

- H_p, the alleged father is the true father,
- H_d, the alleged father is not the true father.

The probability that the alleged father would pass the child's non-maternal alleles is compared to the probability that the alleles would be passed randomly. Thus the value V of the evidence is

$$V = \frac{\text{Probability the alleged father would pass the alleles}}{\text{Probability the alleles would be passed by a random male}}.$$

This may be expressed verbally as 'the alleged father is V times more likely than a randomly selected man to pass this set of alleles'.

Notice the difference between paternity testing and a missing persons case. In paternity testing, the relationship between mother and child is known. The question is whether a particular male is a possible biological father. In missing persons and bloodstain cases the parents are known. The question is whether the stain could have come from one of their offspring.

Evett and Weir (1998) give an example of the use of the likelihood ratio in paternity problems. The ratio in this context has also been called the *paternity index* (PI) by Salmon and Salmon (1980). The example discussed here is an example of the use of the likelihood ratio to include more than one piece of evidence through consideration of more than one DNA marker.

Consider two pieces of evidence E_1 and E_2, where E_1 and E_2 are the DNA profiles of the child, mother and alleged father under the *PentaD* and *VWA* systems, respectively, as shown in Table 9.10. The likelihood ratio or paternity index for $E_i, i = 1, 2$, is

$$PI = \frac{Pr(E_i \mid H_p)}{Pr(E_i \mid H_d)}.$$

Let G_{Ci}, G_{Mi} and G_{AFi} denote the genotypes of the child C, the mother M and the alleged father AF, respectively, for evidence E_i. Let A_{Mi} and A_{Pi} denote the maternal and paternal alleles for evidence E_i. Let $\gamma_{i,j}$ be the frequency of allele j for evidence E_i.

For E_1, the numerator of the likelihood ratio equals $Pr(G_{C1} \mid G_{M1}, G_{AF1}, H_p) = 1/4$. This is because $9 - 13$ crossed with $11 - 13$ will produce $13 - 13$ with probability $1/4$. The denominator equals $Pr(G_{C1} \mid G_{M1}, G_{AF1}, H_d) = Pr(A_{M1} \mid G_{M1}) \times Pr(A_{P1} \mid H_d) = Pr(A_{M1} = 13 \mid G_{M1} = 9 - 13) \times Pr(A_{P1} = 13 \mid H_d) = (1/2) \times \gamma_{1,13}$. The likelihood ratio for *PentaD* is then $1/(2\gamma_{1,13})$.

For E_2, the numerator of the likelihood ratio equals $Pr(G_{C2} \mid G_{M2}, G_{AF2}, H_p) = 1/2$. This is because $16 - 19$ crossed with $18 - 18$ will produce $18 - 19$ with probability $1/2$. The denominator equals $Pr(G_{C2} \mid G_{M2}, G_{AF2}, H_d) = Pr(A_{M2} \mid G_{M2}) \times Pr(A_{P2} \mid H_d) = Pr(A_{M2} = 19 \mid G_{M2} = 16 - 19) \times Pr(A_{P2} = 18 \mid H_d) = (1/2) \times \gamma_{2,18}$. The likelihood ratio for *VWA* is then $1/\gamma_{2,18}$.

Under an assumption of independence, the likelihood ratio for the combination (E_1, E_2) of evidence is

$$\frac{Pr(E_1, E_2 \mid H_p)}{Pr(E_1, E_2 \mid H_d)} = \frac{Pr(E_1 \mid H_p)}{Pr(E_1 \mid H_d)} \times \frac{Pr(E_2 \mid H_p)}{Pr(E_2 \mid H_d)}$$

$$= \frac{1}{2\gamma_{1,13}} \times \frac{1}{\gamma_{2,18}}.$$

(see Section 8.1.3). Assume $\gamma_{1,13} = 0.206$, $\gamma_{2,18} = 0.2274$. Then $1/(2\gamma_{1,13}\gamma_{2,18}) = 10.7 \simeq 11$; the evidence of the two marker systems is 11 times more likely if the alleged father is the true father than if he is not. From Table 3.10 this provides moderate support for the proposition that the alleged father is the true father.

9.8.1 Likelihood of paternity

In the context of paternity, it is appropriate to make a digression from consideration solely of the likelihood ratio and to consider the probability that the alleged father is the true father; that is, the probability that H_p is true. This probability is known as the *likelihood of paternity*.

Consider the two pieces of evidence, E_1 and E_2 of Table 9.10. The odds in favour of H_p, given E_1, may be written, using the odds form of Bayes' theorem (3.6), as

$$\frac{Pr(H_p \mid E_1)}{Pr(H_d \mid E_1)} = \frac{Pr(E_1 \mid H_p)}{Pr(E_1 \mid H_d)} \times \frac{Pr(H_p)}{Pr(H_d)}$$

and

$$Pr(H_d \mid E_1) = 1 - Pr(H_p \mid E_1),$$

so

$$Pr(H_p \mid E_1) = \frac{Pr(E_1 \mid H_p)}{Pr(E_1 \mid H_d)} \times \frac{Pr(H_p)}{Pr(H_d)} \times \{1 - Pr(H_p \mid E_1)\}$$

and

$$Pr(H_p \mid E_1)\left\{1 + \frac{Pr(E_1 \mid H_p)}{Pr(E_1 \mid H_d)} \times \frac{Pr(H_p)}{Pr(H_d)}\right\} = \frac{Pr(E_1 \mid H_p)}{Pr(E_1 \mid H_d)} \times \frac{Pr(H_p)}{Pr(H_d)},$$

so that

$$Pr(H_p \mid E_1) = \left\{1 + \frac{Pr(E_1 \mid H_d)}{Pr(E_1 \mid H_p)} \times \frac{Pr(H_d)}{Pr(H_p)}\right\}^{-1}, \tag{9.13}$$

a result analogous to (4.6). Suppose, rather unrealistically, that the alleged father and only one other man (of unknown blood type) could be the true father and that each possibility is equally likely (Essen-Möller, 1938). Then

$$Pr(H_p) = Pr(H_d) = 0.5$$

and

$$Pr(H_p \mid E_1) = 1/(1 + 2\gamma_{1,13}) = 1/(1 + 0.412) = 0.708.$$

Now include E_2. The posterior odds $Pr(H_p \mid E_1)/Pr(H_d \mid E_1)$ in favour of H_p, given E_1, now replace the prior odds $Pr(H_p)/Pr(H_d)$ (see Section 8.1.3), and the posterior probability for H_p, given E_1 and E_2, is given by

$$
\begin{aligned}
Pr(H_p \mid E_1, E_2) &= \left\{ 1 + \frac{Pr(H_d \mid E_1)}{Pr(H_p \mid E_1)} \times \frac{Pr(E_2 \mid H_d)}{Pr(E_2 \mid H_p)} \right\}^{-1} \quad (9.14) \\
&= \left(1 + \frac{0.292}{0.708} \times \frac{0.227/2}{1/2} \right)^{-1} \\
&= 0.914,
\end{aligned}
$$

where the assumption of the independence of E_1 and E_2 has been made. The probability the alleged father was the true father, the likelihood of paternity, was initially 0.5. After presentation of the *PentaD* evidence (E_1) it became 0.708. After the presentation of the *VWA* evidence (E_2) it became 0.914. Note that this posterior probability is just the ratio of *PI* to $(1 + PI)$; in this case, $10.7/11.7 = 0.914$.

Notice that the assumption $Pr(H_p) = Pr(H_d) = 0.5$ is unrealistic and can lead to breaches of the laws of probability (Berry, 1991b; Allen *et al.*, 1995). If there were two alleged fathers, both of type $(11 - 13, 18 - 18)$ then both would have a posterior probability of 0.914 of being the true father. The probability of one or other being the true father would then be the sum of these two probabilities, 1.828. However, this is greater than 1 and breaches the first law of probability (1.4). There have been many criticisms of this assumption (Ellman and Kaye, 1979; Kaye, 1989; Allen *et al.*, 1995; Taroni and Aitken, 1998a, and references therein). Some courts have been aware of the unrealistic nature of the assumption for a long time. For example:

> Leaving the choice of the prior odds to the legal decision-maker is preferable to presenting or using an unarticulated prior probability. (*Re the Paternity of M.J.B. : T.A.T.*)

Note, though, that in the case with two alleged fathers it could be argued that there are three possibilities, either is the true father or, the third possibility, neither is. The prior probabilities associated with these probabilities could then be 1/3 each. It is better, however, as argued in *Re the Paternity of M.J.B. : T.A.T.*, to leave the choice of prior probabilities to the legal decision maker.

The effect on the posterior probability of altering the prior probability can be determined from (9.13) and (9.14). Some sample results are given in Table 9.11.

This is an example of the general idea expressed in Section 3.5.1. An example specific to paternity is the following.

> The expert's testimony should be required to include an explanation to the jury of what the probability of paternity would be for a varying range of such prior probabilities running from 0.1 to 0.9.' (*State of New Jersey v. J.M. Spann.*)

The probability $Pr(H_p \mid E_1, E_2)$ may be written as

$$\left\{1 + \frac{Pr(E_1 \mid H_d)}{Pr(E_1 \mid H_p)} \times \frac{Pr(E_2 \mid H_d)}{Pr(E_2 \mid H_p)} \times \frac{Pr(H_d)}{Pr(H_p)}\right\}^{-1},$$

and if $Pr(H_d)$ and $Pr(H_p)$ are taken equal to 0.5 then

$$Pr(H_p \mid E_1, E_2) = \left\{1 + \frac{Pr(E_1 \mid H_d)}{Pr(E_1 \mid H_p)} \times \frac{Pr(E_2 \mid H_d)}{Pr(E_2 \mid H_p)}\right\}^{-1}.$$

In general, for n independent DNA markers, giving evidence $E_1, E_2, \ldots, E_n$, with $Pr(H_p) = Pr(H_d)$,

$$Pr(H_p \mid E_1, \ldots, E_n) = \left\{1 + \prod_{i=1}^{n} \frac{Pr(E_i \mid H_d)}{Pr(E_i \mid H_p)}\right\}^{-1}$$

where $\prod_{i=1}^{n} Pr(E_i \mid H_d)/Pr(E_i \mid H_p)$ is the product of the reciprocals of the n likelihood ratios $Pr(E_i \mid H_p)/Pr(E_i \mid H_d)$. This expression is called the *plausibility of paternity* (Berry and Geisser, 1986). Notice that it depends on the assumption $Pr(H_p) = Pr(H_d) = 0.5$, which is unrealistic in many cases. The assumption that $Pr(H_p)$ equals $Pr(H_d)$ may easily be dispensed with to give the following result:

$$Pr(H_p \mid E_1, \ldots, E_n) = \left\{1 + \frac{Pr(H_d)}{Pr(H_p)} \prod_{i=1}^{n} \frac{Pr(E_i \mid H_d)}{Pr(E_i \mid H_p)}\right\}^{-1}.$$

Table 9.11 Posterior probabilities of paternity for various prior probabilities for evidence for alleged father $E_1 = 11 - 13$, $E_2 = 18 - 18$

$Pr(H_p)$	0.5	0.25	0.1	0.01
$Pr(H_p \mid E_1)$	0.708	0.447	0.195	0.024
$Pr(H_p \mid E_1, E_2)$	0.914	0.781	0.516	0.097

Table 9.12   Hummel's likelihood of paternity

Plausibility of paternity	Likelihood of paternity
0.9980–0.9990	Practically proved
0.9910–0.9979	Extremely likely
0.9500–0.9909	Very likely
0.9000–0.9499	Likely
0.8000–0.8999	Undecided
Less than 0.8000	Not useful

The plausibility of paternity has also been transformed into a *likelihood of paternity* (Hummel, 1971, 1983) to provide a verbal scale, given here in Table 9.12. Notice that this verbal scale is one for probabilities. The verbal scale provided by Table 3.10 is for likelihood ratios.

Generally forensic scientists use formulae for calculating the probability of DNA profiles for two related individuals under an assumption of independence of genes. Balding and Nichols (1995) consider paternity indices for the case when the mother, alleged father and alternative father all belong to the same sub-population and estimates of the allele proportions are available only for the total (general) population. Paternity formulae considering parentage or other alleged relationships, when only two individuals are tested (i.e., alleged parent and child), are proposed by Ayres (2000) and Lee *et al.* (2000). The formulae incorporate the coancestry coefficient, F_{ST} (Section 13.5). The effect of incorporating F_{ST} into the equations is, in most situations, to decrease the paternity index for parentage. In fact, uncertainty arises from the fact that matches could be due to allele sharing between the alleged father and the set of alternatives as specified by H_d.

Formulae for the paternity index have also been developed for some of the relationships most often tested for two individuals (alleged full siblings and half-siblings versus unrelated) (Fung *et al.*, 2003). Mutation probabilities have also been incorporated into likelihood ratios. For a discussion on mutation rates and their estimation, see Dawid *et al.* (2001), Dawid (2003) and Vicard and Dawid (2003).

Moreover, it has been shown that it is important to allow for the fact that a close relative of the alleged father may be the true father, in addition to the usual alternative of an unrelated man. Formulae to allow for this are given in Lee *et al.* (1999).

9.8.2   Probability of exclusion in paternity

Weir (1996a) comments that it may be useful to characterise a genetic marker by its ability to exclude a random man from paternity. The so-called *probabilities of exclusion* depend on allele frequencies for the locus but not on the genotype probabilities.

Consider a single autosomal locus with codominant alleles. Let there be K alleles at the locus. Individual alleles will be denoted u, v and y as appropriate. First, assume the mother is of genotype $A_u A_u$. This has probability p_u^2. The child may be either $A_u A_u$ with probability p_u or $A_u A_v$ ($v \neq u$) with probability p_v (given the mother's genotype). If the mother is of genotype $A_u A_v$ with probability $2p_u p_v$ there are five possible genotypes for the child: $A_u A_u$, $A_v A_v$, $A_u A_v$, $A_u A_y$ and $A_v A_y$, where $y \neq u, v$. Weir (1996a) considers the probabilities for excluded men for each of these possible mother–child combinations. The probability of exclusion is then derived by multiplying together the probabilities for mother–child–excluded father combinations and then adding over the seven possibilities. This gives an exclusion probability of

$$Q = \sum_u p_u (1 - p_u)^2 - \frac{1}{2} \sum_u \sum_{v \neq u} p_u^2 p_v^2 (4 - 3p_u - 3p_v).$$

This is maximised when all K alleles at the locus have frequency $1/K$ (see Section 4.5.2 for a similar result associated with discriminating power) and then

$$Q_{\max} = 1 - \frac{2K^3 + K^2 - 5K + 3}{K^4}.$$

The more alleles there are the better the locus is for exclusion. With ten alleles, $Q_{\max} = 0.79$. With 30 equally frequent alleles, $Q_{\max} = 0.9324$.

As discussed before in Section 4.5.4 in the context of the combination of independent systems and discriminating power, the use of several loci will increase the exclusion probabilities. Let Q_l denote the exclusion probability at locus l. The overall probability of exclusion, Q, for independent loci, is

$$Q = 1 - \prod_l (1 - Q_l).$$

For two independent loci, each with ten equally frequent alleles, Q increases from 0.79 to 0.96.

The probability of exclusion when the true father may be a relative of the alleged father is discussed by Fung *et al.* (2002).

10

Continuous Data

10.1 THE LIKELIHOOD RATIO

The previous chapter considered the evaluation of the likelihood ratio where the evidence was represented by discrete data with specific reference to genetic marker systems. The value of the evidence in various different contexts was derived. However, much evidence is of a form in which measurements may be taken and for which the data are continuous. The form of the statistic for the evaluation of the evidence under these circumstances is similar to that for discrete data. Many of the examples in this chapter concern the interpretation of glass evidence; a review of the statistical interpretation of such evidence is given in Curran (2003).

Let the evidence be denoted by E, the two competing propositions by H_p and H_d and background information by I. Then the value V of the evidence is, formally,

$$V = \frac{Pr(E \mid H_p, I)}{Pr(E \mid H_d, I)},$$

as before (see (3.12) and (9.1)). The quantitative part of the evidence is represented by the measurements of the characteristic of interest. Let x denote a measurement on the source evidence and let y denote a measurement on the receptor object. For example, if a window is broken during the commission of a crime, the measurements on the refractive indices of m fragments of glass found at the crime scene will be denoted $x_1, \ldots, x_m$ (denoted $\mathbf{x}^T$). The refractive indices of n fragments of glass found on a suspect will be denoted $y_1, \ldots, y_n$ (denoted $\mathbf{y}^T$). By convention, vectors are printed in bold and the elements of a vector are written in a column. As mentioned in Section 7.3.8, the corresponding row vector is denoted with a superscript T to denote transposition (of a column to a row). See Chapter 11 for further details. The quantitative part of the evidence concerning the glass fragments in this case can be denoted by

$$E = (\mathbf{x}, \mathbf{y}).$$

Statistics and the Evaluation of Evidence for Forensic Scientists: Second Edition
C.G.G. Aitken and F. Taroni © 2004 John Wiley & Sons, Ltd   ISBN: 0-470-84367-5

In the notation of Section 1.6.1, M_c is the broken window at the crime scene, M_s is the set of glass fragments from the suspect, E_c is $\mathbf{x}$, E_s is $\mathbf{y}$, M is (M_c, M_s) and $E = (E_c, E_s) = (\mathbf{x}, \mathbf{y})$. Continuous measurements are being considered and the probabilities Pr are therefore replaced by probability density functions f (see Section 2.4.2) so that

$$V = \frac{f(\mathbf{x}, \mathbf{y} \mid H_p, I)}{f(\mathbf{x}, \mathbf{y} \mid H_d, I)}. \tag{10.1}$$

Bayes' theorem and the rules of conditional probability apply to probability density functions as well as to probabilities. The value, V, of the evidence (10.1) may be rewritten in the following way:

$$\begin{aligned}
V &= \frac{f(\mathbf{x}, \mathbf{y} \mid H_p, I)}{f(\mathbf{x}, \mathbf{y} \mid H_d, I)} \\
&= \frac{f(\mathbf{y} \mid \mathbf{x}, H_p, I)}{f(\mathbf{y} \mid \mathbf{x}, H_d, I)} \times \frac{f(\mathbf{x} \mid H_p, I)}{f(\mathbf{x} \mid H_d, I)}.
\end{aligned}$$

The measurements $\mathbf{x}$ are those on the source, or control, object. Their distribution and corresponding probability density function are independent of whether H_p or H_d is true. Thus

$$f(\mathbf{x} \mid H_p, I) = f(\mathbf{x} \mid H_d, I)$$

and

$$V = \frac{f(\mathbf{y} \mid \mathbf{x}, H_p, I)}{f(\mathbf{y} \mid \mathbf{x}, H_d, I)}.$$

If H_d is true, it is assumed that the measurements ($\mathbf{y}$) on the receptor object and the measurements ($\mathbf{x}$) on the source object are independent. Thus

$$f(\mathbf{y} \mid \mathbf{x}, H_d, I) = f(\mathbf{y} \mid H_d, I),$$

and

$$V = \frac{f(\mathbf{y} \mid \mathbf{x}, H_p, I)}{f(\mathbf{y} \mid H_d, I)}. \tag{10.2}$$

The assumption of independence is relaxed in Chapter 13 in the discussion on DNA profiling. The numerator is a *predictive distribution* (Section 6.3.2). The denominator is the so-called *marginal distribution* of the receptor measurements in the relevant population, the definition of which is assisted by I. This formulation of the expression for V shows that for the numerator the distribution of the receptor measurements, conditional on the source measurements as well as I, is considered. For the denominator, the distribution of the receptor measurements is considered over the distribution of the whole of the relevant population.

The two propositions to be compared are:

- H_p, the receptor sample is from the same source as the bulk sample;
- H_d, the receptor sample is from a different source than the bulk sample.

First, consider H_p. The bulk and receptor measurements are on evidence from the same source. The measurements on this source have a true mean θ, say. For example, if the measurements are of the refractive index of glass then θ denotes the mean refractive index of the window from which the fragments have been taken. For clarity, the conditioning elements H_p and I, the background information, will be omitted in the following argument. The predictive distribution $f(\mathbf{y} \mid \mathbf{x})$ may be expressed as follows.

$$f(\mathbf{y} \mid \mathbf{x}) = \int f(\mathbf{y} \mid \theta) f(\theta \mid \mathbf{x}) d\theta$$

$$= \frac{\int f(\mathbf{y} \mid \theta) f(\mathbf{x} \mid \theta) f(\theta) d\theta}{f(\mathbf{x})}$$

$$= \frac{\int f(\mathbf{y} \mid \theta) f(\mathbf{x} \mid \theta) f(\theta) d\theta}{\int f(\mathbf{x} \mid \theta) f(\theta) d\theta},$$

the ratio of the joint distribution of $\mathbf{x}$ and $\mathbf{y}$ to the marginal distribution of $\mathbf{x}$. Both distributions are independent of θ which is integrated out. The distributions are of measurements over all windows.

For H_d, the situation where the bulk and receptor measurements are from different sources, it is the measurements $\mathbf{y}$ on the receptor object which are the ones of interest. The probability density function for $\mathbf{y}$ is

$$f(\mathbf{y}) = \int f(\mathbf{y} \mid \theta) f(\theta) d\theta.$$

V may then be written as

$$\frac{\int f(\mathbf{y} \mid \theta) f(\mathbf{x} \mid \theta) f(\theta) d\theta}{\int f(\mathbf{x} \mid \theta) f(\theta) d\theta \int f(\mathbf{y} \mid \theta) f(\theta) d\theta}. \tag{10.3}$$

For those unfamiliar with these kinds of manipulations, Bayes' theorem applied to conditional probability distributions is used to write $f(\theta \mid \mathbf{x})$ as $f(\mathbf{x} \mid \theta) f(\theta)/f(\mathbf{x})$. The law of total probability with integration replacing summation is used to write $f(\mathbf{x})$ as $\int f(\mathbf{x} \mid \theta) f(\theta) \, d\theta$.

10.2   NORMAL DISTRIBUTION FOR BETWEEN-SOURCE DATA

The approach to evidence evaluation described above was first proposed by Lindley (1977a) in the context of a problem involving measurements of the refractive index of glass. These measurements may be made on fragments of glass at the scene of a crime and on fragments of window glass found on a suspect's clothing (see Example 1.2). Such measurements are subject to error, and it is this which is incorporated into V.

10.2.1 Sources of variation

Notice that there are often two sources of variation to be considered in measurements. There is variation within a particular source and there is variation between sources.

For example, consider evidence of fragments of glass from a broken window from which refractive index (r.i.) measurements have been made. There is variation in the r.i. measurements amongst the different fragments of glass. These different measurements may be thought of as a sample from the population corresponding to all possible r.i. measurements from that particular window. The population has a mean, θ say, and a variance σ^2. The r.i. measurements of fragments from that window are assumed to be Normally distributed with mean θ and variance σ^2. Secondly, there is variation in the r.i. mean θ between different windows. The mean θ has a probability distribution with its own mean μ, say, and variance τ^2. Typically τ^2 will be much greater than σ^2. Initially it will be assumed that θ has a Normal distribution also. However, a look at Figure 10.1, which is a histogram of r.i. measurements from 2269 examples of float glass from buildings given in Table 10.5 (Lambert and Evett, 1984), shows that this is not a particularly realistic assumption. A more realistic approach will be described in Section 10.4.

Similar considerations apply for other types of evidence. For measurements on the medullary widths of cat hairs, for example, there will be variation amongst hairs from the same cat and amongst hairs from different cats. For human head hairs there will be variations in characteristics between hairs from the same head (and also within the same hair) and between hairs from different heads. For measurements on footprints there is considerable variability between footprints from different people and considerable similarity for multiple impressions taken from the same person (Kennedy *et al.*, 2003).

Thus, when considering the assessment of continuous data at least two sources of variability have to be considered: the variability within the source (e.g., window) from which the measurements were made and the variability between the different possible sources (e.g., windows).

10.2.2 Derivation of the marginal distribution

Let x be a measurement from a particular bulk fragment. Let the mean of measurements from the source of this fragment be θ_1. Let y be one measurement from a particular receptor fragment. Let the mean of measurements from the

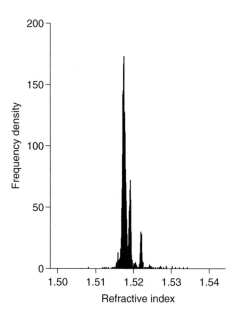

Figure 10.1 Refractive index measurements from 2269 fragments of float glass from buildings (from Lambert and Evett, 1984).

source of this fragment be θ_2. The variance of measurements within a source is assumed constant amongst sources and is denoted σ^2. The dependence of the distribution of these measurements on the source from which they come can be made explicit in the notation. The distributions of X and Y, given θ_1, θ_2 and σ^2, are

$$(X \mid \theta_1, \sigma^2) \sim N(\theta_1, \sigma^2),$$
$$(Y \mid \theta_2, \sigma^2) \sim N(\theta_2, \sigma^2),$$

where the dependence on θ_1 or θ_2 and σ^2 is made explicit. Notice also that variation in X is modelled. Contrast this approach with the coincidence probability approach of Section 4.7 in which the mean of measurements on the bulk fragments was taken as fixed. The conditioning on H_d is implicit. The means θ_1 and θ_2 of these distributions may themselves be thought of as observations from another distribution (that of variation between sources) which for the present is taken to be Normal, with mean μ and variance τ^2. Thus, θ_1 and θ_2 have the same probability density function and

$$(\theta \mid \mu, \tau^2) \sim N(\mu, \tau^2).$$

The distributions of X and of Y, independent of θ, can be determined by taking the so-called *convolutions* of x and of y with θ to give

$$
\begin{aligned}
f(x \mid \mu, \sigma^2, \tau^2) &= \int f(x \mid \theta, \sigma^2) f(\theta \mid \mu, \tau^2) d\theta \\
&= \int \frac{1}{2\pi\sigma^2\tau^2} \exp\left\{ -\frac{1}{2\sigma^2}(x-\theta)^2 \right\} \exp\left\{ -\frac{1}{2\tau^2}(\theta-\mu)^2 \right\} d\theta \\
&= \frac{1}{\sqrt{\{2\pi(\sigma^2+\tau^2)\}}} \exp\left\{ -\frac{1}{2(\sigma^2+\tau^2)}(x-\mu)^2 \right\},
\end{aligned}
$$

using the result that

$$
\frac{1}{2\sigma^2}(x-\theta)^2 + \frac{1}{2\tau^2}(\theta-\mu)^2 = \frac{(\theta-\mu_1)^2}{\tau_1^2} + \frac{(x-\mu)^2}{\sigma^2+\tau^2},
$$

where

$$
\mu_1 = \frac{\sigma^2\mu + \tau^2 x}{\sigma^2+\tau^2},
$$

$$
\tau_1^2 = \frac{\sigma^2\tau^2}{\sigma^2+\tau^2}.
$$

Similarly,

$$
f(y \mid \mu, \sigma^2, \tau^2) = \frac{1}{\sqrt{\{2\pi(\sigma^2+\tau^2)\}}} \exp\left\{ -\frac{1}{2(\sigma^2+\tau^2)}(y-\mu)^2 \right\}.
$$

Notice that τ^2 has been omitted from the distributions of x and y, given θ_1, θ_2 and σ^2. This is because the distributions of x and y, given these parameters, are independent of τ^2. Similarly, the distribution of θ, given μ and τ^2, is independent of σ^2.

The effect of the two sources of variability is that the mean of the r.i. measurements is the overall mean μ and the variance is the sum of the two component variances σ^2 and τ^2. The distribution remains Normal. Thus

$$
(X \mid \mu, \sigma^2, \tau^2) \sim N(\mu, \sigma^2+\tau^2),
$$

$$
(Y \mid \mu, \sigma^2, \tau^2) \sim N(\mu, \sigma^2+\tau^2). \tag{10.4}
$$

10.2.3   Approximate derivation of the likelihood ratio

Consider an application to a broken window as in Example 1.2. A crime is committed in which a window is broken. A suspect is apprehended soon afterwards and a fragment of glass is found on his clothing. Its refractive index is y. A sample of m fragments is taken from the broken window at the scene of the crime and their refractive index measurements are $\mathbf{x} = (x_1, \ldots, x_m)^T$, with mean $\bar{x}$. The two propositions to be compared are:

- H_p, the receptor fragment is from the crime scene window;
- H_d, the receptor fragment is not from the crime scene window.

An approximate derivation of the likelihood ratio may be obtained by replacing θ by $\bar{x}$ in the distribution of y so that $f(y \mid \theta, \sigma^2)$ becomes $f(y \mid \bar{x}, \sigma^2)$. (See (5.11) for a similar result using a uniform prior for a Normal distribution.) This is only an approximate distributional result. A more accurate result is given later to account for the sampling variability of $\bar{x}$. For the present, an approximate result for the numerator is that

$$(Y \mid \bar{x}, \sigma^2, H_p, I) \sim N(\bar{x}, \sigma^2), \tag{10.5}$$

an application of (5.11). Also, from (10.4),

$$(Y \mid \mu, \sigma^2, \tau^2, H_d, I) \sim N(\mu, \tau^2 + \sigma^2). \tag{10.6}$$

For τ^2 much greater than σ^2, assume also that $\tau^2 + \sigma^2$ can be approximated by τ^2. The likelihood ratio is then

$$
V = \left[\frac{1}{\sigma\sqrt{(2\pi)}} \exp\left\{ -\frac{(y - \bar{x})^2}{2\sigma^2} \right\} \right] \bigg/ \left[\frac{1}{\tau\sqrt{(2\pi)}} \exp\left\{ -\frac{(y - \mu)^2}{2\tau^2} \right\} \right]
$$

$$
= \frac{\tau}{\sigma} \exp\left\{ \frac{(y - \mu)^2}{2\tau^2} - \frac{(y - \bar{x})^2}{2\sigma^2)} \right\}
$$

(Evett, 1986). Note that this likelihood ratio depends on an assumption that fragments from a single source are found on the suspect and that these have come from the crime scene.

This result has some intuitively attractive features. The likelihood ratio is larger for values of y which are further from μ and are therefore assumed to be rarer; that is, the rarer the value of the r.i. of the recovered fragment, the larger the likelihood ratio. Also, the larger the value of $\mid y - \bar{x} \mid$, the smaller the value of the likelihood ratio; that is, the further the value of the r.i. of the receptor glass fragment is from the mean of the values of the r.i.s of the source fragments, the smaller the likelihood ratio.

Values for τ equal to 4×10^{-3} and for σ equal to 4×10^{-5} are given by Evett (1986). Values of V for various values of $(y - \mu)/\tau$ and of $(y - \bar{x})/\sigma$, the standardised distances of y from the overall mean and the source mean, are given in Table 10.1. Note that the ratio $\tau/\sigma = 100$, giving ample justification for the approximation τ^2 to the variance of y, given H_d, above. Also, this ratio is a large contributor to the value of V.

Consideration of the probabilities of transfer of fragments (Section 8.3), both from the crime scene and by innocent means from elsewhere (Evett, 1986), is left till later (Section 10.5).

Table 10.1 Likelihood ratio values for varying values of $(y - \bar{x})/\sigma$ and $(y - \mu)/\tau$

$(y - \bar{x})/\sigma$	$(y - \mu)/\tau$		
	0	1	2
0	100	165	739
1	61	100	448
2	14	22	100
3	1	2	8

10.2.4   Lindley's approach

A more detailed analysis was provided by Lindley (1977a). Assume, as before, that the measurements are distributed about the true unknown value, θ, of the refractive index with a Normal distribution and a known constant variance σ^2, and that the propositions H_p and H_d to be compared are as in Section 10.2.3. If m measurements are made at the scene (source measurements, $x_1, \ldots, x_m$) then it is sufficient to consider their mean, $\bar{x} = \sum_{i=1}^{m} x_i/m$. Conditional on θ_1, the mean of the r.i. of the crime window, the mean $\bar{X}$ is Normally distributed about θ_1 with variance σ^2/m. Let $\bar{y}$ denote the mean of n similar measurements (receptor measurements, $y_1, \ldots, y_n$) made on material found on the suspect; conditional on θ_2, $\bar{Y}$ is Normally distributed about θ_2 with variance σ^2/n. In the case H_p, where the source and receptor measurements come from the same source, $\theta_1 = \theta_2$. Otherwise, in the case H_d, $\theta_1 \neq \theta_2$.

The distribution of the true values θ has also to be considered. There is considerable evidence about the distribution of r.i.s; see, for example, Curran *et al.* (2000). First, assume as before that the true values θ are Normally distributed about a mean μ with variance τ^2, both of which are assumed known. Typically τ will be larger, sometimes much larger, than σ (see above, where $\tau/\sigma = 100$). This assumption of Normality is not a realistic one in this context where the distribution has a pronounced peak and a long tail to the right; see Figure 10.1. However, the use of the Normality assumption enables analytic results to be obtained as an illustration of the general application of the method. The unconditional distributions of $\bar{X}$ and $\bar{Y}$ are $N(\mu, \tau + \sigma^2/m)$ and $N(\mu, \tau + \sigma^2/n)$.

A brief derivation of V is given in an Appendix to this chapter (Section 10.7) and follows the arguments of Lindley (1977a). From Section 10.7, if the number of control measurements equals the number of recovered measurements, $m = n$, $z = w = \frac{1}{2}(\bar{x} + \bar{y})$ and $\sigma_1^2 = \sigma_2^2$. Then

$$V \simeq \frac{m^{1/2}\tau}{2^{1/2}\sigma} \exp\left\{-\frac{m(\bar{x} - \bar{y})^2}{4\sigma^2}\right\} \exp\left\{\frac{(z - \mu)^2}{2\tau^2}\right\}. \tag{10.7}$$

Note again that this result assumes implicitly that the fragments are from a single source. Denote this assumption by S. Then the above result is the ratio of the probability density functions $f(\bar{x}, \bar{y} \mid H_p, S)/f(\bar{x}, \bar{y} \mid H_d, S)$. A result including S as one of the uncertain elements and deriving an expression for $f(\bar{x}, \bar{y}, S \mid H_p)/f(\bar{x}, \bar{y}, S \mid H_d)$ was given by Grove (1980). Let T denote the event that fragments were transferred from the broken window to the suspect and persisted there until discovery by the police. Let A be the event that the suspect came into contact with glass from some other source. Assume that $Pr(A \mid H_p) = Pr(A \mid H_d) = p_A$, that $Pr(T \mid H_p) = p_T$, that A and T are independent given H_p and that $Pr(T \mid H_d) \simeq 0$. Grove (1980) shows that

$$
\begin{aligned}
V &= \frac{f(\bar{x}, \bar{y}, S \mid H_p)}{f(\bar{x}, \bar{y}, S \mid H_d)} \\
&= 1 + p_T \left\{ (p_A^{-1} - 1) \frac{f(\bar{x}, \bar{y} \mid H_p)}{f(\bar{x}, \bar{y} \mid H_d)} - 1 \right\} \\
&= (1 - p_T) + \frac{p_T(1 - p_A)}{p_A} \times \frac{f(\bar{x}, \bar{y} \mid H_p)}{f(\bar{x}, \bar{y} \mid H_d)},
\end{aligned}
\tag{10.8}
$$

where $f(\bar{x}, \bar{y} \mid H_p)/f(\bar{x}, \bar{y} \mid H_d)$ is Lindley's (1977a) ratio. The value derived by Grove (1980) takes account of transfer and persistence in a way already derived for discrete data (see Section 8.3.3). Another derivation for continuous data is given by (10.14) in Section 10.5.2.

10.2.5   Interpretation of result

The interpretation of (10.7) is now considered in the particular case $m = n = 1$. V consists of two factors which depend on the measurements. The first is $\exp\{-(\bar{x} - \bar{y})^2/4\sigma^2\}$. This compares the absolute difference $\mid \bar{x} - \bar{y} \mid$ of the control and recovered measurements with their standard deviation $\sigma\sqrt{2}$ on the proposition $(\theta_1 = \theta_2)$ that they come from the same source. Let

$$
\mid \bar{x} - \bar{y} \mid /\sigma\sqrt{2} = \lambda.
$$

Then the value of the first factor is $\exp(-\lambda^2/2)$. A large value of λ favours the hypothesis that the two fragments come from different sources. This factor has an effect like that of a significance test of a null hypothesis of identity $(\theta_1 = \theta_2)$. The second factor, $\exp\{(z - \mu)^2/2\tau^2\}$, with $z = \frac{1}{2}(\bar{x} + \bar{y})$, measures the typicality of the two measurements. This factor takes its smallest value, 1, when $z = \mu$ and increases as $\mid z - \mu \mid$ increases relative to its standard deviation. Thus the more unusual the glass (i.e., the larger the value of $\mid z - \mu \mid$), the greater the value of V and the stronger the inference in favour of a common source for the two measurements. Consider the comment by Parker and Holford (1968) in Section 4.7. The first factor considers similarity. The second factor considers typicality. The assessment of similarity is not by-passed.

Evaluation of the likelihood ratio performance: Tippett plots

The performance of a method of evaluation based on the likelihood ratio may be assessed by comparing the values obtained for comparisons where the two items being compared are known to come from the same source with values obtained for comparisons where the two items being compared are known to come from different sources.

The first step is to estimate the performance of the method used in two situations. The values of evidence for items known to come from the same source are determined. Similarly, the values of evidence for items known to come from different sources are also determined. The results are plotted as follows. The abscissa is graduated in terms of ascending values of likelihood ratio (often on a logarithmic scale) and the ordinate indicates the number of cases when a likelihood ratio value is exceeded. The chart consists of two curves, the first representing the number of cases when the likelihood ratio value is exceeded for within-source comparisons and the second the number of cases when the likelihood ratio value is exceeded for between-source comparisons. The power of the method is measured by comparing the proportions exceeding particular likelihood ratio values for each of the two curves. For example, for within-source comparisons, the proportion exceeding values greater than one should be greater than for between-source comparisons. For between-source comparisons, the proportion less than values less than one should be greater than for within-source comparisons. This represents an alternative way to demonstrate the discriminating power of the analytical method used by the scientist.

In practice, the scientist simulates within-source comparisons. In DNA evidence, for example, and cases in which the offender and the suspect are the same person, the likelihood ratio is determined for each person represented in a database of size n. There are n values for the likelihood ratio for comparisons of profiles from the same person. The next step is to make comparisons between all possible pairs of profiles from different people in the database. There are then $n(n-1)/2$ between-source comparisons and $n(n-1)/2$ corresponding values of the likelihood ratio.

Applications of this procedure, which produces so-called *Tippett plots*, have been described for DNA (Evett *et al.*, 1993; Evett and Buckleton, 1996; Evett and Weir, 1998), for speaker recognition (Meuwly, 2001; Meuwly and Drygajlo, 2001) and for relationships between heroin seizures (Dujourdy *et al.*, 2003).

10.2.6   Examples

Consider a case with only one control ($m = 1$) and one recovered ($n = 1$) fragment of glass. Denote these as $\bar{x}$ and $\bar{y}$, for consistency with (10.7). Suppose the ratio of between-groups standard deviation τ to within-groups standard deviation σ is 100. The control and recovered measurements are found to be two within-group standard deviations apart. Since the variance of $\bar{x} - \bar{y}$ equals

$2\sigma^2$, this separation implies that $|\bar{x}-\bar{y}|/\sigma\sqrt{2}=2$. A conventional significance test (Section 4.6) would reject the hypothesis of a common source at the 5% level of significance. Assume that the mid-point of $\bar{x}$ and $\bar{y}$, which is $(\bar{x}+\bar{y})/2$, the mean, denoted z in (10.7), is the population mean μ. Then, from (10.7),

$$V = \frac{100e^{-2}}{\sqrt{2}} = 9.57.$$

The odds in favour of a common source are increased by a factor of almost 10, a result in contrast to the rejection at the 5% level of significance in a conventional significance test. The values of (10.7) for $\tau/\sigma = 100$ as a function of λ and $\delta = |z-\mu|/\tau$, the deviation of the mean of the two measurements from μ, standardised on the assumption that the hypothesis of a common source is true, are given in Table 10.2.

Consider the more general formula for V, as given in (10.25) in Section 10.7, namely

$$V \simeq \frac{\tau}{a\sigma} \times \exp\left\{-\frac{(\bar{x}-\bar{y})^2}{2a^2\sigma^2}\right\} \times \exp\left\{-\frac{(w-\mu)^2}{2\tau^2} + \frac{(z-\mu)^2}{\tau^2}\right\}. \tag{10.9}$$

The following information is needed in order that V may be evaluated:

- the number of source measurements (m);
- the mean of the source measurements ($\bar{x}$);
- the number of receptor measurements (n);
- the mean of the receptor measurements ($\bar{y}$);
- the variance (assumed known) of the measurements on the source and receptor samples (σ^2);
- the overall mean (assumed known) of the refractive indices (μ);
- the overall variance (assumed known) of the refractive indices (τ^2).

Table 10.2   Value of $\tau(2^{1/2}\sigma)^{-1}\exp(-\frac{1}{2}\lambda^2+\frac{1}{2}\delta^2)$ (10.7) as a function of $\lambda=|\bar{x}-\bar{y}|/(2^{1/2}\sigma)$ and $\delta=|z-\mu|/\tau$ for $\tau/\sigma = 100$

δ	λ				
	0	1.0	2.0	4.0	6.0
0	70.7	42.9	9.57	0.024	1.08×10^{-6}
1.0	117	70.7	15.8	0.039	1.78×10^{-6}
2.0	522	317	70.7	0.175	7.94×10^{-6}
3.0	6370	3860	861	2.14	9.71×10^{-5}

The following values may be derived from the above:

- $z = (\bar{x} + \bar{y})/2$;
- $w = (m\bar{x} + n\bar{y})/(m+n)$;
- $a^2 = 1/m + 1/n$.

In the following numerical example using data from Evett (1977) and Lindley (1977a) we have $\bar{x} = 1.518\,458$, $m = 10$; $\bar{y} = 1.518\,472$, $n = 5$; $\sigma = 0.000\,04$; $\tau = 0.004$. The overall mean μ is taken to be 1.5182 and has been derived from the 2269 measurements for building float glass published by Lambert and Evett (1984); see Figure 10.1. With these figures, $a^2 = 0.3$, $w = 1.518\,463$, $z = 1.518\,465$, and

$$\frac{\tau}{a\sigma} = 182.5742,$$

$$\frac{(\bar{x} - \bar{y})^2}{2a^2\sigma^2} = 0.2042,$$

$$\frac{(w - \mu)^2}{2\tau^2} = 0.00216,$$

$$\frac{(z - \mu)^2}{\tau^2} = 0.00439,$$

$$V = 149.19.$$

The odds in favour of the suspect being at the crime scene are thus increased by a factor of 150.

10.3   ESTIMATION OF A PROBABILITY DENSITY FUNCTION

The estimation of a population mean (μ) and variance (σ^2) by a sample mean ($\bar{x}$) and variance (s^2) of data sampled from the population is a common idea. Also, the probability density function itself may be estimated from data taken from the population. That such a procedure is necessary becomes apparent when it is realised that not all data have a distribution which is readily modelled by a standard distribution. In particular, not all data are unimodal, symmetric and bell-shaped and may be modelled by a Normal distribution. The histogram of the refractive index of glass fragments (Figure 10.1) and a histogram of the medullary width of cat hairs (Figure 10.2, from data in Table 10.3) both illustrate this.

Estimation of a probability density function is not too difficult so long as the distribution is fairly smooth. A procedure known as *kernel density estimation* is used; see Silverman (1986) for technical details. For early applications to

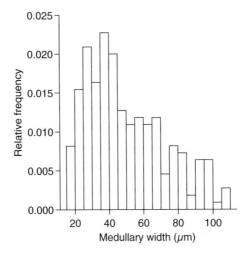

Figure 10.2   Medullary width (in microns) of 220 cat hairs (from Peabody *et al.*, 1983).

forensic science, see Aitken and MacDonald (1979) for an application with discrete data to forensic odontology and Aitken (1986) for an application to the discrimination between cat and dog hairs in which two variables are considered and Evett *et al.* (1992b) for an application in DNA profiling. An example is given here of the application of the technique to the distribution of the medullary width of cat hairs. In Chapter 11, an example involving three variables for the elemental composition of glass is discussed.

Consider data on the medullary widths (in microns) of 220 cat hairs (Peabody *et al.*, 1983). A version of these modified to make the analysis easier is given in Table 10.3, and a histogram to illustrate the distribution is shown in Figure 10.2 from which it can be seen that the data are positively skewed and perhaps not unimodal. The histogram has been constructed from the full data set by selecting intervals of fixed width and fixed boundary points, namely $15.0-20.0$, $20.01-25.00, \ldots, 105.01-110.00$ microns. Individual observations are then allocated to the appropriate interval and a frequency count obtained. Each interval is five units (microns) wide, and there are 220 observations. If each observation is allocated unit height the total area encompassed by the histogram is $5 \times 220 = 1100$ units. Thus if the height of each bar of the histogram is reduced by a factor of 1100, the area under the new diagram is 1. This new histogram may be considered a very naive probability function (with steps at the boundary points of the bars of the histogram).

The method of kernel density estimation may be considered as a development of the histogram. Consider the histogram to be constructed with rectangular blocks, each block corresponding to one observation. The block is

Table 10.3   Medullary widths in microns of 220 cat hairs (Peabody *et al.*, 1983)

17.767	28.600	39.433	52.233	68.467
18.633	28.600	39.867	52.867	69.333
19.067	29.033	39.867	53.300	71.067
19.067	29.033	39.867	53.300	71.500
19.067	29.467	40.300	53.733	71.667
19.133	30.333	40.300	53.733	73.233
19.300	30.767	40.733	54.167	74.533
19.933	31.200	41.167	54.600	75.400
19.933	31.200	41.167	55.033	76.267
20.367	31.300	41.600	55.467	76.267
20.367	31.633	41.600	55.900	77.133
20.367	31.633	42.033	56.767	77.367
20.600	31.807	42.467	57.200	78.000
20.800	32.000	42.467	57.200	78.000
20.800	32.067	42.467	57.200	79.500
21.233	32.067	42.467	57.633	79.733
21.233	32.500	42.467	58.067	80.167
21.400	32.500	42.467	58.067	80.167
22.533	33.800	42.900	58.500	80.167
22.967	33.800	42.900	58.933	81.467
22.967	34.233	42.900	58.933	81.467
23.400	34.667	42.900	60.233	81.900
23.833	34.667	43.333	60.533	82.767
23.833	35.533	44.200	60.667	84.067
24.267	35.533	44.200	60.667	87.100
24.700	35.533	44.300	60.667	87.967
25.133	35.533	45.067	61.100	90.133
25.133	35.533	45.933	61.967	90.267
25.133	36.400	45.933	62.400	91.867
25.133	36.400	45.933	62.400	91.867
25.300	37.267	46.150	63.000	92.733
26.000	37.267	46.583	63.267	93.167
26.000	37.267	46.800	63.700	93.600
26.233	38.567	46.800	65.433	95.333
26.433	38.567	47.167	65.867	96.267
26.433	38.567	48.100	66.300	97.067
26.867	39.000	48.317	66.733	97.500
26.867	39.000	48.967	66.733	97.500
27.133	39.000	48.967	66.733	97.933
27.733	39.000	49.400	67.167	99.667
27.733	39.000	50.267	67.600	100.100
27.733	39.433	51.567	67.600	106.600
28.167	39.433	51.567	68.033	106.600
28.167	39.433	52.000	68.033	107.467

positioned according to the interval in which the observation lies. The method of kernel density estimation used here replaces the rectangular block by a Normal probability density curve, known in this context as the *kernel function*. The curve is positioned by centring it over the observation to which it relates. The estimate of the probability density curve is then obtained by adding the individual curves together over all the observations in the data set and then dividing this sum by the number of observations. Since each component of the sum is a probability density function, each component has area 1. Thus, the sum of the functions divided by the number of observations also has area 1 and is a probability density function.

In the construction of a histogram a decision has to be made initially as to the width of the intervals. If the width is wide, the histogram is very uninformative regarding the underlying distribution. If it is narrow, there is too much detail and general features of the distribution are lost. Similarly, in kernel density estimation, the spread of the Normal density curves has to be determined. The spread of the curves is represented by the variance. If the variance is chosen to be large, the resultant estimated curve is very smooth. If the variance is chosen to be small, the resultant curve is very spiky (see Figure 10.3).

Mathematically, the kernel density estimate of an underlying probability density function can be constructed as follows. The discussion is in the context of estimating the distribution of the medullary widths of cat hairs. There is variation in the medullary width both within hairs from an individual cat and between different cats. Denote the measurement of the mean medullary width of hairs from a particular cat by θ. The corresponding probability density function

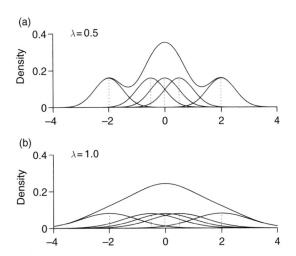

Figure 10.3 Examples of kernel density estimates showing individual kernels. Smoothing parameter values are (a) $\lambda = 0.5$ and (b) $\lambda = 1$.

$f(\theta)$ is to be estimated. A data set $D = \{z_1, \ldots, z_k\}$ is available to enable this to be done. The variance of the width of hairs from different cats is estimated by

$$s^2 = \sum_{i=1}^{k}(z_i - \bar{z})^2/(k-1), \tag{10.10}$$

where $\bar{z}$ is the sample mean. This variance is a mixture of the variances measuring the variability of the medullary width between and within cats and will be used as an approximation to the variance of the medullary width between cats. The sample standard deviation s is then multiplied by a parameter, known as the *smoothing parameter*, denoted here by λ, which determines the smoothness of the density estimate. The kernel density function $K(\theta \mid z_i, \lambda)$ for point z_i is then taken to be a Normal distribution with mean z_i and variance $\lambda^2 s^2$,

$$K(\theta \mid z_i, \lambda) = \frac{1}{\lambda s \sqrt{2\pi}} \exp\left\{ -\frac{(\theta - z_i)^2}{2\lambda^2 s^2} \right\}.$$

The estimate $\hat{f}(\theta \mid D, \lambda)$ of the probability density function is then given by

$$\hat{f}(\theta \mid D, \lambda) = \frac{1}{k} \sum_{i=1}^{k} K(\theta \mid z_i, \lambda). \tag{10.11}$$

Notice here that there is an implicit assumption that a suitable data set D exists and that it is a data set from a relevant population. This latter comment is of particular relevance when considering DNA profiling where there is much debate as to the choice of the relevant population in a particular case. Also, if data were available on variability within groups, an adjustment can be made to the estimate of the between-group variance s^2 (10.10). Consider data of the form $\{z_{ij}, i = 1, \ldots, k, j = 1, \ldots, l\}$, where k is the number of groups and l is the number of members of each group, assumed constant amongst groups. Let $\bar{z}_i$ denote the mean of the ith group and $\bar{z}$ the overall mean. The within-group variance σ^2 is then estimated by

$$\hat{\sigma}^2 = \sum_{i=1}^{k}\sum_{j=1}^{l}(z_{ij} - \bar{z}_i)^2/(kl - k)$$

and the between-group variance τ^2 by

$$s^2 = \sum_{i=1}^{k}(\bar{z}_i - \bar{z})^2/(k-1) - \hat{\sigma}^2/k,$$

an adjustment of $\hat{\sigma}^2/k$ from (10.10).

The smoothing parameter λ has to be chosen. Mathematical procedures exist which enable an automatic choice to be made. For example, a so-called *pseudo-maximum likelihood* procedure (Habbema *et al.*, 1974) was used to determine the value of λ (0.09) used in Figure 10.4. A value of λ equal to 0.50 was used

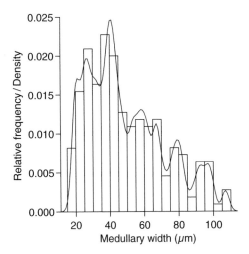

Figure 10.4 Medullary widths, in microns, of cat hairs (Peabody *et al.*, 1983) and associated kernel density estimate with smoothing parameter equal to 0.09.

to produce the curve in Figure 10.5 to illustrate the effect that a larger value of λ produces a smoother curve.

The choice of λ has to be made bearing in mind that the aim of the analysis is to provide a value V for the evidence in a particular case, as represented by the likelihood ratio. Using the kernel density estimation procedure, an expression for V is derived; see (10.13). An investigation of the variation in V as λ varies is worthwhile. If V does not vary greatly as λ varies then a precise value

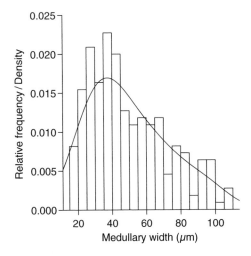

Figure 10.5 Medullary widths, in microns, of cat hairs (Peabody *et al.*, 1983) and associated kernel density estimate with smoothing parameter equal to 0.50.

for λ is not necessary. For example, it is feasible to choose λ subjectively by comparing the density estimate curve $\hat{f}$ obtained for various values of λ with the histogram of the data. The value which provides the best visual fit can then be chosen. Alternatively, from a scientist's personal experience of the distribution of the measurements on the characteristic of interest, it may be thought that certain possible values are not fully represented in the data set D available for estimation. In such a situation a larger value of λ may be chosen in order to provide a smoother curve, more representative of the scientist's experience. The subjective comparison of several plots of the data, produced by smoothing by different amounts, may well help to give a greater understanding of the data than the consideration of one curve, produced by an automatic method.

The choice of λ is also sensitive to outlying observations. The original cat hair data included one hair with a medullary width over 139 microns, the next largest being under 108 microns. The value of λ chosen by the automatic pseudo-maximum likelihood procedure was 0.35, a value which produced a very different estimate of the probability density function from that produced by the value of λ of 0.09 when the data set was modified as has been done by replacing the value of 139 microns by a value of 63 microns. The choice of λ is also difficult if the data are presented in grouped form, as is the case with the glass data (Table 10.5). In this case, the value of λ was chosen subjectively; see Figures 10.6 and 10.7, with values of λ of 0.025 and 0.25.

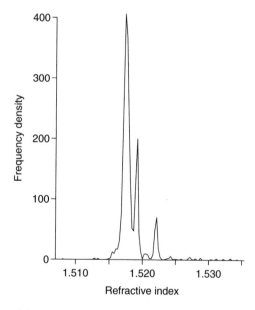

Figure 10.6 Kernel density estimate with smoothing parameter 0.025 of refractive index measurements from 2269 fragments of float glass from buildings (Lambert and Evett, 1984).

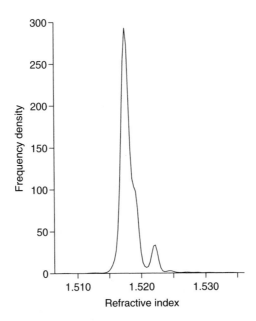

Figure 10.7 Kernel density estimate with smoothing parameter 0.25 of refractive index measurements from 2269 fragments of float glass from buildings (Lambert and Evett, 1984).

10.4 KERNEL DENSITY ESTIMATION FOR BETWEEN-SOURCE DATA

If the assumption of a Normal distribution for θ is thought unrealistic, the argument may be modified for a general distribution for θ using a kernel density estimation as described by Chan and Aitken (1989) for cat hairs, by Berry (1991a) and Berry *et al.* (1992) for DNA profiling and by Aitken and Lucy (2004) for the elemental composition of glass.

An application to the evaluation of fibre evidence in which the marginal distribution of the receptor measurements y was estimated by a kernel density function was given by Evett *et al.* (1987), and a rather more elaborate treatment was given by Wakefield *et al.* (1991). This was a bivariate case involving colour measurements. Further details of these ideas are given in Chapter 11, where they are applied to multivariate data on the elemental composition of glass. The method described here is applicable to situations in which the data are univariate and for which there are two components of variability, that within a particular source (e.g., window or cat) and that between different sources (e.g., windows or cats).

Consider the numerator in the original expression (10.3) for V, namely,

$$\int f(\mathbf{y} \mid \theta) f(\mathbf{x} \mid \theta) f(\theta) d\theta.$$

Given the value for θ, the distribution of $\bar{x} - \bar{y}$ is $N(0, a^2\sigma^2)$ and the distribution of W, given θ, is $N(\theta, \sigma^2/(m+n))$. If a change in the numerator is made from $(\bar{x}, \bar{y})$ to $(\bar{x} - \bar{y}, w)$ then V may be written as

$$\frac{\dfrac{1}{a\sigma}\exp\left\{-\dfrac{(\bar{x}-\bar{y})^2}{2a^2\sigma^2}\right\}\displaystyle\int\dfrac{(m+n)^{1/2}}{\sigma}\exp\left\{-\dfrac{(w-\theta)^2(m+n)}{2\sigma^2}\right\}f(\theta)d\theta}{\displaystyle\int\dfrac{\sqrt{m}}{\sigma}\exp\left\{-\dfrac{(\bar{x}-\theta)^2 m}{2\sigma^2}\right\}f(\theta)d\theta\int\dfrac{\sqrt{n}}{\sigma}\exp\left\{-\dfrac{(\bar{y}-\theta)^2 n}{2\sigma^2}\right\}f(\theta)d\theta} \tag{10.12}$$

(Lindley, 1977a). The probability density function, $f(\theta)$, for θ was previously assumed to be Normal. If this is thought to be unrealistic the probability density function may be estimated by kernel density estimation.

The expression for V in (10.12) may be evaluated when $f(\theta)$ is replaced by the expression in (10.11). Some straightforward, but tedious, mathematics gives the result that

$$V = \frac{K\exp\left\{-\dfrac{(\bar{x}-\bar{y})^2}{2a^2\sigma^2}\right\}\sum_{i=1}^{k}\exp\left\{-\dfrac{(m+n)(w-z_i)^2}{2[\sigma^2+(m+n)s^2\lambda^2]}\right\}}{\sum_{i=1}^{k}\exp\left\{-\dfrac{m(\bar{x}-z_i)^2}{2(\sigma^2+ms^2\lambda^2)}\right\}\sum_{i=1}^{k}\exp\left\{-\dfrac{n(\bar{y}-z_i)^2}{2(\sigma^2+ns^2\lambda^2)}\right\}}, \tag{10.13}$$

where

$$K = \frac{k\sqrt{m+n}\sqrt{\sigma^2+ms^2\lambda^2}\sqrt{\sigma^2+ns^2\lambda^2}}{a\sigma\sqrt{mn}\sqrt{\sigma^2+(m+n)s^2\lambda^2}}.$$

There are four factors, explicitly dependent on the data, in the expression for V which contribute to its overall value:

(a) $\exp\left\{-(\bar{x}-\bar{y})^2/(2a^2\sigma^2)\right\}$;
(b) $\sum_{i=1}^{k}\exp\left\{-(m+n)(w-z_i)^2/2[\sigma^2+(m+n)s^2\lambda^2]\right\}$;
(c) $\sum_{i=1}^{k}\exp\left\{-m(\bar{x}-z_i)^2/2(\sigma^2+ms^2\lambda^2)\right\}$;
(d) $\sum_{i=1}^{k}\exp\left\{-n(\bar{y}-z_i)^2/2(\sigma^2+ns^2\lambda^2)\right\}$.

The first factor, (a), accounts for the difference between the source and receptor evidence. A large difference leads to a smaller value of V, a small difference to a larger value of V.

The second factor, (b), accounts for the location of the combined evidence in the overall distribution from the relevant population. If it is far from the centre of this distribution then V will be smaller than if it were close. This provides a measure of the rarity of the combined evidence.

The third and fourth factors, (c) and (d), account for the rarity or otherwise of the source and receptor evidence, separately. The further these are from the centre of the overall distribution, the smaller the corresponding factor and the larger the value of V.

Notice, also, the difference between σ^2 which measures the variance within a particular source (e.g., window or cat) and s^2 which estimates the overall variance.

10.4.1   Application to medullary widths of cat hairs

Consider a crime in which a cat is involved. For example, in a domestic burglary, there may have been a cat at the crime scene. A suspect is identified who has cat hairs on his clothing. A full assessment of the evidence would require consideration of the suspect's explanation for the presence of these hairs and of the probabilities of transfer of cat hairs from the scene of the crime and from elsewhere. Such issues are not debated here. Measurements are made of the medullary widths, among other characteristics, of these hairs and of a sample of hairs from the domestic cat. Let $\bar{x}$ denote the mean of m hairs from the source (the domestic cat), and let $\bar{y}$ denote the mean of n hairs from the receptor (the suspect's clothing). Some sample results for the value, V, as given by (10.13) of the evidence are given in Table 10.4 for various values of x, y and σ. Variation in σ is given to illustrate the effect of changes in the variation within cats of medullary width on the value of the evidence. The value of the smoothing parameter λ has been taken to be 0.09 and 0.50 to illustrate variation in V with λ. The corresponding density estimate curves are shown in Figures 10.4 and 10.5 which can be used to assess the relative typicality of the evidence.

10.4.2   Refractive index of glass

These data, from Lambert and Evett (1984), are shown in Table 10.5 and illustrated in Figure 10.1. There are many coincident points and an automatic

Table 10.4   Value of the evidence for various values of $\bar{x}$ and $\bar{y}$, the smoothing parameter λ and the within-cat standard deviation σ; $m = n = 10$ throughout; $s = 23$ microns

$\bar{x}$	$\bar{y}$	σ	V	
			$\lambda = 0.09$	$\lambda = 0.50$
15	15	10	16.50	12.01
15	25	10	1.39	0.782
15	35	10	9.81×10^{-4}	4.47×10^{-4}
110	110	10	84.48	53.61
50	50	10	6.97	6.25
50	50	16	3.86	3.93
50	50	5	16.14	12.48
50	55	10	3.75	3.54

Table 10.5   Refractive index of 2269 fragments of float glass from buildings, Lambert and Evett (1984)

r.i.	Count	r.i.	Count	r.i.	Count	r.i.	Count
1.5081	1	1.5170	65	1.5197	7	1.5230	1
1.5119	1	1.5171	93	1.5198	1	1.5233	1
1.5124	1	1.5172	142	1.5199	2	1.5234	1
1.5128	1	1.5173	145	1.5201	4	1.5237	1
1.5134	1	1.5174	167	1.5202	2	1.5240	1
1.5143	1	1.5175	173	1.5203	4	1.5241	1
1.5146	1	1.5176	128	1.5204	2	1.5242	1
1.5149	1	1.5177	127	1.5205	3	1.5243	3
1.5151	1	1.5178	111	1.5206	5	1.5244	1
1.5152	1	1.5179	81	1.5207	2	1.5246	2
1.5153	1	1.5180	70	1.5208	3	1.5247	2
1.5154	3	1.5181	55	1.5209	2	1.5249	1
1.5155	5	1.5182	40	1.5211	1	1.5250	1
1.5156	2	1.5183	28	1.5212	1	1.5254	1
1.5157	1	1.5184	18	1.5213	1	1.5259	1
1.5158	7	1.5185	15	1.5215	1	1.5265	1
1.5159	13	1.5186	11	1.5216	3	1.5269	1
1.5160	6	1.5187	19	1.5217	4	1.5272	2
1.5161	6	1.5188	33	1.5218	12	1.5274	1
1.5162	7	1.5189	47	1.5219	21	1.5280	1
1.5163	6	1.5190	51	1.5220	30	1.5287	2
1.5164	8	1.5191	64	1.5221	25	1.5288	1
1.5165	9	1.5192	72	1.5222	28	1.5303	2
1.5166	16	1.5193	56	1.5223	13	1.5312	1
1.5167	15	1.5194	30	1.5224	6	1.5322	1
1.5168	25	1.5195	11	1.5225	3	1.5333	1
1.5169	49	1.5196	3	1.5226	5	1.5343	1

choice of λ is difficult and perhaps not desirable. Figures 10.6 and 10.7 show the kernel density estimate curves for λ equal to 0.025 and 0.25. The coincidence probabilities, from equation (4.8), and values V of the evidence, using (10.9) and (10.13), are given in Table 10.6.

Notice that, in general, the kernel approach leads to considerably higher values for V than does the Lindley approach. This arises from the more dispersed nature of the Lindley expression. Two examples in Table 10.6 show the failure of the coincidence probability approach. These are examples in which the separation of the source ($\bar{x}$) and receptor ($\bar{y}$) fragments is such that an approach based on coincidence probabilities would declare these two sets of fragments to have come from different windows. However, both the kernel and Lindley approaches give support to the proposition that they come from the same window.

Table 10.6 Coincidence probability and value of the evidence (kernel and Lindley approaches) for various values of $\bar{x}$ and $\bar{y}$ and the smoothing parameter λ (for the kernel approach); $m = 10$, $n = 5$; within-window standard deviation $\sigma = 0.000\,04$, between-window standard deviation $\tau = 0.004$; overall mean $\mu = 1.5182$

$\bar{x}$	$\bar{y}$	Coincidence probability	λ for kernel approach			Lindley
			0.025	0.05	0.25	
1.515 00	1.515 01	2.845×10^{-9}	17 889	7055	2810	226
1.516 00	1.516 01	2.643×10^{-3}	563	489	419	191
1.517 00	1.517 01	2.863×10^{-2}	54.3	52.4	48.9	172
1.518 00	1.518 01	3.083×10^{-2}	53.3	54.4	49.2	164
1.519 00	1.519 01	2.246×10^{-2}	70.0	69.2	102.4	167
1.520 00	1.520 01	8.536×10^{-9}	5524	2297	471.2	182
1.521 00	1.521 01	4.268×10^{-9}	13 083	4381	1397	210
1.522 00	1.522 01	1.321×10^{-2}	128	143	304	259
1.515 00	1.515 05	—	740	519	217	18.4
1.516 00	1.516 05	—	48.4	42.4	32.6	15.6
1.516 00	1.516 10	—	1.76×10^{-2}	1.74×10^{-2}	1.22×10^{-2}	6.30×10^{-3}
1.517 00	1.517 10	—	1.35×10^{-3}	1.42×10^{-3}	1.51×10^{-3}	5.69×10^{-3}

10.5   PROBABILITIES OF TRANSFER

10.5.1   Introduction

Consider transfer of material from the crime scene to the criminal. A suspect is found with similar material on his clothing, say. This material may have come from the crime scene. Alternatively, it may have come from somewhere else under perfectly innocent circumstances. There are two sets of circumstances to consider. First, conditional on the suspect having been present at the crime scene (H_p), there is a probability that material will have been transferred from the scene to the suspect. It has also to be borne in mind that someone connected with the crime may have had no fragments transferred from the scene to his person and have had fragments similar to those found at the crime scene transferred to his person from somewhere else by innocent means. Secondly, there is the probability that a person unconnected with the crime (i.e., this is conditional on a suspect not having been present at the scene, H_d) will have material similar to the crime material on his person.

Consider the case of glass fragments as described by Evett (1986). Let t_i ($i = 0, 1, 2, \ldots$) be the probability that, given H_p, i fragments of glass would have been transferred. More correctly, let t_n be the probability that, given H_p, the presence of the suspect at the crime scene, n fragments would be found on the clothing of the suspect on searching. This allows not only for the mechanism of transfer but also for the mechanisms of persistence and recovery. Let $b_{1,m}$ ($m = 1, 2, \ldots$) denote the probability that a person in the relevant population will have one group of m fragments of glass on their clothing. In general, the proportion of people in the relevant population who have k groups, with $m_1, \ldots, m_k$ fragments of glass from each group, on their clothing can be denoted $b_{k, m_1, \ldots, m_k}$ or $\{b_{k, \mathbf{m}_k}; \mathbf{m}_k = m_1, \ldots, m_k.\}$. Here k will only take values 1 or 2. For $k = 1, b_{k, \mathbf{m}_k}$ will be written as $b_{1,m}$. Also, $b_{1,s}$ and $b_{1,l}$ denote the conditional probabilities that, if one group of fragments is found then it contains a small (s) or large (l) number of fragments. Two cases are considered, one where a single fragment has been found and one where two fragments (one group, two fragments) have been found. A general expression for more than two groups is given in Evett (1986).

10.5.2   Single fragment

The evidence E consists of three parts. The first part is the existence (m_1) of one fragment on the clothing of the suspect. The second part is that its refractive index is y. This is the transferred particle form of the evidence. Let $Pr(m_1 \mid H_d, I)$ correspond to the proportion of people in the relevant population who have a fragment of glass on their clothing. Denote this probability by $b_{1,1}$. Similarly, b_0 denotes the proportion of people in the general population who do not have a fragment of glass on their clothing. The third part of the evidence is the measurements $\mathbf{x}$ on the source material. This is relevant for the determination of the numerator but not the denominator.

Consider the denominator of the likelihood ratio. This is

$$
\begin{aligned}
Pr(E \mid H_d, I) &= Pr(m_1, y \mid H_d, I) \\
&= Pr(m_1 \mid H_d, I) \times f(y \mid H_d, I, m_1) \\
&= b_{1,1} f(y \mid H_d, I, m_1).
\end{aligned}
$$

The probability density function $f(y \mid H_d, I, m_1)$ will be taken to be a Normal density function, with mean μ and variance τ^2 (or, more correctly, $\tau^2 + \sigma^2$), as in Section 10.2.3.

Consider the numerator. If the suspect was present at the crime scene there are two possible explanations for the presence of the glass fragment on the clothing of the suspect. Either the fragment has been acquired by innocent means and no fragment has been transferred from the crime scene (an event with probability t_0) or the fragment was transferred from the crime scene and none was transferred by innocent means, an event with probability t_1 (in Section 14.6.2 these explanations are relaxed to take into account different situations such as the presence of two groups of recovered material, one transferred from the crime scene, the other transferred by innocent reasons). Let $\mathbf{x}$ denote the measurements on the source sample. The numerator is then

$$
t_0 b_{1,1} f(y \mid H_d, I, m_1) + t_1 b_0 \, f(y \mid H_p, \mathbf{x}, I).
$$

Notice the terms $t_0 b_{1,1}$ and $t_1 b_0$. The former is the probability that no particle is transferred from the crime scene and one particle is transferred from the background. The latter is the probability that one particle is transferred from the crime scene and no particle is transferred from the background. Note, also, that in the term involving H_d, the fragment is assumed to have been transferred by innocent means. The probability density function for y in this situation is then the one that holds when the suspect is unconnected with the crime. Hence, the conditioning on H_d is permissible.

The likelihood ratio is then

$$
V = t_0 + \frac{t_1 b_0}{b_{1,1}} \frac{f(y \mid H_p, \mathbf{x}, I)}{f(y \mid H_d, m_1, I)}. \tag{10.14}
$$

There are two comparisons to be made. First, consider (8.5), where $1/\gamma$ is replaced with $f(y \mid H_p, \mathbf{x}, I)/f(y \mid H_d, m_1, I)$. Second, compare this theoretical result with (10.8), derived by Grove (1980), where here t_0 replaces $1 - p_T$, t_1 replaces p_T, b_0 replaces $1 - p_A$ and $b_{1,1}$ replaces p_A. The ratio of the density functions $f(y \mid H_p, \mathbf{x}, I)/f(y \mid H_d, m_1, I)$ was considered earlier. The extension described here accounts for possible different sources of the fragment. For the single-fragment case,

$$V = t_0 + \frac{t_1 b_0}{b_{1,1}} \frac{\sqrt{\sigma^2 + \tau^2}}{\sigma} \exp\left\{ \frac{(y-\mu)^2}{2(\tau^2 + \sigma^2)} - \frac{(y-\bar{x})^2}{2\sigma^2} \right\}.$$

Values for the distributional parameters and for the transfer probabilities from Evett (1986) are given in Tables 10.7 and 10.8. The values for $b_0, b_{1,1}$ and $b_{1,2}$ are suggested by Evett (1986), who cited Pearson *et al.* (1971). Evett refers to $b_2 f_2(y)$ as the probability of a person unconnected with the crime having two fragments of glass of refractive index y on his clothing. It is implicitly assumed in the notation $b_{1,2}$ used here that the two fragments are from one group. The use of b with only one subscript, b_n, indicates n groups but with indeterminate membership numbers. Evett contrasted the probabilities in Tables 10.7 and 10.8 with results from Harrison *et al.* (1985) in which the difference in proportions of people with one and with two fragments on their clothing was smaller. There is a need for a closer investigation of the estimation of these probabilities. Further examples are given in Section 10.4. These transfer probabilities are provided by Evett (1986) citing a personal communication by C.F. Candy. All these probabilities are provided primarily for illustrative purposes. A check will be needed for the relevance of these data from the mid-1980s in the early twenty-first century.

From these values $t_1 b_0 / b_{1,1} = 0.086$. Notice that $\tau/\sigma = 100$ and that $\tau^2 + \sigma^2 \simeq \tau^2$. Thus

$$\frac{\sqrt{\tau^2 + \sigma^2}}{\sigma} \simeq \frac{\tau}{\sigma}.$$

For the single-fragment case

$$V \simeq 8.6 \exp\left\{ \frac{(y-\mu)^2}{2\tau^2} - \frac{(y-\bar{x})^2}{2\sigma^2} \right\}. \tag{10.15}$$

Some values for varying values of $(y-\mu)/\tau$ and $(y-\bar{x})/\sigma$ (standardised differences of y from the mean of the source fragments and from the overall mean)

Table 10.7 Distributional parameters for glass problems

μ	τ	α	σ
1.5186	4×10^{-3}	3×10^{-3}	4×10^{-5}

Table 10.8 Transfer probabilities for glass problems

b_0	$b_{1,1}$	$b_{1,2}$	t_0	t_1	t_2
0.37	0.24	0.11	0	0.056	0.056

Table 10.9 Some values for the likelihood ratio V for the single-fragment case, from Evett (1986)

$(y - \bar{x})/\sigma$	$(y - \mu)/\tau$		
	0.0	1.0	2.0
0.0	9	14	63
1.0	5	9	38
2.0	1	2	9
3.0	0.1	0.2	0.7

are given in Table 10.9. Small values of $(y - \bar{x})/\sigma$ imply similarity between source and recovered fragments. Small values of $(y - \mu)/\tau$ imply a common value of y. Notice the largest value of V is given by a small value of $(y - \bar{x})/\sigma$ and a large value of $(y - \mu)/\tau$.

10.5.3 Two fragments

The evidence E again consists of three parts. The first part is the existence of two fragments (one group, two fragments each), the transferred-particle form, on the clothing of the suspect (m_2). The second part is the measurements y_1, y_2, of their refractive indices. The third part is the bulk form **x**.

The denominator of the likelihood ratio is the probability that a person unconnected with the crime will have two fragments of glass from one group with measurements (y_1, y_2) on his clothing. This is

$$Pr(m_2, y_1, y_2 \mid H_d, I) = Pr(m_2 \mid H_d, I)f(y_1, y_2 \mid H_d, I, m_2)$$
$$= b_{2,1}f(y_1, y_2 \mid H_d, I, m_2).$$

For the numerator there are four possibilities, as given in Table 10.10. Fragments which are categorised as 'not transferred' are assumed to have been acquired by means unconnected with the crime. These four possibilities are exclusive, and the numerator can be expressed as

$$t_0 b_{1,2} f(y_1, y_2 \mid H_d, I, m_2) + t_1 b_{1,1} f(y_1 \mid H_p, \mathbf{x})f(y_2 \mid H_d, I, m_1)$$
$$+ t_1 b_{1,1} f(y_2 \mid H_p, \mathbf{x})f(y_1 \mid H_d, I, m_1) + t_2 b_0 f(y_1, y_2 \mid H_p, \mathbf{x}).$$

Table 10.10 Possible sources of two fragments

Fragment 1	Fragment 2	Probability
Not transferred	Not transferred	$t_0 b_{1,2}$
Not transferred	Transferred	$t_1 b_{1,1}$
Transferred	Not transferred	$t_1 b_{1,1}$
Transferred	Transferred	$t_2 b_0$

The first term corresponds to the transfer of two fragments from the background. The distribution of y_1, y_2 is independent of $\mathbf{x}$, and may be written as shown. The other terms may be derived in a similar way.

The likelihood ratio is then

$$V = t_0 + \phi(1) + \phi(2) \tag{10.16}$$

where

$$\phi(1) = \frac{t_1 b_{1,1} \{f(y_1 \mid H_p, \mathbf{x}, m_1) f(y_2 \mid H_d, I, m_1) + f(y_2 \mid H_p, \mathbf{x}, m_1) f(y_1 \mid H_d, I, m_1)\}}{b_{1,2} f(y_1, y_2 \mid H_d, I, m_2)}$$

and

$$\phi(2) = \frac{t_2 b_0 f(y_1, y_2 \mid H_p, \mathbf{x}), m_2}{b_{1,2} f(y_1, y_2 \mid H_d, I, m_2)}.$$

The term $\phi(1)$ accounts for the case in which one fragment is transferred from the crime scene and one from the background. The term $\phi(2)$ accounts for the case in which both fragments are transferred from the crime scene. Several assumptions are necessary in order to gain numerical insight into the variability of V. Some of these correspond to assumptions made by Evett (1986), but others differ with respect to the distributions of y and (y_1, y_2) in the single-and two-fragment cases, respectively.

First, it is assumed that fragments of glass from the same source have a Normal distribution with a certain mean θ and variance σ^2. This mean θ itself has a Normal distribution with mean μ and variance τ^2 (see the discussion in Section 10.2.1). When considering the two-fragment case with r.i. measurements (y_1, y_2) it is assumed that the mean $\bar{y}$, $(y_1 + y_2)/2$, is Normally distributed, and also that the difference (higher $-$ lower), δ, of the two measurements has an exponential distribution and that these two distributions are independent.

Estimates for parameter values for these distributions based on surveys and casework are given in Evett (1986). The assumptions are as follows:

- The dimensions of the glass fragments are not relevant to the assessment of the evidential value.
- The distributions of the r.i. measurements carried out on fragments from a broken window are independent in magnitude of the number present.
- $(Y \mid H_d, I, m_1) \sim N(\mu, \tau^2 + \sigma^2)$.
- $f(y_1, y_2 \mid H_d, I, m_2) = f(\bar{y}, \delta \mid H_d, I, m_2) = f(\bar{y} \mid H_d, I, m_2) f(\delta \mid H_d, I, m_2)$.
- $(\bar{Y} \mid H_d, I, m_2) \sim N(\mu, \tau^2 + \sigma^2/2)$, a Normal distribution.
- $(\delta \mid H_d, I, m_2) = \alpha^{-1} \exp(-\delta/\alpha)$, an exponential distribution; see Figure 10.8.
- $(Y \mid H_p, \mathbf{x}, m_1) \sim N(\bar{x}, \sigma^2)$.
- $(\bar{Y} \mid H_p, \mathbf{x}, m_1) \sim N(\bar{x}, \sigma^2/2)$.

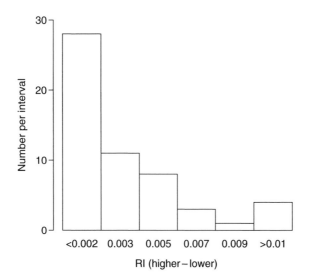

Figure 10.8 The difference in refractive index measurements (higher − lower) for each pair of fragments for individuals who had two fragments of glass on their clothing from Harrison *et al.* (1985). (Reproduced from Evett, 1986, by permission of The Forensic Science Society.)

For the two-fragment case, V may be written as the sum of two terms from (10.16). First, it is assumed that the case in which no fragment is transferred from the scene has negligible probability, so t_0 may be set equal to zero. Then, from (10.16),

$$V = \phi(1) + \phi(2).$$

Assume, as usual, that τ^2 is much greater than σ^2 so that $\tau^2 + \sigma^2$ may be approximated by τ^2. The general expressions for $\phi(1)$ and $\phi(2)$, using the distributional results above, can be shown to be

$$\phi(1) = \frac{t_1 b_{1,1}\alpha\left[\exp\left\{-\dfrac{(y_1-\bar{x})^2}{2\sigma^2}-\dfrac{(y_2-\mu)^2}{2\tau^2}\right\}+\exp\left\{-\dfrac{(y_2-\bar{x})^2}{2\sigma^2}-\dfrac{(y_1-\mu)^2}{2\tau^2}\right\}\right]}{\sigma b_{1,2}\sqrt{2\pi}\exp\left\{-\dfrac{(\bar{y}-\mu)^2}{2\tau^2}\right\}\exp\left\{-\dfrac{|y_1-y_2|}{\alpha}\right\}}$$

and

$$\phi(2) = \frac{t_2 b_0 \alpha\tau\exp\left\{-\dfrac{(y_1-\bar{x})^2}{2\sigma^2}-\dfrac{(y_2-\bar{x})^2}{2\sigma^2}\right\}}{b_{1,2}\sigma^2\sqrt{2\pi}\exp\left\{-\dfrac{(\bar{y}-\mu)^2}{2\tau^2}\right\}\exp\left\{-\dfrac{|y_1-y_2|}{\alpha}\right\}}.$$

Two sets of results are given here in terms of $|\bar{y}-\bar{x}|/\sigma$ when $\bar{y}$ is assumed equal to μ and the parameter values and transfer probabilities are as given in Tables 10.7 and 10.8. Let $k=|\bar{y}-\bar{x}|/\sigma$; $V=\phi(1)+\phi(2)$. If $|y_1-y_2|=\sigma$, then

$$\phi(1) = 3.675\exp\{-(4k^2+1)/8\}(e^{k/2}+e^{-k/2}),$$
$$\phi(2) = 566\exp\{-(4k^2+1)/4\}.$$

If $|y_1-y_2|=4\sigma$, then

$$\phi(1) = 3.825\exp\{-(k^2+4)/2\}(e^{2k}+e^{-2k}),$$
$$\phi(2) = 589\exp(-k^2+4).$$

Graphs of the logarithms to base 10 of $\phi(1)$, of $\phi(2)$ and of $\phi(1)+\phi(2)$ (which is not equal to $\log_{10}(\phi(1))+\log_{10}(\phi(2)))$, plotted against $|\bar{y}-\bar{x}|/\sigma$, are shown in Figure 10.9.

Notice that the graphs of $\log_{10}\phi(2)$ are quadratic. Also, for $|y_1-y_2|=\sigma$ the major contribution to V comes from the transfer of two fragments. For $|y_1-y_2|=4\sigma$, when $|\bar{y}-\bar{x}|/\sigma$ is close to 2, the major contribution to V comes from $\phi(1)$, the term corresponding to the transfer of one fragment only. This is reasonable since for $|y_1-y_2|=4\sigma$ and $|\bar{y}-\bar{x}|=2\sigma$, one of y_1 or y_2 must equal $\bar{x}$.

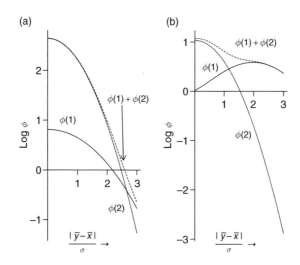

Figure 10.9 Graphs of $\log_{10}\phi(1)$, $\log_{10}\phi(2)$ and $\log_{10}V=\log_{10}(\phi(1)+\phi(2))$ against $|\bar{y}-\bar{x}|/\sigma$ for $\bar{y}=\mu$, for the transfer of two fragments of glass from the scene of the crime to the criminal: (a) $|y_1-y_2|=\sigma$, (b) $|y_1-y_2|=4\sigma$. The value of the evidence is $V=\phi(1)+\phi(2)$. The dotted line is $\log_{10}V$. (Adapted from Evett, 1986.)

10.5.4 A practical approach to glass evaluation

The use of probability density functions and kernel density estimation procedures is very sophisticated and requires considerable skill for their correct implementation. What is called a 'practical approach' to the interpretation of glass evidence is described by Evett and Buckleton (1990). Four scenarios are described and, for each, an expression for a likelihood ratio is derived. Sections 10.5.2 and 10.5.3 gave expressions for the value of the likelihood ratio for cases in which one or two fragments of glass have been transferred to the suspect's clothing. The four scenarios described by Evett and Buckleton (1990) concern the transfer of one or two *groups* of fragments which may or may not have come from one or two windows which have been smashed during the commission of a crime.

The circumstances are as follows. One or two windows have been smashed with criminal intent. A suspect has been apprehended very soon after the crime and one or two groups of glass fragments have been found on his clothing. The two propositions to be compared are

- H_p, the suspect is the man who smashed the window(s) at the scene of the crime;

- H_d, the suspect is not the man who smashed the window(s) at the scene of the crime.

Knowledge of the probabilities of various events is required. These probabilities can be estimated by reference to an appropriate clothing survey (e.g., Pearson *et al.*, 1971; Dabbs and Pearson, 1970, 1972; Pounds and Smalldon, 1978; Harrison *et al.*, 1985; McQuillan and Edgar, 1992; Lambert *et al.*, 1995; Allen and Scranage, 1998; Allen *et al.*, 1998a, b, c, d; Coulson *et al.*, 2001a), with care taken to ensure the relevance of the survey to the case in question, or by the use of personal experience. The probabilities used here are those given by Evett and Buckleton (1990). The various events, with their probabilities, are:

- that a person would have no glass on his clothing by chance alone, probability $p_0 = 0.636$;

- that a person would have one group of fragments on his clothing by chance alone, probability $p_1 = 0.238$;

- that a person would have two groups of fragments on his clothing by chance alone, probability $p_2 = 0.087$;

- that a group of fragments found on members of the population is large, probability $s_l = 0.029$;

- that, in the commission of the crime, no glass is transferred, probability $t_0 = 0.2$;

- that, in the commission of the crime, a large group of fragments is transferred, retained and found, probability $t_l = 0.6$.

This set of probabilities is different from those given in Section 10.5.2, as is the notation. What was previously denoted $b_{i,j}$ has been split into p_i and $s_{i,j}$. No claim is made that either set is definitive. Rather the values are given for illustrative purposes and it is a simple matter to substitute other values where this is thought appropriate. Also, the definition of large is unspecified but again a suitable definition with an appropriate probability may be made for a particular case. If it is not felt possible to choose a particular value for a probability then a range of values may be tried. If V remains relatively stable over the range of probability values, this provides reassurance that an exact value is not crucial. If V does depend crucially on the choice of a probability then careful thought is needed as to the usefulness of the method in the case under consideration.

For both windows the frequency (γ_1, γ_2) of occurrence of glass of the observed r.i.s on clothing is taken to be 3% in both cases, so that $\gamma_1 = \gamma_2 = 0.03$, where γ_1, γ_2 refer to the first and second window, respectively. These frequencies may be obtained from a histogram of r.i. measurements. In a more detailed approach these values would be replaced by probability density estimates. Four cases can be considered.

Case 1. One window is broken, one large group of fragments is found on the suspect and it is similar in properties to the broken window.

The denominator, which is derived assuming the suspect is innocent, is $p_1 s_l \gamma_1$; that is, the product of the probability (p_1) a person has one group of fragments on his clothing, the probability (s_l) that such a group is large and the frequency (γ_1) of glass of the observed properties on clothing.

The numerator is $p_0 t_l + p_1 s_l t_0 \gamma_1$. The first term accounts for the possibility that the suspect has had no glass on his clothing transferred by chance alone (p_0) and has had a large group of items transferred, retained and found in the commission of the crime (t_l). The probability in the circumstances that such a group of fragments has the required properties is 1. The second term accounts for the probability that the suspect has had glass of the required properties transferred by chance alone $(p_1 s_l \gamma_1)$ and no glass transferred in the commission of the crime.

The likelihood ratio is

$$V_1 = t_0 + \frac{p_0 t_l}{p_1 s_l \gamma_1}. \tag{10.17}$$

Case 2. One window is broken and two large groups of fragments are found on the suspect. One group matches the properties of the broken window, the other does not. The likelihood ratio is

$$V_2 = t_0 + \frac{p_1 t_l}{2 p_2 s_l \gamma_1}.$$

The factor 2 appears in the denominator since there are two groups. The characteristics (size and frequency of occurrence) of the second group of fragments occur in both the numerator and denominator and cancel out. Hence they do not appear in the final expression.

Case 3. Two windows are broken and one large group of fragments is found on the suspect, the properties of which match one of the broken windows. The likelihood ratio is

$$V_3 = t_0^2 + \frac{p_0 t_0 t_l}{p_1 s_l \gamma_1},$$

where it has been assumed that the transfer probabilities (t_0, t_l) are the same for both windows.

Case 4. Two windows are broken and two large groups of glass are identified, one of which matches one broken window and one of which matches the other. The likelihood ratio is

$$V_4 = t_0^2 + \frac{p_0 t_l^2}{2 p_2 s_l^2 \gamma_1 \gamma_2} + \frac{p_1 t_0 t_l}{2 p_2 s_l \gamma_1} + \frac{p_1 t_0 t_l}{2 p_2 s_l \gamma_2}.$$

Using the probability figures given above, it is easy to verify that in these four cases the second term is the dominant one. Thus, the following approximate results are obtained

$$V_1 \simeq \frac{p_0 t_l}{p_1 s_l \gamma_1}, \tag{10.18}$$

$$V_2 \simeq \frac{p_1 t_l}{2 p_2 s_l \gamma_1}, \tag{10.19}$$

$$V_3 \simeq \frac{p_0 t_0 t_l}{p_1 s_l \gamma_1}, \tag{10.20}$$

$$V_4 \simeq \frac{p_0 t_l^2}{2 p_2 s_l^2 \gamma_1 \gamma_2}. \tag{10.21}$$

The results, substituting the probability values listed into (10.18) to (10.21), are

$$V_1 \simeq 1843,$$

$$V_2 \simeq 943,$$

$$V_3 \simeq 369,$$

$$V_4 \simeq 1\,738\,000.$$

It would be wrong to place too much worth on the exact numerical values of these results. There are many imponderables, such as the specifications of the transfer probabilities to be considered. However, a comparison of the orders of magnitude provides a useful qualitative assessment of the relative worth of these results. For example, consider a comparison of V_3 with V_4. The latter, V_4, is bigger than the former, V_3, by a factor of about 5000. The effect on the value of the evidence when two windows have been broken of discovering two groups of fragments, with similar properties to the broken windows, rather than just one, on the clothing of a suspect is considerable. An approximate general formula is given by Curran *et al.* (2000).

10.5.5   Graphical models for the assessment of transfer probabilities

Bayesian interpretations of transfer evidence which address propositions at the activity level require estimates of transfer probabilities. For example, in a case in which a window has been broken, consideration needs to be given to the probability of transfer, persistence and recovery of glass fragments. These three components of the process may be referred to, generally, as the *transfer process*.

A crime is committed. It is expected by the investigators that there will have been transfer of evidence in both directions between the criminal and the crime scene. A suspect is apprehended and evidence is found on his person which can be associated with the crime scene. The factors in the transfer process to be considered include, among others, transfer itself, persistence and recovery.

Transfer will depend on the nature of the contact between the criminal and the scene. For example, if a window has been broken to gain entry, then the type of window and the distance from the window at which the criminal stood in order to break it will be factors in assessing the quantity of glass transferred.

Persistence will depend on the elapsed time between the commission of the crime and the apprehension of the suspect and on the nature of the clothes that the suspect may be thought to have been wearing at the time of the crime. This will also depend on the nature of the contact, as in transfer. The persistence of glass fragments from a broken window may have different characteristics from the persistence of blood following a prolonged assault.

Recovery will depend on the previous two factors. It will also depend on the quality of the resources available for the detection and collection of the evidence.

The relationships amongst these three factors of transfer, persistence and recovery may be illustrated graphically with nodes (circles) to represent the factors and arrows (links) joining the nodes to indicate a relationship between the nodes thus joined. In addition, the factors which contribute to the three factors in the transfer process may also be represented by nodes with links to other nodes to indicate relationships as appropriate. The resulting diagram is known as a *graph* and the associated models are known as *graphical models*.

It is possible to include probabilistic relationships in such a graph. For example, the probability of transfer of evidence in a particular case may be dependent on several factors. This dependence can then be represented with a conditional probability distribution, the conditioning being on the values of the factors for which the transfer is dependent. The use of graphical models requires decisions to be made concerning not only the values of parameters in probability distributions but also the type of probability distributions themselves. For example, in the case of a broken window, the distance, D, from the window at which the criminal was standing when it was broken is unknown. Because there is uncertainty associated with D, it may be modelled probabilistically. One suggestion (Curran *et al.*, 1998a) is to model D with a gamma distribution (5.1).

The distribution $Pr(N = n)$ of the number N of fragments transferred is dependent on D and on other factors, such as those mentioned above. This distribution is not expressible as a formula from which probabilities may be simply obtained by the replacement of the values of the contributing factors. Instead, the distribution has to be derived empirically through a process known as *simulation*, which is beyond the scope of this book. Further details of the simulation process used for the modelling of the transfer of glass fragments are available in Curran *et al.* (2000), and from whom the appropriate software is available. Curran *et al.* (2000, p. 124) comment that the 'simulation process can be thought of as generating thousands of cases where the crime details are approximately the same and observing the number of fragments recovered'. For each of these 'thousands of cases' a value of n is obtained. A histogram of the n is then derived and used as an approximation to the distribution $Pr(N = n)$. From this approximation, an estimate of the probability at the particular value of N for the case in hand may be obtained.

Consider a case in which a suspect was apprehended about an hour or two after the commission of the crime. There is eyewitness evidence that the criminal was about 1 metre from the window when it was broken. The scientist expects that for a criminal standing 1 metre from the window in the circumstances of this crime about 60 fragments of glass would have been transferred. The scientist expects about 80% to 90% of any fragments transferred to the clothing of the criminal to be lost in the first hour and 50% to 70% of the fragments remaining at the beginning of an hour to be lost in the subsequent hour. The scientist also expects to recover about 90% to 95% of the fragments remaining on the clothing at the time of inspection.

The scientist inspects the clothing of the suspect and finds $(N = 4)$ fragments of glass. The simulation process of Curran *et al.* (2000) indicates a probability $Pr(N = 4)$ of 0.08 and $Pr(N = 0)$ of 0.104. The values 0.08 and 0.104 are values for t_n in (10.17), with $n = 4$ and $n = 0$, respectively.

A similar graphical approach using nodes and arrows for discrete data only is described in greater detail in Chapter 14.

10.6   APPROACH BASED ON *t*-DISTRIBUTION

The case in which a window is broken, a large group of fragments is found on the suspect and the group is similar in properties to the broken window was discussed earlier (Case 1 in Section 10.5.4) and the likelihood ratio given by (10.17).

The factor $1/\gamma_1$ in (10.17) is an approximation to the ratio given in (10.2). Another approach to the evaluation of V in (10.2) is to consider the summary statistics for the recovered and control data. Let n_y and n_x denote the number of recovered and control measurements, respectively, and denote the recovered data $\mathbf{y}$ as $(y_1, \ldots, y_{n_y})$ and the control data $\mathbf{x}$ as $(x_1, \ldots, x_{n_x})$. These data may be replaced by summary statistics for the means and variances, with

$$\bar{x} = \sum_{i=1}^{n_x} x_i / n_x \quad \text{and} \quad \bar{y} = \sum_{j=1}^{n_y} y_j / n_y,$$

and

$$s_x^2 = \sum_{i=1}^{n_x} (x_i - \bar{x})^2 / (n_x - 1) \quad \text{and} \quad s_y^2 = \sum_{j=1}^{n_y} (y_j - \bar{y})^2 / (n_y - 1).$$

Then, following the argument of Walsh *et al.* (1996), the ratio (10.2) may be written as

$$V = \frac{f(\bar{x} - \bar{y} \mid \bar{x}, s_x, s_y, H_p)}{f(\bar{y} \mid \bar{x}, s_x, s_y, H_d)}. \tag{10.22}$$

The numerator of (10.22) may be taken to be a Student t-density with a Welch modification (Welch, 1937) when the data $\mathbf{x}$ and $\mathbf{y}$ are Normally distributed and the population variances σ_x^2 and σ_y^2, of which s_x^2 and s_y^2 are estimates, are not assumed equal. The statistic for the numerator which is to be referred to the t-density with the Welch modification is

$$t_W = \frac{\bar{x} - \bar{y}}{\sqrt{\dfrac{s_x^2}{n_x} + \dfrac{s_y^2}{n_y}}}. \tag{10.23}$$

The statistic t_W does not have a t-distribution but may be approximated by a t-distribution with ν degrees of freedom, where ν may be estimated from the data as

$$\nu = \frac{\left(\dfrac{s_x^2}{n_x} + \dfrac{s_y^2}{n_y}\right)}{\left(\dfrac{s_x^4}{n_x^2(n_x - 1)} + \dfrac{s_y^4}{n_y^2(n_y - 1)}\right)}, \tag{10.24}$$

which need not necessarily be an integer. Density values for t_W are provided by many readily available statistical software packages.

The denominator of (10.22) is the value of the probability density for the relevant population of glass at $\bar{y}$. This is usually obtained from a kernel density estimate, such as (10.11).

Consider as an example the data in Table 10.11 with $\bar{y} = 1.519\,507\,3$, $\bar{x} = 1.519\,573\,0$, $s_y = 5.24 \times 10^{-5}$, $s_x = 4.55 \times 10^{-5}$, $n_x = 10$ and $n_y = 11$. The value of the t-statistic using a pooled standard deviation, with 19 degrees of freedom, is 3.06. The 99.5% point of a t-distribution with 19 degrees of freedom is 2.86, so the null hypothesis that the recovered (Johnston) and control data are samples from populations with the same mean is rejected in favour of the two-sided alternative of samples from populations with different means at the 1% level. The significance probability (Section 4.6) for the two-sided test is 0.0064. Thus, the conclusion of a scientist who used this approach would be to reject the hypothesis that the glass fragments found on Johnston came from the crime scene window and this evidence would be discarded. However, one of the problems associated with the use of significance probabilities is that there is a dichotomy between data for which the null hypothesis is rejected and data for which the null hypothesis is not rejected, the 'fall-off-the-cliff' effect of Section 1.3.3.

The value of the numerator is obtained from a Student t-density function (Section 2.4.3) with (in the notation of that section) x equal to the Johnston mean 1.519 507 3, μ equal to the control mean, 1.519 573 0, ν equal to the (non-integer) degrees of freedom 18.97 (derived from (10.24)) and λ equal to $\sqrt{\frac{s_x^2}{n_x} + \frac{s_y^2}{n_y}}$ which equals 2.14×10^{-5} in this example. Transformation to $y = (x - \mu)/\lambda$ gives a value for the t-statistic of 3.07 with 18.97 degrees of freedom.

Table 10.11 Refractive indices of glass fragments for Johnston, recovered, and a control set with means, separate and pooled standard deviations (s.d.). The number of recovered fragments $n_y = 11$ and the number of control fragments $n_x = 10$. Example presented in Walsh *et al.* (1996)

	Johnston	Control
	1.519 40	1.519 50
	1.519 46	1.519 52
	1.519 47	1.519 53
	1.519 48	1.519 56
	1.519 50	1.519 57
	1.519 52	1.519 59
	1.519 52	1.519 60
	1.519 53	1.519 60
	1.519 56	1.519 62
	1.519 57	1.519 64
	1.519 57	
Mean	1.519 507 3	1.519 573 0
s.d.	5.24×10^{-5}	4.55×10^{-5}
Pooled s.d.		4.92×10^{-5}

Reference to appropriate statistical software gives a value for the central t-density function of 0.007. An adjustment by the factor $1/\lambda$ gives a value for the non-central t-density (Section 2.4.3) of $0.007/2.14 \times 10^{-5}$ or 328. This is the value for the numerator.

The value of the denominator is obtained from population data and a kernel density estimate. For this example, the value of the density estimate at the Johnston mean is taken to be 109 (Walsh *et al.*, 1996). The likelihood ratio is then $328/109 = 3.0$. This provides slight support for the proposition that the fragments found on Johnston's clothing come from the crime scene window. This conclusion is in contrast to the rejection of this proposition at the 1% level using a two-tailed test.

Goldmann *et al.* (2004) describe another application of the t-distribution and the Welch modification, in which the determination of the source of illicit pills through examination of the dye present in the pills is assisted. The dye considered is CI 14720. There is a sample Y with 5 pills to be compared with a specific batch X with 20 pills and with another batch containing 100 pills attributed to the same producer, Z, as produced X. The measurement of interest is the concentration of the dye, expressed as a percentage. Two pairs of propositions are compared. The first is:

- H_{p1}, sample Y comes from batch X;
- H_{d1}, sample Y does not come from batch X.

The second is:

- H_{p2}, sample Y comes from a batch produced by Z;
- H_{d2}, sample Y does not come from a batch produced by Z.

The summary statistics (Goldmann *et al.* 2004) are presented in Table 10.12. No sample size is given for the general population. The population is characterised by illicit pills coloured with dye CI 14720. The percentage concentration of the dye in the pills is Normally distributed with mean 0.300% and standard deviation 0.06%. This contrasts with the denominator in the previous example in which a kernel density estimate was used. The degrees of freedom and pooled standard deviations of Y with X and of Y with the batch produced by Z are $\nu_{y.x} = 4.50, s_{\text{pooled},y.x} = 0.0092$ and $\nu_{y.z} = 5.75, s_{\text{pooled},y.z} = 0.0098$.

Table 10.12 Summary statistics for concentration of dye CI 14720 in illicit pills (Goldmann *et al.*, 2004)

Sample	Sample Y	Batch X	Producer Z	General population
Size	5	20	100	-
Mean (%)	0.165	0.140	0.180	0.300
Standard deviation	0.02	0.01	0.04	0.06

Consider the first pair of propositions. The value of the evidence is obtained by comparing the probability density of $\bar{y}$ in batch X and the probability density of $\bar{y}$ in the general population. This is 5.4.

Consider the second pair of propositions. The value of the evidence is obtained by comparing the probability density of $\bar{y}$ in batches Z and the probability density of $\bar{y}$ in the general population. This is 23.3.

10.7 APPENDIX: DERIVATION OF V WHEN THE BETWEEN-SOURCE MEASUREMENTS ARE ASSUMED NORMALLY DISTRIBUTED

The marginal distributions of the means of the source and receptor measurements (with m measurements on the source and n on the receptor), $\bar{X}$ and $\bar{Y}$, in the denominator are independent and are, respectively, $N(\mu, \tau^2 + \sigma^2/m)$ and $N(\mu, \tau^2 + \sigma^2/n)$.

Let $\sigma_1^2 = \tau^2 + \sigma^2/m$ and $\sigma_2^2 = \tau^2 + \sigma^2/n$, where τ^2 is the between-source variance. Then $(\bar{X} - \bar{Y}) \sim N(0, \sigma_1^2 + \sigma_2^2)$ and $Z = (\sigma_2^2\bar{X} + \sigma_1^2\bar{Y})/(\sigma_1^2 + \sigma_2^2)$ is distributed as $N(\mu, \sigma_1^2\sigma_2^2/(\sigma_1^2 + \sigma_2^2))$, and $(\bar{X} - \bar{Y})$ and Z are also independent. The denominator may then be written as

$$\frac{1}{2\pi\sigma_1\sigma_2} \exp\left\{-\frac{(\bar{x} - \bar{y})^2}{2(\sigma_1^2 + \sigma_2^2)}\right\} \exp\left\{-\frac{(z - \mu)^2(\sigma_1^2 + \sigma_2^2)}{2\sigma_1^2\sigma_2^2}\right\}.$$

In the numerator, it can be shown that the joint unconditional distribution of $\bar{X}$ and $\bar{Y}$ is bivariate Normal with means μ, variances σ_1^2 and σ_2^2 and covariance τ^2.

The distribution of $\bar{X} - \bar{Y}$ is $N\{0, \sigma^2(\frac{1}{m} + \frac{1}{n})\}$. Let $W = (m\bar{X} + n\bar{Y})/(m+n)$. The distribution of W is $N(\mu, \tau^2 + \sigma^2/(m+n))$. Also, $(\bar{X} - \bar{Y})$ and W are independent. Let $a^2 = 1/m + 1/n$ and $\sigma_3^2 = \tau^2 + \sigma^2/(m+n)$. Then the numerator may be written as

$$\frac{1}{2\pi a\sigma\sigma_3} \exp\left\{-\frac{(\bar{x} - \bar{y})^2}{2a^2\sigma^2}\right\} \exp\left\{-\frac{(w - \mu)^2}{2\sigma_3^2}\right\}.$$

The value, V, of the evidence is the ratio of the numerator to the denominator; after some simplification, this is

$$\frac{\sigma_1\sigma_2}{a\sigma\sigma_3} \exp\left\{-\frac{(\bar{x} - \bar{y})^2\tau^2}{a^2\sigma^2(\sigma_1^2 + \sigma_2^2)}\right\} \exp\left\{-\frac{(w - \mu)^2}{2\sigma_3^2} + \frac{(z - \mu)^2(\sigma_1^2 + \sigma_2^2)}{2\sigma_1^2\sigma_2^2}\right\}.$$

Large values of this provide good evidence that the suspect was at the crime scene.

This expression may be simplified. Typically, τ is much larger than σ. Then $\sigma_1^2 = \sigma_2^2 = \sigma_3^2 = \tau^2$, $Z = (X + Y)/2$ and

$$V \simeq \frac{\tau}{a\sigma} \exp\left\{-\frac{(\bar{x} - \bar{y})^2}{2a^2\sigma^2}\right\} \exp\left\{-\frac{(w - \mu)^2}{2\tau^2} + \frac{(z - \mu)^2}{\tau^2}\right\}. \tag{10.25}$$

If the number of control measurements equals the number of recovered measurements then $m = n$, $Z = W = \frac{1}{2}(X + Y)$, $\sigma_1^2 = \sigma_2^2$ and

$$V \simeq \frac{m^{1/2}\tau}{2^{1/2}\sigma} \exp\left\{-\frac{m(\bar{x} - \bar{y})^2}{4\sigma^2}\right\} \exp\left\{\frac{(z - \mu)^2}{2\tau^2}\right\}.$$

11

Multivariate Analysis

11.1 INTRODUCTION

Often more than one characteristic, or variable, is recorded for a piece of evidence. For example, the concentrations of various elements in fragments of glass may be recorded. These various characteristics may not be independent, and it is necessary to allow for the dependence in the evaluation of the evidence. An example of the importance of allowing for the dependence between characteristics was given in Section 4.6.3, where the characteristics are the refractive index and density of glass. The product of two separate significance probabilities was 0.0016, which may be thought to be highly significant. It was shown in Section 4.6.3 that when the dependence between the two variables was included in the analysis the significance probability was only 0.1225. The two characteristics were significant individually at the 5% level but together were not significant at the 10% level. It is possible to transform multivariate data, for example using a method known as principal component analysis (Jolliffe, 1986), to derive statistically independent variables from correlated measurements. Such an approach in the context of footprints is discussed in Kennedy *et al.* (2003).

A method for the evaluation of evidence through the derivation of an appropriate likelihood ratio will be developed in this chapter. The method will be compared with a significance test using Hotelling's T^2 (Section 11.4). The likelihood ratio generalises the likelihood ratio (10.7) and assumes that the data are Normally distributed. As for the development of (10.7), there are assumed to be two sources of variation, that within sources and that between sources. The ideas may be extended using kernel density estimation to situations in which the between-source distribution is not Normally distributed, as in Section 10.4 for univariate data, but this extension is not discussed here. There may also be data for which it is possible to develop specialist distributions, but again this is not discussed here. One such example is a bivariate distribution developed to model measurements of complementary chromaticity coordinates taken from fibres from a number of garments (Hoggart *et al.*, 2003).

The ideas will be illustrated with an example concerning elemental concentrations of glass. The arithmetic involved is considerable. In order to have data

Statistics and the Evaluation of Evidence for Forensic Scientists: Second Edition
C.G.G. Aitken and F. Taroni © 2004 John Wiley & Sons, Ltd   ISBN: 0-470-84367-5

which can be presented easily a rather unrealistic example will be considered in which only two variables are considered and in which there are only two control items and two recovered items. Two sets of results will be calculated, one for when the control and recovered items come from the same source and one for when the control and recovered items come from different sources. Results will be derived for significance tests using Hotelling's T^2 (Section 11.4), for univariate likelihood ratios using (10.7) and for the multivariate generalisation. A brief derivation of the multivariate likelihood ratio is given in an Appendix.

An application of multivariate analysis for the classification and discrimination of glass fragments is described in Hicks *et al.* (2003). An application to a different evidential type, that of the discrimination of ball-point pen inks, is described in Thanasoulias *et al.* (2003).

11.2   DESCRIPTION OF EXAMPLE

The example which is used to illustrate these methods is based on an example of elemental concentrations for glass fragments from bottles described in Curran *et al.* (1997a). The calculations are minimised by considering only two elements, aluminium and barium. In the notation of the Appendix, $p = 2$. There are two control and two recovered items in each case ($n_1 = n_2 = 2$). There are two groups, glass from brown bottles and glass from colourless bottles.

The overall mean elemental concentration μ, the within-group covariance matrix $\mathbf{U}$ and the between-group covariance matrix $\mathbf{C}$ are taken to be

$$\mu = \begin{pmatrix} 0.805 \\ 0.016 \end{pmatrix},$$

$$\mathbf{U} = \begin{pmatrix} 0.002 & 0.00004 \\ 0.00004 & 0.000002 \end{pmatrix}, \tag{11.1}$$

$$\mathbf{C} = \begin{pmatrix} 0.011 & 0.0006 \\ 0.0006 & 0.00004 \end{pmatrix}.$$

These three parameters μ, $\mathbf{U}$ and $\mathbf{C}$ can be obtained from population data of glass fragments. The first component of μ is the population mean of the elemental concentration of aluminium and the second component of μ is the population mean of the elemental concentration of barium. The top left-hand figure in $\mathbf{U}$ is the population within-source variance for aluminium. The bottom right-hand figure in $\mathbf{U}$ is the population within-source variance for barium. The top left-hand figure in $\mathbf{C}$ is the population between-source variance for aluminium. The bottom right-hand figure in $\mathbf{C}$ is the population between-source variance for barium. The off-diagonal terms (top right and bottom left) provide a measure of the correlation between the concentrations of the two elements.

The measurements $\mathbf{y}_{11}, \mathbf{y}_{12}$ on the two control fragments are taken to be the same throughout. These are

$$\mathbf{y}_{11} = \begin{pmatrix} 0.929 \\ 0.022 \end{pmatrix}, \; \mathbf{y}_{12} = \begin{pmatrix} 0.859 \\ 0.018 \end{pmatrix}.$$

The mean of these two vectors is obtained by taking the mean of the first component and the mean of the second component. These two means together give the mean vector. This may be denoted by $\bar{\mathbf{y}}_1$ and is

$$\bar{\mathbf{y}}_1 = \begin{pmatrix} 0.894 \\ 0.020 \end{pmatrix}.$$

There will be two sets of recovered fragments. One will be used for the evaluation of the evidence when the control and recovered fragments come from the same source, and one will be used for the evaluation of the evidence when the control and recovered fragments come from different sources.

The measurements $\mathbf{y}_{21}, \mathbf{y}_{22}$ on the two recovered fragments taken to be from the same source as the control fragments are

$$\mathbf{y}_{21} = \begin{pmatrix} 0.845 \\ 0.018 \end{pmatrix}, \quad \mathbf{y}_{22} = \begin{pmatrix} 0.931 \\ 0.020 \end{pmatrix},$$

with mean

$$\bar{\mathbf{y}}_2 = \begin{pmatrix} 0.888 \\ 0.019 \end{pmatrix}.$$

The measurements $\mathbf{y}_{31}, \mathbf{y}_{32}$ on the two recovered fragments taken to be from a different source from the control fragments are

$$\mathbf{y}_{31} = \begin{pmatrix} 0.751 \\ 0.011 \end{pmatrix}, \quad \mathbf{y}_{32} = \begin{pmatrix} 0.659 \\ 0.009 \end{pmatrix},$$

with mean

$$\bar{\mathbf{y}}_3 = \begin{pmatrix} 0.705 \\ 0.010 \end{pmatrix}.$$

Note that these are denoted with a subscript 3 to distinguish them from those recovered fragments, with a subscript 2, deemed to be from the same source as the control fragments, with a subscript 1. Note also that, for multivariate data, $\mathbf{y}$ is used to denote control as well as recovered data.

Various measures for the evaluation of the evidence will be considered:

- univariate t-tests for aluminium and for barium separately;
- Hotelling's T^2, which combines the information provided by the two elements;
- univariate likelihood ratios using (10.7);
- multivariate likelihood ratios using the results in the Appendix.

11.3   UNIVARIATE t-TESTS

There are two propositions to consider:

- H_p, the control and recovered fragments of glass come from the same source;
- H_d, the control and recovered fragments of glass come from different sources.

Measurements $\mathbf{y}$ are taken of elemental concentrations of glass fragments from the crime scene (n_1 measurements) and from a suspect (n_2 measurements). Denote the means of the populations from which the glass may have come as $\boldsymbol{\mu}_1$ and $\boldsymbol{\mu}_2$. If H_p is true, $\boldsymbol{\mu}_1$ and $\boldsymbol{\mu}_2$ are equal. If H_d is true, $\boldsymbol{\mu}_1$ and $\boldsymbol{\mu}_2$ are not equal. The variance σ^2 is taken to be equal for the (possibly) two populations. The mean of the measurements from the crime scene is denoted $\bar{\mathbf{y}}_1$. The mean of the measurements from the suspect is denoted $\bar{\mathbf{y}}_2$ or $\bar{\mathbf{y}}_3$, depending on whether the fragments have been chosen to come from the same source or from different sources.

The test statistic is then

$$t = \frac{|\bar{y}_{1k} - \bar{y}_{lk}|}{\sigma\sqrt{\dfrac{1}{n_1} + \dfrac{1}{n_l}}}, \tag{11.2}$$

where $k = 1$ or 2 to denote aluminium ($k = 1$) or barium ($k = 2$) and $l = 2$ or 3, depending on the scenario. The numbers n_1, n_2 and n_3 of fragments in each group are all 2 in these examples. Note, also, that a t-test is being advocated even though the standard deviation σ has been estimated from some population. The use of σ is to relate this analysis to the fuller analysis described later in which the within-group covariance matrix $\mathbf{U}$ is used. The value for σ^2 is either the top left-hand cell of $\mathbf{U}$ for aluminium or the bottom right-hand cell for barium.

Four situations will be considered:

(a) measurements of aluminium concentrations for fragments from the same source;
(b) measurements of aluminium concentrations for fragments from different sources;
(c) measurements of barium concentrations for fragments from the same source;
(d) measurements of barium concentrations for fragments from different sources.

For the two situations for aluminium, the denominator of (11.2) is $(0.002/2 + 0.002/2)^{1/2} = 0.045$. For the two situations for barium, the denominator of (11.2) is $(0.000\,002/2 + 0.000\,002/2)^{1/2} = 0.0014$. The results for the four situations are respectively:

(a) $|\bar{y}_{11} - \bar{y}_{21}| = 0.894 - 0.888 = 0.006$, and $t = 0.006/0.045 = 0.133$;

(b) $|\bar{y}_{11} - \bar{y}_{31}| = 0.894 - 0.705 = 0.189$, and $t = 0.189/0.045 = 4.2$;

(c) $|\bar{y}_{12} - \bar{y}_{22}| = 0.020 - 0.019 = 0.001$, and $t = 0.001/0.0014 = 0.71$;

(d) $|\bar{y}_{12} - \bar{y}_{32}| = 0.020 - 0.010 = 0.010$, and $t = 0.010/0.0014 = 7.14$.

There are 2 degrees of freedom in all situations ($n_1 + n_2 - 2 = n_1 + n_3 - 2 = 2$). The 2.5% point of the t-distribution with 2 degrees of freedom is 4.3027 and the 0.5% point is 9.9248. Thus, for a two-sided test, there is insufficient evidence in (a) and (c) to reject the proposition that the two sets of fragments come from the same source. For (b), there is almost sufficient evidence at the 5% level to reject the proposition that the two sets of fragments come from the same source. For (d) there is sufficient evidence at the 5% level to reject the proposition that the two sets of fragments come from the same source. The significance probability in a two-sided test for the outcome in (d) is 2%.

11.4 HOTELLING'S *T*²

This distribution is a multivariate analogue of the univariate t-distribution.
As before, there are two propositions to consider:

H_p: The control and recovered fragments of glass come from the same source.
H_d: The control and recovered fragments of glass come from different sources.

Bivariate measurements **y** are taken of elemental concentrations of aluminium and of barium for glass fragments from the crime scene (n_1 measurements) and from a suspect (n_2 or n_3 measurements). Denote the bivariate means of the populations from which the glass may have come as μ_1 and μ_2. If H_p is true, μ_1 and μ_2 are equal. (Equality of a vector follows from equality of each component separately.) If H_d is true, μ_1 and μ_2 are not equal. The covariance matrix to be used in the test statistic is the within-group covariance matrix **U** and is taken to be equal for the (possibly) two populations. The mean of the measurements from the crime scene is denoted $\bar{\mathbf{y}}_1$. The mean of the measurements from the suspect is denoted $\bar{\mathbf{y}}_2$ or $\bar{\mathbf{y}}_3$, depending on whether the fragments have been chosen to come from the same source or from different sources.

The test statistic is then

$$T^2 = (\bar{\mathbf{y}}_1 - \bar{\mathbf{y}}_l)^T \left[\left(\frac{1}{n_1} + \frac{1}{n_l} \right) \mathbf{U} \right]^{-1} (\bar{\mathbf{y}}_1 - \bar{\mathbf{y}}_l) \tag{11.3}$$

for $l = 2, 3$. As for the univariate t-statistics, the covariance matrix should be estimated from the control and recovered fragments, but the population matrix $\mathbf{U}$ is being used instead for ease of comparison with the multivariate Normal results to follow. The covariance matrix $\mathbf{U}$ is given in (11.1). The statistic T^2 has a scaled F-distribution such that

$$T^2 \sim \frac{(n_1 + n_l - 2)p}{n_1 + n_l - p - 1} F_{p, n_1 + n_l - p - 1}$$

where in this case $l = 2$ or 3, $p = 2$ and $n_1 = n_2 = n_3 = 2$. Thus $(n_1 + n_l - 2)p/(n_1 + n_l - p - 1) = 4$ and the F-distribution has $(2, 1)$ degrees of freedom. The significance of the test can then be determined by evaluating T^2, dividing the value obtained by 4 and referring the result to the F-distribution with $(2,1)$ degrees of freedom (using statistical tables or software).

Two situations will be considered: (a) measurements of aluminium and barium concentrations together for fragments from the same source and (b) measurements of aluminium and barium concentrations together for fragments from different sources.

(a) The difference in the means $\bar{\mathbf{y}}_1$ and $\bar{\mathbf{y}}_2$ is $(0.006, 0.001)^T$. Also $(1/n_1 + 1/n_2)\mathbf{U}^{-1} = (1/2 + 1/2)\mathbf{U}^{-1} = \mathbf{U}^{-1}$. The value of T^2 (11.3) is thus

$$(0.006, 0.001)\, \mathbf{U}^{-1} \begin{pmatrix} 0.006 \\ 0.001 \end{pmatrix} = 0.6633. \tag{11.4}$$

(b) The difference in the means $\bar{\mathbf{y}}_1$ and $\bar{\mathbf{y}}_3$ is $(0.189, 0.010)^T$. Also $(1/n_1 + 1/n_3)\mathbf{U}^{-1} = (1/2 + 1/2)\mathbf{U}^{-1} = \mathbf{U}^{-1}$. The value of T^2 (11.3) is thus

$$(0.189, 0.010)\, \mathbf{U}^{-1} \begin{pmatrix} 0.189 \\ 0.010 \end{pmatrix} = 50.1008.$$

These values divided by 4 (the appropriate value of $(n_1 + n_l - 2)p/(n_1 + n_l - p - 1)$ for this example), that is, 0.168 and 12.525, should then be referred to $F_{2,1}$. The 5% point of the $F_{2,1}$ distribution is 199.5, so these results are very insignificant. There is very little evidence in either case of the control and recovered fragments coming from different sources.

11.5　LIKELIHOOD RATIO FOR UNIVARIATE NORMALITY WITH TWO SOURCES OF VARIATION

The methods described in Sections 11.3 and 11.4 consider only the within-group variance, and the evidence is evaluated with a significance probability.

A univariate likelihood ratio which allows for between-group variance as well as within-group variance for univariate Normally distributed data is given in (10.7). The within-group variances (σ^2 in the notation of (10.7)) are given by the diagonal terms in **U**, namely 0.002 for aluminium and 0.000 002 for barium. The between-group variances (τ^2 in the notation of (10.7)) are given by the diagonal terms in **C**, namely 0.011 for aluminium and 0.000 04 for barium. The number of control and recovered fragments m corresponds to n_1, n_2 and n_3 in this chapter. The difference $(\bar{x} - \bar{y})$ in control and recovered means in the notation of (10.7) is now denoted $(\bar{\mathbf{y}}_1 - \bar{\mathbf{y}}_l)$, where $l = 1, 2$ for recovered fragments which come from the same source as or from different sources than the control fragments. The final notational comment is that z in (10.7) is $\frac{1}{2}(\bar{\mathbf{y}}_1 + \bar{\mathbf{y}}_l)$, the mean of the control and recovered fragments.

As in Section 11.3, four situations are considered: (a) measurements of aluminium concentrations for fragments from the same source; (b) measurements of aluminium concentrations for fragments from different sources; (c) measurements of barium concentrations for fragments from the same source; (d) measurements of barium concentrations for fragments from different sources. The expression $m^{1/2}\tau/(2^{1/2}\sigma)$ has two possible values, one when considering aluminium and one when considering barium. For aluminium it is equal to $(0.011/0.002)^{1/2} = 5.5^{1/2} \simeq 2.35$, and for barium it is equal to $(0.000 04/0.000 002)^{1/2} = 20^{1/2} \simeq 4.47$.

(a) Here, $\sigma^2 = 0.002, \tau^2 = 0.011, \mu = 0.805, z = 0.891$, and $\bar{y}_1 - \bar{y}_2 = 0.006$. The value of the evidence is then, from (10.7)

$$\left(\frac{0.011}{0.002}\right)^{1/2} \exp\left\{-\frac{0.006^2}{0.004} + \frac{0.086^2}{0.022}\right\} = 3.25.$$

(b) Here, $\sigma^2 = 0.002$, $\tau^2 = 0.011$, $\mu = 0.805$, $z = 0.7995$, and $\bar{y}_1 - \bar{y}_3 = 0.189$. The value of the evidence is then, from (10.7)

$$\left(\frac{0.011}{0.002}\right)^{1/2} \exp\left\{-\frac{0.189^2}{0.004} + \frac{0.0055^2}{0.022}\right\} = 0.00031 \simeq 1/3200.$$

(c) Here, $\sigma^2 = 0.000 002$, $\tau^2 = 0.000 04$, $\mu = 0.016$, $z = 0.0195$, and $\bar{y}_1 - \bar{y}_2 = 0.001$. The value of the evidence is then, from (10.7)

$$\left(\frac{0.000 04}{0.000 002}\right)^{1/2} \exp\left\{-\frac{0.001^2}{0.000 004} + \frac{0.0035^2}{0.000 08}\right\} = 4.06.$$

(d) Here, $\sigma^2 = 0.000002$, $\tau^2 = 0.00004$, $\mu = 0.0.016$, $z = 0.015$, and $\bar{y}_1 - \bar{y}_3 = 0.01$. The value of the evidence is then, from (10.7)

$$\left(\frac{0.00004}{0.000002} \right)^{1/2} \exp \left\{ -\frac{0.01^2}{0.000004} + \frac{0.001^2}{0.00008} \right\} = 6.3 \times 10^{-11}.$$

Let $V_{a,1}$ and $V_{a,2}$ denote the values of the evidence for the elemental concentrations for aluminium in glass fragments from the same and different sources, respectively, and let $V_{b,1}$ and $V_{b,2}$ denote the values of the evidence for the elemental concentrations for barium in glass fragments from the same and different sources, respectively. Assume, for the moment, that the measurements of the aluminium and barium elemental concentrations are independent. Then, for the scenario where the two groups of fragments have been chosen to come from the same source, the overall value of the evidence is

$$V_{a,1} \times V_{b,1} = 3.25 \times 4.06 = 13.2.$$

For the scenario where the two groups of fragments have been chosen to come from different sources, the overall value of the evidence is

$$V_{a,2} \times V_{b,2} = (3.1 \times 10^{-4}) \times (6.3 \times 10^{-11}) = 2.0 \times 10^{-14}.$$

11.6   LIKELIHOOD RATIO FOR MULTIVARIATE NORMALITY WITH TWO SOURCES OF VARIATION

The previous three sections have discussed, with reference to a particular data set, examples where the within-group variance only is considered and the evidence is evaluated with a significance probability in Sections 11.3 and 11.4, the first of which considered univariate data, the second of which considered multivariate data. A univariate likelihood ratio which allows for between-group variance as well as within-group variance for univariate Normally distributed data was then discussed in Section 11.5. For completeness, the data are now analysed using a multivariate likelihood ratio which allows for between-group variance as well as within-group variance for multivariate Normally distributed data. A technical derivation of the component parts of the likelihood ratio is given in the Appendix. Here a numerical example will be given and the results can be compared with those of the previous three sections. As in Section 11.4, there are two scenarios. The first is where the control and recovered fragments are chosen to come from the same source. The second is where the control and recovered fragments are chosen to come from different sources. The data are the same as in the previous sections. Numerical values will be given for the various components of the formulae and then put together at the end to give the value of the evidence for each of the two scenarios.

First, consider the case where the control and recovered fragments are chosen to come from the same source.

The overall mean of the two sets of fragments is given by

$$\bar{\mathbf{y}} = (n_1\bar{\mathbf{y}}_1 + n_2\bar{\mathbf{y}}_2)/(n_1 + n_2) = (\bar{\mathbf{y}}_1 + \bar{\mathbf{y}}_2)/2 = \begin{pmatrix} 0.891 \\ 0.0195 \end{pmatrix}.$$

The matrices $\mathbf{S}_1$ and $\mathbf{S}_2$ measure the variability within the control and recovered fragments, respectively, and are derived from (11.19):

$$\mathbf{S}_1 = \begin{pmatrix} 0.002\,45 & 0.000\,14 \\ 0.000\,14 & 0.000\,008 \end{pmatrix}, \tag{11.5}$$

$$\mathbf{S}_2 = \begin{pmatrix} 0.014\,792 & 0.000\,172 \\ 0.000\,172 & 0.000\,002 \end{pmatrix}. \tag{11.6}$$

Define the matrices $\mathbf{D}_1 = \mathbf{U}/n_1$ and $\mathbf{D}_2 = \mathbf{U}/n_2$. For the example under discussion, $n_1 = n_2 = 2$ and so $\mathbf{D}_1 + \mathbf{D}_2 = \mathbf{U}$.

Consider the determination of the numerator.

The term H_3 (11.17) measures the difference between the means of the control and recovered fragments. In this example,

$$H_3 = (\bar{\mathbf{y}}_1 - \bar{\mathbf{y}}_2)^T(\mathbf{D}_1 + \mathbf{D}_2)^{-1}(\bar{\mathbf{y}}_1 - \bar{\mathbf{y}}_2) = (\bar{\mathbf{y}}_1 - \bar{\mathbf{y}}_2)^T(\mathbf{U}^{-1})(\bar{\mathbf{y}}_1 - \bar{\mathbf{y}}_2) = 0.6633,$$

as in (11.4).

The term H_2 (11.16) measures the difference between the overall mean of the control and recovered fragments and the overall population mean. In this example,

$$H_2 = (\bar{\mathbf{y}} - \boldsymbol{\mu})^T \left(\frac{\mathbf{U}}{n_1 + n_2} + \mathbf{C} \right)^{-1} (\bar{\mathbf{y}} - \boldsymbol{\mu}) = 0.7816.$$

The term H_1 (11.15) measures the variability in the measurements on the control and recovered fragments with respect to the within-group covariance matrix. In this example,

$$H_1 = \sum_{l=1}^{2} \text{trace}(\mathbf{S}_l\mathbf{U}^{-1}) = 12.2975,$$

where $\mathbf{S}_1$ and $\mathbf{S}_2$ are given in (11.5) and (11.6), respectively, and the trace of a matrix is as defined in the Appendix (Section 11.9.1).

The sum $H_1 + H_2 + H_3 = 13.7424$.

The other terms in (11.14) involve determinants, inverses and square roots of determinants of functions of the two covariance matrices $\mathbf{U}$ and $\mathbf{C}$. For this example, they take values:

$$| 2\pi\mathbf{U} |^{-\frac{1}{2}(n_1+n_2)} = 1.11 \times 10^{14},$$

$$| 2\pi\mathbf{C} |^{-1/2} = 5.62 \times 10^2,$$

$$| 2\pi\{(n_1+n_2)\mathbf{U}^{-1}+\mathbf{C}^{-1}\}^{-1} |^{1/2} = 7.11 \times 10^{-5}.$$

All these terms can be put together to give the numerator of the likelihood ratio, which is

$$(1.11 \times 10^{14}) \times (5.62 \times 10^2) \times (7.11 \times 10^{-5}) \times \exp(-13.7424/2)$$

$$= 4.6 \times 10^9. \tag{11.7}$$

Now, consider the denominator. The first term is given by (11.20). The expression

$$\frac{1}{2}(\bar{\mathbf{y}}_1 - \boldsymbol{\mu})^T \left(\frac{\mathbf{U}}{n_1} + \mathbf{C}\right)^{-1} (\bar{\mathbf{y}}_1 - \boldsymbol{\mu})$$

measures the distance of the mean of the control group from the overall mean. In this example it equals 0.35. The expression

$$\frac{1}{2}\text{trace}(\mathbf{S}_1\mathbf{U}^{-1})$$

compares the internal variation for the control group with the within-group covariance matrix. In this example, it equals 2.02.

The various terms involving determinants of the covariance matrices take the following values:

$$| 2\pi\mathbf{U} |^{-\frac{1}{2}n_1} = 1.05 \times 10^7,$$

$$| 2\pi\mathbf{C} |^{-1/2} = 5.62 \times 10^2,$$

$$| 2\pi(n_1\mathbf{U}^{-1}+\mathbf{C}^{-1})^{-1} |^{1/2} = 1.33 \times 10^{-4},$$

All these terms can be put together to give the first term in the denominator of the likelihood ratio, which is

$$(1.05 \times 10^7) \times (5.62 \times 10^2) \times (1.33 \times 10^{-4}) \times \exp(-2.37)$$

$$= 7.33 \times 10^4. \tag{11.8}$$

The second term is given by an analogous expression to (11.20). The expression

$$\frac{1}{2}(\bar{\mathbf{y}}_2 - \boldsymbol{\mu})^T \left(\frac{\mathbf{U}}{n_1} + \mathbf{C}\right)^{-1} (\bar{\mathbf{y}}_2 - \boldsymbol{\mu}) \tag{11.9}$$

measures the distance of the mean of the recovered group from the overall mean. In this example it equals 0.38. The expression

$$\frac{1}{2}\text{trace}(\mathbf{S}_2\mathbf{U}^{-1})$$

compares the internal variation for the recovered group with the within-group covariance matrix. In this example, it equals 4.13.

The various terms involving determinants of the covariance matrices are the same as for the first term in the denominator.

All these terms can be put together to give the second term in the denominator of the likelihood ratio, which is

$$(1.05 \times 10^7) \times (5.62 \times 10^2) \times (1.33 \times 10^{-4}) \times \exp(-4.51) = 8.63 \times 10^3.$$

The value of the evidence is then

$$V = \frac{4.6 \times 10^9}{7.33 \times 10^4 \times 8.63 \times 10^3} \simeq 7.3.$$

Now, consider the case where the control and recovered fragments are chosen to come from different sources. Many of the terms and calculations will be the same as or similar to the calculations done for the example where the control and recovered fragments were chosen to come from the same sources. This is because the control group has been chosen to be the same in both cases and because the covariance matrices $\mathbf{U}$ and $\mathbf{C}$ and the sample sizes n_1 and n_3 are the same.

The differences are the second within-group variability matrix $\mathbf{S}_3$, the overall mean $\bar{\mathbf{y}}$, and H_3 which measures the distance between the means of the control and recovered measurements.

The overall mean of the two sets of fragments is given by

$$\bar{\mathbf{y}} = (n_1\bar{\mathbf{y}}_1 + n_3\bar{\mathbf{y}}_3)/(n_1 + n_3) = \frac{1}{2}(\bar{\mathbf{y}}_1 + \bar{\mathbf{y}}_3) = \begin{pmatrix} 0.7995 \\ 0.0150 \end{pmatrix}.$$

The matrix $\mathbf{S}_3$ is given by

$$\mathbf{S}_3 = \begin{pmatrix} 0.00423 & 0.000092 \\ 0.000092 & 0.000002 \end{pmatrix}. \tag{11.10}$$

Consider the evaluation of the numerator. In this example,

$$H_3 = (\bar{\mathbf{y}}_1 - \bar{\mathbf{y}}_3)^T(\mathbf{D}_1 + \mathbf{D}_2)^{-1}(\bar{\mathbf{y}}_1 - \bar{\mathbf{y}}_3) = (\bar{\mathbf{y}}_1 - \bar{\mathbf{y}}_3)^T\mathbf{U}^{-1}(\bar{\mathbf{y}}_1 - \bar{\mathbf{y}}_3) = 50.1008.$$

This is much larger than H_3 for the example where the control and recovered fragments were taken to be from the same source. As the negative of this value is used in the exponential term, there will be a considerable reduction in the value of the evidence arising from this. Also, let $\bar{\mathbf{y}} = (n_1\bar{\mathbf{y}}_1 + n_3\bar{\mathbf{y}}_3)/(n_1 + n_2) = (\bar{\mathbf{y}}_1 + \bar{\mathbf{y}}_3)/2$. Then

$$H_2 = (\bar{\mathbf{y}} - \boldsymbol{\mu})^T \left(\frac{\mathbf{U}}{n_1 + n_3} + \mathbf{C} \right)^{-1} (\bar{\mathbf{y}} - \boldsymbol{\mu}) = 0.0642$$

and

$$H_1 = \text{trace}(\mathbf{S}_1\mathbf{U}^{-1}) + \text{trace}(\mathbf{S}_3\mathbf{U}^{-1}) = 6.164\,16.$$

Thus $H_1 + H_2 + H_3 = 56.329\,16$.

The other terms which involve determinants, inverses and square roots of determinants of functions of the two covariance matrices $\mathbf{U}$ and $\mathbf{C}$ are the same as before, namely,

$$|\,2\pi\mathbf{U}\,|^{-\frac{1}{2}(n_1+n_2)} = 1.11 \times 10^{14},$$

$$|\,2\pi\mathbf{C}\,|^{-1/2} = 5.62 \times 10^2,$$

$$|\,2\pi\{(n_1 + n_2)\mathbf{U}^{-1} + \mathbf{C}^{-1}\}^{-1}\,|^{1/2} = 7.11 \times 10^{-5}.$$

The numerator of the likelihood ratio is then

$$(1.11 \times 10^{14}) \times (5.62 \times 10^2) \times (7.11 \times 10^{-5}) \times \exp(-56.329\,16/2) = 2.60.$$

Now, consider the denominator. The first term is the same as (11.8) and is equal to 7.33×10^4.

The second component of the exponential part of the second term of the denominator is given by

$$\frac{1}{2}(\bar{\mathbf{y}}_3 - \boldsymbol{\mu})^T \left(\frac{\mathbf{U}}{n_3} + \mathbf{C} \right)^{-1} (\bar{\mathbf{y}}_3 - \boldsymbol{\mu}).$$

In this example it equals 0.45540. The expression

$$\frac{1}{2}\text{trace}(\mathbf{S}_3\mathbf{U}^{-1})$$

equals $1.063\,33$. The sum of these two terms is $1.518\,73$. As before, the various terms involving determinants of the covariance matrices are the same as for the first term in the denominator.

All these terms can be put together to give the second term in the denominator of the likelihood ratio, which is

$$(1.05 \times 10^7) \times (5.62 \times 10^2) \times (1.33 \times 10^{-4}) \times \exp(-1.518\,73) = 1.72 \times 10^5.$$

The value of the evidence is then

$$V = \frac{2.60}{7.33 \times 10^4 \times 1.72 \times 10^5} = 2.1 \times 10^{-10}.$$

11.7   CAVEAT LECTOR

Multivariate data are becoming more prevalent in forensic science. One common occurrence is the elemental composition of glass. A simple, indeed simplistic, example has been given in this chapter as an illustration of the ideas. Discrete data are not analysed here. The analysis of discrete data requires consideration of interactions in a multi-dimensional table and careful judgement is needed to decide amongst the possible margins of the table which are to be fixed for assessing the appropriate model to fit. Each dimension of the table corresponds to a different characteristic. Discrete characteristics may be qualitative or quantitative, and qualitative characteristics may be *nominal* or *ordinal*. It is also possible to have a multivariate problem in which some of the characteristics are discrete and some are continuous. This is not discussed here either.

The advantage of a multivariate analysis is that the associations amongst the various characteristics may be properly considered without making an assumption of independence which may be unwarranted.

Other methods for evaluating measurements of elemental concentrations developed at the turn of the century include a likelihood ratio approach involving the probability density of Hotelling's T^2 to a univariate kernel density estimate based on a transformation of multivariate data to a univariate function (Curran *et al.*, 1997a, b). Information content (Section 7.4) also provides for discrimination amongst glass fragments through the use of elemental concentrations (Koons and Buscaglia, 1999a). Koons and Buscaglia (1999a, b) criticise the use of databases for the post-data analysis of glass measurements because the characteristics vary over location and time. They are also concerned that the data are multivariate and thus require large numbers of samples to predict accurately probabilities for events with low frequencies of occurrence.

The capability of discrimination may be assessed through the use of type 1 and type 2 errors in hypothesis tests (Koons and Buscaglia, 2002). This is done through the use of *t*-tests on each variable and assessing the significance of each test. There is a Welch (1937) modification (Section 10.6) for unequal variances, and a Bonferroni correction to adjust the significance level for multivariate comparisons. The Bonferroni correction allows for the multiple comparisons which are made if a series of independent tests are conducted. For such comparisons, if an individual type 1 error probability α is used and p variables are compared, the overall type 1 error probability is approximately $p\alpha$. The overall error probability may be reduced to α by reducing the individual type 1 error probability to α/p. This reduction is known as the Bonferroni correction. Excellent discrimination with a type 2 error probability of 0.009% when using a type 1 error probability of 5% over ten variables is reported by Koons and Buscaglia (2002). This provides a very good answer to the pre-data question and shows that elemental concentrations provide excellent discrimination. The approach could be criticised for the use of a definite cut-off point of 5% for the type 1 error, but this would be unduly critical for a method which is obviously so successful.

A concern with the use of multivariate data is that the precision of the estimation of the parameters ($\boldsymbol{\mu}$, $\mathbf{U}$, $\mathbf{C}$) and associated probabilities and likelihood ratios may not be very precise. For example, if the concentrations of ten elements are to be considered, the number of parameters to be estimated is 10 (for $\boldsymbol{\mu}$) plus 55 variances and covariances for each of $\mathbf{U}$ and $\mathbf{C}$. This is 120 in total. Many data are needed to do this with high precision. For this reason, it is recommended that the likelihood ratio approach described here be restricted to only a few variables. For example, if only three variables are considered then 15 parameters are to be estimated.

Another concern is that some of the estimated probability densities are extreme, either extremely large or extremely small. Probability densities, unlike probabilities, may be greater than one. For example, in the likelihood ratio for the comparison of multivariate control and recovered data from the same source, the numerator (11.7) is 4.6×10^9. This is an extremely large number obtained by multiplying together several other numbers, one of which, 1.11×10^{14}, is even larger. The two terms in the denominator, (11.8) and (11.9), 7.33×10^4 and 8.63×10^3, are moderately large. The result, 7.3, is the result of multiplications and divisions of several large numbers, and great care has to be taken in checking the precision of the result of such operations. Analogously, for the calculation of the likelihood ratio for the comparison of multivariate control and recovered data from different sources, the example provides a very small number, 2.1×10^{-10}.

The precision of these results should be checked before too much reliance is placed on them. The data used to motivate the example on which this chapter relies (Curran *et al.*, 1997a) were presented to three significant figures. The final results, 7.3 and 2.1×10^{-10}, are presented to only two significant figures. It is suggested that a check on the precision be made by adjusting the input data by a small amount, say 1%, and working through the analysis again. If the result is changed by a similar amount, say 1%, then one may be reasonably confident in its precision. However, if the result is changed by a large amount then one's confidence will be correspondingly reduced.

11.8 SUMMARY

Four methods of analysing multivariate data have been compared. Two use significance probabilities. These are not recommended because of the artificial nature of the significance level and the dichotomy between results which are just significant and those that are just not. There is also a problem in that the result is not easily assimilated into an overall assessment of the value of the evidence in total which has been presented. In addition, the first method used two independent *t*-tests without consideration as to whether the variables were independent.

The other two methods used likelihood ratios. This approach does not have an artificial cut-off and the result is easily assimilated. The first method, however, only considered within-group variability. The second method considers both within-group and between-group variability. An extension to the second method allows for the between-group distribution to be non-Normal and models the distribution with a kernel density estimate (Aitken and Lucy, 2004).

11.9 APPENDIX

11.9.1 Matrix terminology

A brief introduction to matrices is given here. If further details are desired, good references are Mardia *et al.* (1979) and Graybill (1969).

A matrix **A** is a rectangular array of numbers. If **A** has r rows and c columns it is said to be of *order* $r \times c$ (read as 'r by c'). For example, r measurements on c characteristics (or variables) may be arranged in this way. Another example is the matrix of variances and covariances Σ, introduced in Section 2.4.6 and known as the *covariance matrix*. If $r = c$, the matrix is said to be *square*, and an example of a square matrix is the covariance matrix. The matrices **U** and **C** in the example discussed in this chapter are 2×2 matrices as the number of variables in the example is 2. The diagonal terms are the variances and the off-diagonal terms (those in the top right-hand and bottom left-hand corners) are the covariances. The covariances are equal since the covariance between variables 1 and 2 is the same as the covariance between variables 2 and 1. For the purpose of discussion within this Appendix, a 2×2 matrix, denoted **A**, will be used with cell entries given by

$$\mathbf{A} = \begin{pmatrix} a_{11} & a_{12} \\ a_{21} & a_{22} \end{pmatrix}. \tag{11.11}$$

The subscripts denote the cell in which the item is located. Thus a_{ij} is the member of the (i, j)th cell, the cell in row i and column j of the matrix. A matrix **A** is sometimes denoted $\{a_{ij}\}$.

The trace of a matrix

The *trace* of a matrix is the sum of the terms in the leading diagonal, that is the diagonal which runs from the top left-hand corner to the bottom right-hand corner of the matrix. Thus, the trace of **A** is $a_{11} + a_{22}$. This is denoted trace(**A**). It only exists if the matrix is square.

The transpose of a matrix

The *transpose* of an $r \times c$ matrix $\mathbf{A} = \{a_{ij}\}, (i = 1, \ldots, r; j = 1, \ldots, c)$ is a $c \times r$ matrix $\mathbf{B} = \{b_{ji}\}, (j = 1, \ldots, c; i = 1, \ldots, r)$ such that $b_{ji} = a_{ij}$. The element in the jth row and ith column of **B** is the element in the ith row and jth column of A. The transpose of **A** is denoted $\mathbf{A}^T$.

Let **A** be a 3×2 matrix

$$\mathbf{A} = \begin{pmatrix} a_{11} & a_{12} \\ a_{21} & a_{22} \\ a_{31} & a_{32} \end{pmatrix}.$$

Then

$$\mathbf{A}^T = \begin{pmatrix} a_{11} & a_{21} & a_{31} \\ a_{12} & a_{22} & a_{32} \end{pmatrix}.$$

A matrix **A** is symmetric if it is equal to its transpose, $\mathbf{A} = \mathbf{A}^T$. Such a matrix is of necessity square, with $r = c$. An $r \times 1$ matrix is a column vector and its transpose is a $1 \times r$ row vector. For $r = 3$,

$$\mathbf{x} = \begin{pmatrix} x_1 \\ x_2 \\ x_3 \end{pmatrix}$$

and

$$\mathbf{x}^T = \begin{pmatrix} x_1 & x_2 & x_3 \end{pmatrix}.$$

Addition of two matrices

Two matrices of the same order may be added together to give a third matrix of the same order. Let **B** be a matrix

$$\mathbf{B} = \begin{pmatrix} b_{11} & b_{12} \\ b_{21} & b_{22} \end{pmatrix}.$$

Then the matrix $\mathbf{A} + \mathbf{B}$ is obtained by adding corresponding cell entries together.

$$\mathbf{A} + \mathbf{B} = \begin{pmatrix} a_{11} + b_{11} & a_{12} + b_{12} \\ a_{21} + b_{21} & a_{22} + b_{22} \end{pmatrix}.$$

Note that $\mathbf{A} + \mathbf{B} = \mathbf{B} + \mathbf{A}$.

Determinant of a matrix

The determinant of a square matrix **A** is a non-negative number, denoted $| \mathbf{A} |$. For the 2×2 matrix **A** in (11.11), the determinant is the difference between the product of the leading diagonal terms and the product of the off-diagonal terms. Thus

$$| \mathbf{A} | = a_{11}a_{22} - a_{12}a_{21}.$$

Care has to be taken in understanding the notation. For a matrix, the symbols $|\cdot|$ denote the determinant. For a real number, the symbols $|\cdot|$ denote the positive value of the number. Multiplication of a matrix by a constant, c say, results in a matrix in which every cell is multiplied by c. Thus

$$c\mathbf{A} = \begin{pmatrix} ca_{11} & ca_{12} \\ ca_{21} & ca_{22} \end{pmatrix}.$$

and $|c\mathbf{A}| = c^2 |\mathbf{A}|$. In general, for a $p \times p$ matrix, $|c\mathbf{A}| = c^p |\mathbf{A}|$.

Matrix multiplication

Examples are given here of how matrices, row and column vectors may be multiplied together for 2×2 matrices, and row and column vectors with two components. First, consider the multiplication of a vector and a matrix. The column vector is $\mathbf{x}$, where

$$\mathbf{x} = \begin{pmatrix} x_1 \\ x_2 \end{pmatrix},$$

and the matrix $\mathbf{A}$ is as in (11.11). The order of multiplication of a pair of matrices is important and the number of columns in the first member of the pair must equal the number of rows in the second member. The outcome is a matrix in which the number of rows equals the number of rows in the first member of the pair of matrices and the number of columns equals the number of columns in the second member of the pair of matrices. Thus, multiplication of an $r \times c$ matrix and a $c \times p$ matrix results in an $r \times p$ matrix.

Multiplication of the 1×2 row vector $\mathbf{x}^T$ and the 2×2 matrix $\mathbf{A}$, written as $\mathbf{x}^T\mathbf{A}$ gives a 1×2 row vector. Note that a row or column vector may be thought of as a matrix with only one row or one column. The row vector $\mathbf{x}^T\mathbf{A}$ is

$$(x_1 a_{11} + x_2 a_{21}, x_1 a_{12} + x_2 a_{22}).$$

The members of the row vector multiply the corresponding members of the columns of $\mathbf{A}$ and the resultant products are then summed.

Multiplication of the 2×2 matrix $\mathbf{A}$ and the 2×1 column vector $\mathbf{x}$ written as $\mathbf{Ax}$ gives a (2×1) column vector. The column vector $\mathbf{Ax}$ is

$$\mathbf{Ax} = \begin{pmatrix} a_{11}x_1 + a_{12}x_2 \\ a_{21}x_1 + a_{22}x_2 \end{pmatrix}.$$

This is not equal to $\mathbf{x}^T\mathbf{A}$. The expression $\mathbf{x}^T\mathbf{Ax}$ is

$$(x_1 a_{11} + x_2 a_{21}, x_1 a_{12} + x_2 a_{22}) \begin{pmatrix} x_1 \\ x_2 \end{pmatrix},$$

which is equal to

$$x_1^2 a_{11} + x_1 x_2 a_{21} + x_1 x_2 a_{12} + x_2^2 a_{22} = x_1^2 a_{11} + x_1 x_2 (a_{21} + a_{12}) + x_2^2 a_{22}.$$

This is simply a number, not a matrix. A 2×2 symmetric matrix has $a_{12} = a_{21}$. Thus, for such a matrix

$$\mathbf{x}^T \mathbf{A} \mathbf{x} = x_1^2 a_{11} + 2 x_1 x_2 a_{12} + x_2^2 a_{22}.$$

For the multiplication of two matrices, the rows of the first matrix and the columns of the second are multiplied together component by component. Thus

$$\mathbf{AB} = \begin{pmatrix} a_{11}b_{11} + a_{12}b_{21} & a_{11}b_{12} + a_{12}b_{22} \\ a_{21}b_{11} + a_{22}b_{21} & a_{21}b_{12} + a_{22}b_{22} \end{pmatrix},$$

$$\mathbf{BA} = \begin{pmatrix} a_{11}b_{11} + a_{21}b_{12} & a_{12}b_{11} + a_{22}b_{12} \\ a_{11}b_{21} + a_{12}b_{22} & a_{12}b_{21} + a_{22}b_{22} \end{pmatrix}. \tag{11.12}$$

In general, $\mathbf{AB} \neq \mathbf{BA}$. However if $\mathbf{A}$ and $\mathbf{B}$ are symmetric then $a_{12} = a_{21}$, $b_{12} = b_{21}$ and $\mathbf{AB} = \mathbf{BA}$. For the product $\mathbf{AB}$, $\mathbf{A}$ is said to *pre-multiply* $\mathbf{B}$ and $\mathbf{B}$ is said to *post-multiply* $\mathbf{A}$.

The inverse of a matrix

The square matrix $\mathbf{I}$ defined as

$$\mathbf{I} = \begin{pmatrix} 1 & 0 \\ 0 & 1 \end{pmatrix}$$

is known as the *identity matrix*. This is because pre- or post-multiplication of another square matrix $\mathbf{A}$ by $\mathbf{I}$ leaves $\mathbf{A}$ unchanged: $\mathbf{AI} = \mathbf{IA} = \mathbf{A}$. This may be checked from (11.12). The existence of an identity matrix then leads naturally to the concept of an inverse of a matrix. The inverse of a square matrix $\mathbf{A}$ is defined as that matrix, denoted $\mathbf{A}^{-1}$, which when used to pre- or post-multiply $\mathbf{A}$ gives as a product the identity matrix $\mathbf{I}$. Thus $\mathbf{AA}^{-1} = \mathbf{A}^{-1}\mathbf{A} = \mathbf{I}$. For $\mathbf{A}$ as in (11.11), the inverse of $\mathbf{A}$ is given by

$$\mathbf{A}^{-1} = \frac{1}{|\mathbf{A}|} \begin{pmatrix} a_{22} & -a_{12} \\ -a_{21} & a_{11} \end{pmatrix}. \tag{11.13}$$

Matrix multiplication $\mathbf{AA}^{-1}$ and $\mathbf{A}^{-1}\mathbf{A}$ verifies that the products are $\mathbf{I}$ and that the matrix given in (11.13) satisfies the definition of an inverse.

Note that an inverse only exists if the determinant $|\mathbf{A}|$ is non-zero. A matrix for which the determinant is zero is said to be *singular* and an inverse does not exist. This is the matrix equivalent of the non-existence of the reciprocal of the number 0. The two rows and the two columns of a singular 2×2 matrix are equal or proportional. Note, also, that it makes no sense to consider division by a matrix. The operation with matrices which is equivalent to division by a number is multiplication by the inverse of a matrix.

11.9.2 Determination of a likelihood ratio with an assumption of Normality

The model as discussed here assumes two sources of variation, that between replicates within the same group or source (known as within-group variation) and that between groups or sources (known as between-group variation). It is assumed that both the within-group variation and the between-group variation are Normally distributed. The number of variables to be considered is denoted p. These need not be independent. The number of groups is m and the number of items within groups is n, constant over all groups. (The mathematics is more complicated but feasible if there are different numbers of measurements within groups.)

Within-group. Denote the mean vector within group i by $\boldsymbol{\theta}_i$ and the within-group covariance matrix by $\mathbf{U}$. The subscript i is omitted from the covariance matrix to indicate that it is assumed that the within-group variability is constant over all groups. This is an extension of the assumption made in standard univariate analysis of variance techniques. Then, given $\boldsymbol{\theta}_i$ and $\mathbf{U}$, the distribution of $\mathbf{X}_{ij}$ (a $p \times 1$ column vector indicating the p characteristics of the jth member of the ith group) is taken to be Normal, where the notation is the same as in Section 2.4.6:

$$(\mathbf{X}_{ij} \mid \boldsymbol{\theta}_i, \mathbf{U}) \sim N(\boldsymbol{\theta}_i, \mathbf{U}), \quad i = 1, \ldots, m; \ j = 1, \ldots, n.$$

Between-group. Denote the mean vector between groups by $\boldsymbol{\mu}$ and the between-group covariance matrix by $\mathbf{C}$. The distribution of the $\boldsymbol{\theta}_i$, as measures of between-source variability, is taken to be Normal:

$$(\boldsymbol{\theta}_i \mid \boldsymbol{\mu}, \mathbf{C}) \sim N(\boldsymbol{\mu}, \mathbf{C}).$$

This distribution arises as the model is assumed to be a so-called *random effects* model. The different groups in the population database are thought of as a *random* sample from a larger population (or super-population). Thus, measurements on the groups in the (population) database are a random sample from a larger population and thus have variability. The term 'multivariate random effects' arises because the data are multivariate.

There are multivariate data from a crime scene, which are assumed to come from one source, and multivariate data from a suspect which are assumed to come from one source. These sources may or may not be the same source. A likelihood ratio is derived to measure the support for the proposition that the sources are the same, as opposed to the proposition that the sources are different.

Let there be n_c observations at the crime scene, with vectors of measurements $\mathbf{Y}_{11}, \ldots, \mathbf{Y}_{1n_c}$ and n_s observations from a suspect, with vectors of measurements

$\mathbf{Y}_{21}, \ldots, \mathbf{Y}_{2n_s}$, where n_c is not necessarily equal to n_s. The data will be referred to as crime and suspect data, respectively. This terminology then covers the two situations in which the control data may or may not be those at the crime scene and the recovered data may or may not be those associated with the suspect. The means of these two sets of observations are

$$\bar{\mathbf{Y}}_1 = \sum_{j=1}^{n_c} \mathbf{Y}_{1j}/n_c \quad \text{and} \quad \bar{\mathbf{Y}}_2 = \sum_{j=1}^{n_s} \mathbf{Y}_{2j}/n_s.$$

The distributions of the means ($\bar{\mathbf{Y}}_l$; $l = 1, 2$) of the measurements on the crime and suspect data, conditional on the source (crime or suspect), are also taken to be Normal, with means $\boldsymbol{\theta}_l$ and covariance matrix $\mathbf{D}_l$, where $\mathbf{D}_1 = n_c^{-1}\mathbf{U}$ and $\mathbf{D}_2 = n_s^{-1}\mathbf{U}$. Thus

$$(\bar{\mathbf{Y}}_l \mid \boldsymbol{\theta}_l, \mathbf{D}_l) \sim N(\boldsymbol{\theta}_l, \mathbf{D}_l); \qquad l = 1, 2.$$

Then it can be shown that

$$(\bar{\mathbf{Y}}_l \mid \boldsymbol{\mu}, \mathbf{C}, \mathbf{D}_l) \sim N(\boldsymbol{\mu}, \mathbf{C} + \mathbf{D}_l); \qquad l = 1, 2.$$

This is the multivariate generalisation of (10.6).

The value of the evidence ($\bar{\mathbf{y}}_l$; $l = 1, 2$) is then the ratio of two probability density functions, evaluated at the point ($\bar{\mathbf{y}}_1, \bar{\mathbf{y}}_2$). For the calculations in the numerator it is assumed that the crime and suspect items come from the same source and the means $\boldsymbol{\theta}_1$ and $\boldsymbol{\theta}_2$ are equal. For the calculations in the denominator it is assumed that the crime and suspect items come from different sources and the means $\boldsymbol{\theta}_1$ and $\boldsymbol{\theta}_2$ are not equal.

First, consider the numerator, where $\boldsymbol{\theta}_1 = \boldsymbol{\theta}_2 = \boldsymbol{\theta}$ say, which is unknown. The parameter $\boldsymbol{\theta}$ can be eliminated by integration, analogous to the approach of Section 10.1, to obtain a probability density function $f_0(\bar{\mathbf{y}}_1, \bar{\mathbf{y}}_2 \mid \boldsymbol{\mu}, \mathbf{D}_1, \mathbf{D}_2, \mathbf{C})$ given by

$$\int_{\boldsymbol{\theta}} f(\bar{\mathbf{y}}_1 \mid \boldsymbol{\theta}, \mathbf{D}_1).f(\bar{\mathbf{y}}_2 \mid \boldsymbol{\theta}, \mathbf{D}_2)f(\boldsymbol{\theta} \mid \boldsymbol{\mu}, \mathbf{C})d\boldsymbol{\theta}.$$

The component probability density functions are multivariate Normal. The expressions for $f(\bar{\mathbf{y}}_1 \mid \boldsymbol{\theta}, \mathbf{D}_1), f(\bar{\mathbf{y}}_2 \mid \boldsymbol{\theta}, \mathbf{D}_2)$ and $f(\boldsymbol{\theta} \mid \boldsymbol{\mu}, \mathbf{C})$ above are obtained by appropriate substitutions in the general formula (2.25).

The integral can then be shown to be equal to

$$f_0(\mathbf{y}_1, \mathbf{y}_2 \mid \boldsymbol{\mu}, \mathbf{U}, \mathbf{C}) = \mid 2\pi\mathbf{U} \mid^{-\frac{1}{2}(n_1+n_2)} \mid 2\pi\mathbf{C} \mid^{-1/2} \mid 2\pi\{(n_1+n_2)\mathbf{U}^{-1} + \mathbf{C}^{-1}\}^{-1} \mid^{1/2}$$

$$\times \exp\left\{-\frac{1}{2}(H_1 + H_2 + H_3)\right\} \tag{11.14}$$

where

$$H_1 = \sum_{l=1}^{2} \text{trace}(\mathbf{S}_l\mathbf{U}^{-1}), \tag{11.15}$$

$$H_2 = (\bar{\mathbf{y}} - \boldsymbol{\mu})^T \left(\frac{\mathbf{U}}{n_1 + n_2} + \mathbf{C} \right)^{-1} (\bar{\mathbf{y}} - \boldsymbol{\mu}), \tag{11.16}$$

$$H_3 = (\bar{\mathbf{y}}_1 - \bar{\mathbf{y}}_2)^T (\mathbf{D}_1 + \mathbf{D}_2)^{-1} (\bar{\mathbf{y}}_1 - \bar{\mathbf{y}}_2), \tag{11.17}$$

$$\bar{\mathbf{y}} = (n_1\bar{\mathbf{y}}_1 + n_2\bar{\mathbf{y}}_2)/(n_1 + n_2), \tag{11.18}$$

$$\mathbf{S}_l = \sum_{j=1}^{n_l} (\mathbf{y}_{lj} - \bar{\mathbf{y}}_l)(\mathbf{y}_{lj} - \bar{\mathbf{y}}_l)^T \tag{11.19}$$

(Aitken and Lucy, 2004). The exponential term is a combination of three terms: H_3, which accounts for the difference $(\bar{\mathbf{y}}_1 - \bar{\mathbf{y}}_2)$ between the means of the measurements on the control and recovered items; H_2, which accounts for their rarity (as measured by the distance of the mean weighted by sample sizes from $\boldsymbol{\mu}$); and H_1, which accounts for internal variability.

Second, consider the denominator, where $\boldsymbol{\theta}_1 \neq \boldsymbol{\theta}_2$. The probability density function $f_1(\mathbf{y}_1, \mathbf{y}_2 \mid \boldsymbol{\mu}, \mathbf{U}, \mathbf{C})$ is given by

$$\int_{\boldsymbol{\theta}} \{f(\mathbf{y}_1 \mid \boldsymbol{\theta}, \mathbf{U}) \times f(\boldsymbol{\theta} \mid \boldsymbol{\mu}, \mathbf{C})\} d\boldsymbol{\theta} \times \int_{\boldsymbol{\theta}} \{f(\mathbf{y}_2 \mid \boldsymbol{\theta}, \mathbf{D}_2) \times f(\boldsymbol{\theta} \mid \boldsymbol{\mu}, \mathbf{C})\} d\boldsymbol{\theta},$$

where $\mathbf{y}_1$ and $\mathbf{y}_2$ are taken to be independent as the data are assumed to be from different sources. The integral

$$\int_{\boldsymbol{\theta}} \{f(\mathbf{y}_1 \mid \boldsymbol{\theta}, \mathbf{U}) \times f(\boldsymbol{\theta} \mid \boldsymbol{\mu}, \mathbf{C})\} d\boldsymbol{\theta}$$

can be shown to be equal to

$$f(\mathbf{y}_1 \mid \boldsymbol{\mu}, \mathbf{U}, \mathbf{C}) = | 2\pi\mathbf{U} |^{-\frac{1}{2}n_1} | 2\pi\mathbf{C} |^{-1/2} | 2\pi(n_1\mathbf{U}^{-1} + \mathbf{C}^{-1})^{-1} |^{1/2}$$

$$\times \exp\left\{ -\frac{1}{2}\text{trace}(\mathbf{S}_1\mathbf{U}^{-1}) - \frac{1}{2}(\bar{\mathbf{y}}_1 - \boldsymbol{\mu})^T \left(\frac{\mathbf{U}}{n_1} + \mathbf{C} \right)^{-1} (\bar{\mathbf{y}}_1 - \boldsymbol{\mu}) \right\}, \tag{11.20}$$

with an analogous result for

$$\int_{\boldsymbol{\theta}} \{f(\mathbf{y}_2 \mid \boldsymbol{\theta}, \mathbf{U}) \times f(\boldsymbol{\theta} \mid \boldsymbol{\mu}, \mathbf{C})\} d\boldsymbol{\theta}.$$

The value of the evidence is the ratio of $f_0(\mathbf{y}_1, \mathbf{y}_2 \mid \boldsymbol{\mu}, \mathbf{U}, \mathbf{C})$ to the product of $\int_{\boldsymbol{\theta}} \{f(\mathbf{y}_1 \mid \boldsymbol{\theta}, \mathbf{U}) \times f(\boldsymbol{\theta} \mid \boldsymbol{\mu}, \mathbf{C})\} d\boldsymbol{\theta}$ and $\int_{\boldsymbol{\theta}} \{f(\mathbf{y}_2 \mid \boldsymbol{\theta}, \mathbf{U}) \times f(\boldsymbol{\theta} \mid \boldsymbol{\mu}, \mathbf{C})\} d\boldsymbol{\theta}$.

12

Fibres

12.1 INTRODUCTION

The *transfer* and *presence by chance* of material are fundamental factors when the evidence is assessed under *activity level* propositions (see Section 7.2.1; see also and Cook *et al.*, 1993; Evett, 1984). Various technical information that the scientist collects during the analysis, such as (a) the number of fibres recovered, (b) the materials involved (the type of material of the receptor and of the potential source) and (c) the intensity of the *action* under consideration, are essential for the estimation of the factors appearing in the likelihood ratio. Procedures for their estimation have been reviewed by Chabli (2001).

The influence of the factors is easily shown by the values of the likelihood ratio which can be obtained in different scenarios. Values vary in a wide range (from support for the prosecutor's proposition to support for that of the defence). It has been shown (Champod and Taroni, 1999) that likelihood ratios can attain values greater than the 'classical' $1/\gamma$, where γ represents the relative frequency of the features of interest in a relevant population, and values less than 1.

12.2 LIKELIHOOD RATIOS IN SCENARIOS INVOLVING FIBRES

Various scenarios, which are principally assessed under propositions described at the activity level, were developed in Champod and Taroni (1999). Propositions of interest were:

- H_p, the suspect sat on the driver's seat of the stolen car;
- H_{d1}, the suspect never sat on the driver's seat of the stolen car,

or

- H_p, the suspect sat on the driver's seat of the stolen car;
- H_{d2}, the suspect sat on this seat one week ago for legitimate reasons.

Statistics and the Evaluation of Evidence for Forensic Scientists: Second Edition
C.G.G. Aitken and F. Taroni © 2004 John Wiley & Sons, Ltd   ISBN: 0-470-84367-5

In another context, the propositions could be:

- H_p, the victim sat on the passenger's seat of the suspect's car;
- H_d, the victim never sat on the passenger's seat of the suspect's car.

Interestingly, these kinds of propositions allow the scientist to illustrate their impact on the evaluation of the evidence by determining the influence of probabilities such as the *transfer* and *background probabilities* on the likelihood ratio and showing that the value of the likelihood ratio at the activity level can (under certain assumptions) become $1/\gamma$. Consider the following examples.

12.2.1 Fibres evidence left by an offender

Imagine the following scenario: a stolen car is used in a robbery on the day of its theft. One hour later, the car is abandoned. During the night the stolen vehicle is found by the police. On the polyester seats (lower and upper back), n extraneous textile fibres are collected. The day following the robbery, a suspect is apprehended. His red woollen pullover is confiscated and submitted to the laboratory. On the driver's seat one group of $n = 170$ foreign fibres has been collected.

In the notation of Section 3.5.1, the evidence Ev is the material from the car's seat M_c (where c denotes *crime scene*) and from the suspect's pullover M_s (where s denotes *suspect*) and the characteristics E_c and E_s of the material. These characteristics will be denoted y (for recovered E_c in this context) and x (for control E_s in this context) respectively. Thus, the group of $n = 170$ red woollen fibres recovered is described by a set y of extrinsic (physical attributes such as quantity and position) and intrinsic characteristics (chemical or physical descriptors such as analytical results), and the suspect's red woollen pullover generates known fibres described by a set x of intrinsic characteristics. (For a discussion on extrinsic and intrinsic characteristics, see also Section 8.3.3.)

The likelihood ratio is expressed as follows:

$$V = \frac{Pr(y, x \mid H_p, I)}{Pr(y, x \mid H_d, I)}$$

where

- H_p, the suspect sat on the driver's seat of the stolen car;
- H_d, the suspect never sat on the driver's seat of the stolen car.

Note that H_d implies that *another person* sat on the driver's seat of the stolen car. This point is important in the assessment of the *transfer* probabilities, as will be seen later. The previous equation can be expanded using the third law of probability (1.7):

$$V = \frac{Pr(y, x \mid H_p, I)}{Pr(y, x \mid H_d, I)} = \frac{Pr(y \mid x, H_p, I)}{Pr(y \mid x, H_d, I)} \times \frac{Pr(x \mid H_p, I)}{Pr(x \mid H_d, I)}. \tag{12.1}$$

It is reasonable to assume that the probability of the characteristics of the suspect's pullover, x, does not depend on whether or not the suspect sat on the driver's seat of the stolen car. So, the second ratio of the right-hand side of (12.1) equals 1 and the likelihood ratio is reduced to

$$V = \frac{Pr(y \mid x, H_p, I)}{Pr(y \mid x, H_d, I)}.$$

It is commonly accepted that the denominator of the likelihood ratio is reduced to $Pr(y \mid H_d, I)$, independent of the characteristics of the control object (the suspect's pullover). Note that this is not so in DNA evidence (Sections 13.4 and 13.6), where the fact that a person is known to share the stain characteristics with those of a crime stain influences the estimate of the conditional probability called *random match probability*. For a comment on this point applied to transfer evidence other than DNA, see Aitken and Taroni (1997).

The scientist has to assess (a) the probability of the observed characteristics of the recovered fibres, y, given that the suspect sat on the driver's seat of the stolen car and his pullover shares the same forensic characteristics as the fibres found on the car seat (the numerator of V), and (b) the probability of the observed characteristics of the recovered fibres, y, given that the suspect never sat on the driver's seat of the stolen car (the denominator of V). In order to assess the evidence under these two propositions, it is important to note that the scientist is interested in propositions which imply an activity (the act of sitting on a driver's seat) and then considers the logical consequence of this activity.

Imagine a person (the suspect or the offender) who sat in the driver's seat. He (or rather his clothes) had contact with the seat, so that fibres from the clothes will have been transferred to the seat. For the successful recovery and analysis of these fibres, it is necessary for them to have persisted on the seat. There are two main explanations for the presence of the evidence on the seat:

(a) The recovered group of $n = 170$ fibres was transferred, has persisted and was successfully recovered from the driver's seat. Under this explanation, the driver's seat did not have this group of fibres before the commission of the crime. Call this event T_n (or T_{170} in this case).

(b) The recovered group of $n = 170$ fibres was not transferred, has not persisted nor was recovered from the driver's seat. Under this explanation, the recovered fibres are unconnected with the action under investigation: the fibres were on the driver's seat before. Call this event T_0.

These explanations may be considered as association propositions (see Section 9.5.3). Inclusion of these two association propositions and omission of the background information I for the sake of simplicity leads to:

$$V = \frac{Pr(y \mid x, H_p, T_{170})Pr(T_{170} \mid x, H_p) + Pr(y \mid x, H_p, T_0)Pr(T_0 \mid x, H_p)}{Pr(y \mid H_d, T_{170})Pr(T_{170} \mid H_d) + Pr(y \mid H_d, T_0)Pr(T_0 \mid H_d)}. \quad (12.2)$$

Consideration then needs to be given to the eight conditional probabilities forming the likelihood ratio. $Pr(y \mid x, H_p, T_{170})$ represents the probability of observing a group of 170 red woollen fibres on the car seat given that the suspect wore a red woollen pullover, that he sat on the driver's seat of the stolen car and that the group of fibres was transferred, persisted and was recovered successfully during the activity. If the suspect sat on the driver's seat and the group has been transferred, this means that it was not there before the activity. So, this probability is $1 \times b_0$, where 1 is the probability of a match and b_0 is the probability of the presence by chance of no groups.

$Pr(T_{170} \mid x, H_p)$ represents the probability that a group of 170 red woollen fibres was transferred, persisted and was recovered successfully from the driver's seat given that the suspect sat on the driver's seat of the stolen car. This represents the probability, say t_{170}, that the fibres had been transferred, had remained and were recovered from the suspect's pullover. This probability depends on physical characteristics (e.g., sheddability and construction) of the suspect's pullover (Roux *et al.*, 1996). It is assumed that the characteristics are from the control group because the scientist assesses the probability under H_p.

$Pr(y \mid x, H_p, T_0)$ is the probability that a group of 170 red woollen fibres are observed on the driver's seat given that the suspect wore a red woollen pullover, that he sat on the driver's seat of the stolen car and that this group of fibres was not transferred, did not persist nor was recovered successfully during the activity. If the group has not been transferred, this means that it was present on the seat before the activity. Let $b_{1,m} \times \gamma$ represent the probability of the chance occurrence of a group of m (a comparable number of) fibres on the driver's seat $(b_{1,m})$ linked to the estimated frequency γ of the characteristics y.

$Pr(T_0 \mid x, H_p)$ represents the probability that no group of fibres was transferred, persisted or was recovered successfully from the suspect's pullover to the driver's seat. This probability, t_0, is estimated given that the suspect sat on the driver's seat, H_p.

The numerator of the likelihood ratio is then $b_0 t_{170} + b_{1,m} \gamma t_0$.

Consider now the terms in the denominator of (12.2). $Pr(y \mid H_d, T_{170})$ represents the probability of observing a group of 170 red woollen fibres given that the suspect never sat on the driver's seat of the stolen car and that the group of fibres was transferred, persisted and was recovered successfully during the activity. If the suspect never sat on the driver's seat and the group has been transferred, this means the driver's seat did not have this group of fibres before the commission of the crime and the event of the shared characteristics is one of chance. This probability is $b_0 \times \gamma$.

$Pr(T_{170} \mid H_d)$ represents the probability that a group of 170 red woollen fibres was transferred, persisted and was recovered successfully from the driver's seat given that the suspect never sat on the driver's seat of the stolen car. This means that the probability, say t'_{170}, has to be estimated given that the fibres have been transferred from the *offender's* garment (not the suspect's pullover). This probability depends on physical characteristics of an *unknown garment*, the one wore by the offender. This probability is different from the probability p of Section 9.5.4 (d) in which p denotes the probability the material (the stain in Section 9.5.4) would have been left by the suspect even though the suspect was innocent of the offence. The suspect is a known person. For a probability like t'_{170}, the person, or garment here, is not known.

$Pr(y \mid H_d, T_0)$ is the probability that a group of 170 red woollen fibres is observed on the driver's seat given that the suspect never sat on the driver's seat and that this group of fibres was not transferred, did not persist nor was recovered successfully during the activity. If the group was not transferred, it was present on the seat before the commission of the crime. The probability of the chance occurrence of a group of foreign fibres on the driver's seat linked to the estimated frequency of the characteristics y is then $b_{1,m} \times \gamma$.

$Pr(T_0 \mid H_d)$ represents the probability that no group of fibres was transferred, persisted or was recovered from the offender's garments to the driver's seat. This probability, t'_0, is estimated given that the suspect never sat on the driver's seat and, thus, another individual sat in the stolen car, H_d.

The denominator of the likelihood ratio is $b_0 \gamma t'_{170} + b_{1,m} \gamma t'_0$. The likelihood ratio (12.2) is then

$$V = \frac{b_0 t_{170} + b_{1,m} \gamma t_0}{b_0 \gamma t'_{170} + b_{1,m} \gamma t'_0}. \tag{12.3}$$

This expression illustrates that, theoretically, $Pr(T_{170} \mid x, H_p) \neq Pr(T_{170} \mid H_d)$ and $Pr(T_0 \mid x, H_p) \neq Pr(T_0 \mid H_d)$. Practically, as the probabilities are estimated throughout controlled experiments using the garments involved under propositions H_p and H_d it is reasonable that the estimates are different. Multiple and complex variables involved in transfer, persistence and recovery phenomena should directly be taken into account for the estimation of t_{170}, t_0, t'_{170} and t'_0. The aim is to try to reproduce the alleged case. A modelling technique for the assessment of transfer probabilities in glass fragments scenarios has been suggested (Curran *et al.*, 1998a; see also Section 8.3.3). The technique can also be easily used to estimate probabilities in fibres scenarios as given in Champod and Taroni (1999). The use of surveys to obtain estimates for different scenarios involving fibres has been proposed by Siegel (1997) and by Roux *et al.* (1999).

Background probabilities, $b_{g,m}$ may be estimated from data obtained in surveys where groups of extraneous fibres were recovered from surfaces of interest. The probabilities derived depend on the types of fibres considered. In fact, the estimates are influenced by the conditions of transfer and sheddability of the potential garments involved (Roux and Margot, 1997). Sometimes 'target fibres studies' are performed. These studies enable the estimation of $b_{1,m} \gamma$ directly (Palmer and Chinherende, 1996). The probabilities b_0 and $b_{1,m}$ are considered as mutually exclusive parts of the event 'having 0, 1 or more groups of extraneous fibres which can be distinguished from the garments of the habitual user(s) of the car' (Champod and Taroni, 1997) such that

$$\sum_{g=1}^{\infty} b_{g,m_g} = 1.$$

Then:

$$b_{1,m_1} = 1 - b_0 - \sum_{g=2}^{\infty} b_{g,m_g} \leq 1 - b_0.$$

For practical reasons, discussed in Champod and Taroni (1997), $b_{1,m}$ is set as the strict complement of b_0 and values for b_{2,m_2} to b_{∞,m_∞} are supposed to be equal to 0.

In the scenario described above, it is reasonable to assume that, on average, a large number of fibres are transferred, persist and are recovered from the driver's seat of the stolen car. This implies that t_{170} is much greater than t_0. A similar assumption applies for t'_{170} and t'_0, which are estimated through controlled experiments involving woollen garments potentially worn by the offender.

The likelihood ratio (12.3) can then be reduced to

$$V = \frac{t_{170}}{\gamma t'_{170}}$$

because $b_{1,m}\gamma t_0$ and $b_{1,m}\gamma t'_0$ are negligible in this case. Assuming (unrealistically) that transfer characteristics of the suspect's pullover do not differ from those of the offender's garment, the likelihood ratio may be reduced further to $1/\gamma$. Lists of references reporting values for γ are given in Chabli (2001) and in Cantrell *et al.* (2001).

Imagine two slight modifications of the previous scenario, notably in the number of recovered fibres on the driver's seat. Assume also that transfer probabilities differ under propositions H_p and H_d, so that $Pr(T_n \mid x, H_p) \neq Pr(T_n \mid H_d)$ and that $Pr(T_0 \mid x, H_p) \neq Pr(T_0 \mid H_d)$. In such situations Champod and Taroni (1999) have noted the following:

(a) If a group of 10 fibres has been recovered on the driver's seat, if 10 fibres is the average number expected under the proposition of implication of the suspect and if, on average, a potential offender's garment will transfer 60 fibres, then the likelihood ratio exceeds the reciprocal of the relative frequency, $1/\gamma$, without any significant influence of the background probabilities b_0 and $b_{1,m}$ (calculations have been performed where b_0 equals 0.01, 0.5 and 0.99).

(b) If a group of 10 fibres has been recovered on the driver's seat, if the 10 fibres do not correspond to the average number expected under the proposition of implication of the suspect (e.g., the suspect's pullover transfers on average

30 fibres) and if, on average, a potential offender's garment will transfer a comparable number of fibres (e.g., 5 fibres), then the likelihood ratio may be lower than 1 so that the findings support the defence proposition.

Transfer probabilities may be estimated (Curran *et al.*, 1998a; see also Section 10.5.5). Given that the number of fibres transferred does not decrease as a function of time, and the quality of the recovery technique is extremely high (between 90% and 95% of the fibres shed are recovered; see Chable *et al.*, 1994), the following results are obtained:

- For modification (a), $Pr(T_{10} \mid x, H_p) = t_{10} = 0.098$, $Pr(T_0 \mid x, H_p) = t_0 = 0.005$, $Pr(T_{10} \mid H_d) = t'_{10} = 0.0001$ and $Pr(T_0 \mid H_d) = t'_0 = 0.0001$. Using conservative background probabilities, $b_0 = 0.01$ and $b_{1,m} = 0.99$ (these figures imply that it is very probable that a group of extraneous fibres will be found on the driver's seat; note also that $b_0 + b_{1,m} \leq 1$), and a relative frequency γ for the fibre characteristics of 0.01, the following likelihood ratio is obtained.

$$V = \frac{b_0 t_{10} + b_{1,m} \gamma t_0}{b_0 \gamma t'_{10} + b_{1,m} \gamma t'_0} = \frac{0.01 \times 0.098 + 0.99 \times 0.01 \times 0.005}{0.01 \times 0.01 \times 0.0001 + 0.99 \times 0.01 \times 0.0001}$$
$$\approx 1030.$$

This value strongly support proposition H_p.

- For modification (b), $Pr(T_{10} \mid x, H_p) = t_{10} = 0.006$, $Pr(T_0 \mid x, H_p) = t_0 = 0.0001$, $Pr(T_{10} \mid H_d) = t'_{10} = 0.017$ and $Pr(T_0 \mid H_d) = t'_0 = 0.021$. Using conservative background probabilities, $b_0 = 0.01$ and $b_{1,m} = 0.99$, and a relative frequency of 0.01, the following likelihood ratio is obtained:

$$V = \frac{b_0 t_{10} + b_{1,m} \gamma t_0}{b_0 \gamma t'_{10} + b_{1,m} \gamma t'_0} = \frac{0.01 \times 0.006 + 0.99 \times 0.01 \times 0.0001}{0.01 \times 0.01 \times 0.017 + 0.99 \times 0.01 \times 0.021}$$
$$= 0.29.$$

This likelihood ratio supports H_d, with a value of $0.29^{-1} = 3.4$.

These conclusions are reasonable. In fact the number of recovered fibres is in agreement with transfer characteristics of the suspect's pullover and of the potential garments of the offender, respectively.

12.2.2 Comments on the fibres scenario

Various comments may be made following these examples. First of all, extraneous fibres have been collected on the upright part of the driver's seat. There is no information at all on potential fibres which may be on the bottom part of the seat. It is interesting to combine observations made at different parts of the seat. This could be done more easily through the use of Bayesian networks (see Chapter 14).

Secondly, the background probability $b_{g,m}$ represents the probability of the chance occurrence of g group(s) of m foreign fibres on the driver's seat. This probability can be divided into two probabilities, p_i and $s_{i,j}$, as currently done in glass analysis (Section 10.5.4) or as proposed in the fibre pre-assessment scenario in Section 12.3. The probability p_i denotes the probability of the presence of i (≥ 0) groups of fibres on the surface of the seat, and $s_{i,j}$ denotes the probability that group i of recovered material is of size j, where j may take a positive integer value or be replaced by the letter l or s to denote the fact the group is large or small, respectively. Therefore $b_{g,m}$ can be replaced in formulae by $p_i s_{i,j}$. For the assessment of glass fragments, it is assumed (as mentioned by Curran *et al.*, 2000) that (a) there is no association between the number of groups found on surfaces of interest and the sizes of those groups, and (b) there is no association between the frequency of a given type of item and either the number of groups or the size of the groups. These assumptions are questionable in the context of fibres evidence.

Thirdly, the trousers of the suspect have not been seized. It is of interest to look at these in association with the driver's seat and consider the evidence potentially found on the two items. Suggestions on how to approach such a situation can be found in Chapter 14, where Bayesian networks are presented.

Fourthly, a pullover has had contact with the seat, so that fibres from the pullover will be transferred to the seat. But note that fibres from the seat can also be transferred to the pullover. The scientist should be interested in any possible cross-transfer; see Sections 12.2.4 and 14.8.

Finally, a grouping approach is adopted for the recovered foreign fibres (see Section 8.4 for a formal approach to how grouping may be done). Here the group was defined as a set of materials which share the same forensic attributes. The scientist declares the presence of a group of fibres if there is sufficient specificity in the shared features to link these traces reasonably with a unique source. However, this declaration is no more than a logically qualified opinion.

12.2.3   Fibres evidence not left by the offender

We now consider a different situation, taken from Champod and Taroni (1999). Here, the scientist is interested in a car belonging to a man who is suspected of abducting a woman. The victim was wearing a red woollen pullover. According to the suspect, nobody sat in the passenger's seat of his car. On this seat, an extraneous group of fibres – a group of 170 red wool fibres – has been collected. In his defence the suspect denied that the victim has ever sat on the passenger's seat of his car. This defence implies that there is no offence at all, so the recovered fibres are not related to the activity (sitting in the passenger's seat of the car) so that they are on the passenger's seat by chance alone. This represents an important point for the understanding of the development of the likelihood ratio equation (see also the development presented in Sections 8.3.2 and 8.3.3). In fact, even if the numerator of the likelihood ratio is still the same as the previous one (12.2), the denominator changes. There is no reason to develop $Pr(y|H_d)$ using association hypotheses T_n and T_0 because fibres cannot be transferred (persist and be recovered) during the action (because there is no action!). Therefore, the likelihood ratio becomes:

$$V = \frac{b_0 t_{170} + b_{1,m}\gamma t_0}{b_{1,m}\gamma}. \qquad (12.4)$$

The consequence of this is that if b_0 is close to 1 (this means, for example, that the seats are cleaned regularly), and if, as stated before, $b_{1,m} \simeq 1 - b_0$, then $b_{1,m}$ (the probability that the recovered group is present on the seat by chance alone) is close to 0 and the likelihood ratio (12.4) is increased. A numerical example is offered in Section 8.3.3 in the context of DNA evidence. The likelihood ratio (12.4) is the method of evaluation currently presented in glass fragments scenarios (see Section 10.5.4). This happens because the presence of glass fragments on a suspect's pullover (for example) is generally explained by chance alone (in fact, H_d normally specifies that the suspect has not broken the window, so an alternative action which explains the presence of the fragments is not given; their presence is explained by chance alone). Thus (12.4) may be rewritten as

$$V = \frac{b_0 t_{170} + b_{1,m}\gamma t_0}{b_{1,m}\gamma} = \frac{b_0 t_{170}}{b_{1,m}\gamma} + t_0 \simeq \frac{b_0 t_{170}}{b_{1,m}\gamma},$$

as t_0 is negligible compared with the first term. An extension of this scenario is the one given by Buckleton and Evett (1989) where the number of extraneous groups, g, of fibres (a number of fibres consistent with the nature of the activity) is greater than 1. Recovered material is denoted $y_1, y_2, \ldots, y_g$. Only one group is compatible with the victim's pullover, x_1.

The likelihood ratio developed by Buckleton and Evett (1989) shows that it is important not only to focus on the fibres that match the victim's garments (for example) but also to consider other groups of fibres compatible with the alleged action:

$$V = \{Pr(y_1, y_2, \ldots, y_g \mid x_1, T_n, H_p)Pr(T_n \mid x_1, H_p)$$

$$+ Pr(y_1, y_2, \ldots, y_g \mid x_1, T_0, H_p)Pr(T_0 \mid x_1, H_p)\}/Pr(y_1, y_2, \ldots, y_g \mid H_d).$$

$$(12.5)$$

Likelihood ratio (12.5) can reasonably be reduced to $t_n/\gamma_1 g$. Details of the development are also given in Champod and Taroni (1999).

12.2.4   Cross-transfer

A stolen vehicle is used in a robbery on the day of its theft. An hour later it is abandoned. The vehicle is found by the police a few hours later. On the polyester seats, which were recently cleaned with a car vacuum cleaner, extraneous textile fibres are collected. The car owner lives alone and has never lent his vehicle to anyone. The owner wears nothing but cotton. The day following the robbery a suspect is apprehended, and his red woollen pullover and his denim jeans are confiscated.

On the driver's seat one group of relevant foreign fibres is collected (this is a group other than cotton). It consists of a large number of red wool fibres. The evidence E_1 is (y_1, x_1), where y_1 refers to the recovered fibres on the car seat and x_1 refers to known (control) material from the suspect's red woollen pullover. The fibres on the driver's seat are assumed to have been transferred from the clothing of the offender to the seat.

Foreign fibre groups are fibres which can be distinguished from fibres from a known source (either associated with the suspect or associated with an object such as a car).

On the suspect's pullover and denim jeans (together) there are many foreign fibre groups. One consists of 20 extraneous black fibres. They are in agreement (in some sense) with the fibres of the driver's seat. The evidence E_2 is (y_2, x_2), where y_2 refers to the 20 recovered fibres on the suspect's clothes and x_2 refers to known material from the driver's seat.

The competing propositions (at the activity level) could be:

- H_p, the suspect sat on the driver's seat of the stolen car;
- H_d, the suspect never sat on the driver's seat of the stolen car.

When two individuals or an individual and an object, such as a car seat, are in contact a reciprocal transfer of material is usually involved. The two sets of recovered traces then have to be considered as dependent. If a transfer has occurred in one direction and the expert has recovered traces characterising this transfer, then the expert would, in general, expect to find trace evidence characterising the transfer in the other direction. The presence of evidence transferred in one direction gives information about the presence of evidence transferred in the other direction. From (8.2), the value of the evidence is

$$V = \frac{Pr(E_2 \mid H_p, E_1)}{Pr(E_2 \mid H_d, E_1)} \times \frac{Pr(E_1 \mid H_p)}{Pr(E_1 \mid H_d)}. \qquad (12.6)$$

The second ratio in (12.6) is equal to $1/\gamma_1$, where γ_1 is the estimated frequency of the compared characteristics from y_1 in extraneous groups of fibres of similar sizes found on stolen car seats. The first ratio has to account for the transfer of 20 recovered fibres (y_2) to the suspect's clothing from the driver's seat of the stolen car. This is done by multiplying the unconditional value $1/\gamma_2$, where γ_2 is the estimated frequency of the compared characteristics from y_2 in extraneous groups of fibres of similar sizes found on the clothing of potential offenders, by a probability u_{20} which is the *conditional transfer probability* that the 20 recovered fibres on the suspect's clothing have been transferred (have remained and been recovered) from the driver's seat of the stolen car to the suspect's upper garment. The probability is conditional because it is estimated given that the suspect has sat on the seat (proposition H_p is believed true in the numerator) and that E_1 is true (e.g., 170 red wool fibres have been found on the seat as in Section 12.2.1). It is also subjective and its value has to be considered a new for each individual case. The first ratio is then equal to u_{20}/γ_2. The value of the evidence is then

$$V = \frac{u_{20}}{\gamma_1 \gamma_2} \qquad (12.7)$$

(Champod and Taroni, 1999). This expression can be generalized. For example, if there were k fibres found on the suspect's clothing in a group which matched the car seat, u_{20} would be replaced by u_k. If there were g_1 groups of fibres on the driver's seat of which one matched the fibres of the suspect's clothing, $1/\gamma_1$ would be replaced by $1/(g_1\gamma_1)$. Similarly, if there were g_2 groups of fibres on the suspect's clothes of which one matched the fibres on the driver's seat, $1/\gamma_2$ would be replaced by $1/(g_2\gamma_2)$. The value of the evidence is then

$$V = \frac{u_k}{(g_1\gamma_1)(g_2\gamma_2)}. \qquad (12.8)$$

It is of interest to examine the behaviour of V for different values of the components of (12.8). For example, consider $\gamma_1 = \gamma_2 = 0.05, g_1 = 1, g_2 = 6$ and $k = 20$. Then

$$V = 400u_{20}/6 = 66.7u_{20}.$$

If the cross-transfer evidence was not taken into consideration, the value of the evidence of the fibres on the driver's seat would be $1/\gamma_1 = 20$. The cross-transfer evidence *increases* the value of the evidence if

$$\frac{400u_{20}}{6} > 20.$$

This is only so if $u_{20} > 6/20$ or 0.3. If the two transfers are not thought compatible, then it may be that the scientist's subjective view is that $u_{20} < 0.3$ and the presence of the evidence on the suspect's clothing is supportive of the defence proposition rather than the prosecution proposition. A cross-transfer does not necessarily increase the strength of the evidence. The following observation is pertinent here:

> It is often common to focus on evidence regarding the occurrence of events and easy to overlook evidence regarding the non-occurrence of events. In any inferential context it is just as important to inquire about what did not happen as it is to inquire about what did happen. (Schum, 1994, p. 96)

12.3 PRE-ASSESSMENT IN FIBRES SCENARIOS

The theory of the pre-assessment approach proposed by Cook *et al.* (1998a) was presented in Section 7.2.2. The aim of this section is to develop a practical example in a fibres scenario. The scenario is taken from Champod and Jackson (2000), who work through the following steps: identification of the information the scientist may need, identification of the relevant propositions used to assess the findings, progress through the pre-assessment of the case, determination of the examination strategy, assessment of the likelihood ratio and its sensitivity, and identification of the effect of a change in the propositions.

12.3.1 The case history

Two armed and masked men burst into a post office, threatened the staff, the takings for the day were handed over and the men left. Witnesses said that one of the men was wearing a dark green balaclava mask and the other man was wearing a knotted stocking mask. They said also that the two men drove away from the scene in a car driven by a third man. Some way along the presumed getaway route, a dark green balaclava was found. Mr U was arrested the following day. He denied all knowledge of the incident. Reference samples of his head hairs and blood were taken as well as combings from his head hair. Mr U has not yet been charged with the robbery because there is very little evidence against him.

12.3.2 Formulation of the pairs of propositions and events

Investigators are interested in knowing if Mr U has worn the mask because the intended charge is robbery. The scientist is at first able to define propositions at the *source* level, such as:

- H_{p1}, hairs in the mask came from Mr U,
- H_{d1}, hairs in the mask came from someone else;
- H_{p2}, saliva in the mask came from Mr U,
- H_{d2}, saliva in the mask came from someone else;
- H_{p3}, fibres in Mr U's hair combings came from the mask,
- H_{d3}, fibres in Mr U's hair combings came from some other garment or fabric.

If the scientist wishes to assess the findings under *activity* level propositions (e.g., H_p, Mr U wore the mask at the time of the robbery, and H_d, Mr U has never worn the mask) – propositions that are more relevant for the court because they link more strongly a person to an offence – they need (a) good background information on the offence (the time of the offence, time of arrest, time the sample is taken, etc.) to be able to estimate precisely the factors that relate to transfer and persistence of the evidence of interest, and (b) data (i.e., literature) on transfer, persistence and recovery of hairs, fibres and saliva when someone wears a mask, as well as survey data on masks. Published data on hairs and saliva are very limited; data on fibre transfer to hair and on persistence are available (see, for example, Ashcroft *et al.*, 1988; Salter and Cook, 1996; Cook *et al.*, 1997). If criteria (a) and (b) are satisfied, the scientist can consider fibres first. If the scientist has no sufficient background information to consider the evidence (fibres, saliva, hairs) under *activity* level propositions, then they are obliged to stay at the *source* level. In the scenario described above, the strategy would be to offer an assessment at the activity level for fibres but to consider the other evidence at the *source* level.

The second step in the pre-assessment is to determine the possible *findings*. Considering fibres evidence on the hair combings, the scientist could be presented with the following results: no fibres have been observed, a small number of fibres have been observed (say one to three) or a large number of fibres have been observed (i.e., more than three). Note that the definitions of these categories can be flexible and may depend on the available data. Note also that more than one group of fibres could be present.

Then, *events* involved in the activity level assessment are defined. To help the scientist to determine which events are relevant for the pre-assessment of the findings in such a scenario, it is useful to consider what could happen if Mr U wore the mask at the time of the robbery. If Mr U wore the mask, then three possibilities may be envisaged:

- no fibres have been transferred, have persisted and have been recovered (call this event T_0);
- a small number of fibres have been transferred, have persisted and have been recovered (event T_s);
- a large number of fibres have been transferred, have persisted and have been recovered (event T_l).

The presence of fibres has two main explanations: transfer during the course of the crime or presence beforehand by chance. The events linked to this *background* presence are:

- no group of fibres is present by chance (event P_0);
- one group of fibres is present by chance (event P_1).

When such a chance group occurs, it may be a small or a large group. The events are:

- the group of fibres present by chance is small (event S_s);
- the group of fibres present by chance is large (event S_l).

(Note that in Section 12.2.2, probabilities values for P_i and $S_{i,j}$ are grouped into $b_{g,m}$).

Finally, when a recovered group of fibres of unknown origin is compared with the control, two outcomes are possible:

- the recovered fibres match the control with respect to the features analysed (event M);

- the recovered fibres do not match the control with respect to the features analysed (event $\bar{M}$).

If the scientist takes into account discrete variables and makes the assumptions that a sample known to come from a particular source is compared with the source, the probability of a match equals 1.

12.3.3 Assessment of the expected value of the likelihood ratio

The scientist, analysing the hair combings, could observe one of the four situations given in Table 12.1. This list of outcomes does not take into account other possibilities such as the observation of a group of fibres coming from the transfer and a second group of fibres from the background. This point is discussed in Section 14.6. The arithmetical approach proposed here needs clear assumptions. These could include, for example, that there is one and only one group present.

The events (and the related probabilities) occurring under the two propositions of interest, notably H_p and H_d as previously specified, are given in Table 12.2.

The next step in the pre-assessment process is to estimate these probabilities using data from published literature, case–specific experiments or subjective judgements based on the scientist's own experience.

- *Transfer probabilities.* To assess these probabilities, t_0, t_s, t_l, it is useful to answer questions such as 'if the suspect wore a mask, what is the probability that no/a small/a large number of fibres are transferred, persisted and are then recovered?' Note that information on the suspect (i.e., type and length of the suspect's hair), on the material involved (i.e., sheddability), on the

Table 12.1 Findings from the analysis of the hair combings

Outcome	Number of groups	Number of non-matching groups	Number of matching groups	Size of matching groups
A	0	0	0	
B	1	1	0	
C	1	0	1	Small
D	1	0	1	Large

Table 12.2 Events and probabilities relating to findings under H_p and H_d

Outcome from Table 12.1	Events (Ev) to occur if H_p is true	$Pr(Ev\|H_p)$	Events (Ev) to occur if H_d is true	$Pr(Ev\|H_d)$
A	T_0, P_0	t_0p_0	P_0	p_0
B	$T_0, P_1, \bar{M}$	$t_0p_1(1-m)$	$P_1, \bar{M}$	$p_1(1-m)$
C	T_0, P_1, S_s, M or T_s, P_0	$t_0p_1s_sm + t_sp_0$	P_1, S_s, M	p_1s_sm
D	T_0, P_1, S_l, M or T_l, P_0	$t_0p_1s_lm + t_lp_0$	P_1, S_l, M	p_1s_lm

methods used to search and collect fibres, on the circumstances of the case (i.e., alleged activities, time delays) are all relevant information for a correct assessment of the probabilities.

- *Background probabilities.* These probabilities, p_0, p_1, s_s, s_l, assess the likelihood of having by chance no fibres or one group of fibres (which could be small or large, as previously defined) on the hair if no fibres have been transferred or if the suspect denied wearing the mask. (Note that H_d specifies that the suspect never wore any mask. If the alternative proposition changes, for example, if the suspect stated that he wore a similar mask two days before the alleged facts, background probabilities change and new assessments are required.)

- *Match probability.* This probability, m, represents – in some way – an assessment of the rarity of the extraneous fibres found on the head of a person innocently accused of wearing a mask (fibres which match by chance the reference fibres coming from the mask). This rarity may be assessed in various ways. The scientist can refer to literature where fibres have been recovered on hairs of individuals and consider the relative proportions of fibres presenting the features of interest. He can use a 'target fibre' study, stressing that such studies offer a different probability estimate, notably the probabilities of observing by chance one target group of fibres that match the control, $Pr(p_1, s_s, m|H_d)$ and $Pr(p_1, s_l, m|H_d)$, and not an estimate of m. Databases can also be used, assuming that the potential source of fibres are hats, neckwear, bedding and jumpers, so that the scientist will be able to assess the relative frequency of the matching fibres in this population.

Likelihood ratios using probabilities proposed by Champod and Jackson (2000) are given in Table 12.3.

Likelihood ratios obtained in this pre-assessment of fibres evidence offer an answer to the legitimate question 'is it useful to proceed in fibres analysis?'. It has been shown that all situations offer a likelihood ratio different from the 'inconclusive' value of 1. If no fibres at all are recovered, or if a group of fibres is recovered and this group does not match the control object, likelihood ratios supporting the defence's proposition are obtained. On the other hand, if one group (small or large) of fibres is recovered and the group matches the control group, then a likelihood ratio greater than 1 is obtained, and the prosecution's proposition is supported.

Table 12.3 Likelihood ratios for the outcomes from Table 12.1 with $t_0 = 0.01$, $t_s = 0.04$, $t_l = 0.95$; $p_0 = 0.78$, $p_1 = 0.22$; $s_s = 0.92$, $s_l = 0.08$; $m = 0.05$, as proposed by Champod and Jackson(2000)

Outcome	V
A	0.01
B	0.01
C	3.09
D	842.05

Pre-assessment can be applied (as an approach for decision making) in more sophisticated cases. Imagine a cross-transfer case (also called a 'two-way transfer case'). Pre-assessment can be updated when a staged approach is taken (e.g., the victim's pullover is analysed first, then the suspect's pullover; Cook *et al.*, 1999). The results of the examination of one of the garments are used to inform the decision about whether the second garment should be examined. As specified by Cook *et al.* (1999), it is easy to see how the principles could be extended to other kinds of cases. For example, if the crime involves the smashing of a sheet of glass and if clothing is submitted from a suspect, then the phased approach could be applied to the order in which the garments are examined. If examination of the jacket reveals no glass, then how does formal pre-assessment inform the decision about examining the trousers or shoes?

12.4   THE RELEVANT POPULATION OF FIBRES

Consider the following example. An offender attempted to enter the rear of a house through a hole which he cut in a metal grille, but failed when a security alarm went off. He left the scene. About 10 minutes after the offence, a suspect wearing a red pullover was apprehended in the vicinity of the house following information from an eyewitness who testified that he saw a man wearing a red pullover running away from the scene. At the scene, a tuft of red fibre was found on the jagged end of one of the cut edges of the grille.

If the propositions from the prosecution and defence are that the fibres at the crime scene came from the suspect (H_p) and the fibres at the crime scene did not come from the suspect (H_d), respectively, following the argument of Section 9.2, the value of the evidence is given by:

$$V = \frac{Pr(y \mid x, H_p, I)}{Pr(y \mid H_d, I)},$$

where y is the evidence of the red fibres on the grille and x is the evidence of fibres on the suspect's red pullover.

If H_d is true, the probability in the denominator is the probability of finding the characteristics of the tuft of fibre in the grille in a population of potential sources (a relevant population). Assume that a survey has been made of the characteristics in a relevant population and the frequency of the characteristics is γ. As in Section 9.2, the value of the evidence is then

$$V = \frac{1}{\gamma}.$$

As given in Section 8.5, the relevant population is defined by H_d and I. Some considerations for the definition in this example are given by Champod and Taroni (1999):

- If H_d is that the suspect had never been present at the scene, then the relevant population is defined as that of red upper garments worn by burglars (accepting that the eyewitness had seen the burglar and was correct in the report that the burglar was wearing a red upper garment).

- If H_d is that the suspect has been correctly identified by the witness but had never been in contact with the grille, then the relevant population is defined by the potential sources of red fibres, without any distinction in respect of the colour of the garment worn by the burglar.

- In the absence of an eyewitness, if the suspect has been apprehended because he was wearing a red pullover and the fibres found at the crime scene were red, then the relevant population is defined as that of potential perpetrators wearing red garments (further discussion of the implications of such a so-called search strategy is given in Chapter 13).

- In the absence of an eyewitness, if the suspect has been apprehended independently of the forensic attributes of the tuft, then the relevant population is defined by the potential perpetrators, without any distinction in respect to the colour of the garment.

Analogously, an example of the influence of the defence's strategy for the definition of the relevant population is given in Robertson and Vignaux (1995a). Blood which did not belong to the murder victim was found at the scene of the crime. A man of Maori appearance was seen running away. Subsequently, a Maori person was arrested and identified by an eyewitness as the man seen running away. The proposition of the prosecution, H_p, is that this man is the criminal. The defence has two possible alternatives:

- H_{d1}, the accused was the person seen running away but was not the murderer;

- H_{d2}, the accused person was not the person seen running away and the eyewitness identification was wrong.

Under H_{d1} there is no information about the murderer, so the murderer is to be considered as a randomly selected person in New Zealand. The scientist will then use the population as a whole as the relevant population for the value of γ. Under H_{d2} there is information about the murderer. He was of Maori appearance, so the murderer is to be considered as a randomly selected person of Maori appearance. The scientist will then use the Maori population as the relevant population for the value of γ.

13

DNA Profiling

13.1 INTRODUCTION

A summary of the general principles and techniques of DNA typing is given in
the two National Research Council reports (NRC, 1992, 1996; Kaye, 1997a,
Lempert, 1997) and a review of current developments is given in Foreman *et al.*
(2003). A brief summary is given here. More details are to be found in Rudin
and Inman (2003). The following two paragraphs are taken from the entry
for DNA in the Micropaedia of the 15th edition of Encyclopaedia Britannica
(1993). These paragraphs are then followed by a description from Chapter 2 of
the second NRC (1996) report.

Deoxyribonucleic acid (DNA) is an organic chemical of complex molecular
structure. It codes genetic information for the transmission of inherited traits.
The structure of DNA consists of two strands of a phosphoryl-deoxyribose
polymer that are connected in a double spiral, or helix, by hydrogen bonds
between complementary nitrogenous bases attached to the polymer. The
sequence of nitrogenous bases dictates what specific genetic information the
molecule encodes; a segment of DNA that codes for the cell's synthesis of a
specific protein is called a *gene*. Only four bases, adenine, guanine, cytosine and
thymine (abbreviated to A, G, C and T) occur in DNA. The repeated sequences
of these encode all inherited traits.

The nitrogenous bases on each strand of the DNA molecule bind to specific
complementary bases on the opposing strand. DNA replicates by separation into
two single strands, each of which serves as a template for a new strand built
up by bonding of new bases to the bases of the original. Adenine binds only
to thymine and guanine binds only to cytosine, so the new strand is precisely
complementary to the original strand.

In higher organisms, the genetic material is organized into microscopic structures called *chromosomes*. A fertilized human egg has 46 chromosomes (23
pairs). A set of 23 chromosomes with the genetic information they contain is
termed the *genome*. A chromosome is a very thin thread of DNA, surrounded
by other materials, mainly protein. The DNA thread is actually double, two

---

Statistics and the Evaluation of Evidence for Forensic Scientists: Second Edition
C.G.G. Aitken and F. Taroni © 2004 John Wiley & Sons, Ltd   ISBN: 0-470-84367-5

strands coiled round each other like a twisted rope ladder with stiff wooden steps. The basic chemical unit of DNA is the nucleotide, consisting of a base (a half-step in the ladder) and a sugar-phosphate complex (the adjacent section of the rope). The total DNA in a genome amounts to about 3 billion nucleotide pairs. A gene is a segment of DNA, ranging from a few thousand to more than a hundred thousand nucleotide pairs. The specific sequence of nucleotides in a gene acts as an encoded message that is translated into the specific amino acid sequence of a polypeptide or protein. The gene product might be detected only chemically or might lead to a visible trait such as eye pigment. The position on the chromosome where a particular gene resides is its *locus*.

Different forms of a gene are called *alleles*. If the same allele is present in both chromosomes of a pair, the person is *homozygous*. If the two alleles are different, the person is *heterozygous*. The genetic makeup of a person is the *genotype*. Genotype can refer to a single gene locus with two alleles denoted, say, A and a. In such a case, there are three possible genotypes AA, Aa and aa. Genotype can also refer to several loci or to the entire set of genes. In forensic analysis, the genotype for the group of analysed loci is called the DNA *profile*. There is a large number of alleles at a given locus so the number of possible genotypes is very large. For example (NRC, 1996), at a locus with 20 alleles, there are 20 homozygous genotypes and $(20 \times 19)/2 = 190$ heterozygous ones, making a total of 210 genotypes. With four such loci, the number of genotypes is 210^4 or about 2 billion.

An alternative approach to DNA analysis uses a laboratory process known as the polymerase chain reaction (PCR) in which a chosen short segment of DNA can be copied millions of times. A description of the process is given in NRC (1996). Because the amplification process is almost unlimited, PCR-based methods make possible the analysis of very tiny amounts of DNA. Also, PCR-methods usually permit an exact identification of each allele, and thus there are no measurement uncertainties. They do not have as many alleles per locus as VNTRs (variable number of tandem repeats), but that is compensated for by the very large number of loci that are potentially usable.

Another class of genetic marker is mitochondrial DNA. Mitochondria are microscopic particles found in the cell, but outside the nucleus, so they are not associated with chromosomes. The transmission of mitochondria is from mother to child. All the children of one woman have identical mitochondrial DNA. Thus it is particularly useful for associating persons related through their maternal lineage (Gill *et al.*, 1994). Similar results apply for the Y-chromosome and male lineage.

For the assessment of the probability that DNA from a randomly selected person has the same profile as that of DNA found at a crime scene, say, it is necessary to know the frequency of that profile in the population. The frequency is determined by comparison with a reference data set. These reference data sets are very small in comparison with the set of all possible profiles which may be found. Thus, the frequency of individual alleles are used to estimate the frequency of a given profile. In order to do this, assumptions need to be made about the mating structure of the population.

It is conventional in genetics to designate each gene or marker locus with a letter and each allele at that locus with a numeric subscript. Thus, the tenth allele at locus A would be denoted A_{10} and the fifth allele at locus B would be denoted B_5. In general, the ith allele at locus A would be denoted A_i. The frequency of A_i will be denoted p_i, and these frequencies will be such that $\sum p_i = 1$, where the summation is over all the alleles at the locus.

13.2 HARDY–WEINBERG EQUILIBRIUM

Allelic frequencies may be calculated from the observed frequencies in a sample. Once the allelic frequencies have been calculated it is then possible to determine the expected frequencies, assuming Hardy–Weinberg equilibrium. A comparison of the observed and expected frequencies provides a measure of how good the equilibrium assumption is. Consider a locus in a population with k alleles, indexed by i and j, and $m = k(k+1)/2$ genotypes. A sample of size n is taken from the population and the number of members of the sample for each of the genotypes is counted, with x_{ij} members with genotype $\{i, j\}$ (denoting component alleles i, j for $1 \le i \le j \le k$), where the individual is heterozygous if $i \ne j$ and homozygous if $i = j$. The relative frequencies x_{ij}/n are denoted P_{ij}. The proportion p_i of alleles of type i is then given by

$$p_i = P_{ii} + \frac{1}{2}\sum_{\substack{j \ne i}}^{j=k} P_{ij}. \tag{13.1}$$

The division by 2 is because in heterozygotes only half the alleles are i.

The assumption that the alleles on a locus, one from the father and one from the mother, are independent of each other leads to an equilibrium distribution for the relative frequencies of alleles in a population. This is known as *Hardy–Weinberg* equilibrium or *random mating*. The relative frequencies for the results of mating for a locus with two alleles are given in Table 13.1. Thus, the genotype frequency of AA is p^2, of Aa is $2pq$, and of aa is q^2, and $p^2 + 2pq + q^2 = (p+q)^2 = 1$. In general, let p_i and p_j be the population proportions of alleles A_i and A_j for $i, j = 1, \ldots, k$, where k is the number of alleles at the locus in question. The expected genotypic frequencies P_{ij} are obtained from the following equations assuming Hardy–Weinberg equilibrium:

$$P_{ij} = \begin{cases} 2p_i p_j, & i \ne j, \\ p_i^2, & i = j. \end{cases} \tag{13.2}$$

Note that in what follows population frequencies are denoted with Latin letters, usually p. This is counter to the convention used in the rest of the book of

Table 13.1 Hardy–Weinberg proportions for a locus with two alleles, A and a, with frequencies p and q such that $p+q = 1$

Alleles (and frequencies) in eggs	Alleles (and frequencies) in sperm	
	$A(p)$	$a(q)$
$A(p)$	p^2	pq
$a(q)$	pq	q^2

Greek letters denoting population frequencies. However, the use of Latin letters, such as p (and q for $1-p$), is so widespread that it is felt that the lesser confusion would be caused by following this practice. The allelic relative frequencies and genotype frequencies for *HumTHO1* genotypes based on 95 unrelated Turkish individuals (Çakir *et al.*, 2001) are given in Table 13.2. At the *HumTHO1* locus, six alleles are used. Consider allele 6. This has an allelic frequency of 0.295 in this population. This is computed from the observed genotypic frequencies in the right-hand portion of the table. The first six genotypes include the allele 6. Combining their frequencies, using (13.1), and dividing by 95 the relative frequency is obtained.

$$\{10+(6+8+14+2+6)/2\}/95 = 0.295.$$

The expected genotypic frequencies are derived from the allelic frequencies (which themselves are derived from the *observed* genotypic frequencies) from (13.2).

The value of DNA evidence is often determined using a likelihood ratio which is just the reciprocal of the profile frequency, say $1/2p_ip_j$ for a heterozygous suspect matching a bloodstain found on a crime scene. Although this is expedient, it glosses over some issues and does not accommodate population structure (Section 13.5), relatives (Section 13.6) or the application to mixtures (Section 13.10; Weir, 1996b).

The discriminating power described in Section 4.5 can also be applied here to determine the so-called *exclusion power* of a locus. The probability that two randomly chosen people have a particular genotype is the square of its frequency in the population. The probability Q that two randomly chosen people have the same unspecified genotype is the sum of the squares of the frequencies of all the genotypes. Thus, for the general case above with n alleles,

$$Q = \sum_{i=1}^{n} P_{ii}^2 + \sum_{i=2}^{n} \sum_{j<i} P_{ij}^2.$$

Table 13.2 Observed and expected frequencies of *HumTHO1* genotypes based on 95 unrelated Turkish individuals (from Çakir *et al.*, 2001)

Alleles		Genotypes		
Allele	Frequency (%)		Observed	Expected
			Frequency	
6	0.295	6–6	10	8.3
7	0.147	6–7	6	8.2
8	0.184	6–8	8	10.3
9	0.232	6–9	14	13.0
9.3	0.026	6–9.3	2	1.5
10	0.116	6–10	6	6.5
		7–7	2	2.1
		7–8	4	5.1
		7–9	7	6.5
		7–9.3	0	0.7
		7–10	7	3.2
		8–8	4	3.2
		8–9	9	8.1
		8–9.3	2	0.9
		8–10	4	4.1
		9–9	4	5.1
		9–9.3	1	1.1
		9–10	5	5.1
		9.3–9.3	0	0.1
		9.3–10	0	0.6
		10–10	0	1.3
		Total	95	95
		Homozygotes	21.1%	21.2 %
		Heterozygotes	78.9 %	78.8 %

The discriminating or exclusion power for the genotype is then $1 - Q$. The larger the discriminating power, the better the genotype is at discriminating amongst people.

As in Section 4.5.4, suppose there are m independent systems with corresponding values $Q_1, \ldots, Q_m$. The probability DP_m of being able to distinguish between two individuals using the m tests is, as before (Weir, 1996a),

$$DP_m = 1 - \prod_{l=1}^{m} Q_l.$$

This chapter gives an overview of various issues associated with the evaluation of evidence from DNA profiles. Further details are given in more specialist books such as Evett and Weir (1998), Buckleton *et al.* (2004) and Balding (2005).

13.3   DNA LIKELIHOOD RATIO

Imagine the likelihood ratio calculation in a common situation involving DNA where there is the profile E_c of the crime sample and the profile E_s of the suspect. Let I represent the background information and let the propositions, at the source level for example, be:

- H_p, the suspect is the source of the stain;
- H_d, another person, unrelated to the suspect, is its source (i.e., the suspect is not the source of the stain).

Both profiles are of type A, say. The likelihood ratio can then, from (9.2), be expressed as

$$\frac{Pr(E_c = A \mid E_s = A, H_p, I)}{Pr(E_c = A \mid E_s = A, H_d, I)}$$

Assume that the DNA typing system is sufficiently reliable that two samples from the same person will be found to match when the suspect is the donor of the stain (proposition H_p), and that there are no false negatives. The recovered sample is of type A if it is known that the suspect is of type A, if H_p is assumed true. So, $Pr(E_c = A \mid E_s = A, H_p, I) = 1$.

It is widely assumed that the DNA profiles from two different people (the suspect and the donor of the stain when proposition H_p is true) are independent. Then $Pr(E_c = A \mid E_s = A, H_d, I) = Pr(E_c = A|I)$. In such a case only the so-called *profile probability*, ($2p_ip_j$ for a heterozygote stain/suspect and p_i^2 for a homozygote) with which an unknown person would have the profile A is needed. This is a widely accepted simplification. In reality, the evidential value of a match between the profile of the recovered sample and that of the suspect needs to take into account the fact that there is a person (the suspect) who has already been seen to have that profile (type A). So, the probability of interest is $Pr(E_c = A \mid E_s = A, H_d, I)$ and this can be quite different from $Pr(E_c = A \mid H_d, I)$ (see Sections 13.5 and 13.10; Weir, 2000a).

In fact, observing one gene in the sub-population increases the chance of observing another of the same type. Hence, within a sub-population, DNA profiles with matching allele types are more common than suggested by the independence assumption, even when two individuals are not directly related (see Section 13.6 on relatives).

The conditional probability (also called *match probability* or *random match probability*) incorporates the effect of population structure or other dependencies between individuals, such as that imposed by family relationships (Weir, 2000b).

13.4 UNCERTAINTY

The role of uncertainty in the determination of the allelic frequencies and the frequency of a particular DNA profile in the population is discussed by Balding and Nichols (1994). Suppose that the relative frequency of a DNA profile in a particular population is known to be γ. If two individuals, J and K say, are taken at random from the population then the probability that K has the profile, given that J has the profile, is p.

In practice, the profile relative frequency, $\hat{p}$ say, from a sample is not known exactly. The expected value is p. Let σ^2 denote the variance of the profile relative frequency about its expected value. Let P denote the random variable corresponding to $\hat{p}$. The variance of a random variable is related to the expectation of the random variable through the result that the variance is equal to the expectation of the square of the random variable minus the square of the expectation. For the profile relative frequency P, this result may be represented symbolically by $Var(P) = E(P^2) - \{E(P)\}^2$. This may be rewritten as $E(P^2) = Var(P) + \{E(P)\}^2 = \sigma^2 + p^2$. The probability of a match is given by

$$Pr(K \text{ has the profile } | J \text{ has the profile }) = \frac{Pr(J \text{ and } K \text{ have it })}{Pr(J \text{ has it })}$$

$$= \frac{E(P^2)}{E(P)}$$

$$= \frac{p^2 + \sigma^2}{p}$$

$$= p + \frac{\sigma^2}{p}.$$

Thus, the probability K has the profile, given J has it, is greater than the profile frequency p. Information that one person has a profile frequency increases the probability that another person has it.

13.5 VARIATION IN SUB-POPULATION ALLELE FREQUENCIES

How many other individuals among the population of possible culprits might be expected also to share this DNA type? The answer to this question is complicated by the phenomenon of genetic correlations due to shared ancestry (Balding, 1997). So, simple calculation of profile frequencies (using the 'product rule', where allele proportions are multiplied together within and across loci to give an estimate of the proportion for the complete multi-locus matching profile) is not sufficient when there are dependencies between different individuals involved in the case under examination: the suspect and the perpetrator (the real source of the recovered sample) featured in the alternative proposition, H_d.

The more common source of dependency is a result of membership in the same population and having similar evolutionary histories. The mere fact that populations are finite in size means that two people taken at random from a population have a non-zero chance of having relatively recent common ancestors. So, disregarding this correlation of alleles in the calculation of the weight of the evidence calculation results in an exaggeration of the strength of the evidence against the compared person (e.g., the suspect in a criminal case or the alleged father in a civil paternity case) even though it is not as important as the relatedness in the same population (see Section 13.6).

A measure F_{ST} of inter-population variation in allele frequencies was introduced by Wright (1922). It can be considered as a measure of population structure. Extensive studies have been made of allele frequencies in many human populations to estimate values of F_{ST} (Balding *et al.*, 1996; Foreman *et al.*, 1997b, 1998; Balding and Nichols, 1997; Lee *et al.*, 2002), suggesting that it is prudent to use F_{ST} values from the large end of the currently observed range.

Evett and Weir (1998) discuss three so-called *F-statistics* which provide a measure of relationship between a pair of alleles, relative to some background level of relationship. These are as follows. The notation is that used by Wright (1951, 1965). Evett and Weir (1998) use the notation of Cockerham (1969, 1973).

- The extra extent to which two alleles within one individual are related when compared to pairs of alleles in different individuals but within the same sub-population is denoted F_{IS}.

- The extent of relatedness of alleles within an individual compared to alleles of different individuals in the whole population is denoted F_{IT}.

- The relationship between alleles of different individuals in one sub-population when compared to pairs of alleles in different sub-populations, also known as the *coancestry coefficient*, is denoted F_{ST}.

Slight differences between the definitions of Wright (1951, 1965) and Cockerham (1969, 1973) are pointed out by Evett and Weir (1998). Wright defined his quantities for alleles identified by the gametes carrying them; Cockerham defined his statistics for alleles defined by the individuals carrying them. For random mating sub-populations, Evett and Weir (1998) comment that the distinction can be ignored.

The following argument for deriving the probability of a match, taking account of inbreeding, is taken from Balding and Nichols (1994). Let p_A and p_B denote the population proportions of alleles A and B. Interpret the value of F_{ST} as the probability that two alleles are identical through inheritance from a common ancestor in the same sub-population. With reasonable assumptions, two alleles drawn from the sub-population are identical through a common ancestor in the sub-population with probability F_{ST} and the ancestor is of type A with probability p_A. If not identical by descent, two alleles are of type A with probability p_A^2. Thus, the probability of drawing allele A in both of two random draws from the sub-population is

$$Pr(A^2 \mid p_A) = p_A\{F_{ST} + (1 - F_{ST})p_A\}. \tag{13.3}$$

The observation of one A allele in the sub-population makes it likely that A is more common in the sub-population than in the general population and hence $Pr(A^2 \mid p_A)$ is larger than the probability p_A^2 of drawing two consecutive A alleles in the general population. The probability of drawing first an A followed by a B allele is

$$Pr(AB \mid p_A, p_B) = p_A p_B(1 - F_{ST}). \tag{13.4}$$

In general, let $Pr(A^r B^s)$ denote the probability that amongst $r + s$ alleles drawn randomly from the sub-population, the first r are of type A and the following s are of type B then

$$Pr(A^{r+1} B^s \mid p_A, p_B) = Pr(A^r B^s \mid p_A, p_B) \frac{rF_{ST} + p_A(1 - F_{ST})}{1 + (r+s-1)F}. \tag{13.5}$$

Special cases of (13.5) with $s = 0, r = 3$ and then $s = 0, r = 2$ gives

$$Pr(A^4 \mid p_A) = Pr(A^2 \mid p_A) \frac{(2F_{ST} + (1 - F_{ST})p_A)(3F_{ST} + (1 - F_{ST})p_A)}{(1 + F_{ST})(1 + 2F_{ST})} \tag{13.6}$$

and with $s = 2, r = 1$ and then $s = 1, r = 1$,

$$Pr(A^2 B^2 \mid p_A, p_B) = Pr(AB \mid p_A, p_B) \frac{(F_{ST} + (1 - F_{ST})p_A)(F_{ST} + (1 - F_{ST})p_B)}{(1 + F)(1 + 2F)}. \tag{13.7}$$

Assume that an innocent suspect, with profile G_s, is drawn from the same sub-population as the criminal, who has profile G_c, but that the two people are not closely related. Let A and B be the two observed alleles. Then $Pr(G_c = AB \mid G_s = AB) = Pr(G_c = AB, G_s = AB)/Pr(G_s = AB) = Pr(A^2 B^2)/Pr(AB)$. Assume also that p_A and p_B are available only for a collection of sub-populations. Then, from (13.7), the probability that the criminal has a particular genotype given that the suspect has been found to have that type is

$$Pr(G_c = AB \mid G_s = AB) = 2 \frac{\{F_{ST} + (1 - F_{ST})p_A\}\{F_{ST} + (1 - F_{ST})p_B\}}{(1 + F_{ST})(1 + 2F_{ST})}, \tag{13.8}$$

known as the *conditional match probability*. Note that when $F_{ST} = 0$, the probability reduces to $2p_A p_B$, the basic result assuming Hardy–Weinberg equilibrium.

Similarly, the homozygote match probability may be obtained from (13.6), to give

$$Pr(G_c = A^2 \mid G_s = A^2) = \frac{\{2F_{ST} + (1 - F_{ST})p_A\}\{3F_{ST} + (1 - F_{ST})p_A\}}{(1 + F_{ST})(1 + 2F_{ST})}.$$

These are the equations referred to in recommendation 4.2 of the NRC (1996) report. They allow the scientist to obtain match probabilities for complete profiles. Single-locus probabilities from Balding and Nichols' (1994) formula are multiplied across loci. It should be stressed that the results hold for two people in the same sub-population, but are an average over sub-populations. Allele frequencies are an average over sub-populations and are not those in a particular sub-population. The last two equations allow population-wide allele frequencies to be used for sub-populations for which F_{ST} applies. A simple derivation of Balding and Nichols' (1994) formula for heterozygotes and homozygotes is presented in Harbison and Buckleton (1998). Logical implications of applying the principles of population genetics to the evaluation of DNA evidence are presented in Triggs and Buckleton (2002).

It is possible to assess the effect of population sub-structure on forensic calculations. For heterozygotes between alleles with equal frequencies p, the likelihood ratio is assessed at the source level as the reciprocal of the conditional match probabilities (an analogous result to (9.3)). For various values of F_{ST}, the likelihood ratios are presented Table 13.3. The effect of F_{ST} decreases as allele frequencies increase and is not substantial when $p = 0.1$ even for F_{ST} as high as 0.01 (Weir, 1998).

It is important to distinguish between *profile probabilities* and *random match probabilities*. It is very helpful to use the term *profile probability* for the chance of a single individual having a particular profile, in distinction to *random match probability* for the chance of a person having the profile when it is known that another person has the profile. The random match probability, therefore, explicitly requires statements about two profiles. Profile probabilities are of some interest but are unlikely to be relevant in forensic calculations (Weir, 2001b). It is of little consequence that the profile is rare in the population – what is

Table 13.3 Effects of population structure, as represented by F_{ST}, on the likelihood ratio, the reciprocal of the conditional match probability (13.8) for heterozygotes between alleles with equal frequency p

Allele frequency	F_{ST}			
	0	0.001	0.01	0.05
$p = 0.01$	5000	4152	1301	163
$p = 0.05$	200	193	145	61
$p = 0.1$	50	49	43	27

relevant is the rarity of the profile, given that one person (e.g., the perpetrator) has the profile. In other words, it is relevant to know the probability that the defendant would have the profile given that the perpetrator has the profile and that these are different people (Balding and Donnelly, 1995a). In practical cases, the pool of possible culprits usually contains individuals with differing levels of ancestry shared with the defendant and therefore differing between-person correlations. The task is then to present to the court the plausible range of match probabilities in a helpful and fair way. Further developments for mixed racial populations that avoid the approach of reporting separate estimates for each race are proposed by Triggs *et al.* (2000).

The distinction between profile and match probabilities is rarely made by practicing forensic scientists, and this is most likely because the two quantities have the same value in the simple case when 'product rule' calculations are valid. If there is no relatedness in a large population, due to either immediate family membership or common evolutionary history, and there is completely random mating and population homogeneity, and an absence of linkage, selection, mutation and migration, then all the alleles in a DNA profile are independent. The profile probability and the match probability are both just the product of the allele probabilities, together with factors of 2 for each heterozygous locus. So, when $F_{ST} = 0$, the probability reduces to p_A^2, the basic result assuming Hardy–Weinberg equilibrium for homozygotes.

13.6   RELATED INDIVIDUALS

The discussion in Section 13.5 has assumed that the reference population contains no related individuals. Unrelated individuals have a very low probability of sharing the same profile, but the probability increases for related individuals. In fact, relatives have the possibility of receiving the same genetic material from their common ancestors and therefore having the same DNA profile (Balding, 2000). So, the largest effect of dependencies between the DNA profiles of two individuals is when they are related. Other than for an identical twin, relationships such as brothers or fathers or cousins have very large effects on the likelihood ratio when their DNA profiles are not available. If it is possible that a sibling or a close relative could have been the contributor of the stain recovered at a crime scene, the likelihood ratio has to reflect that. Brothers, for example, have at least a 25% probability of sharing the same genotype at any locus. Consider, for example, the following two propositions:

- H_p, the suspect is the source of the crime sample;
- H_d, a relative of the suspect left the crime sample.

Let the evidence be the observation of alleles $A_i A_j$ of the source sample and of the receptor sample with population frequencies p_i and p_j. Assume that, if H_p

is true, then the numerator of the likelihood ratio is 1, an assumption which will not necessarily be true but will serve to illustrate the point. If there is no familial relationship between the source and the receptor sample then the denominator would be $2p_ip_j$ for $i \neq j$, and p_i^2 for $i = j$. The effects of different familial relationships are given in Table 13.4 from Weir and Hill (1993), where numerical values are given assuming allelic frequencies of 0.1. The numerator is assumed to have the value 1. See also Brookfield (1994) and Belin *et al.* (1997) for further examples.

Sjerps and Kloosterman (1999) develop further the scenarios involving relatives in analysing cases where the DNA profile of a crime sample exonerated the suspect as the origin of the recovered sample. In some situations, such a result can suggest that a close relative of the suspect might match the stain, in particular when the profiles share rare alleles.

Weir (2001a) completes the conditional probabilities of Table 13.4 to account for the population structure parameter F_{ST} (see Table 13.5, where F_{ST} is denoted θ for clarity).

Match probability values for a variety of specified alternatives (possible sources of the stain other than the suspect) that correspond to individuals who exhibit different degrees of relatedness to the suspect when there are full matching profiles are considered by Foreman and Evett (2001). The most common SGM-plus (second generation multiplex-plus, 10-locus short tandem repeat (STR) profiling system) profile was determined using databases routinely used in forensic casework. They took account of sampling error using the size-bias correction (Curran *et al.*, 2002) and recommended general match probability

Table 13.4 Match probability $Pr(G_c \mid G_s, H_d, I)$ that a relative has the same genotype as the suspect and the corresponding value for V assuming allelic frequencies of 0.1, from Weir and Hill (1993). (Reproduced by permission of The Forensic Science Society.)

Suspect	Relative	$Pr(G_c \mid G_s, H_d, I)$	Likelihood ratio, V
A_iA_j	Father or son	$(p_i + p_j)/2$	10
	Full brother	$(1 + p_i + p_j + 2p_ip_j)/4$	6.67
	Half brother	$(p_i + p_j + 4p_ip_j)/4$	16.67
	Uncle or nephew	$(p_i + p_j + 4p_ip_j)/4$	16.67
	First cousin	$(p_i + p_j + 12p_ip_j)/8$	25
	Unrelated	$2p_ip_j$	50
A_iA_i	Father or son	p_i	10
	Full brother	$(1 + p_i)^2/4$	3.3
	Half brother	$p_i(1 + p_i)/2$	18.2
	Uncle or nephew	$p_i(1 + p_i)/2$	18.2
	First cousin	$p_i(1 + 3p_i)/4$	30.8
	Unrelated	p_i^2	100

Table 13.5   Effects of family relatedness on match probability, $Pr(G_c \mid G_s, H_d, I)$, from Weir (2001a). Note the use of θ for F_{ST} for clarity. ©2001. John Wiley & Sons, Ltd. Reproduced with permission

Suspect	Relationship	$Pr(G_c \mid G_s, H_d, I)$
A_iA_j	Full sibs	$\dfrac{(1+p_i+p_j+2p_ip_j)+(5+3p_i+3p_j-4p_ip_j)\theta+2(4-2p_i-2p_j+p_ip_j)\theta^2}{4(1+\theta)(1+2\theta)}$
	Parent and child	$\dfrac{2\theta+(1-\theta)(p_i+p_j)}{2(1+\theta)}$
	Half sibs	$\dfrac{(p_i+p_j+4p_ip_j)+(2+5p_i+5p_j-8p_ip_j)\theta+(8-6p_i-6p_j+4p_ip_j)\theta^2}{4(1+\theta)(1+2\theta)}$
	First cousins	$\dfrac{(p_i+p_j+12p_ip_j)+(2+13p_i+13p_j-24p_ip_j)\theta+2(8-7p_i-7p_j+6p_ip_j)\theta^2}{8(1+\theta)(1+2\theta)}$
	Unrelated	$\dfrac{2[\theta+(1-\theta)p_i][\theta+(1-\theta)p_j]}{(1+\theta)(1+2\theta)}$
A_iA_i	Full sibs	$\dfrac{(1+p_i)^2+(7+7p_i-2p_i^2)\theta+(16-9p_i+p_i^2)\theta^2}{4(1+\theta)(1+2\theta)}$
	Parent and child	$\dfrac{2\theta+(1-\theta)p_i}{1+\theta}$
	Half sibs	$\dfrac{[2\theta+(1-\theta)p_i][2+4\theta+(1-\theta)p_i]}{2(1+\theta)(1+2\theta)}$
	First cousins	$\dfrac{[2\theta+(1-\theta)p_i][1+11\theta+3(1-\theta)p_i]}{4(1+\theta)(1+2\theta)}$
	Unrelated	$\dfrac{[2\theta+(1-\theta)p_i][3\theta+(1-\theta)p_i]}{(1+\theta)(1+2\theta)}$

Table 13.6   General match probability values recommended for use when reporting full SGM-plus profile matches, with an F_{ST} value of 0 in situation 6 and a value of 0.02 for situations 1–5, from Foreman and Evett (2001). (Reproduced by permission of Springer-Verlag.)

Situation	Relatedness with the suspect	Match probability
1	Sibling	1 in 10 000
2	Parent/child	1 in 1 million
3	Half-sibling or uncle/nephew	1 in 10 million
4	First cousin	1 in 100 million
5	Unrelated (sub-population)	1 in a billion
6	Unrelated (population)	1 in a billion

values for use when reporting full profile matches with an F_{ST} value of 0 in situation 6 and a value of 0.02 for situations 1–5 (see Table 13.6). A general discussion of this topic is presented in Foreman *et al.* (1997a) and Evett *et al.* (2000e).

13.7 MORE THAN TWO PROPOSITIONS

Consider a situation in which the evidence E is that of a DNA profile of a stain of body fluid found at the scene of a crime and of the DNA profile from a suspect which matches in some sense the crime stain. There are three propositions to be considered:

- H_p, the suspect left the crime stain;
- H_{d1}, a random member of the population left the crime stain;
- H_{d2}, a brother of the suspect left the crime stain.

This situation has been discussed by Evett (1992). Let θ_0, θ_1 and θ_2 denote the prior probabilities for each of these three propositions ($\theta_0 + \theta_1 + \theta_2 = 1$). Assume that $Pr(E \mid H_p) = 1$. Denote $Pr(E \mid H_{d1})$ by ϕ_1 and $Pr(E \mid H_{d2})$ by ϕ_2. Also, H_d, the complement of H_p, is assumed to be the conjunction of H_{d1} and H_{d2}. Then, using Bayes' theorem (3.3) and the law of total probability (1.10), a result analogous to (8.4) is obtained:

$$Pr(H_p \mid E) = \frac{Pr(E \mid H_p)\theta_0}{Pr(E \mid H_p)\theta_0 + Pr(E \mid H_{d1})\theta_1 + Pr(E \mid H_{d2})\theta_2}$$

$$= \frac{\theta_0}{\theta_0 + \phi_1\theta_1 + \phi_2\theta_2},$$

$$Pr(H_d \mid E) = \frac{\phi_1\theta_1 + \phi_2\theta_2}{\theta_0 + \phi_1\theta_1 + \phi_2\theta_2},$$

and hence the posterior odds in favour of H_p are

$$\frac{\theta_0}{\phi_1\theta_1 + \phi_2\theta_2}.$$

Let the relevant population size be N and let the number of siblings be n, where $N \gg n$. It can be assumed that $\theta_0 = 1/N, \theta_1 = (N - n)/N$ and $\theta_2 = (n - 1)/N$. The posterior odds in favour of H_p are approximately equal to

$$\frac{1}{\phi_1 N + \phi_2(n - 1)}.$$

Example 13.1 Consider a situation where two single-locus probes are used, each giving two bands which match the suspect (Evett, 1992). Each of the four bands has a frequency of 0.01 so the probability (ϕ_1) of a match from a random individual is $(4 \times 0.01^4) = 1/25\,000\,000$. If $N = 100\,000$, and there are no siblings, so that $n = 0$, then the posterior odds in favour of H_p are

$1/(N\phi_1) = 250$. If there is a brother, then $\phi_2 \simeq (1/4)^2, n = 2$ and the posterior odds are

$$1/\left\{\frac{1}{250} + \left(\frac{1}{4}\right)^2\right\} = \frac{1}{0.004 + 0.063} \simeq 15.$$

The existence of the brother has reduced the posterior odds by a factor of over 15 from 250 to 15.

Example 13.2 A more general approach involving a brother is discussed by Balding (1997, 2000). The existence of just one brother among the possible culprits can outweigh the effect of very many unrelated men for realistic values of the match probabilities. The possible culprits are the defendant, one brother of the defendant (note that the brother may, however, be missing, or refuse to collaborate with investigators, or it may not even be known whether or not the defendant has any brothers) and 100 unrelated men. Only the DNA profile of the defendant is available. Consider that the match probability for the brother is $1/100$ and the match probability for the other men is $1/1\,000\,000$. Suppose that the probability $Pr(I \mid H_p)$ of the non-DNA evidence I was the same for all the possible culprits (102 individuals). Then

$$Pr(H_p \mid E, I) = 1/\{1 + 0.01 + (100 \times 0.000\,001)\} = 0.99.$$

Thus $Pr(H_d \mid E, I) = 0.01$.

If the brother is ignored, then

$$Pr(H_p \mid E, I) = 1/\{1 + (100 \times 0.000\,001)\} = 0.9999$$

and $Pr(H_d \mid E, I) = 0.0001$. Consideration of the presence of the brother has increased the probability of the defence proposition by a factor of 100. See Section 13.9 for further details.

The consequences are more dramatic if other brothers, cousins or other relatives are members of the relevant population of culprits.

It may in some cases be plausible that the non-DNA evidence has approximately the same weight for the defendant as for some or all of the siblings. In this case, it can be assumed that the probability, $Pr(H_p \mid E, I)$, that the defendant is guilty is at most $1/(1 + nq)$, where n denotes the number of such siblings and q represents the sibling match probability (Balding and Donnelly, 1995a).

Note that the consideration of three propositions has led to consideration of posterior odds in favour of one of the propositions. The determination of a likelihood ratio under such circumstances is discussed in Section 8.1.3. The likelihood ratio is used for comparing propositions in pairs. For comparing more than two propositions, propositions have to be combined in some meaningful way to provide two propositions for comparison.

13.8   DATABASE SEARCHING

When a scientific expert has to assess the weight of DNA results, the manner in which the suspect was selected is crucial. The evaluation of evidence for the identification of a suspect which came from the use of a database has been a matter of debate for some time (Thompson and Ford, 1989). The compilation of DNA databases could enable police forces to collect samples taken during investigations of unsolved criminal cases, as well as samples from convicted felons, in order that such stored information could be used to select suspects in a way similar to the collection and storage of fingerprint records (see, for example, *Rise v. State of Oregon*). Databases are now established in all 50 states of the USA, in the UK and in other European countries.

Confusion surrounding the interpretation of the outcome of such a search can arise because the probability of a match increases as the database gets larger. Robertson and Vignaux (1995b) explain this confusion, stating:

> It is commonly claimed that the evidential value of a match, when a suspect is selected through a search in a database, is affected by the number of comparisons one has made. Certainly, the larger the database the more likely we are to find a match.

This leads to the erroneous conclusion that the larger the database the weaker the evidence. This is one reason for believing that the evidential value of a DNA match under such circumstances may be of little or no evidential value: the evidence is relevant if it is more (or less) likely to exist if the defendant is guilty than if he is innocent. In some cases, a match between the suspect and the aggressor is certain if the defendant is guilty. But a match between the suspect and the criminal is also certain when the match follows from a search of a database under the hypothesis he is not guilty, because the suspect was chosen based on the fact his DNA profile matches the criminal's profile. Therefore, the consequent likelihood ratio of 1 suggests that this evidence has no probative value: the evidence is as likely if the suspect is innocent as if he is guilty (Thompson and Ford, 1989).

It has long been recognised that it could be fallacious to apply the formula for posterior odds of identity in cases of selection of the suspect by the evidence of a match in a database without recognising that the probability of matching evidence is necessarily 1 in these cases, regardless of the identity or non-identity of the suspect with the criminal (Fairley, 1975). Confusion arises because it is not clear whether the scientist is concerned with the probability of finding a match or with the increase that arises, through the discovery of the match, in the probability that it was the accused who left the trace (Robertson and Vignaux, 1995b). Analyses of the second point have been provided by Balding and Donnelly (1995a, 1996), by Dawid and Mortera (1996) and by Evett and Weir (1998).

These analyses showed that the likelihood ratio is higher following a search than in a case where the size of the potential criminal population is known and no sequential searches have been performed. In fact, each person who does not match with the DNA profile of the recovered trace is excluded. Therefore, the exclusion of these individuals from the potential culprits increases the probability of involvment of the individual who matches. Although a database search is useful, it has to be emphasised that the strength of the overall case against the suspect can be much weaker than in the *probable cause setting*, defined by Balding and Donnelly (1996) as the setting in which the suspect has been identified on other grounds and subsequently subjected to DNA profiling. This is because of a lack of supporting evidence; no further incriminating evidence has been obtained (Balding and Donnelly, 1996; Donnelly and Friedman, 1999). Therefore, the discovery of a match in a database does not necessarily mean that the perpetrator of the crime has been found.

The fact that the likelihood ratio is greater than the reciprocal of the match probability may be justified through the arguments developed by Balding and Donnelly (1996) and Evett and Weir (1998).

Let H_p be the proposition that the suspect is the source of DNA found at the crime scene, H_d the proposition that the suspect is not the source of the DNA found at the crime scene, and E be the evidence that the profile of the DNA found at the crime scene and the profile of the DNA of the suspect match. Then the value of the evidence is given by

$$V = \frac{Pr(E \mid H_p)}{Pr(E \mid H_d)}.$$

A search has been made of a database which contains the DNA profiles of N named individuals. Exactly one of the profiles in the database matches that of the DNA found at the crime scene and that individual becomes the suspect. Note that V does not depend on the probability that a search through the database would find a matching profile. The evidence is not that at least one (or exactly one) profile of the individuals in the database matches the crime stain profile. Other information does not affect the value of the evidence. Other information will be heard at any trial and will be accounted for there. To assess the value of the evidence, including the outcome of the search, let O denote the event that no other individual in the database matches the profile of the crime stain. Also, E may be separated into two components, E_c, the profile of the crime stain and E_s, the profile of the suspect. Then V can be written as

$$V = \frac{Pr(E_c, E_s, O \mid H_p)}{Pr(E_c, E_s, O \mid H_d)}$$

$$= \frac{Pr(E_c, E_s, \mid H_p, O)}{Pr(E_c, E_s \mid H_d, O)} \frac{Pr(O \mid H_p)}{Pr(O \mid H_d)}. \tag{13.9}$$

In the probable cause setting, the value of the evidence is

$$\frac{Pr(E_c, E_s, \mid H_p, O)}{Pr(E_c, E_s \mid H_d, O)}$$

which is the first ratio in (13.9). Here the conditioning is extended to the information O. The numerator equals p, the profile frequency; the fact that there has been a database search does not affect this probability. The denominator expresses the probability that two individuals chosen at random match. Information O increases confidence in the rarity of the match probability. So,

$$\frac{Pr(E_c, E_s \mid H_p, O)}{Pr(E_c, E_s \mid H_d, O)} = \frac{p}{p^2} = \frac{1}{p}.$$

It is necessary to determine if the second ratio $Pr(O \mid H_p)/Pr(O \mid H_d)$ is smaller or greater than 1. Consider $\bar{O}$. This is the event that at least one of the other individuals in the database matches the profile of the crime stain. If H_d is true there are two ways in which $\bar{O}$ may occur. One of the other individuals may be the source of the stain, or none may be the source but at least one happens by chance to match E_c. If H_p is true, then only the second of these is possible. Thus $Pr(\bar{O} \mid H_p) < Pr(\bar{O} \mid H_d)$. Hence $Pr(O \mid H_p) > Pr(O \mid H_d)$ and the second ratio in (13.9) is greater than unity. A more extended development which involves the general discriminating power of the profiling system for the assessment of the second ratio is proposed by Evett and Weir (1998). Their approach also shows that the second ratio is greater than unity. Thus, the value of the evidence when there has been a database search is greater than when the probable cause setting applies.

Balding and Donnelly (1996) and Evett and Weir (1998) argue that, although the difference in value is difficult to quantify in general, it seems likely that the database search value will be only slightly greater than the simple likelihood ratio and that it is therefore convenient, and beneficial to the defendant, to calculate and report the simple value.

The above argument may be used to counter the following possible defence strategy: 'The DNA profile found at the scene of the crime occurs in the population with a frequency of 1 in a million. The police database contains 10 000 profiles. The probability that, on searching the database, a match will be found is thus $10\,000 \times (1/1\,000\,000) = 1/100$. This figure, rather than 1 in a million, is the relevant match probability. This is not nearly small enough to be regarded as convincing evidence against the defendant.' From this point of view, the effect of the database search is to weaken, very dramatically, the strength of the evidence against the defendant.

The argument also contradicts the following proposal from the NRC (1996) report. Consider a crime profile which has a profile frequency of p. Consider a database with N unrelated individuals in it. The probability that the profile of an individual from the database does not match the crime profile is $1 - p$ and, assuming independence, the probability that no profiles from the database match the crime profile is $(1 - p)^N$. Hence, the probability that at least one profile from the database matches the crime scene profile, by chance alone, is $1 - (1 - p)^N$ which, for small p such as occurs with DNA profiles, is approximately equal to Np, hence the figure $1/100$ in the previous paragraph. This result gives rise to the simple rule that, to determine the probability of a match in a search of a database, one should take the match probability p and multiply it by the size N of the database.

This simple rule is concerned with the probability that there is at least one match in the database. An extreme case can illustrate why the rule approximates to the answer for which it is designed and why it is not the right answer. Assume p, the match probability, is extremely small and N is extremely large, such that Np is close to one. It can be argued that the probability of finding at least one match increases as N increases, and may even become close to 1 as N approaches the population of the world. Note that the correct result is $1 - (1 - p)^N$, which will never be greater than 1. The simple rule, Np, cannot be used if it will give an answer greater than 1. As Np becomes larger, so the evidence reduces in value. However, the counter-argument (Balding and Donnelly, 1996; Balding 1997) is that the evidence becomes stronger as N becomes larger. This latter argument is the correct one. It makes more sense to attach greater value to the outcome of the search of a large database in which only one match has been found when all other members of the database have been eliminated from the enquiry. The suspect so identified is now one of a smaller overall population (smaller by the elimination of $N - 1$ members).

Balding and Donnelly (1996) make an interesting comment, in the light of later discussion (Stockmarr, 1999; Dawid, 2001; Devlin, 2000; Evett *et al.*, 2000c, d; Balding, 2002; Meester and Sjerps, 2003). An alternative pair of propositions is that the source of the crime stain is or is not in the database. The probability of the evidence of exactly one match in the database given the source is in the database is 1. The probability of the evidence of exactly one match in the database, given the source is not in the database, is $Np(1 - p)^{N-1}$. The likelihood ratio is then (assuming $(1 - p)^{N-1}$ to be 1) $1/Np$, which is the value given by the NRC (1996). This result assumes that each of the individuals in the database is, without the DNA evidence, equally likely to be the source. Note that it is possible for Np to be greater than 1, as has been pointed out by Donnelly and Friedman (1999). Such a result would imply that the evidence favours the defendant and may be possible if a sufficiently large number of people are profiled. Thus one could have the rather bizarre situation where the large the number of people who were profiled and found not to match the crime scene profile, the more support there would be for the defendant's case.

The flaw in this argument is explained very well by Balding (2002). He notes that many statisticians will instinctively feel that a database search weakens the evidence against the suspect because one is conducting multiple comparisons. He comments that in a database search there is a 'crucial distinction', namely that it is known in advance that exactly one proposition of the form 'X is the culprit' is true. Consider two scenarios. In the first, there is evidence that a particular proposition is true. This would be the case if there had been no database search. In the second, there is evidence that a particular proposition is true and that many other propositions are false. This would be the case if there had been a database search. Put in this way, the evidence of a database search in which no other matches were found strengthens the case against the defendant. The bigger the search which results in only a single match, the more reason there is to be convinced that the observed match is unique in the population.

13.8.1   Search and selection effect (double counting error)

Database searches are currently used by scientists in other areas of forensic science such as shoeprints and fingerprints. As above, there may be a concern that the fact that the mark was retrieved as a result of a database search in some way weakens the evidence from the comparison with the shoe, for example. But this is not the case (Evett *et al.*, 1998a). The likelihood ratio summarises all of the evidence which derives from the comparison. The fact that the mark (of the shoe) was found from a database search is relevant to the forming of prior odds in favour of the proposition that the suspect is the offender for that case. If there is no evidence other than the geography of the incident and arrest then the prior odds would be small, but in this case they would presumably be increased by the evidence of the property that was found at the suspect's home (Evett *et al.*, 1998a). Each piece of evidence must be considered only once in relation to each issue, otherwise its effect is unjustifiably doubled. However, this does not mean that once an item of evidence has been used by one decision maker for one purpose it cannot be used by another decision maker for another purpose. Thus, the fact that the police have used an item of evidence to identify a suspect does not mean that the court cannot use it to determine guilt. Of course, the court must not use the fact that the accused is in the dock as evidence of guilt and then also consider the evidence produced, since to do so would be to double-count the evidence which led to the arrest and which is also used in court.

The following example was presented by Robertson and Vignaux (1995a). A man might be stopped in the street because he was wearing a bloodstained shirt and the value of this has to be considered. It may be thought that because this was the reason for selecting this particular suspect the value of the evidence should be changed; it is less useful than if the suspect was arrested on the basis of other evidence. This is not correct. The power of the evidence is still determined by the ratio of the two probabilities of the accused having a bloodstained shirt if he is guilty and if he is not guilty. It is just that there happens to be less evidence in one case than the other. When the suspect is stopped because of a bloodstained shirt there may be no other evidence. When the suspect is arrested on the basis of other evidence and then found to have a bloodstained shirt, the likelihood ratio for the bloodstained shirt is to be combined with a prior which has already been raised by other evidence. The value of one item of evidence is not to be confused with the value of the evidence as a whole.

13.9 ISLAND PROBLEM

Consider an island on which there are $N + 2$ people, one of whom is murdered. Of the remaining $N + 1$ people, N are innocent and one is the murderer. Label these from 0 to N. There is trace evidence (e.g., a DNA profile frequency) which links the murderer to the crime. The frequency of this evidence in the relevant population is γ. The value of the evidence is then

$$1/(1 + N\gamma)$$

(Section 3.5.6).

The arguments in this section are based on those expressed in Balding and Donnelly (1995b) and Balding (2000). A discussion of extensions beyond the scope of this book are given in Balding and Donnelly (1995b) and Balding (2005). Other relevant articles are Balding (1995) and Dawid and Mortera (1996). However, the initial extension can be discussed here and provides a useful insight into ideas underlying inferences for forensic identification. There are two propositions to be considered:

- H_p, the suspect is the criminal;
- H_d, the suspect is not the criminal.

The background information is denoted I, and this is assumed to be independent of the evidence. If H_d is true, one of the other members of the island population is the criminal. Let C be the random variable denoting the criminal and s the identity of the suspect (s is one of $0, 1, \ldots, N$). The expression $C = s$ denotes that the suspect is the criminal. The expression $C = x$ denotes that individual x is the criminal.

The evidence, E, is that the criminal is known to have the DNA profile and a suspect s has been apprehended and observed to have the DNA profile. Proposition H_d is that the criminal is not the suspect, and this can be denoted $C \neq s$.

The probability of guilt, given $C = s$ and that the criminal and the suspect both have the trait, and assuming independence between E and I, may be written as

$$Pr(C = s \mid E, I)$$
$$= \frac{Pr(E \mid C = s)Pr(C = s \mid I)}{Pr(E \mid C = s)Pr(C = s \mid I) + \sum_{x \neq s} Pr(E \mid C = x)Pr(C = x \mid I)}.$$

$$(13.10)$$

Let $V_s(x)$ denote the likelihood ratio for x versus s,

$$V_s(x) = \frac{Pr(E \mid C = x)}{Pr(E \mid C = s)}.$$

The notation with s as a subscript is indicative of the asymmetry of the context in that the evidence is being considered with relation to propositions H_p and H_d that s is or is not the criminal. Let $w_s(x)$ be defined by the ratio

$$w_s(x) = \frac{Pr(C = x \mid I)}{Pr(C = s \mid I)}.$$

This ratio is neither an evidential value, since I is the conditioning, nor an odds, since the propositions $C = s$ and $C = x$ are not complementary. It is interesting to consider the relationships between different situations and the values of $w_s(x)$, as discussed in Balding (2000). If the case against s rests primarily on DNA evidence, there may be many x for which $w_s(x) \simeq 1$. For most sexual or violent crimes $w_s(x) \simeq 0$ when x refers to women, children and invalids. If there is strong alibi evidence or the victim has not been able to identify s then $w_s(x) \gg 1$. With this notation, (13.10) may be written as

$$Pr(C = s \mid E, I) = \frac{1}{1 + \sum_{x \neq s} V_s(x) w_s(x)},$$

a result which has already been presented in (8.13) using different notation. An example of this result was given in Section 13.7.

Partition the evidence E into E_c, the crime scene profile, and E_s, the suspect's profile, with the additional knowledge that $E_c = E_s$. Then

$$V_s(x) = \frac{Pr(E_c = E_s \mid C = x)}{Pr(E_c = E_s \mid C = s)}.$$

This may be thought of as the *random match probability* under four assumptions:

- the crime sample DNA is that of the perpetrator;
- the matches are unequivocal;
- if the defendant were the perpetrator then the defendant and crime sample DNA profiles would be certain to match; and
- the fact the defendant's DNA profile was investigated is not, in itself, informative about the profile

(Balding and Nichols, 1995). Balding and Nichols also explain why it is inappropriate to ignore the conditioning on the observed profile and take the match probability to be equivalent to the relevant frequency of the defendant's profile.

• A match involves two profiles, not one; there is no logical framework for linking profile frequencies with the defendant's guilt or innocence.

• Allowance has to be made for the possibility the perpetrator is related to the defendant.

• There is no logical framework for combining the DNA evidence, quantified by a profile frequency, with the non-DNA evidence.

13.10 MIXTURES

A mixed sample in this context is a stain (trace) which contains a mixture of genetic material from more than one person (Weir, 1995). Notably, formulae for mixed samples were proposed by Weir *et al.* (1997) based on Evett *et al.* (1991). The formulae are based on the assumption of independence of all the alleles in the mixture. This assumption implies Hardy–Weinberg and linkage equilibrium, as well as independence between individuals. Therefore, the level of dependence among individuals within the same population was ignored (or set equal to zero, $F_{ST} = 0$, Section 13.6). More recent treatments of the mixture problem involve the effect of a structured population and develop approaches dealing with situations where contributors to a DNA mixture are of different ethnic groups and there are different numbers of contributors (known and unknown). Examples are given by Harbison and Buckleton (1998), Buckleton *et al.* (1998), Curran *et al.* (1999b), Fukshansky and Baer (1999), Triggs *et al.* (2000), Fung and Hu (2000b, 2002) and Hu and Fung (2003).

A further development was proposed by Fukshansky and Baer (2000) in cases where one suspect who is unavailable for testing is not unknown, but is a known person whose relatives are available for genetic testing. Moreover, the value of the extra information offered by the peak area resulting from the electrophoretic analysis has also been considered (Evett *et al.*, 1998b; Gill *et al.* 1998a, b; Clayton *et al.*, 1998; Gill, 2001) . Some limitations on the evaluation of the formulae proposed for the evaluation of DNA profiles with more than one contributor have been solved. For example, it is not necessary that all unknown persons should belong to the same ethnic group or that there should be no relationship between the non-tested (unknown) persons subjected to the statistical analysis of the stain or between the unknown and tested persons, irrespective of whether these tested persons have contributed to the stain or not. An extension of these ideas to apply to paternity cases is described by Liao *et al.* (2002).

As a simple example, Evett and Weir (1998) show how the evidential value of a DNA profile can be substantially reduced when there is clearly more than one contributor. Assume the evidence is from a rape committed by one individual.

The mixture obtained from a vaginal swab has a profile with alleles *abc*. The victim has alleles *ab* and the suspect has allele *c*. The prosecution proposition H_p is that the contributors of the mixture were the victim and the suspect. The defence proposition H_d is that the contributors of the stain were the victim and an unknown person. If H_p is true, the numerator of the likelihood ratio equals 1 because the mixture profile is exactly as expected under this situation. If H_d is true, the donor of part of the mixture (if not the suspect) could be an individual having profile *ac*, *bc* or *cc*. The denominator of the likelihood ratio is given by the product of profile probabilities of all these potential donors. The likelihood ratio is then

$$V = \frac{1}{2p_a p_c + 2p_b p_c + p_c^2}.$$

The presence of the victim's alleles (*ab*) has weakened the value of the evidence against the suspect because it has increased the number of potential contributors of the mixed sample.

It has been noted (Section 13.5) that the profile probability generally does not represent the correct estimate of the conditional probability in the denominator of the likelihood ratio. The scientist is interested, for example, in $Pr(\text{offender} = A_i A_j \mid \text{suspect} = A_i A_j)$ not in $Pr(\text{offender} = A_i A_j)$. The denominator of the previous example should then be rewritten as

$$Pr(A_a A_c \mid A_c A_c) + Pr(A_b A_c \mid A_c A_c) + Pr(A_c A_c \mid A_c A_c),$$

where the first term of each conditional probability refers to the DNA profile of the offender (the real donor of the mixed sample), and the second to the suspect's profile. Therefore, a likelihood ratio of

$$\frac{1}{Pr(A_a A_c \mid A_c A_c) + Pr(A_b A_c \mid A_c A_c) + Pr(A_c A_c \mid A_c A_c)}$$

can be rewritten as

$$\frac{Pr(A_c A_c)}{Pr(A_a A_c, A_c A_c) + Pr(A_b A_c, A_c A_c) + Pr(A_c A_c, A_c A_c)}. \tag{13.11}$$

If there are x alleles of type a out of a total of n alleles sampled from a sub-population, then the probability that the next allele sampled will be of type a is

$$\frac{x F_{ST} + (1 - F_{ST}) p_a}{1 + (n - 1) F_{ST}}$$

where F_{ST} represents the coancestry coefficient (see Section 13.5). The observation of one allele a in the sub-population makes it likely that a is more common in the sub-population than in the general population. The likelihood ratio, V becomes

$$V = \frac{(1+F_{ST})(1+2F_{ST})}{[2F_{ST}+(1-F_{ST})p_c]\{2(1-F_{ST})(p_a+p_b)+[3F_{ST}+(1-F_{ST})p_c]\}} \quad (13.12)$$

(Harbison and Buckleton, 1998). The same scenario but with additional information that the contributors of a DNA mixture are of different ethnic groups is discussed in Fung and Hu (2002) and Hu and Fung (2003), and some results are given in Table 13.7. For example, Buckleton *et al.* (1998) mention that in the Superior Court of the State of California for the County of Los Angeles (case number BA097211) the suspect was an African-American, the victims were Caucasian-Americans and the true perpetrator(s) could have been from any ethnic group(s) (Weir, 1995). Situation (c) in Table 13.7 corresponds to the one previously developed by Harbison and Buckleton (1998) for one ethnic group (13.12). Scenarios with three- and four-allele mixtures and propositions involving more than one unknown contributor shared between two different sub-populations are also presented in Fung and Hu (2002). The calculations consider indistinguishable contributors. Sometimes the major and minor contributors of a mixed stain can be inferred by considering the peak areas (intensities) of the alleles, and it may be possible to determine which alleles are from the same contributors. If so, the number of possible contributors would be reduced and the problem would be simplified. Analogously, Curran *et al.* (1999b) proposed a flexible approach to the same problem allowing the evaluation of many different mixed stain DNA profiles. All these treatments assumed a specific number of unknown contributors.

The calculation of likelihood ratios under plausible ranges of numbers of contributors may provide a solution to this. The scientist could then report the more conservative results (Curran *et al.*, 1999b). Through a real case

Table 13.7 Likelihood ratio for three-allele mixed sample, heterozygous victim and homozygous suspect. The victim has alleles a, b, the suspect has allele c. There are three ethnic groups under consideration, labelled 1, 2 and 3. The co-ancestry coefficient F_{ST} for the ethnic group, 1, of the unknown donor is labelled θ_1 for clarity. The allelic frequencies are p_{a1}, p_{b1} and p_{c1} in ethnic group 1. Ethnicities of the unknown donor, the victim and the suspect are considered. From Fung and Hu (2002). (Reproduced by permission of Springer-Verlag.)

Case	Unknown	Victim	Suspect	Likelihood ratio
(a)	1	1	1	$\dfrac{(1+3\theta_1)(1+4\theta_1)}{[2\theta_1+(1-\theta_1)p_{c1}][7\theta_1+(1-\theta_1)(p_{c1}+2p_{a1}+p_{b1})]}$
(b)	1	1	2	$\dfrac{(1+\theta_1)(1+2\theta_1)}{(1-\theta_1)p_{c1}[5\theta_1+(1-\theta_1)(p_{c1}+2p_{a1}+p_{b1})]}$
(c)	1	2	1	$\dfrac{(1+\theta_1)(1+2\theta_1)}{[2\theta_1+(1-\theta_1)p_{c1}][3\theta_1+(1-\theta_1)(p_{c1}+2p_{a1}+p_{b1})]}$
(d)	1	3	3	$\dfrac{1}{p_{c1}[\theta_1+(1-\theta_1)(p_{c1}+2p_{a1}+p_{b1})]}$

example (Superior Court of the State of California for the County of Los Angeles, case number BA097211; three-allele mixture with homozygous victim and heterozygous suspect), Buckleton *at al.* (1998) varied the number of unknown contributors, r, from 2 to 10 to determine the range of likelihood ratio values, suggesting that r could not credibly exceeed 10 so that the maximum reduction of the likelihood ratio is a factor of 4 with the two extreme values. An alternative approach to solve the same problem has recently been proposed by Lauritzen and Mortera (2002).

13.11 ERROR RATE

Proper consideration of the role of error rates in the evaluation of DNA profiles is important. The need to consider errors in the assessment of forensic scientific evidence in general has been mentioned by Gaudette (1986, 1999). The possibility of events which are 'outrageous', with very small probabilities, is mentioned by Meier and Zabell (1980) in connection with a forged document (Section 7.3.6) in the context of the assessment of handwriting evidence.

When evaluating the strength of DNA evidence for proving that two samples have a common source, one must consider two factors. One factor is the random match probability. A coincidental match occurs when two different people have the same DNA profile. The second factor is the probability of a false positive. A false positive occurs when a laboratory erroneously reports a DNA match between two samples that actually have different profiles. A false positive may occur due to error in the collection or handling of samples, misinterpretation of test results, or incorrect reporting of test results (Thompson, 1995). Either a coincidental match or a false positive could cause a laboratory to report a DNA match between samples from different people. Thus the random match probability and the false positive probability should both be considered in order to make a fair evaluation of DNA evidence. Laboratory error rate, as determined, for example, in proficiency testing, does not necessarily equate to the false positive probability in a particular case. The unique circumstances of each case may make various types of error more or less likely than a conjectured error rate. Nevertheless, data on the rate of various types of error in proficiency testing can provide insight into the likely range of values for a particular case (Thompson, 1997; Koehler, 1997b). When DNA evidence is presented in court, juries typically receive data on the probability of a coincidental match only (Kaye and Sensabaugh, 2000). A second practical difficulty is the presentation of a logical framework which takes account of both the probability of a match and of an error. Various suggestions have been made (Robertson and Vignaux, 1995a; Balding and Donnelly, 1995a; Balding, 2000). For example:

- In order to achieve a satisfactory conviction based primarily on DNA evidence, the prosecution needs to persuade the jury that the relevant error probabilities are small.

- If the probability of an error, such that the DNA profile of one person x is distinct from that of s and the observation of matching profiles is due to an error in one or both recorded profiles, is much greater than the probability of matching profiles between x and s, then the latter probability is effectively irrelevant to the weight of the evidence. Extremely small match probabilities can therefore be misleading unless the relevant error probabilities are also extremely small.

- What matters are not the probabilities of any profiling or handling errors but only the probabilities of errors that could have led to the observed DNA profile match.

An alternative framework for considering the role that error may play in determining the value of forensic DNA evidence in a particular case is presented in Thompson *et al.* (2003). Even a small false positive probability can, in some circumstances, be highly significant so serious consideration has to be given to its estimation. Accurate estimates of the false positive probabilities can be crucial for assessing the value of DNA evidence.

Consider two propositions:

- H_p, the crime scene stain came from a suspect;
- H_d, the crime scene stain did not come from a suspect.

The evidence E is a report of a DNA match between the suspect's profile and the profile of the sample. (See also Chapter 14 for consideration of reported matches.) The probability of a random match and the probability of a false positive both contribute to $Pr(E \mid H_d)$. Let M denote a true match. It is assumed that either the suspect and the crime scene stain have matching DNA profiles (M) or the suspect and the crime scene stain do not have matching DNA profiles ($\bar{M}$). From the law of total probability (1.10),

$$Pr(E \mid H_p) = Pr(E \mid M, H_p)Pr(M \mid H_p) + Pr(E \mid \bar{M}, H_p)Pr(\bar{M} \mid H_p)$$

and

$$Pr(E \mid H_d) = Pr(E \mid M, H_d)Pr(M \mid H_d) + Pr(E \mid \bar{M}, H_d)Pr(\bar{M} \mid H_d).$$

The value of the evidence is then

$$\frac{Pr(E \mid H_p)}{Pr(E \mid H_d)} = \frac{Pr(E \mid M, H_p)Pr(M \mid H_p) + Pr(E \mid \bar{M}, H_p)Pr(\bar{M} \mid H_p)}{Pr(E \mid M, H_d)Pr(M \mid H_d) + Pr(E \mid \bar{M}, H_d)Pr(\bar{M} \mid H_d)}.$$

Assume that $Pr(E \mid M)$ is independent of H_p and H_d; that is, the probability that a match will be reported if there really is a match is not affected by whether the match is coincidental. Consequently, $Pr(E \mid M, H_p) = Pr(E \mid M, H_d) = Pr(E \mid M)$.

The suspect and crime scene stain will necessarily have matching DNA profiles if the suspect is the source of the stain so $Pr(M \mid H_p) = 1$ and $Pr(\bar{M} \mid H_p) = 0$. Finally, because $\bar{M}$ can only arise under H_d, $Pr(E \mid \bar{M}, H_d)$ can be simplified to $Pr(E \mid \bar{M})$. So, the likelihood ratio becomes:

$$\frac{Pr(E \mid H_p)}{Pr(E \mid H_d)} = \frac{Pr(E \mid M)}{Pr(E \mid M)Pr(M \mid H_d) + Pr(E \mid \bar{M})Pr(\bar{M} \mid H_d)}.$$

In this version of the likelihood ratio, the term $Pr(E \mid M)$ is the probability that the laboratory will report a match if the suspect and the crime scene stain have matching DNA profiles, and is assumed to be 1.

The term $Pr(M \mid H_d)$ is the probability of a coincidental match. For a comparison between single-source samples, $Pr(M \mid H_d)$ is the random match probability, denoted γ, and $Pr(\bar{M} \mid H_d)$ is the complement of the random match probability. The term $Pr(E \mid \bar{M})$ is the false positive probability, denoted ϵ. Thus

$$\frac{Pr(E \mid H_p)}{Pr(E \mid H_d)} = \frac{1}{\gamma + \{\epsilon(1 - \gamma)\}}.$$

The influence of variations in γ, ϵ and the prior odds in favour of H_p on the posterior odds that the suspect was the source of the crime scene stain is shown in Table 13.8.

The prior odds presented in Table 13.8 are designed to correspond to two distinct case types that vary in how strongly the suspect is implicated as the source of the specimen by evidence other than the DNA match. Prior odds of 2:1 describe a case in which the other evidence is fairly strong but not sufficient, by itself, for conviction. It has been reported that DNA testing leads to the exclusion of approximately one-third of suspects in sexual assault cases. Hence, prior odds of 2:1 might describe a typical sexual assault case submitted for DNA testing.

Table 13.8 Posterior odds that a suspect is the source of a sample that reportedly has a matching DNA profile, as a function of prior odds, random match probability, and false positive probability. Extracted from Thompson *et al.* (2003). (Reprinted with permission from ASTM international)

Prior odds	Random match probability	Probability of a false positive	Posterior odds
2:1	10^{-9}	0	2 000 000 000
2:1	10^{-9}	0.0001	20 000
2:1	10^{-6}	0	2 000 000
2:1	10^{-6}	0.0001	19 802
1:1000	10^{-9}	0	1 000 000
1:1000	10^{-9}	0.0001	10
1:1000	10^{-6}	0	1 000
1:1000	10^{-6}	0.0001	9.9

Prior odds of 1:1000 describe a case in which there is almost no evidence apart from the DNA match. The random match probabilities presented are chosen to represent two values that may plausibly arise in actual cases. Random match probabilities on the order of 1 in 1 billion are often reported when laboratories are able to match two single-source samples over ten or more STR loci. Random match probabilities closer to 1 in 1 million are common when fewer loci are examined, when the laboratory can obtain only a partial profile of one of the samples. The probability of a false positive in any particular case will depend on a variety of factors. Some years ago there was a suggestion that the overall rate of false positives was between 1 in 100 and 1 in 1000 (Koehler, 1995). Of course, for cases in which special steps, such as repeat testing, have been taken to reduce the chance of error the false positive probability will be reduced. If two independent tests comparing the same samples each had a false positive probability of one in 100, then the probability of a false positive on both tests would be 1 in 10 000. A false positive probability of zero is also included for purposes of comparison. More results are available in Thompson *et al.* (2003). Finally, it can be shown that the probability of a false negative is not relevant, at least to a first approximation, for the probability of guilt (Balding, 2000).

14

Bayesian Networks

14.1 INTRODUCTION

Methods of formal reasoning have been proposed to assist forensic scientists and jurists to understand all of the dependencies which may exist between different aspects of evidence and to deal with the formal analysis of decision making. One of the more prevalent is a diagrammatic approach that uses graphical probabilistic methods, such as Bayesian networks (BNs), also called Bayes nets. These have been found to provide a valuable aid in the representation of relationships between characteristics of interest in situations of uncertainty, unpredictability or imprecision.

The use of graphical models to represent legal issues is not new. Charting methods developed by Wigmore (1937) can be taken as a predecessor of modern graphical methods such as BNs. Examples of the use of such charts, which were developed to provide formal support for the reaching of conclusions based on many pieces of evidence, can be found in Robertson and Vignaux (1993c), Schum (1994) and Anderson and Twining (1998).

The use of such probabilistic networks has been revived with the analyses of complex and famous cases such as the Collins case (Edwards, 1991; see also Section 4.4), and the Sacco–Vanzetti case (Kadane and Schum, 1996) with an emphasis on the credibility and relevance of testimonial evidence. More recently, the Omar Raddad case (Levitt and Blackmond Laskey, 2001) and the O.J. Simpson trial (Thagard, 2003) have also been analysed using graphical models.

Graphical models in the assessment of scientific evidence are described here with particular reference to the role of missing evidence, error rates (false positives), transfer evidence, combination of evidence, and cross-transfer evidence. The issues involved in the determination of the factors (nodes), associations (links) and probabilities to be included are discussed.

Pre-assessment is also approached through the use of BNs. An example is presented using fibres as transfer evidence. The aspects developed here using graphical methods have been previously discussed in Section 7.2.1 and Chapter 12.

Statistics and the Evaluation of Evidence for Forensic Scientists: Second Edition
C.G.G. Aitken and F. Taroni © 2004 John Wiley & Sons, Ltd   ISBN: 0-470-84367-5

14.2 BAYESIAN NETWORKS

A common set of issues has begun to emerge surrounding the representation of problems which are structured with belief networks (Cowell *et al.*, 1999). Bayesian networks are a widely applicable formalism for a compact representation of uncertain relationships among parameters in a domain (in this case, forensic science).

These graphical probabilistic models combine probability theory and graph theory. They provide a natural tool for dealing with two of the problems, uncertainty and complexity, that occur throughout applied mathematics and engineering (Jordan, 1999).

Fundamental to the idea of a graphical model is the combination of simpler parts. Probability theory provides the glue whereby the parts are combined, ensuring that the system as a whole is coherent and that inferences can be made. The task of specifying relevant equations can be made invisible to the user and arithmetic can be almost completely automated. Most important, the intellectually difficult task of organising and arraying complex sets of evidence to exhibit their dependencies and independencies can be made visual and intuitive.

Bayesian networks are a method for discovering valid, novel and potentially useful patterns in data where uncertainty is handled in a mathematically rigorous, but simple and logical, way. A BN is a collection of nodes, representing uncertain state variables, linked by arrows (also called arcs or edges) that represent either causal or evidential relationships. For the purposes of this discussion, each variable has a finite number of mutually exclusive states. Extensions to continuous variables are possible but not discussed here. A BN represents relationships amongst uncertain events by means of nodes and arrows. These are combined to form what is known as a *directed acyclic graph* (DAG), that is, one in which no loops or double-headed arrows are permitted.

If a node A has no entering arrows, it is called a *source* or *parent* node and a table containing unconditional probabilities $Pr(A)$ will be required. On the other hand, if A receives arrows from other variables $B_1, \ldots, B_n$, then A is called a *child* node and the variables $B_1, \ldots, B_n$ are the parent nodes. The table of probabilities (*node probability table*) of node A will contain *conditional* node probabilities $Pr(A \mid B_1, \ldots, B_n)$.

The combination of nodes and arrows constitutes paths through the net. Therefore, a net can be taken as a compact graphical representation of an evolution of all possible stories related to a scenario. Attention is concentrated on BNs essentially because they are easy to develop. Examples of the use of BNs in forensic science have been given in several papers (Aitken and Gammerman, 1989; Dawid and Evett, 1997; Dawid *et al.*, 2002; Evett *et al.*, 2002; Garbolino and Taroni, 2002; Aitken *et al.*, 2003; and Mortera *et al.*, 2003).

In summary, the use of BNs has some key advantages that could be described as follows:

- the ability to structure inferential processes, permitting the consideration of problems in a logical and sequential fashion;
- the requirement to evaluate all possible stories;
- the communication of the processes involved in the inferential problems to others in a succinct manner, illustrating the assumptions made at each node;
- the ability to focus the discussion on probability and underlying assumptions.

14.2.1 The construction of Bayesian networks

For the construction of BNs, it is important to note that they do not represent the flow of information, but serve as a direct representation of a part of the real world (Jensen, 2001). This means that, through the use of a BN, an expert can articulate their subjective view of a real-world system both graphically and numerically. Therefore, the model which is obtained as a result of the modelling process will principally be influenced by the *properties* and the expert's *individual view*, *perception*, and ultimately, extent of *understanding*, of the domain of interest. The problem is well posed in Dawid *et al.* (2002), where the authors argue that finding an appropriate representation of a case under examination is crucial for several reasons (viability, computational routines, etc.), and that the graphical construction is to some extent an art-form, but one which can be guided by scientific and logical considerations. The search for good representations for specific problems therefore is an important task for continuing research in this area, and was addressed by Taroni *et al.* (2004).

Given this, the question of the appropriateness of a given BN should always be regarded with respect to the context in which its construction took place. For example, there may be situations in which the knowledge about a domain of interest is severely limited. In addition, there may be processes taking place which are incompletely understood and apparently random,. Furthermore, the imperfect domain knowledge may be impossible to improve, or may only be improved at an unacceptably high cost.

Nevertheless, there is a prospect of evolution, and a BN should be considered within a continuous process of development. A BN can be taken as an instant representation of a given state of knowledge about a problem of interest. As new knowledge becomes available, the qualitative and/or quantitative specifications may be adapted in order to account for the newly acquired understanding of domain properties.

It has also been reported that different models can be used to represent the questions surrounding the same problem, because the same problem can be approached at different levels of detail, and because existing opinions about domain properties are diverging:

> either you agree with me that E is relevant for H, but our likelihoods are different, or you believe that E is directly relevant for H, and I believe that it is only indirectly relevant, or you believe that it is relevant and I believe it is not. These disagreements explain why we can offer different Bayesian networks models for the same hypothesis. (Garbolino, 2001)

Each node represents a random variable that can assume either discrete or continuous values, though only discrete nodes which take a finite number of states are considered here.

There are three basic types of connections among nodes in a BN: *serial, diverging* and *converging* connections. These are illustrated in Figure 14.1.

There is a *serial* connection linking three nodes A, B, and C when there is an arrow from A to B, another one from B to C and no arrows from A to C (Figure 14.1(a)). A serial connection is appropriate when we judge that knowledge of the truth state of A provides relevant information about the occurrence of B and knowledge of the truth state of B in turn provides relevant information about C but, when the truth state of B is known, then knowledge of the state of A does not provide any more relevant information about C. That

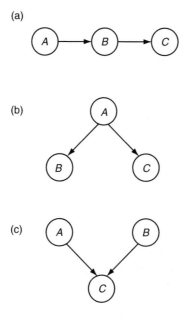

Figure 14.1 Basic connections in Bayesian networks: (a) serial, (b) diverging and (c) converging.

is, A influences C through B but only B directly influences C or, in other words, B *screens off* C from A. If the value of B is known, then A and C are probabilistically independent, that is, $Pr(A \mid B, C) = Pr(A \mid B)$. A serial connection among three nodes is the simplest example of what is known as a *Markov chain*.

As an example, let A be the proposition that the suspect is the offender, B that the bloodstain found on the crime scene comes from the suspect, and C that the suspect's blood sample and the bloodstain from the crime scene share the same DNA profile. Then A is relevant for B and B for C but, given B, the cause of the presence of blood could be different from A.

An example of a *diverging* connection linking A, B and C obtains when there are two arrows originating from A and pointing to B and C, and no arrow between B and C (Figure 14.1(b)). It is said that A *separates* B from C. If the value of A is known, then B and C are probabilistically independent, that is, $Pr(B \mid A, C) = Pr(B \mid A)$ and $Pr(C \mid A, B) = Pr(C \mid A)$. A diverging connection is the graphical representation of what may be called a *spurious correlation* .

Nodes B and C are correlated because they both depend on a third factor, A. When A is fixed, the correlation vanishes. There are many examples of such spurious correlations. For example, a positive correlation may be shown between the number of doctors in a town and the number of deaths in a town. As the number of doctors (B) increases, so does the number of deaths (C). This does not mean that doctors are bad for one's health. Rather it is the case that both factors are correlated positively with the population of the town (A). Another example is where A is the proposition that the suspect has been in contact with the victim, B is that the bloodstain on the suspect's clothes comes from the victim and C is that the bloodstain on the victim comes from the suspect.

An example of a *converging* connection linking A, B and C is when there is an arrow from A pointing to C and another one from B pointing to C, and no arrow between A and B (Figure 14.1(c)). It is said that A and B are probabilistically independent unless either the value of C or the value of any of the children of C is known. Another way of expressing the same idea is to say that A and B are conditionally dependent given the value of C (Jensen, 2001). Thus $Pr(AB) = Pr(A)Pr(B)$ but $Pr(AB \mid C)$ may not be equal to $Pr(A \mid C)Pr(B \mid C)$. Contrast this with the discussion in Section 1.6.7 where there are events which are conditionally independent but not unconditionally independent.

For example, let A be the proposition that the suspect is the offender and B that the bloodstain found on the scene of the crime comes from the offender: knowledge that one of these events occurred would not provide information about the occurrence of the other, but if it is true that the bloodstain found at the crime scene comes from the suspect (proposition C), then A and B become related. Converging connections in BNs are particularly important because they represent a very common pattern of reasoning: conditional dependence or *explaining away*.

d-separation properties

The criterion of *d*-separation, where *d* denotes *directional*, is a graphical criterion (Pearl, 1988) that designates the blocking (or stopping) of the flow of information (or of dependencies) between variables that are connected through a sequence of consecutive arrows (called a *path*). Consider this concept in case of the three basic connections (serial, diverging and converging) that are possible in BNs (as in Figure 14.1):

- In serial and diverging connections, a path is said to be *d*-separated if the middle variable is *instantiated* (a variable is called 'instantiated' if its state is changed from unknown to known).

- In converging connections, on the otherhand, a path is called *d*-separated as long as the intermediate variable, or one of its descendants, is *not* instantiated.

In other words, if two variables in a causal network are *d*-separated, then changes in the truth state of one variable will have no impact on the truth state of the other variable. If two variables are not *d*-separated, they are called *d*-connected (Jensen, 2001).

Chain rule for Bayesian networks

Through the use of Bayes' theorem, every joint probability distribution can be decomposed as a product of conditional probabilities. However, the joint probability table grows exponentially with the number of parameters. This complexity can be reduced when working with BNs, where it is supposed that a variable, given knowledge of its parents, is independent of all the variables which are not its descendants. If the conditional relationships implied by the structure of a Bayesian network hold for a set of variables $A_1, \ldots, A_n$, then the joint probability distribution $Pr(A_1, \ldots, A_n)$ is given by the product of all specified conditional probabilities

$$Pr(A_1, \ldots, A_n) = \prod_{i=1}^{n} Pr(A_i \mid \text{par}(A_i)),$$

where $\text{par}(A_i)$ denotes the set of parental variables of A_i.

Consider the chain rule for the three basic sequential connections that are possible in BNs (Figure 14.1). For a serial connection from A to C via B, $Pr(A, B, C) = Pr(A)Pr(B \mid A)Pr(C \mid A, B)$ can be reduced to $Pr(A, B, C) = Pr(A)Pr(B \mid A)Pr(C \mid B)$. For a diverging connection, the joint probability can be written as $Pr(C, A, B) = Pr(A)P(B \mid A)Pr(C \mid A)$, whereas for a converging connection it would be $Pr(A, B, C) = Pr(A)Pr(B)Pr(C \mid A, B)$.

Bayesian network formalism

The key feature of a BN is the fact that it provides a method for decomposing a joint probability distribution of many variables into a set of local distributions of a few variables within each set. This facilitates the investigation of relationships amongst the variables in the context of a particular case. The network discussed here consists only of nodes which are binary (more than two states in a variable are used in the fibres examples); they represent events and take only one of two values 'true' and 'false'.

A BN formalism is very appropriate in the context of criminal investigations since it is the development which is most responsible for progress in the construction of systems which are capable of handling uncertain information in a practical manner.

Bayesian networks are a useful tool for forensic scientists as they assist in the construction of a logical framework in complex situations, and the need for support for such a construction has been already suggested (Friedman, 1986a, b; Edwards, 1991; Aitken and Gammerman, 1989; Schum, 1994, 1999; Aitken *et al.*, 1996 a, b; Kadane and Schum, 1996; Dawid and Evett, 1997; Tillers 2001; Evett *et al.*, 2002; Mortera *et al.*, 2003).

The graphical nature of the approach facilitates formal discussion of the structure of proposed models. The BN enables the description of the uncertain relationships amongst factors of concern in a criminal investigation. On a quantitative level, the approach enables the user's subjective knowledge to be incorporated into the model on the same basis as more objectively derived data (e.g., data from surveys of DNA profile frequencies). These features allow the creation of a model which may contain mathematical relationships as well as subjective elements. The subjective elements are provided by the experience of the people who contribute to the modelling of the system.

Moreover, the graph enables a study of the sensitivity of the outcome to changes in the truth state of other variables of interest. Evett *et al.* (2002) suggest such an approach and show that it can be used for case pre-assessment too.

Two simple examples are presented to illustrate how a probabilistic model may be represented graphically.

Example 14.1 This relates evidence E, a match between characteristics of the recovered stain and DNA profile of a suspect, and the proposition that the suspect is the donor of the stain (H_p). The relationship is shown in Figure 14.2. The directed arrow from H (where H denotes either H_p or H_d) to E illustrates that probabilities are known for $Pr(E \mid H_p)$ and $Pr(E \mid H_d)$. It is desired to determine $Pr(H_p \mid E)$. Given values for $Pr(H_p), Pr(E \mid H_p)$ and $Pr(E \mid H_d)$, it is possible, using Bayes' theorem, to determine $Pr(H_p \mid E)$:

$$Pr(H_p \mid E) = \frac{Pr(E \mid H_p) \times Pr(H_p)}{Pr(E \mid H_p) \times Pr(H_p) + Pr(E \mid H_d) \times Pr(H_d)}.$$

Figure 14.2 Bayesian network for evidence E and proposition H.

Example 14.2 This relates evidence of a reported match (*RM*) between the DNA profile of a bloodstain found on the clothing of a victim of a crime and the DNA profile of a suspect, the event of a true match (*M*) between these two profiles and the proposition that the suspect is the source of the bloodstain on the victim's clothing (H_p). The relationship is shown in Figure 14.3, and is an example of a serial connection.

The directed arrows from *H* to *M* and from *M* to *RM* show that probabilities are known or are required for $Pr(M \mid H_p), Pr(M \mid H_d), Pr(RM \mid M)$ and $Pr(RM \mid \bar{M})$, where $\bar{M}$ is the complement of *M* and denotes no match. Also, and importantly, the separation of the node for *RM* from the node for H_p by the node for *M* shows that *RM* is conditionally independent of H_p, given *M*. Analogously to Example 14.1, given values for $Pr(M \mid H_p)$, $Pr(M \mid H_d)$, $Pr(RM \mid M)$, $Pr(RM \mid H_d)$ and $Pr(H_p)$, it is possible, using Bayes' theorem, to determine $Pr(H_p \mid RM)$:

$$Pr(H_p \mid RM) = \frac{Pr(RM \mid H_p) \times Pr(H_p)}{Pr(RM \mid H_p) \times Pr(H_p) + Pr(RM \mid H_d) \times Pr(H_d)},$$

$$Pr(RM \mid H_p) = Pr(RM \mid M, H_p) \times Pr(M \mid H_p) + Pr(RM \mid \bar{M}, H_p) \times Pr(\bar{M} \mid H_p)$$

$$= Pr(RM \mid M) \times Pr(M \mid H_p) + Pr(RM \mid \bar{M}) \times Pr(\bar{M} \mid H_p),$$

with a similar expression for $Pr(RM \mid H_d)$. More details on this practical example are presented in Section 14.5.

More complicated diagrams can be analysed in a similar manner known as propagation, though the procedures become more complicated with the diagrams. Analysis may be done with software packages such as HUGIN.

From the previous examples, the following two intuitive principles can be inferred:

- The cause produces the effect: knowing that the cause happened, it can be foreseen that the effect will or might probably occur. This is a *predictive* line of reasoning.

- The effect does not produce the cause; but knowing that the effect occurred, it may be inferred that the cause probably occurred. This is a line of reasoning against the causal direction that is termed *diagnostic*.

Various applications of BNs are discussed in Sections 14.3 to 14.8.

Figure 14.3 Bayesian network for a serial connection for a reported match *RM* in a DNA profile, where *M* denotes a match and *H* a proposition.

14.3   EVIDENCE AT THE CRIME LEVEL

14.3.1   Preliminaries

Early studies of the use of BNs in forensic science (Aitken and Gammermann, 1989; Dawid and Evett, 1997) discussed scientific evidence in the context of individual case scenarios. It has been shown that a standard analysis of patterns of inference concerning scientific evidence is also possible without a primary focus on a particular scenario (Garbolino and Taroni, 2002). The authors discussed some of the main issues that forensic scientists should account for if they assess scientific evidence in the light of propositions that are of judicial interest, for example, that the suspect is the offender.

As described earlier (see Section 9.5.2), to solve the problem probabilistically, a link is needed between the stain at the crime scene and the main proposition, that the suspect is the offender. The link is made in two steps. The first is the consideration of the proposition that the crime stain came from the offender (*the association proposition*). Then, assuming that the crime stain came from the offender, the second step is the consideration of a proposition that the crime stain came from the suspect (*the intermediate association proposition*).

Four nodes enable the scientist to solve the judicial question of interest, as shown in Figure 14.4.

14.3.2   Description of probabilities required

It is assumed that the four nodes are all binary. The two possible values for each of the nodes are as follows:

- H, the suspect is or is not the offender, denoted H_p or H_d;
- B, the crime stain did or did not come from the offender, denoted B or $\bar{B}$;

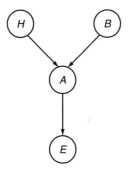

Figure 14.4 Four-node network for evaluation of evidence at the crime level.

- A, the crime stain did or did not come from the suspect, denoted A or $\bar{A}$;
- E, the suspect and crime stain have the same DNA profile, denoted E or $\bar{E}$.

Nodes H and B are parent nodes and are independent (there is no link between them). Knowing that the stain comes from the offender does not tell us anything about the probability that the suspect is (or is not) the offender. Thus, only one probability needs to be specified for each: $Pr(H_p)$, the probability the suspect is the offender, and $Pr(B)$, the probability the crime stain came from the offender (i.e., the 'relevance' term). The complementary probabilities $Pr(H_d)$ and $Pr(\bar{B})$ follow automatically. The outcome of node A is dependent on the values of H and B. Four probabilities are needed:

- $Pr(A \mid H_p, B)$, the probability that the crime stain came from the suspect, conditional on the suspect being the offender and the crime stain coming from the offender; this probability equals 1.
- $Pr(A \mid H_p, \bar{B})$, the probability that the crime stain came from the suspect, conditional on the suspect being the offender and the crime stain not coming from the offender; here the stain certainly did not come from the suspect, so the probability equals 0.
- $Pr(A \mid H_d, B)$, the probability that the crime stain came from the suspect, conditional on the suspect not being the offender and the crime stain coming from the offender; here the stain certainly did not come from the suspect, so the probability equals 0.
- $Pr(A \mid H_d, \bar{B})$, the probability that the crime stain came from the suspect, conditional on the suspect not being the offender and the crime stain not coming from the offender; this is the probability that the stain would have been left by the suspect even though he was innocent of the offence (this probability is denoted p in Section 9.5.4).

For the fourth node E, there are two probabilities to determine. The first is $Pr(E \mid A)$, the probability the suspect and the crime stain have the same profile, given the crime stain came from the suspect; this is 1. The second is $Pr(E \mid \bar{A})$, the probability the suspect and the crime stain have the same profile, given the crime stain did not come from the suspect; this is the profile frequency in the relevant population, γ.

The probability of interest is $Pr(H_p \mid E)$. The probabilities given above are what is provided. There is then an observation that E takes the value 'suspect and crime stain have the same DNA profile'. Then

$$
\begin{aligned}
Pr(H_p \mid E) &= \frac{Pr(E \mid H_p)Pr(H_p)}{Pr(E)} \\
&= \frac{Pr(E \mid H_p)Pr(H_p)}{Pr(E \mid A)Pr(A) + Pr(E \mid \bar{A})Pr(\bar{A})}.
\end{aligned}
$$

The probability of A can be determined as

$$
\begin{aligned}
Pr(A) &= Pr(A \mid H_p, B)Pr(H_p, B) + Pr(A \mid H_p, \bar{B})Pr(H_p, \bar{B}) \\
&\quad + Pr(A \mid H_d, B)Pr(H_d, B) + Pr(A \mid H_d, \bar{B})Pr(H_d, \bar{B}) \\
&= Pr(A \mid H_p, B)Pr(H_p)Pr(B) + Pr(A \mid H_p, \bar{B})Pr(H_p)Pr(\bar{B}) \\
&\quad + Pr(A \mid H_d, B)Pr(H_d)Pr(B) + Pr(A \mid H_d, \bar{B})Pr(H_d)Pr(\bar{B}),
\end{aligned}
$$

making use of the independence of H and B. Then $Pr(E \mid H_p)$ can be determined as

$$
\begin{aligned}
Pr(E \mid H_p) &= Pr(E \mid H_p, A)Pr(A \mid H_p) + Pr(E \mid H_p, \bar{A})Pr(\bar{A} \mid H_p) \\
&= Pr(E \mid A)Pr(A \mid H_p) + Pr(E \mid \bar{A})Pr(\bar{A} \mid H_p),
\end{aligned}
$$

and

$$
Pr(A \mid H_p) = Pr(A \mid H_p, B)Pr(B) + Pr(A \mid H_p, \bar{B})Pr(\bar{B})
$$

with $Pr(\bar{A} \mid H_p) = 1 - Pr(A \mid H_p)$.

If it is assumed that $Pr(B) = r$, and $Pr(A \mid \bar{B}, H_d) = p$, a simplified version of the equation presented in Section 9.5.4 (with $k = 1$) is obtained. This emphasises the appropriateness of the graphical structure and the associated probabilistic assessments. If it also assumed further that p equals 0 and that the relevance reaches its maximum ($r = 1$), then the likelihood ratio is reduced to its simplest form, $1/\gamma$.

The work of Garbolino and Taroni (2002) can be taken as a demonstration that BNs can be used to represent accurately existing and accepted probabilistic solutions for forensic inferential problems.

14.4 MISSING EVIDENCE

14.4.1 Preliminaries

This is an example where a Bayesian network structure can be elicited from a likelihood ratio formula which has already been given (Taroni *et al.*, 2004). This is the problem of missing evidence, for which Lindley and Eggleston (1983) have provided a general Bayesian formula. According to Schum (1994), evidence is called *missing* if it is expected, but is neither found nor produced on request. The example presented in Lindley and Eggleston (1983) relates to a collision between two motor cars. The scenario is as follows:

> The plaintiff sues the defendant, claiming that it was his car that collided with the plaintiff's. The evidence of identification is weak, and the defendant relies on the

fact that, his car being red, the plaintiff has produced no evidence that any paint, red or otherwise, was found on the plaintiff's car after the collision.

14.4.2   Determination of a structure for a Bayesian network

A likelihood ratio to assist the court in the examination of the effect that the evidence is missing (M) has on the truth or otherwise of the variable of interest H is presented by Lindley and Eggleston (1983):

$$
\frac{Pr(M \mid H_p)}{Pr(\bar{M} \mid H_d)}
$$
$$
= \frac{Pr(M \mid E_1)Pr(E_1 \mid H_p) + Pr(M \mid E_2)Pr(E_2 \mid H_p) + Pr(M \mid E_3)Pr(E_3 \mid H_p)}{Pr(\bar{M} \mid E_1)Pr(E_1 \mid H_d) + Pr(\bar{M} \mid E_2)Pr(E_2 \mid H_d) + Pr(\bar{M} \mid E_3)Pr(E_3 \mid H_d)}.
$$
$$(14.1)$$

It is easily seen that the construction of a BN based upon an existing formula has the advantage that the number and definition of the nodes are already given. From the above example on missing evidence, there are three variables that can be derived:

1. The variable H represents the event that the defendant is guilty of the offence for which he has been charged. This event may be either true or false. It has two states, H_p and H_d.
2. The variable M represents the event that evidence is missing. This variable can take the value true or false, denoted M and $\bar{M}$, respectively.
3. The variable E designates the form of the evidence that is missing. Three states were proposed for E:

   - E_1, there was red paint on the plaintiff's car;
   - E_2, there was paint on the plaintiff's car, but it was not red;
   - E_3, there was no paint on the plaintiff's car.

In order to find a graphical representation that correctly represents the conditional dependencies as specified by the likelihood ratio (14.1), it is helpful to follow a two-stage approach. Eggleston and Lindley (1983) assert that (14.1) contains all the relevant considerations for the paint scenario, notably conditional probabilities for:

- the various forms of the evidence given that the prosecution hypothesis H_p is true and given the defence hypothesis H_d is true,
- the evidence being missing were it E_1, E_2 and then E_3, respectively.

Consider the first of the two points mentioned above. Acceptance that the probability of the evidence is *conditioned* on the truth state of the variable H means, graphically, that H is chosen as a parental variable for E (see Figure 14.5(a)).

The situation is similar for the second point. If the event that evidence is missing (M) is *conditioned* on the form the missing evidence can take (E), then E can be chosen as a parental variable for M (see Figure 14.5(b)).

Since the variable E shown in Figure 14.5(a) is the same as in Figure 14.5(b), the two network fragments combine to give the Bayesian network structure shown in Figure 14.5(c).

Whilst searching for an appropriate structure for a BN, based on the three variables M, E and H, it would be legitimate to ask whether there could be an arrow pointing from H to M. Consideration of the proposed network structure as shown in Figure 14.5(c), indicates that there should be none. As with any other graphical element employed in BN structures, the absence of an arrow must be justified as well. In the current example, the absence of a directed edge between H and M can be justified by the indications given by Eggleston and Lindley (1983), who assume that 'were the form of the missing evidence known, then the view of the defendant's guilt would not be altered by knowing that this evidence had, or had not, been produced in court'. In other words, 'the actual evidence suppresses any importance being attached to its omission'.

In a formal notation, this corresponds to $Pr(H_p \mid E, M) = Pr(H_p \mid E)$, where E may take any of its three possible states E_1, E_2 or E_3, and M either M or $\bar{M}$. The proposed BN correctly encodes this property through its serial connection, where H and M are *conditionally independent* given E is known. It may also be said that the transmission of evidence between the nodes H and M is blocked whenever E is instantiated , or that the node E screens off M from H. This is a practical example of d-separation .

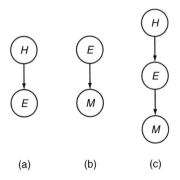

(a)   (b)   (c)

Figure 14.5 Bayesian network fragments representing the relation between (a) the variables E and H, and (b) the variables M and E; (c) Bayesian network for missing evidence.

The main aim of the current example is the determination of ways to find appropriate qualitative structures for BNs. Thus, the numerical specification of the BN for the current example will not be treated here. However, the implementation of the BN in a suitable computer program would provide further means to validate the proposed network structure. Notably, the effect of different probabilities for the suppression of favourable and unfavourable evidence for the defendant on the odds of guilt can be examined and compared to the indications given in Lindley and Eggleston (1983).

14.4.3 Comments

The example on missing evidence has been chosen not only to illustrate that the construction of a BN can be oriented and guided by existing likelihood ratio formulae, but also to show that a BN can be useful for improving the understanding of likelihood ratios that may not otherwise be transparent. (This has already been pointed out by Garbolino and Taroni (2002) who propose Bayesian networks to justify Evett's (1993a) likelihood ratio formula that included the concept of 'relevance'.) Moreover, the current example clearly has shown that the use of BNs is not restricted to the formal discussion of evidence that is present, which in fact is only one form amongst others, but is also applicable to problems such as missing evidence.

The BN derived not only reflects but also supports graphically the understanding of the concept of *conditional independence*. Note also another important structural aspect of BNs; the absence of an arrow is just as informative as the presence of an arrow.

14.5 ERROR RATES

14.5.1 Preliminaries

A Bayesian framework for the assessment of the influence of false positives in the analysis of DNA evidence has been considered (Thompson *et al.*, 2003; see also Section 13.11).

In analogy to (3.6), the traditional approach for considering the hypothesis S, that a specimen came from a particular suspect, in the light of the evidence (R) of the forensic scientist's report of a DNA match between the suspect's profile and the profile of the specimen, is given by Bayes' theorem:

$$\frac{Pr(S \mid R)}{Pr(\bar{S} \mid R)} = \frac{Pr(R \mid S)}{Pr(R \mid \bar{S})} \times \frac{Pr(S)}{Pr(\bar{S})}. \tag{14.2}$$

In order to account for the possibility of error (notably false positive), an intermediate proposition M was introduced. Proposition M denotes a true match, as distinct from R, a reported match. Analogously to S, M must be taken as an *unobserved* variable, since its truth state cannot be known with certainty, but can be revised based on new information, such as R. A modified likelihood ratio is presented in Thompson *et al.* (2003).

$$\frac{Pr(R \mid S)}{Pr(R \mid \bar{S})} = \frac{Pr(R \mid M) \times Pr(M \mid S) + Pr(R \mid \bar{M}) \times Pr(\bar{M} \mid S)}{Pr(R \mid M) \times Pr(M \mid \bar{S}) + Pr(R \mid \bar{M}) \times Pr(\bar{M} \mid \bar{S})}. \tag{14.3}$$

14.5.2 Determination of a structure for a Bayesian network

The relevant number and definitions of the nodes for modelling the error rate problem can, analogously to the previous example, be based on the existing probabilistic solution. Formally, the definitions of the three variables R, M, and S, all binary, are as follows:

- S, the specimen did or did not come from the suspect, denoted S or $\bar{S}$;
- R, the forensic scientist reports a match or a non-match between the suspect's profile and the profile of the sample, denoted R or $\bar{R}$;
- M, the suspect and the specimen have or do not have matching DNA profiles, denoted M or $\bar{M}$.

In order to construct a Bayesian network with the three nodes R, M and S, the dependencies amongst the corresponding variables have to be determined.

For the derivation of (14.3), it was assumed that $Pr(R \mid M, S) = Pr(R \mid M, \bar{S}) = Pr(R \mid M)$. Therefore, the joint probability of R, M and S can be written as

$$Pr(RMS) = Pr(R \mid M)Pr(M \mid S)Pr(S). \tag{14.4}$$

The appropriate Bayesian network structure that correctly represents the dependencies defined in (14.4) is serial ($S \rightarrow M \rightarrow R$), as shown in Figure 14.6. This can be understood by considering the chain rule for BNs, which yields

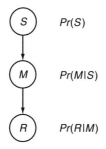

Figure 14.6 Bayesian network for error rates. The probabilistic dependencies are indicated on the right-hand side of each node.

(14.4) for the variables R, M, and S, and for which the structural dependencies are such as specified in Figure 14.6.

14.6 TRANSFER EVIDENCE

14.6.1 Preliminaries

Chapter 12 approaches scenarios involving the evaluation of group(s) of recovered fibres associated with the investigation of an offence. Two main categories of situations have been presented where the evidence was assessed under propositions at the *activity* level, notably situations in which the recovered fibres come from the offender and situations in which the recovered fibres are present by chance alone. It has been shown that (12.3) is an extended version of (12.4). The formal development has shown that parameters of interest are the *transfer, persistence* and *recovery* (transfer probabilities), the *presence by chance* (background probabilities) and the *relative frequency*, γ, of the relevant features of the recovered fibres in a population.

A pre-assessment of the potential value of findings (classified into three categories: no fibres, a small amount of fibres, a large amount of fibres) has also been presented following recent literature. All the scenarios presented in Chapter 12 can easily be translated into BNs allowing the scientist to avoid the development and calculation of elaborate formulae.

14.6.2 Determination of a structure for a Bayesian network

In the fibres scenario described in Section 12.2.1, the variables of interest are as follows:

- H represents the event that the suspect has sat on the driver's seat of the stolen car. This event may be either true or false (if false, another person sat on the seat). This node has two states, H_p and H_d.

- T represents the event that fibres have been transferred to, have persisted on and have successfully been recovered from the driver's seat. This event may be either true or false. It has two states, t and $\bar{t}$.

- X represents the features of the control object (i.e., the suspect's pullover). It has two states, x and $\bar{x}$ (other features).

- Y represents the features of the recovered fibres. It has three states, y, $\bar{y}$ (other features) and *two groups*. The third state takes into account for the possibility (not expressed in the formulae presented in Chapter 12) of recovering fibres transferred from the offender and fibres already present on the surface on interest. This situation is avoided in the formulae presented in Chapter 12, where it is assumed that only one group of fibres has been observed.

- B represents the event that a compatible group of fibres is present on the driver's seat by chance alone. This event may be either true or false. It has two states, b and $\bar{b}$.

The BN structure that represents the dependencies expressed in (12.2) is a combination of a serial and a divergent connections as presented in Figure 14.7.

A table of conditional probabilities for node T allows the scientist to take into account values related to transfer phenomena: transfer from the suspect's pullover and transfer from an offender's garments, $Pr(T \mid x, H_p)$ and $Pr(T \mid H_d)$, respectively (called t_n and t'_n in Chapter 12). In the scenario developed in Section 12.2.3, the presence of the recovered fibres is explained by chance alone, so $Pr(T \mid H_d)$ is set equal to 0. The probabilities linked to node Y are given in Table 14.1.

The numerator of the likelihood ratio is obtained when states H_p and x are instantiated and a value for y registered. The denominator is obtained when state H_d is instantiated.

A slight modification of this network allows the scientist to approach the pre-assessment of findings. To describe the fibre case developed in Section 12.3.3, a redefinition of nodes is necessary. New nodes, P (presence of a group) and S

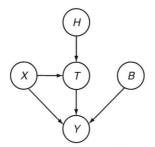

Figure 14.7 Five-node network: no outcome.

Table 14.1 Conditional probabilities for transfer with one group. The features of the control object are X, and the features of the recovered fibres are Y, which may be from $0(\bar{y})$, $1(y)$ or 2 groups. These recovered fibres may be present by chance alone (b_1) or not (b_0), or have been transferred from the victim (t_1) or not (t_0)

X	Type $= x$				Type $= \bar{x}$			
T	t_1		t_0		t_1		t_0	
B	b_0	b_1	b_0	b_1	b_0	b_1	b_0	b_1
y	1	0	0	0.01	0	0	0	0.01
$\bar{y}$	0	0	1	0.99	1	0	1	0.99
2 groups	0	1	0	0	0	1	0	0

(size of the group), substitute for node B, and a node O, say, is introduced. Nodes X and Y are eliminated. The final node O is characterized by five states:

1. no groups of fibres observed,
2. a group of non-matching fibres,
3. a small group of matching fibres,
4. a large group of matching fibres,
5. two groups of fibres.

The last state is not considered by Champod and Jackson (2000) in their arithmetical approach (see Section 12.3.3). The 'two groups of fibres' state enables consideration of situations in which fibres coming from the transfer are combined with fibres coming from the background.

   All other nodes are binary, except the *transfer* one. An extension of states of this variable should be introduced:

1. no transfer,
2. transfer of a small amount of fibres,
3. transfer of a large amount of fibres.

The corresponding BN is presented in Figure 14.8.

   Probability values proposed in Section 12.3.3 and conditional probabilities linked to node O presented in Table 14.2 confirm the likelihood ratios presented in Table 12.3.

14.6.3 Comment on the *transfer* node

More elaborate BNs can be developed to take into account the amount of information coming from transfer, persistence and recovery aspects. For the sake of illustration, the fibres example is generalised to include a scenario involving glass fragments. A burglar has smashed the window of a house. A suspect is arrested. A quantity, Q_r, of glass fragments matching, in some sense, the type of glass of the house's window are recovered from the suspect's pullover.

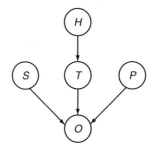

Figure 14.8 Five-node network: outcome, converging and serial in Bayesian networks.

Table 14.2 Conditional probabilities for transfer with many groups, as illustrated in Section 12.3.3. The transfer may be of no fibres (t_0), a small amount (t_s) or a large amount (t_l). A group may be present (p_1) or not (p_0), and the group may be small (s_s) or large (s_l). The outcome O has five categories: no groups, one not-matching group, one small matching group, one large matching group and two groups.

T	t_0				t_s				t_l			
P	p_0		p_1		p_0		p_1		p_0		p_1	
S	s_s	s_l	s_s	s_l	s_s	s_l	s_s	s_l	s_s	s_l	s_s	s_l
0 groups	1	1	0	0	0	0	0	0	0	0	0	0
1 not-matching group	0	0	0.95	0.95	0	0	0	0	0	0	0	0
1 small matching group	0	0	0.05	0	1	1	0	0	0	0	0	0
1 large matching group	0	0	0	0.05	0	0	0	0	1	1	0	0
2 groups	0	0	0	0	0	0	1	1	0	0	1	1

Instead of a single transfer node, T, which takes account of all the phenomena linked to the transfer, it is possible to extend the net using the following information:

- The number, Q_r, of glass fragments that are recovered from the suspect's pullover. This quantity depends on the number, Q_p, of glass fragments that have persisted on the pullover and on the quality of the technique used for the search, which can be assessed by the proportion of glass fragments lifted from the pullover. This measure is denoted P_l.

- The number, Q_p, of glass fragments that have persisted on the pullover is dependent on the number, Q_t, of glass fragments transferred and on the proportion, P_s, of fragments that were shed between the time of transfer and the time of the examination of the pullover.

- The number, Q_t, of transferred fragments depends on the proposition, H.

The states of the events can be categorical in the form of none, few, many, or none, small, large, as previously described in the pre-assessment of the scenario for fibres in Section 7.2.

14.7 COMBINATION OF EVIDENCE

The aim of the examples in this chapter is to draw attention to considerations that relate to the dependence and independence properties of the basic building blocks (see Figure 14.1) of Bayesian networks as well as to illustrate methods of construction. Examples so far have involved serial connections. An example is now given for which a diverging connection is appropriate.

The general expression of the likelihood ratio for a single piece of evidence E is given in (3.12). The corresponding representation of this relation in terms of a BN is shown in Figure 14.5(a).

Suppose there was a second piece of evidence. Denote the two pieces of evidence A and B. The combined effect of A and B has to be evaluated in order to revise the belief in a proposition of interest H (Section 8.1.3). Bayes' theorem can again be applied:

$$\frac{P(H_p)}{P(H_d)} \times \overbrace{\frac{P(A, B \mid H_p)}{P(A, B \mid H_d)}}^{V_{comb}} = \frac{P(H_p \mid A, B)}{P(H_d \mid A, B)} \tag{14.5}$$

According to the product rule (1.7), the likelihood ratio for the combined evidence (V_{comb}) can be written as:

$$\frac{P(A \text{ and } B \mid H_p)}{P(A \text{ and } B \mid H_d)} = \frac{P(A \mid B, H_p)}{P(A \mid B, H_d)} \times \frac{P(B \mid H_p)}{P(B \mid H_d)}. \tag{14.6}$$

Suppose that the two pieces of evidence A and B are independent so that $Pr(A \mid B, H_p) = Pr(A \mid \bar{B}, H_p) = Pr(A \mid H_p)$ (with an analogous expression for $P(A \mid B, H_d)$). Consequently, with the use of (1.6), V_{comb} reduces to:

$$\frac{P(A \text{ and } B \mid H_p)}{P(A \text{ and } B \mid H_d)} = \frac{P(A \mid H_p)}{P(A \mid H_d)} \times \frac{P(B \mid H_p)}{P(B \mid H_d)}. \tag{14.7}$$

In terms of a BN for the three variables A, B and H, these relations translate to the structural dependencies as shown in Figure 14.1, where H replaces A, and A and B replace B and C. The diverging connection in Figure 14.1 allows consideration of the following:

- The cause H can produce each of the two effects A and B. Events A and B *depend* on H.

- Knowledge about A provides relevant information for H, which in turn will provide relevant information for judging the truth state of B. In other words, A and B are d-connected, given H is not instantiated.

- Similarly, the probability that A given H is true will not be affected by knowing in addition that B is true; A and B are d-separated, knowing H.

Imagine a situation in which a suspect has been found as a result of a search of the DNA profile of a crime stain against a database of N suspects (Section 13.8). The suspect's profile was the only profile found to match, and all the other $(N-1)$ profiles did not match. This scenario of a database search consists essentially of two pieces of information:

- E, a match between the genotype of the suspect G_s and the genotype of the crime stain G_c, that is, $G_s = G_c$;
- D, the information that the other $(N-1)$ profiles of the database do not match.

The likelihood ratio for the combined pieces of evidence E and D is

$$V = \frac{P(E,D \mid H_p)}{P(E,D \mid H_d)} = \frac{P(E \mid D, H_p)}{P(E \mid D, H_d)} \times \frac{P(D \mid H_p)}{P(D \mid H_d)}. \qquad (14.8)$$

Following the analysis of Balding and Donnelly (1996), the first ratio of the right-hand side of (14.8) reduces to $1/\gamma$ approximately, where γ is the random match probability. The second ratio reduces to $1/(1-\phi)$, where ϕ is the probability that the source of the crime stain is among the other $N-1$ suspects.

A BN structure analogous to Figure 14.1(c) can then be constructed, with H replacing A, E replacing B and D replacing C. The numerical values of the relevant conditional probabilities are in the conditional node probability tables of the variables E and D.

Note that this BN is an explicit representation of the assumption that, knowing that H is true, the probability of a match between the suspect's genotype and the genotype of the crime stain is not influenced by the fact that there has been a search of a database.

It may further be noted that the use of a BN for representing the probabilistic approach proposed by Balding and Donnelly (1996) underlines the view that the result of the database search has the character of an additional piece of information. The database search has increased the likelihood ratio. There is a match and also the additional information that the suspect has been chosen from a database and $(N-1)$ other people have been excluded.

14.8 CROSS-TRANSFER EVIDENCE

A scenario involving cross-transfer and the construction of a BN to reflect the dependencies amongst various parts is described in Aitken *et al.* (2003). Particular variations from the general scenario can be allowed for in the network.

An assault has been committed. There is one victim (V) and one criminal (C). There has been contact between C and V. The evidence under consideration is such as to yield DNA profiles of the victim and the criminal. This evidence could be semen and vaginal fluids (in a rape case) or blood (in an assault). There could be transfer from the victim to the criminal (vaginal fluids in a rape) or from the criminal to the victim (semen in a rape) or both. It is possible there may be transfer in one direction only. Suppose the victim had been killed with a knife and there is no evidence of a fight. The probability of a transfer of blood from the criminal to the victim is low. The probability of a transfer of blood from the victim to the criminal is high.

Generally, the two sets of recovered traces (bloodstains, for example) have to be considered as dependent. In fact, if a transfer has occurred in one direction (for example, from V to C), and the expert has recovered traces characterising this transfer, then the expert would generally expect to find trace evidence characterising the transfer in the other direction (from C to V). The presence of one set of transfer evidence gives information about the presence of the other set of transfer evidence. The absence of the other set of transfer evidence would in itself be significant (Lindley and Eggleston, 1983).

A suspect (S) is identified. Consider, first, the study of the suspect for evidence likely to yield a DNA profile of the victim. Several bloodstains are found on some of the suspect's clothing. There are various subjective probabilities in the model, and these are described in general in Chapter 9. The bloodstains could have arisen because of a transfer from V to S, either innocently (and this consideration is accounted for with the parameter denoted p and referred to as *innocent acquisition* in the analysis) or because S committed the crime (and this consideration is accounted for with the parameter denoted t for *transfer* in the analysis). Alternatively, the stains could be there because of the suspect's lifestyle, either innocently through the nature of their occupation or, not so innocently, through the nature of other activities which may be violent. This consideration is accounted for with the parameter denoted b and referred to as *innocent presence* in the analysis. Some of the stains may be there innocently and some because the suspect committed the crime, and it may not be known which is which. One or more stains are chosen for analysis. These may be *relevant* or not and the parameter for consideration of this is denoted r. The analysis of evidence found on the victim can be considered analogously to that found on the suspect.

Thus, there are four features to account for in the analysis: innocent presence (b), innocent acquisition (p), transfer (t) and relevance (r). These apply to both S and V. The transfer probabilities (t) depend on whether or not there has been transfer in the other direction. It is an important feature of this analysis that it models cross-transfer evidence. Often, consideration of transfer evidence does not account for the presence or absence of evidence which may have been transferred in the opposite direction from that under consideration. Examples of consideration of cross-transfer in the evaluation of evidence are given in Champod and Taroni (1999) and Cook *et al.* (1999). It is demonstrated that a simple multiplication of likelihood ratios for the two directions of transfer is not valid and a solution to the problem is described. The probabilities for the four features listed at the beginning of the paragraph depend also on whether it is the victim or suspect who is being considered. The relative importance for the probability of guilt of these four features may be studied through variation of the probabilities and the changes in the calculated probability of guilt that arise as a result.

Once a stain has been chosen on the suspect (victim) for analysis, it is analysed and found to match the victim's (suspect's) profile. It is possible to draw a distinction between a reported match and a true match and consider a non-zero probability of a false positive (Thompson *et al.*, 2003; see also Section 14.5). It is assumed that there are no false negatives. The network illustrated in Figure 14.9 allows for the probability of guilt of the suspect to be determined, given that a match is reported between the DNA profile of the stains analysed from the suspect and that of the victim and/or a match is reported between the DNA profile of the stains analysed from the victim and that of the suspect, with other considerations taken into account: relevance, transfer, innocent presence and innocent acquisition. These other considerations are determined given the contextual information, and the corresponding probabilities are subjective probabilities assessed by the scientist (Taroni *et al.*, 2001).

14.8.1 Description of nodes

For a BN it is necessary to have a clear description of each of the nodes. The example in Aitken *et al.* (2003) and represented in Figure 14.9 has 14 nodes, the factors of which include the transfer of material between the suspect and the victim, the choice of the stains on the suspect and victim, the background

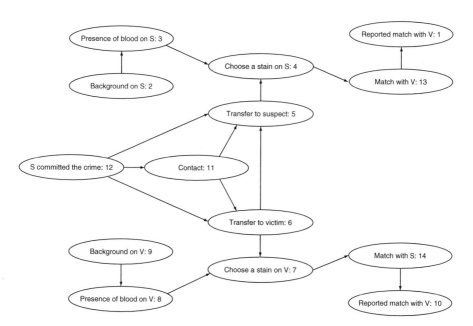

Figure 14.9 Complete network for evaluation of cross-transfer evidence in DNA profiles. (Reprinted from Aitken *et al.*, 2003, with permission from Elsevier.)

activities, the matches and reported matches of stains, and the ultimate issue of whether the suspect committed the crime or not.

14.8.2   Probabilities for nodes

These may be unconditional or conditional probabilities, depending on whether nodes are parent or child nodes. It is assumed there are no false negatives: the probability of reporting no match, conditional on their being a true match, is zero. Thus, the probability of reporting a match, conditional on their being a true match, is one. Probabilities are also required for the event that there is no true match. It is assumed that there is a non-zero probability of a false positive: the probability of reporting a match, conditional on their being no match, is non-zero. It is assumed to be small. For example, assume it were 0.001; then the probability of a true positive is 0.999. Node 1 in Figure 14.9 is an example of a binary node; the values are *report a match* and *report no match*. Node 13 is a node in which the response can be a continuous variable. The value is conditional on the relationship of the stain chosen on the suspect and the crime. The probability of a match between the DNA profile of the stain chosen from S and the DNA profile of V, if the chosen stain relates to the crime, is 1. The probability of a match between the DNA profile of the stain chosen from S and the DNA profile of V, if the chosen stain does not relate to the crime, is the DNA profile frequency of the victim.

Node 5 is one of the nodes for the event of transfer of biological material. Node 6 is the other. These two nodes represent the cross-transfer. Node 5 is associated with the transfer of biological material from V to S. The structure of the BN reflects the beliefs of the investigators in the relationships amongst the factors which may affect the probability of the outcome of node 5. This probability depends on the values of three other nodes: S *committed the crime* (node 12), *Contact between S and V* (node 11), and *Transfer of biological material from S to V* (node 6). Further details are available in Aitken *et al.* (2003).

Note that absence of evidence or missing evidence is taken into account in the framework. If the transfer probabilities are not equal, this means that the correlation between transfer to V from S and transfer to S from V is not spurious. These two transfers do not become independent, when the outcome (contact or not) of node 11 is known, because the presence, or the absence (missing evidence) of one is relevant for the other. This is recognised through the provision of a direct link joining the two nodes.

14.9   FACTORS TO CONSIDER

The ideas here are extracted from Puch and Smith (2002) in a discussion on Bayesian networks for the assessment of fibre evidence in which a balaclava has been used in the commission of a crime (see Section 14.6 for a simplified

BN, and Section 12.3.1). The defence is that the suspect owns a mask similar to the one collected by the police but has not worn it for t hours. Factors modelled include the following (most of the distributions are described in Chapter 2):

- The average number L of fibres transferred to the offender's head hair from the balaclava; Puch and Smith model this with a gamma distribution, (5.1), (Evans *et al.*, 2000), with parameters a_l and b_l chosen based on characteristics of mask sheddability and transfer reception.
- The actual number X transferred; Puch and Smith model this with a Poisson distribution with mean L.
- The proportion of fibres Q that have persisted in the offender's head up to time t without considering physical and head disturbance; this is modelled with a beta distribution with parameters a_q and b_q.
- The number X_t of fibres that persisted with success rate pdQ from the initially transferred fibres X_0; this is modelled with a binomial distribution .
- The proportion S of fibres that are retrieved in the laboratory; this is a beta distribution with parameters a_s, b_s.
- The number Y_t of fibres that are actually retrieved from the suspect's hair given that the proportion of recoverable fibres is S and that the number of fibres on the offender's hair is X_t.

14.9.1   Parameter choice

The choice of values for the parameters in the above model may be obtained from look-up tables constructed from experimental data. The parameters are a_l, b_l, a_q, b_q, p and q. For example, parameters a_l and b_l are chosen based on mask sheddability and transfer reception. These take values very low, low, medium, high and very high. Experimental data are classified into these categories, and for each category a_l and b_l are computed. These values are then recorded in a look-up table which the software package FINDS uses to determine the gamma distribution that is displayed to the forensic scientist. Persistence is also classified as very low, low, medium, high and very high. Head disturbance choices are none, combed and washed. Physical disturbance choices are low, medium and high. 'Retrieval recipient' refers to whether the hair type facilitates the retrieval; the choices are very high, high, medium, low and very low.

14.10   SUMMARY

The construction of a Bayesian network requires the determination of:

- the factors to include;
- the links between factors;

- the conditional probabilities among factors;
- the number of levels and the associated probability distribution if a factor represents a discrete or categorical variable,
- the probability distribution if a node represents a continuous variable (though this is not discussed here),

Certain distributions will require values for their parameters. These values can be determined from experimental data and inserted into look-up tables. This can be done for various levels of the appropriate variables.

For a particular case, appropriate values can be inserted into the BN. This can be done as a pre-assessment procedure to determine whether an analysis will support the proposition of the scientist's client. If it is decided that such an analysis is worth pursuing, an analysis can be done to determine the likelihood ratio. For examples of fibres described by Puch and Smith (2002) the computation of the likelihood ratios incorporates the probability distribution of the number of fibres retrieved from the suspect's hair and the uncertainty over the presence of fibres in people's head hair by chance, taking account of the rarity of the fibres.

The BN described in Aitken *et al.* (2003) uses binary nodes only and the software package HUGIN. Puch and Smith (2002) determine a likelihood ratio for evidence of the number of fibres retrieved from the suspect's head. Aitken *et al.* (2003) include a node for the ultimate issue: the suspect committed the crime. The evidence is in two parts: first, the reported match between the DNA profile of a stain on the suspect and the DNA profile of the victim; and second, the reported match between the DNA profile of a stain on the victim and the DNA profile of the suspect. It is possible to use this BN to determine a likelihood ratio for the the combination of these two parts of the evidence. This can be done by conditioning, in turn, on what may be thought of as the prosecution and the defence propositions, namely that the suspect did or did not commit the crime. By propagation of these conditions through the network, the probabilities of the two reported matches may be determined for each proposition. The likelihood ratio is then the ratio of the product of the probabilities under the two propositions. The numerator is the probability assuming the prosecution proposition is true; the denominator is the probability assuming the defence proposition is true. The relationship between the two factors – the reported match between the DNA profile of a stain on the suspect and the DNA profile of the victim, and the reported match between the DNA profile of a stain on the victim and the DNA profile of the suspect – is allowed for by the separation of the two nodes.

References

Adams, B.J. (2003) The diversity of adult dental patterns in the United States and the implications for personal identification. *Journal of Forensic Sciences*, **48**, 497–503.

Aitchison, J. and Dunsmore, I.R. (1975) *Statistical Prediction Analysis*, Cambridge University Press, Cambridge.

Aitchison, J., Habbema, J.D.F. and Kay, J.W. (1977) A critical comparison of two methods of statistical discrimination. *Applied Statistics*, **26**, 15–25.

Aitken, C.G.G. (1986) Statistical discriminant analysis in forensic science. *Journal of the Forensic Science Society*, **26**, 237–247.

Aitken, C.G.G. (1991) Report on International Conference on Forensic Statistics. *Journal of the Royal Statistical Society, Series A*, **154**, 45–48. Selected papers included on pp. 49–130.

Aitken, C.G.G. (1993) Statistics and the law: report of a discussion session at the Royal Statistical Society Conference, Sheffield, September 1992. *Journal of the Royal Statistical Society, Series A*, **156**, 301–304.

Aitken, C.G.G. (1999) Sampling – how big a sample? *Journal of Forensic Sciences*, **44**, 750–760.

Aitken, C.G.G. (2000) Interpretation of evidence and sample size determination. In *Statistical Science in the Courtroom* (Gastwirth, J.L., ed.), Springer-Verlag, New York, pp. 1–24.

Aitken, C.G.G. (2003) Conviction by probability. *New Law Journal*, **153**, 1153–1154.

Aitken, C.G.G. and Gammerman, A. (1989) Probabilistic reasoning in evidential assessment. *Journal of the Forensic Science Society*, **29**, 303–316.

Aitken, C.G.G. and Lucy, D. (2002) Estimation of the quantity of a drug in a consignment from measurements on a sample. *Journal of Forensic Sciences*, **47**, 968–975.

Aitken, C.G.G. and Lucy, D. (2004) Evaluation of trace evidence in the form of multivariate data. *Applied Statistics*, **53**, 109–122.

Aitken, C.G.G. and MacDonald, D.G. (1979) An application of discrete kernel methods to forensic odontology. *Applied Statistics*, **28**, 55–61.

Aitken, C.G.G. and Robertson, J. (1987) A contribution to the discussion of probabilities and human hair comparisons. *Journal of Forensic Sciences*, **32**, 684–689.

Aitken, C.G.G. and Stoney, D.A. (eds) (1991) *The Use of Statistics in Forensic Science*, Ellis Horwood, Chichester.

Statistics and the Evaluation of Evidence for Forensic Scientists: Second Edition
C.G.G. Aitken and F. Taroni © 2004 John Wiley & Sons, Ltd   ISBN: 0-470-84367-5

Aitken, C.G.G. and Taroni, F. (1997) A contribution to the discussion on 'Bayesian analysis of deoxyribonucleic acid profiling data in forensic identification applications', Foreman *et al. Journal of the Royal Statistical Society, Series A,* **160**, 463.

Aitken, C.G.G. and Taroni, F. (1998) A verbal scale for the interpretation of evidence (Letter to the Editor). *Science & Justice,* **38**, 279–281.

Aitken, C.G.G., Gammerman, A., Zhang, G., Connolly, T., Bailey, D., Gordon, R. and Oldfield, R. (1996a) Bayesian belief networks with an application in specific case analysis. In *Computational Learning and Probabilistic Reasoning* (Gammerman A., ed.), John Wiley & Sons Ltd, London, pp. 169–184.

Aitken, C.G.G., Connolly, T., Gammerman, A., Zhang, G., Bailey, D., Gordon, R. and Oldfield, R. (1996b) Statistical modelling in specific case analysis. *Science & Justice,* **36**, 245–255.

Aitken, C.G.G., Bring, J., Leonard, T. and Papasouliotis, O. (1997) Estimation of quantities of drugs handled and the burden of proof. *Journal of the Royal Statistical Society, Series A,* **160**, 333–350.

Aitken, C.G.G., Taroni, F. and Garbolino, P. (2003) A graphical model for the evaluation of cross-transfer evidence in DNA profiles. *Theoretical Population Biology,* **63**, 179–190.

Allen, R.J., Balding, D.J., Donnelly, P., Friedman, R., Kaye, D., LaRue, H., Park, R.C., Robertson, B. and Stein, A. (1995) Probability and proof in *State v. Skipper*: an internet exchange. *Jurimetrics Journal,* **35**, 277–310.

Allen, T.J., and Scranage, J.K. (1998) The transfer of glass. Part I – Transfer of glass to individuals at different distances. *Forensic Science International,* **93**, 167–174.

Allen, T.J., Hoefler, K. and Rose, S.J. (1998a) The transfer of glass. Part II – A study of the transfer of glass to a person by various methods. *Forensic Science International,* **93**, 175–193.

Allen, T.J., Hoefler, K. and Rose, S.J. (1998b) The transfer of glass. Part III – The transfer of glass from a contaminated person to another uncontaminated person during a ride in a car. *Forensic Science International,* **93**, 195–200.

Allen, T.J., Cox, A.R., Barton, S., Messam, P. and Lambert, J.A. (1998c) The transfer of glass. Part IV – The transfer of glass fragments from the surface of an item to the person carrying it. *Forensic Science International,* **93**, 201–208.

Allen, T.J., Locke, J. and Scranage, J.K. (1998d) Breaking of flat glass. Part V – Size and distribution of fragments from vehicle windscreens. *Forensic Science International,* **93**, 209–218.

Anderson, T. and Twining, W. (1998) *Analysis of Evidence: How to Do Things with Facts Based on Wigmore's Science of Judicial Proof.* Northwestern University Press, Evanston, IL.

Anderson, T.W. (1984) *An Introduction to Multivariate Analysis,* John Wiley & Sons, Inc., New York.

Antelman, G. (1997) *Elementary Bayesian Statistics,* Edward Elgar, Cheltenham.

Ashcroft, C.M., Evans, S. and Tebbett, I.R. (1988) The persistence of fibres in head hair. *Journal of the Forensic Science Society,* **28**, 289–293.

Ayres, K.L. (2000) Relatedness testing in subdivided populations. *Forensic Science International,* **114**, 107–115.

Balding, D.J. (1995) Estimating products in forensic identification using DNA profiles. *Journal of the American Statistical Association,* **90**, 839–844.

Balding, D.J. (1997) Errors and misunderstandings in the second NRC report. *Jurimetrics Journal,* **37**, 469–476.

Balding, D.J. (1999) When can a DNA profile be regarded as unique? *Science & Justice*, **39**, 257–260.

Balding, D.J. (2000) Interpreting DNA evidence: can probability theory help? In: *Statistical Science in the Courtroom* (Gastwirth, J.L. ed.), Springer-Verlag, New York, pp. 51–70.

Balding, D.J. (2002) The DNA database search controversy. *Biometrics*, **58**, 241–244.

Balding, D.J. (2005) *Assessing Weight-of-Evidence for DNA Profiles*, John Wiley & Sons Ltd, Chichester. (Working title.)

Balding, D.J. and Donnelly, P. (1994a) How convincing is DNA evidence? *Nature* 368, 285–286.

Balding, D.J. and Donnelly, P. (1994b) The prosecutor's fallacy and DNA evidence. *Criminal Law Review*, 711–721.

Balding, D.J. and Donnelly, P. (1995a) Inferring identity from DNA profile evidence. *Proceedings of the National Academy of Sciences USA*, **92**, 11741–11745.

Balding, D.J. and Donnelly, P. (1995b) Inference in forensic identification (with discussion). *Journal of the Royal Statistical Society, Series A*, **158**, 21–53.

Balding, D.J. and Donnelly, P. (1996) Evaluating DNA profile evidence when the suspect is identified through a database search. *Journal of Forensic Sciences*, **41**, 603–607.

Balding, D.J. and Nichols, R.A. (1994) DNA profile match probability calculation: how to allow for population stratification, relatedness, database selection and single bands. *Forensic Science International*, **64**, 125–140.

Balding, D.J. and Nichols, R.A. (1995) A method for quantifying differentiation between populations at multi-allelic loci and its implications for investigating identity and paternity. In *Human Identification: The Use of DNA Markers* (Weir, B.S., ed.), Kluwer Academic, Dordrecht, pp. 3–12.

Balding, D.J. and Nichols, R.A. (1997) Significant genetic correlations among Caucasians at forensic DNA loci. *Heredity*, **78**, 583–589.

Balding, D.J., Greenhalgh, M. and Nichols, R.A. (1996) Population genetics of STR loci in Caucasians. *International Journal of Legal Medicine*, **108**, 300–305.

Balthazard, V. (1911) De l'identification par les empreintes digitales. *Comptes Rendus des Séances de l'Académie des Sciences*, **152**, 1862–1864.

Bar-Hillel, M. and Falk, R. (1982) Some teasers concerning conditional probabilities. *Cognition*, **11**, 109–122.

Barnard, G.A. (1958) Thomas Bayes – a biographical note (together with a reprinting of Bayes, 1763). *Biometrika*, **45**, 293–315. Reprinted in Pearson and Kendall (1970), 131–153.

Barnett, P.D. and Ogle R.R. (1982) Probabilities and human hair comparison. *Journal of Forensic Sciences*, **27**, 272–278.

Bates, J.W. and Lambert, J.A. (1991) Use of the hypergeometric distribution for sampling in forensic glass comparison. *Journal of the Forensic Science Society*, **31**, 449–455.

Bayes, T. (1763) An essay towards solving a problem in the doctrine of chances. *Philosophical Transactions of the Royal Society of London for 1763*, **53**, 370–418. Reprinted with Barnard (1958) in Pearson and Kendall (1970), 131–153.

Belin, T.R., Gjertson, D.W. and Hu, M. (1997) Summarizing DNA evidence when relatives are possible suspects. *Journal of the American Statistical Association*, **92**, 706–716.

Bentham, J. (1827) *Rationale of Judicial Evidence, Specially Applied to English Practice* (Mill, J.S., ed.), Hunt and Clarke, London.

Berger, J.O. (1985) *Statistical Decision Theory and Bayesian Analysis*, Springer-Verlag, New York.

Berger, J.O. and Sellke, T. (1987) Testing a point null hypothesis: the irreconcilability of *P* values and evidence. *Journal of the American Statistical Association*, **82**, 112–139.

Bernardo, J.M. and Smith, A.F.M. (1994) *Bayesian Theory*. John Wiley & Sons, Ltd, Chichester.

Bernoulli, J. (1713) *Ars conjectandi*, Basle, Switzerland.

Bernoulli, N. (1709) *Specimina artis conjectandi ad quaestiones juris applicatae*, Basle, Switzerland.

Berry, D.A. (1990) DNA fingerprinting: what does it prove? *Chance*, **3**, 15–25.

Berry, D.A. (1991a) Inferences using DNA profiling in forensic identification and paternity cases. *Statistical Science*, **6**, 175–205.

Berry, D.A. (1991b) Probability of paternity. In *The Use of Statistics in Forensic Science* (Aitken, C.G.G. and Stoney, D.A., eds), Ellis Horwood, Chichester, pp. 150–156.

Berry, D.A. (1991c) Bayesian methods in Phase III trials. *Drug Information Journal*, **25**, 345–368.

Berry, D.A. (1993) A case for Bayesianism in clinical trials. *Statistics in Medicine*, **12**, 1377–1393.

Berry, D.A. (1996) *A Bayesian Perspective*, Duxbury Press, Belmont, CA.

Berry, D.A. and Geisser, S. (1986) Inference in cases of disputed paternity. In *Statistics and the Law* (DeGroot, M.H., Fienberg, S.E. and Kadane, J.B., eds), John Wiley & Sons, Inc., New York, pp. 353–382.

Berry, D.A., Evett, I.W. and Pinchin, R. (1992) Statistical inference in crime investigations using deoxyribonucleic acid profiling (with discussion). *Applied Statistics*, **41**, 499–531.

Bertillon, A. (1897/1898) La comparaison des écritures et l'identification graphique. *Revue Scientifique*, 18.12.1897–1.1.1898. Reprint published by Typographie Chamerot et Renouard, Paris.

Bertillon, A. (1899) Déposition Bertillon (du 18 janvier, 2 février, 4 février et 6 février 1899. In *La révision du procès Drefus – Enquête de la Cour de Cassation*, P.-V. Stock, Paris, pp. 482–500.

Bertillon, A. (1905) *Le Réquisitoire de M. Baudoin, Procureur Général*, Imprimerie Nationale, Paris.

Biasotti, A.A. (1959) A statistical study of the individual characteristics of fired bullets. *Journal of Forensic Scientists*, **4**, 34–50.

Biasotti, A.A. and Murdock, J. (1997) Firearms and toolmark identification: the scientific basis of firearms and toolmark identification. In *Modern Scientific Evidence: The Law and Science of Expert Testimony* (Faigman, L., Kaye, D.H., Saks, M.J. and Sanders, J., eds), West Publishing, St. Paul, MN, pp. 144–150.

Booth, G., Johnston, F. and Jackson, G. (2002) Case assessment and interpretation – application to a drugs supply case. *Science & Justice*, **42**, 123–125.

Brenner, C.H. and Weir, B.S. (2003) Issues and strategies in the DNA identification of World Trade Center victims. *Theoretical Population Biology*, **63**, 173–178.

Briggs, T.J. (1978) The probative value of bloodstains on clothing. *Medicine, Science and the Law*, **18**, 79–83.

Bring, J. and Aitken, C.G.G. (1997) Burden of proof and estimation of drug quantities under the Federal Sentencing Guidelines. *Cardozo Law Review*, **18**, 1987–1999.

Brookfield, J.F.Y. (1994) The effect of relatives on the likelihood ratio associated with DNA profile evidence in criminal cases. *Journal of the Forensic Science Society*, **34**, 193–197.

Brown, G.A. and Cropp, P.L. (1987) Standardised nomenclature in forensic science. *Journal of the Forensic Science Society*, **27**, 393–399.

Buckleton, J. (1999) What can the 90's teach us about good forensic science? Paper presented at the First International Conference on Forensic Human Identification in The Millenium, London, 24–26 October.

Buckleton, J.S. and Evett, I.W. (1989) Aspects of the Bayesian interpretation of fibre evidence. CRSE Report 684, Home Office Forensic Science Service, 1–17.

Buckleton, J.S., Triggs, C.M. and Walsh, S.J. (2004) *DNA evidence*, CRC Press, Boca Raton, Florida, USA. (To appear.)

Buckleton, J.S. and Walsh, K.A.J. (1991) Knowledge-based systems. In *The Use of Statistics in Forensic Science* (Aitken, C.G.G. and Stoney, D.A., eds), Ellis Horwood, Chichester, pp. 186–206.

Buckleton, J., Walsh, K.A.J., Seber, G.A.F. and Woodfield, D.G. (1987) A stratified approach to the compilation of blood group frequency surveys. *Journal of the Forensic Science Society*, **27**, 103–112.

Buckleton, J.S., Walsh, K.A.J. and Evett, I.W. (1991) Who is 'random man'? *Journal of the Forensic Science Society*, **31**, 463–468.

Buckleton, J.S., Evett, I.W. and Weir, B.S. (1998) Setting bounds for the likelihood ratio when multiple hypotheses are postulated. *Science & Justice*, **38**, 23–26.

Budowle, B., Chakraborty, R., Carmody, G. and Monson, K.L. (2000) Source attribution of a forensic DNA profile. *Forensic Science Communications*, **2**. Available at http://www.fbi.gov/hq/lab/fsc/backissu/july2000/index.htm

Bunch, S.G. (2000) Consecutive matching striation criteria: a general critique. *Journal of Forensic Sciences*, **45**, 955–962.

Çakir, A.H., Şimşek, F., Açik, L. and Taşdelen, B. (2001) Distribution of *HumTPOX*, *HumvWA*, *HumTHO1* alleles in a Turkish population sample. *Journal of Forensic Sciences*, **46**, 1257–1259.

Calman, K.C. (1996) Cancer: science and society and the communication of risk. *British Medical Journal*, **313**, 799–802.

Calman, K.C. and Royston, G.H.D. (1997) Risk language and dialects. *British Medical Journal*, **315**, 939–941.

Cantrell, S., Roux, C., Maynard, P. and Robertson, J. (2001) A textile fibre survey as an aid to the interpretation of fibres evidence in the Sydney region. *Forensic Science International*, **123**, 48–53.

Carracedo, A., Barros, F., Lareu, M.V., Pestoni, C. and Rodríguez-Calvo, M.S. (1996) Focusing the debate on forensic genetics. *Science & Justice*, **36**, 204–205.

Carracedo, A., Bär, W., Lincoln, P.J., Mayr, W., Morling, N., Olaisen, B., Schneider, P., Budowle, B., Brinkmann, B., Gill, P., Holland, M., Tully, G. and Wilson, M. (2000) DNA Commission of the International Society for Forensic Genetics: guidelines for mitochondrial DNA typing. *Forensic Science International*, **110**, 79–85.

Ceci, S.J. and Friedman, R.D. (2000) The suggestibility of children: scientific research and legal implications. *Cornell Law Review*, **86**, 33–108.

Chable, J., Roux, C. and Lennard, C.J. (1994) Collection of fiber evidence using water-soluble cellophane tape. *Journal of Forensic Sciences*, **39**, 1520–1527.

Chabli, S. (2001) Scene of crime evidence: fibres. In *Proceedings of the 13th INTERPOL Forensic Science Symposium* (Tontarski Jr, R.E., ed.), National Forensic Science Technology Centre, Largo, FL, pp. 106–119.

Champod, C. (1995) Locard, numerical standards and 'probable' identification. *Journal of Forensic Identification*, **45**, 132–159.

Champod, C. (1996) *Reconnaissance automatique et analyse statistique des minuties sur les empreintes digitales*. Doctoral thesis, Institut de Police Scientifique et de Criminologie, Université de Lausanne, Imprimerie Evard, Concise, Switzerland.

Champod, C. (1999) The inference of identity of source: theory and practice. Paper presented at the First International Conference on Forensic Human Identification in The Millenium, London, 24–26 October.

Champod, C. (2000) Identification/individualization. In *Encyclopedia of Forensic Sciences* (Siegel, J.A., Saukko, P.J. and Knupfer, G.C., eds) Academic Press, San Diego, CA, pp. 1077–1084.

Champod, C. and Evett, I.W. (2000) Commentary on Broeders, A.P.A. (1999) 'Some observations on the use of probability scales in forensic identification', *Forensic Linguistics*, **6**, 228–241; *Forensic Linguistics*, **7**, 238–243.

Champod, C. and Evett, I.W. (2001) A probabilistic approach to fingerprint evidence. *Journal of Forensic Identification*, **51**, 101–122.

Champod, C. and Jackson, G. (2000) European Fibres Group Workshop: Case assessment and Bayesian interpretation of fibres evidence. In *Proceedings of the 8th Meeting of European Fibres Group*, Kraków, Poland, 33–45.

Champod, C. and Jackson, G. (2002) Comments on the current debate on the Bayesian approach in marks examination. *Information Bulletin for Shoeprint/Toolmark Examiners*, **8**(3), 22–25.

Champod, C. and Margot, P. (1996) Computer assisted analysis of minutiae occurrences on fingerprints. In *Proceedings of the International Symposium on Fingerprint Detection and Identification* (Almog, J. and Springer, E., eds), Israel National Police, Jerusalem, pp. 305–318.

Champod, C. and Margot, P. (1997) Analysis of minutiae occurrences in fingerprints – the search for non-combined minutiae. In *Current Topics in Forensic Science – Proceedings of the 14th Meeting of the International Association of Forensic Sciences* (Takatori, T. and Takasu, A.), Shunderson Communications, Ottawa, pp. 55–58.

Champod, C. and Meuwly, D. (2000) The inference of identity in forensic speaker recognition. *Speech Communication*, **31**, 193–203.

Champod, C. and Taroni, F. (1997) Bayesian framework for the evaluation of fibre transfer evidence. *Science & Justice*, **37**, 75–83.

Champod, C. and Taroni, F. (1999) Interpretation of evidence: the Bayesian approach. In *Forensic Examination of Fibres* (Robertson, J. and Grieve, M., eds). Taylor and Francis, London, pp. 379–398.

Champod, C., Taroni, F. and Margot, P. (1999) The Dreyfus case – an early debate on experts' conclusions (an early and controversial case on questioned document examination). *International Journal of Forensic Document Examiners*, **5**, 446–459.

Champod, C., Evett, I.W., Jackson, G. and Birkett, J. (2000) Comments on the scale of conclusions proposed by the *ad hoc* committee of the ENFSI marks working group. *Information Bulletin for Shoeprint/Toolmark Examiners*, **6**(3), 11–18.

Champod, C., Evett, I.W. and Kuchler, B. (2001) Earmarks as evidence: a critical review. *Journal of Forensic Sciences*, **46**, 1275–1284.

Champod, C., Baldwin, D., Taroni, F. and Buckleton, J.S. (2003) Firearms and tool marks identification: the Bayesian approach. *AFTE (Association of Firearm and Toolmark Examiners) Journal*, **35**, 307–316.

Champod, C., Lennard, C.J., Margot, P. and Stoilovic, M. (2004) , *Fingerprints and Other Ridge Skin Impressions*. CRC Press, Boca Raton, FL.

Chan, K.P.S. and Aitken, C.G.G. (1989) Estimation of the Bayes factor in a forensic science problem. *Journal of Statistical Computation and Simulation*, **33**, 249–264.

Charpentier, A. (1933) *Historique de l'affaire Dreyfus*, Fasquelle, Paris.

Clayton, T.M., Whitaker, J.P., Sparkes, R. and Gill, P. (1998) Analysis and interpretation of mixed forensic stains using DNA STR profiling. *Forensic Science International*, **91**, 55–70.

Cochran, W.G. (1977) *Sampling Techniques*, 3rd edition. John Wiley & Sons, Ltd, Chichester.

Cockerham, C.C. (1969) Variance of gene frequencies. *Evolution*, **23**, 72–84.

Cockerham, C.C. (1973) Analysis of gene frequencies. *Genetics*, **74**, 679–700.

Cohen, L.J. (1977) *The Probable and the Provable*, Clarendon Press, Oxford.

Cohen, L.J. (1988) The difficulty about conjunction in forensic proof. *The Statistician*, **37**, 415–416.

Coleman, R.F. and Walls, H.J. (1974) The evaluation of scientific evidence. *Criminal Law Review*, 276–287.

Colón, M., Rodríguez, G. and Díaz, R.O. (1993) Representative sampling of 'street' drug exhibits. *Journal of Forensic Sciences*, **38**, 641–648.

Condorcet (de Caritat, M.J.A.N.), Marquis de (1785) *Essai sur l'application de l'analyse à la probabilité des décisions rendues à la pluralité des voix*, Imprimerie Royale, Paris.

Cook, R., Evett, I.W., Jackson, G. and Rogers, M. (1993) A workshop approach to improving the understanding of the significance of fibres evidence. *Science & Justice*, **33**, 149–152.

Cook, R., Webb-Salter, M.T. and Marshall, L. (1997) The significance of fibres found in head hair, *Forensic Science International*, **87**, 155–160.

Cook, R., Evett, I.W., Jackson, G., Jones, P.J. and Lambert, J.A. (1998a) A model for case assessment and interpretation. *Science & Justice*, **38**, 151–156.

Cook, R., Evett, I.W., Jackson, G., Jones, P.J. and Lambert, J.A. (1998b) A hierarchy of propositions: deciding which level to address in casework. *Science & Justice*, **38**, 231–239.

Cook, R., Evett, I.W., Jackson, G., Jones, P.J. and Lambert, J.A. (1999) Case pre-assessment and review of a two-way transfer case. *Science & Justice*, **39**, 103–122.

Coulson, S.A., Buckleton, J.S., Gummer, A.B. and Triggs, C.M. (2001a) Glass on clothing and shoes of members of the general population and people suspected of breaking crimes. *Science & Justice*, **41**, 39–48.

Coulson, S.A., Coxon, A. and Buckleton, J.S. (2001b) How many samples from a drug seizure need to be analyzed? *Journal of Forensic Sciences*, **46**, 1456–1461.

Cournot, A.A. (1838) Sur les applications du calcul des chances à la statistique judiciaire. *Journal des Mathématiques Pures et Appliquées*, **3**, 257–334.

Cowell, R.C., Dawid, A.P., Lauritzen, S.L. and Spiegelhalter, D. (1999) *Probabilistic Networks and Expert Systems*, Springer-Verlag, New York.

Crow, E.L., Davis, F.A. and Maxfield, M.W. (1960) *Statistics Manual*, Dover, New York, 51–52.

Cullison, A.D. (1969) Probability analysis of judicial fact-finding: a preliminary outline of the subjective approach. *University of Toledo Law Review*, 538–598.

Curran, J.M. (2003) The statistical interpretation of forensic glass evidence. *International Statistical Review*, **71**, 497–520.

Curran, J.M., Triggs, C.M., Almirall, J.R., Buckleton, J.S. and Walsh, K.A.J. (1997a) The interpretation of elemental composition measurements from forensic glass evidence: I. *Science & Justice*, **37**, 241–244.

Curran, J.M., Triggs, C.M., Almirall, J.R., Buckleton, J.S. and Walsh, K.A.J. (1997b) The interpretation of elemental composition measurements from forensic glass evidence: II. *Science & Justice*, **37**, 245–249.

Curran, J.M., Triggs, C.M., Buckleton, J.S., Walsh, K.A.J. and Hicks, T. (1998a) Assessing transfer probabilities in a Bayesian interpretation of forensic glass evidence. *Science & Justice*, **38**, 15–21.

Curran, J.M., Triggs, C.M. and Buckleton, J.S. (1998b) Sampling in forensic comparison problems. *Science & Justice*, **38**, 101–107.

Curran, J.M., Triggs, C.M., Buckleton, J.S. and Coulson, S. (1998c) Combining a continuous Bayesian approach with grouping information. *Forensic Science International*, **91**, 181–196.

Curran, J.M., Buckleton, J.S. and Triggs, C.M. (1999a) Commentary on Koons, R.D. and Buscaglia, J., The forensic significance of glass composition and refractive index measurements, *J. Forensic Sci.*, 1999, **44**, 496–503. *Journal of Forensic Sciences*, **44**, 1324–1325.

Curran, J.M., Triggs, C.M., Buckleton, J.S and Weir, B.S. (1999b) Interpreting DNA mixtures in structured populations. *Journal of Forensic Sciences*, **44**, 987–995.

Curran, J.M., Hicks, T.N. and Buckleton, J.S. (2000) *Forensic Interpretation of Glass Evidence*. CRC Press, Boca Raton, FL.

Curran, J.M., Buckleton, J.S., Triggs, C.M. and Weir, B.S. (2002) Assessing uncertainty in DNA evidence caused by sampling effects. *Science & Justice*, **42**, 29–37.

Dabbs, M.G.D. and Pearson, E.F. (1970) Heterogeneity in glass. *Journal of the Forensic Science Society*, **10**, 139–148.

Dabbs, M.G.D. and Pearson, E.F. (1972) Some physical properties of a large number of window glass specimens. *Journal of Forensic Sciences*, **17**, 70–78.

Darboux, J.G., Appell, P.E. and Poincaré, J.H. (1908) Examen critique des divers systèmes ou études graphologiques auxquels a donné lieu le bordereau. In *L'affaire Drefus – La révision du procès de Rennes – enquête de la chambre criminelle de la Cour de Cassation*, Ligue française des droits de l'homme et du citoyen, Paris, pp. 499–600.

Darroch, J. (1985) Probability and criminal trials. *Newsletter of the Statistical Society of Australia*, **30**, 1–7.

Darroch, J. (1987) Probability and criminal trials; some comments prompted by the Splatt trial and The Royal Commission. *Professional Statistician*, **6**, 3–7.

Davis, R.J. (ed.) (1986) The Splatt case. *Journal of the Forensic Science Society*, **26**, 219–221.

Davis, R.J. and DeHaan, J.D. (1977) A survey of men's footwear. *Journal of the Forensic Science Society*, **17**, 271–285.

Davison, A.C. and Hinkley, D.V. (1997) *Bootstrap Methods and Their Application*, Cambridge University Press, Cambridge.

Dawid, A.P. (1987) The difficulty about conjunction. *The Statistician*, **36**, 91–97.

Dawid, A.P. (1994) The island problem: coherent use of identification evidence. In *Aspects of Uncertainty: A Tribute to D.V. Lindley* (Freeman, P.R. and Smith, A.F.M., eds), John Wiley & Sons, Ltd, Chichester, pp. 159–170.

Dawid, A.P. (2001) Comment on Stockmarr (Likelihood ratios for evaluating DNA evidence when the suspect is found through a database search. *Biometrics*, 1999, **55**, 671–677), *Biometrics*, **57**, 976–980.

Dawid, A.P. (2002) Bayes's theorem and the weighing of evidence by juries. In *Bayes's theorem: Proceedings of the British Academy, Vol. 113* (Swinburne, R., ed.), Oxford University Press, Oxford, pp. 71–90.

Dawid, A.P. (2003) An object-oriented Bayesian network for evaluating mutation rates. In *Proceedings of the Ninth International Workshop on Artificial Intelligence and Statistics* (Bishop, C.M. and Frey, B.J., eds), Key West (available at http://research.microsoft.com/conferences/aistats2003/proceedings/188.pdf).

Dawid, A.P. and Evett, I.W. (1997) Using a graphical model to assist the evaluation of complicated patterns of evidence. *Journal of Forensic Sciences*, **42**, 226–231.

Dawid, A.P. and Mortera, J. (1996) Coherent analysis of forensic identification evidence. *Journal of the Royal Statistical Society, Series B*, **58**, 425–443.

Dawid, A.P., Mortera, J. and Pascali, V.L. (2001) Non-fatherhood or mutation? A probabilistic approach to parental exclusion in paternity testing. *Forensic Science International*, **124**, 55–61.

Dawid, A.P., Mortera, J., Pascali, V.L. and van Boxel, D. (2002) Probabilistic expert systems for forensic inference from genetic markers. *Scandinavian Journal of Statistics*, **29**, 577–595.

de Finetti, B. (1930) Fondamenti logici del ragionamento probabilistico. *Bollettino della Unione Matematica Italiana*, **9**, 258–261.

de Finetti, B. (1931) Sul significato soggettivo delle probabilita. *Fundamenta Mathematicae*, **17**, 298–329.

de Finetti, B. (1952) La nozione di evento. *Atti del Congresso di Metodologia*, Centro Studi Metodologici, Ramella, Turin, pp. 170–174. Translated in B. de Finetti, *Probabilità e induzione*, Clueb, Bologna (1993).

de Finetti, B. (1968) Probability: the subjectivistic approach. In *La philosophie contemporaine. Vol. 2* (Klibansky, R., ed.), La Nuova Italia, Florence, pp. 45–53.

Decorte, R. and Cassiman, J.J. (1993) Forensic medicine and the polymerase chain reaction technique. *Journal of Medical Genetics*, **30**, 625–633.

DeGroot, M.H. (1970) *Optimal Statistical Decisions*, McGraw-Hill, New York.

Devlin, B. (2000) The evidentiary value of a DNA database search. *Biometrics*, **56**, 1276.

Diaconis, P. and Freedman, D. (1981) The persistence of cognitive illusions. *Behavioural and Brain Sciences*, **4**, 333–334.

Dickson, D. (1994) As confusion leads to retrial in UK. *Nature*, **367**, 101–102.

Donnelly, P. and Friedman, R. (1999) DNA database searches and the legal consumption of scientific evidence. *Michigan Law Review*, **97**, 931–984.

Dujourdy, L., Barbati, G., Taroni, F., Guéniat, O., Esseiva, P., Anglada, F. and Margot, P. (2003) Evaluation of links in heroin seizures. *Forensic Science International*, **131**, 171–183.

Edwards, A.W.F. (1992) *Likelihood*, expanded edition, John Hopkins University Press, Baltimore, MD.

Edwards, W. (1986) Comment. *Boston University Law Review*, **66**, 623–626.

Edwards, W. (1991) Influence diagrams, Bayesian imperialism, and the *Collins* case: an appeal to reason. *Cardozo Law Review*, **13**, 1025–1079.

Edwards, W., Lindman, H. and Savage, L.J. (1963) Bayesian statistical inference for psychological research. *Psychological Review*, **70**, 193–242. Reprinted in *Robustness of Bayesian Analyses* (Kadane, J., ed.), Elsevier, Amsterdam, 1984.

Encyclopaedia Britannica (1993) DNA. In *Micropaedia, Vol. 4, Encyclopaedia Britannica* (15th edition), pp. 140–141.

Eggleston, R. (1983) *Evidence, Proof and Probability*, 2nd edition. Weidenfeld and Nicolson, London.

Ellman, I.M. and Kaye, D. (1979) Probabilities and proof: can HLA and blood group testing prove paternity? *New York University Law Review*, **54**, 1131–1162.

Engel, E. and Venetoulias, A. (1991) Monty Hall's probability puzzle. *Chance*, **4**, 6–9.

Essen-Möller, E. (1938) Die Beweiskraft der Ähnlichkeit im Vaterschaftsnachweis: Theoretische Grundlagen. *Mitteilungen der Anthropologischen Gesellschaft*, **68**, 9–53.

Evans, M., Hastings, N. and Peacock, B. (2000) *Statistical distributions*, 3rd edition. John Wiley & Sons, Inc., New York.

Evett, I.W. (1977) The interpretation of refractive index measurements. *Forensic Science International*, **9**, 209–217.

Evett, I.W. (1978) The interpretation of refractive index measurements, II. *Forensic Science International*, **12**, 34–47.

Evett, I.W. (1983) What is the probability that this blood came from that person? A meaningful question? *Journal of the Forensic Science Society*, **23**, 35–39.

Evett, I.W. (1984) A quantitative theory for interpreting transfer evidence in criminal cases. *Applied Statistics*, **33**, 25–32.

Evett, I.W. (1986) A Bayesian approach to the problem of interpreting glass evidence in forensic science casework. *Journal of the Forensic Science Society*, **26**, 3–18.

Evett, I.W. (1987a) Bayesian inference and forensic science: problems and perspectives. *The Statistician*, **36**, 99–105.

Evett, I.W. (1987b) On meaningful questions: a two-trace transfer problem. *Journal of the Forensic Science Society*, **27**, 375–381.

Evett, I.W. (1990) The theory of interpreting scientific transfer evidence. In *Forensic Science Progress, Volume 4*, Springer-Verlag, Berlin, pp. 141–179.

Evett, I.W. (1992) Evaluating DNA profiles in the case where the defence is 'It was my brother'. *Journal of the Forensic Science Society*, **32**, 5–14.

Evett, I.W. (1993a) Establishing the evidential value of a small quantity of material found at a crime scene. *Journal of the Forensic Science Society*, **33**, 83–86.

Evett, I.W. (1993b) Criminalistics: the future of expertise. *Journal of the Forensic Science Society*, **33**, 173–178.

Evett, I.W. (1995) Avoiding the transposed conditional. *Science & Justice*, **35**, 127–131.

Evett, I.W. (1998) Towards a uniform framework of reporting opinions in forensic science casework. *Science & Justice*, **38**, 198–202.

Evett, I.W. and Buckleton, J.S. (1990) The interpretation of glass evidence. A practical approach. *Journal of the Forensic Science Society*, **30**, 215–223.

Evett, I.W. and Buckleton, J.S. (1996) Statistical analysis of STR data. In *Advances in Forensic Haemogenetics, Volume 6* (Carracedo, A., Brinkmann, B. and Bär, W., eds), Springer-Verlag, Berlin, pp. 79–86.

Evett, I.W. and Lambert, J.A. (1982) The interpretation of refractive index measurements, III. *Forensic Science International*, **20**, 237–245.

Evett, I.W. and Lambert, J.A. (1984) The interpretation of refractive index measurements, IV. *Forensic Science International*, **26**, 149–163.

Evett, I.W. and Lambert, J.A. (1985) The interpretation of refractive index measurements, V. *Forensic Science International*, **27**, 97–110.

Evett, I.W. and Weir, B.S. (1992) Flawed reasoning in court. *Chance*, **4**, 19–21.

Evett, I.W. and Weir, B.S. (1998) *Interpreting DNA Evidence*. Sinauer Associates, Sunderland, MA.

Evett, I.W. and Williams, R. (1996) A review of the sixteen point fingerprint standard in England and Wales. *Journal of Forensic Identification*, **46**, 49–73.

Evett, I.W., Cage, P.E. and Aitken, C.G.G. (1987) Evaluation of the likelihood ratio for fibre transfer evidence in criminal cases. *Applied Statistics*, **36**, 174–180.

Evett, I.W., Werrett, D.J. and Buckleton, J.S. (1989a) Paternity calculations from DNA multilocus profiles. *Journal of the Forensic Science Society*, **29**, 249–254.

Evett, I.W., Werrett, D.J. and Smith, A.F.M. (1989b) Probabilistic analysis of DNA profiles. *Journal of the Forensic Science Society*, **29**, 191–196.

Evett, I.W., Buffery, C., Willott, G. and Stoney, D.A. (1991) A guide to interpreting single locus profiles of DNA mixtures in forensic cases. *Journal of the Forensic Science Society*, **31**, 41–47.

Evett, I.W., Pinchin, R. and Buffery, C. (1992a) An investigation of the feasibility of inferring ethnic origin from DNA profiles. *Journal of the Forensic Science Society*, **32**, 301–306.

Evett, I.W., Scranage, J. and Pinchin, R. (1992b) An efficient statistical procedure for interpreting DNA single locus profiling data in crime cases. *Journal of the Forensic Science Society*, **32**, 307–326.

Evett, I.W., Scranage, J. and Pinchin, R. (1993) An illustration of the advantages of efficient statistical methods for RFLP analysis in forensic science. *American Journal of Human Genetics*, **52**, 498–505.

Evett, I.W., Lambert, J.A. and Buckleton, J.S. (1995) Further observations on glass evidence interpretation. *Science & Justice*, **35**, 283–289.

Evett, I.W., Lambert, J.A. and Buckleton, J.S. (1998a) A Bayesian approach to interpreting footwear marks in forensic casework. *Science & Justice*, **38**, 241–247.

Evett, I.W., Gill, P.D. and Lambert, J.A. (1998b) Taking account of peak areas when interpreting mixed DNA profiles. *Journal of Forensic Sciences*, **43**, 62–69.

Evett, I.W., Jackson, G., Lambert, J.A. and McCrossan, S. (2000a) The impact of the principles of evidence interpretation and the structure and content of statements. *Science & Justice*, **40**, 233–239.

Evett, I.W., Jackson, G. and Lambert, J.A. (2000b) More on the hierarchy of propositions: exploring the distinction between explanations and propositions. *Science & Justice*, **40**, 3–10.

Evett, I.W., Foreman, L.A. and Weir, B.S. (2000c) Letter to the Editor. *Biometrics*, **56**, 1274–1275.

Evett, I.W., Foreman, L.A. and Weir, B.S. (2000d) A response to Devlin (The evidentiary value of a DNA database search. *Biometrics*, **56**, 1276). *Biometrics*, **56**, 1277.

Evett, I.W., Foreman, L.A., Jackson, G. and Lambert, J.A (2000e) DNA profiling: a discussion of issues relating to the reporting of very small match probabilities. *Criminal Law Review*, 341–355.

Evett, I.W., Gill, P.D., Jackson, G., Whitaker, J. and Champod, C. (2002) Interpreting small quantities of DNA: the hierarchy of propositions and the use of Bayesian networks. *Journal of Forensic Sciences*, **47**, 520–530.

Faber, N.M., Sjerps, M., Leijenhorst, H.A.L. and Maljaars, S.E. (1999) Determining the optimal sample size in forensic casework – with application to fibres. *Science & Justice*, **39**, 113–122.

Fairley, W.B. (1973), Probabilistic analysis of identification evidence. *Journal of Legal Studies*, **II**, 493–513.

Fairley W.B. (1975), Probabilistic analysis of identification evidence. In *Utility, Probability and Human Decision Making* (Wendt, D. and Vlek, C., eds), Reidel, Dordrecht, p. 251.

Fairley, W.B. and Mosteller, W. (1974) A conversation about Collins. *University of Chicago Law Review*, **41**, 242–253.

Fairley, W.B. and Mosteller, W. (1977) *Statistics and Public Policy*, Addison-Wesley, London, pp. 355–379.

Falk, R. (1992) A closer look at the probabilities of the notorious three prisoners. *Cognition*, **43**, 197–223.

Fienberg, S.E. (ed.) (1989) *The Evolving Role of Statistical Assessments as Evidence in the Courts*, Springer-Verlag, New York.

Fienberg, S.E. and Finkelstein, M.P. (1996) Bayesian statistics and the law. In *Bayesian Statistics 5* (Bernardo, J.M., Berger, J.O., Dawid, A.P. and Smith, A.F.M., eds), Oxford University Press, Oxford, pp. 129–146.

Fienberg, S.E. and Kadane, J.B. (1983) The presentation of Bayesian statistical analyses in legal proceedings. *The Statistician*, **32**, 88–98.

Fienberg, S.E. and Kaye, D.H. (1991) Legal and statistical aspects of some mysterious clusters. *Journal of the Royal Statistical Society, Series A*, **154**, 265–270.

Fienberg, S.E. and Schervish, M.J. (1986) The relevance of Bayesian inference for the presentation of statistical evidence and for legal decision making. *Boston University Law Review*, **66**, 771–798.

Fienberg, S.E., Krislov, S.H. and Straf, M.L. (1996) Understanding and evaluating statistical evidence in litigation. *Jurimetrics Journal*, **36**, 1–32.

Finkelstein, M.O. and Fairley, W.B. (1970) A Bayesian approach to identification evidence. *Harvard Law Review*, **83**, 489–517.

Finkelstein, M.O. and Fairley, W.B. (1971) A comment on 'Trial by mathematics'. *Harvard Law Review*, **84**, 1801–1809.

Finkelstein, M.O. and Levin, B. (2001) *Statistics for Lawyers*, 2nd edition, Springer-Verlag, New York.

Finney, D.J. (1977) Probabilities based on circumstantial evidence. *Journal of the American Statistical Association*, **72**, 316–318.

Fisher, R.A. (1951) Standard calculations for evaluating a blood-group system. *Heredity*, **5**, 51–102.

Fleming, P., Blair, P., Bacon, C. and Berry, J. (2000) *Sudden Unexpected Deaths in Infancy*, Her Majesty's Stationery Office, London.

Fong, W. and Inami, S.H. (1986) Results of a study to determine the probability of chance match occurrences between fibres known to be from different sources. *Journal of Forensic Sciences*, **31**, 65–72.

Foreman, L.A. and Evett, I.W. (2001) Statistical analyses to support forensic interpretation for a new ten-locus STR profiling system. *International Journal of Legal Medicine*, **114**, 147–155.

Foreman, L.A., Smith, A.F.M. and Evett, I.W. (1997a) Bayesian analysis of deoxyribonucleic acid profiling data in forensic identification applications (with discussion). *Journal of the Royal Statistical Society, Series A*, **160**, 429–469.

Foreman, L.A., Smith, A.F.M. and Evett, I.W. (1997b) A Bayesian approach to validating STR multiplex databases for use in forensic casework. *International Journal of Legal Medicine*, **110**, 244–250.

Foreman, L.A., Lambert, J.A. and Evett, I.W. (1998) Regional genetic variation in Caucasians. *Forensic Science International*, **95**, 27–37.

Foreman, L.A., Champod, C., Evett, I.W., Lambert, J.A. and Pope, S. (2003) Interpreting DNA evidence: a review. *International Statistical Review*, **71**, 473–495.

Frank, R.S., Hinkley, S.W. and Hoffman, C.G. (1991) Representative sampling of drug seizures in multiple containers. *Journal of Forensic Sciences*, **36**, 350–357.

Freeling, A.N.S. and Sahlin, N.-E. (1983) Combining evidence. In *Evidentiary Value* (Gardenfors, P., Hansson, B. and Sahlin, N.-E., eds), C.W.K. Gleerup, Lund, Sweden, pp. 58–74.

Friedman, R.D. (1986a) A diagrammatic approach to evidence. *Boston University Law Review*, **66**, 571–622.

Friedman, R.D. (1986b) A close look at probative value. *Boston University Law Review*, **66**, 733–759.

Friedman, R.D. (1996) Assessing evidence. *Michigan Law Review*, **94**, 1810–1838.

Friedman, R.D., Kaye, D.H, Mnookin, J., Nance, D. and Saks, M. (2002) Expert testimony on fingerprints: an Internet exchange. *Jurimetrics*, **43**, 91–98.

Fukshansky, N. and Baer, W. (1999) Biostatistical evaluation of mixed stains with contributors of different ethnic origin. *International Journal of Legal Medicine*, **112**, 383–387.

Fukshansky, N. and Baer, W. (2000) Biostatistics for mixed stains: the case of tested relatives of a non-tested suspect. *International Journal of Legal Medicine*, **114**, 78–82.

Fung, W.K. (2003) User-friendly programs for easy calculations in paternity testing and kinship determinations. *Forensic Science International*, **136**, 22–34.

Fung, W.K. and Hu, Y.-Q. (2000a) Interpreting DNA mixtures based on the NRC-II recommendation 4.1. *Forensic Science Communications*, **2**. Available at http://www.fbi.gov/hq/lab/fsc/backissu/oct2000/fung.htm

Fung, W.K. and Hu, Y.-Q. (2000b) Interpreting forensic DNA mixtures, allowing for uncertainty in population substructure and dependence. *Journal of the Royal Statistical Society, Series A*, **163**, 241–254.

Fung, W.K. and Hu, Y.-Q. (2002) The statistical evaluation of DNA mixtures with contributors from different ethnic groups. *International Journal of Legal Medicine*, **116**, 79–86.

Fung, W.K., Chung, Y. and Wong, D. (2002) Power of exclusion revisited: probability of excluding relatives of the true father from paternity. *International Journal of Legal Medicine*, **116**, 64–67.

Fung, W.K., Carracedo, A. and Hu, Y.Q. (2003) Testing for kinship in a subdivided population. *Forensic Science International*, **135**, 105–109.

Gaensslen, R.E., Bell, S.C. and Lee, H.C. (1987a) Distribution of genetic markers in United States populations: 1. Blood group and secretor systems. *Journal of Forensic Sciences*, **32**, 1016–1058.

Gaensslen, R.E., Bell, S.C. and Lee, H.C. (1987b) Distribution of genetic markers in United States populations: 2. Isoenzyme systems. *Journal of Forensic Sciences*, **32**, 1348–1381.

Gaensslen, R.E., Bell, S.C. and Lee, H.C. (1987c) Distribution of genetic markers in United States populations: 3. Serum group systems and haemoglobin variants. *Journal of Forensic Sciences*, **32**, 1754–1774.

Garber, D. and Zabell, S. (1979) On the emergence of probability. *Archive for History of Exact Sciences*, **21**, 33–53.

Garbolino, P. (2001) Explaining relevance. *Cardozo Law Review*, **22**, 1503–1521.

Garbolino, P. and Taroni, F. (2002) Evaluation of scientific evidence using Bayesian networks. *Forensic Science International*, **125**, 149–155.

Gastwirth, J.L. (1988a) *Statistical Reasoning in Law and Public Policy, Volume 1: Statistical Concepts and Issues of Fairness*, Academic Press, San Diego, CA.

Gastwirth, J.L. (1988b) *Statistical Reasoning in Law and Public Policy, Volume 2: Tort Law, Evidence and Health,* Academic Press, San Diego, CA.

Gastwirth, J.L. (ed.) (2000) *Statistical Science in the Courtroom,* Springer-Verlag, New York.

Gastwirth, J.L., Freidlin, B. and Miao, W. (2000) The *Shonubi* case as an example of the legal system's failure to appreciate statistical evidence. In *Statistical Science in the Courtroom* (Gastwirth, J.L., ed.), Springer-Verlag, New York, pp. 405–413.

Gaudette, B.D. (1982) A supplementary discussion of probabilities and human hair comparisons. *Journal of Forensic Sciences,* **27**, 279–289.

Gaudette, B.D. (1986) Evaluation of associative physical evidence. *Journal of the Forensic Science Society,* **26**, 163–167.

Gaudette, B.D. (1999) Evidential value of hair examination. In *Forensic Examination of Hair* (Robertson, J., ed.), Taylor & Francis, London, pp. 243–260.

Gaudette, B.D. (2000) Comparison: significance of hair evidence In *Encyclopedia of Forensic Sciences* (Siegel, J.A., Saukko, P.J. and Knupfer, G.C., eds), Academic Press, San Diego, pp. 1018–1024.

Gaudette, B.D. and Keeping, E.S. (1974) An attempt at determining probabilities in human scalp hair comparison. *Journal of Forensic Sciences,* **19**, 599–606.

Geisser, S. (1993) *Predictive Inference: An Introduction.* Chapman & Hall, London.

Gelfand, A.E. and Solomon, H. (1973) A study of Poisson's models for jury verdicts in criminal and civil trials. *Journal of the American Statistical Association,* **68**, 271–278.

Gettinby, G. (1984) An empirical approach to estimating the probability of innocently acquiring bloodstains of different ABO groups in clothing. *Journal of the Forensic Science Society,* **24**, 221–227.

Gill, P. (2001) An assessment of the utility of single nucleotide polymorphisms (SNPs) for forensic purposes. *International Journal of Legal Medicine,* **114**, 204–210.

Gill, P., Ivanov, P.L., Kimpton, C., Piercy, R., Benson, N., Tully, G., Evett, I., Hagelberg, E. and Sullivan, K. (1994) Identification of the remains of the Romanov family by DNA analysis. *Nature Genetics,* **6**, 130–135.

Gill, P., Sparkes, R., Pinchin, R., Clayton, T.M., Whitaker, J. and Buckleton, J.S. (1998a) Interpreting simple STR mixtures using allele peak areas. *Forensic Science International,* **91**, 41–53.

Gill, P. Sparkes, R. and Buckleton, J.S. (1998b) Interpretation of simple mixtures of when artefacts such as stutters are present – with special reference to multiplex STRs used by the Forensic Science Service. *Forensic Science International,* **95**, 213–224.

Gill, P., Brenner, C., Brinkmann, B., Budowle, B., Carracedo, A., Jobling, M.A., de Knijff, P., Kayser, M., Krawczak, M., Mayr, W., Morling, N., Olaisen, B., Pascali, V., Prinz, M., Roewer, L., Schneider, P., Sajantila, A. and Tyler-Smith, C. (2001) DNA Commission of the International Society for Forensic Genetics: recommendations on forensic analysis using Y-chromosome STRs. *Forensic Science International,* **124**, 5–10.

Goldmann, T., Taroni, F. and Margot, P. (2004) Analysis of dyes in illicit pills (amphetamine and derivatives). *Journal of Forensic Sciences.*

Good, I.J. (1950) *Probability and the Weighing of Evidence,* Griffin, London.

Good, I.J. (1956) Discussion of paper by G. Spencer Brown. In *Information Theory: Third London Symposium, 1955* (Cherry, C., ed.), Butterworths, London, pp. 13–14.

Good, I.J. (1959) Kinds of probability. *Science,* **129**, 443–447.

Good, I.J. (1983) A correction concerning my interpretation of Peirce and the Bayesian interpretation of Neyman–Pearson hypothesis determination. *Journal of Statistical Computation and Simulation,* **18**, 71–74.

Good, I.J. (1991) Weight of evidence and the Bayesian likelihood ratio. In *The Use of Statistics in Forensic Science* (Aitken, C.G.G. and Stoney, D.A., eds), Ellis Horwood, Chichester, pp. 85–106.

Goodman, J. (1992) Jurors' comprehension and assessment of probabilistic evidence. *American Journal of Trial Advocacy*, **16**, 361.

Graybill, F.A. (1969) *Introduction to Matrices with Applications in Statistics.* Wadsworth, Belmont, CA.

Grieve, M.C. (2000a) A survey on the evidential value of fibres and on the interpretation of the findings in fibre transfer cases. Part 1 – Fibre frequencies. *Science & Justice*, **40**, 189–200.

Grieve, M.C. (2000b) A survey on the evidential value of fibres and on the interpretation of the findings in fibre transfer cases. Part 2 – Interpretation and reporting. *Science & Justice*, **40**, 201–209.

Grieve, M.C. and Biermann, T.W. (1997) The population of coloured textile fibres on outdoor surfaces. *Science & Justice*, **37**, 231–239.

Grieve, M.C. and Dunlop, J. (1992) A practical aspect of the Bayesian interpretation of fibre evidence. *Journal of the Forensic Science Society*, **32**, 169–175.

Grieve, M.C., Biermann, T.W. and Davignon, M. (2001) The evidential value of black cotton fibres. *Science & Justice*, **41**, 245–260.

Groom, P.S. and Lawton, M.E. (1987) Are they a pair? *Journal of the Forensic Science Society*, **27**, 189–192.

Grove, D.M. (1980) The interpretation of forensic evidence using a likelihood ratio. *Biometrika*, **67**, 243–246.

Grove, D.M. (1981) The statistical interpretation of refractive index measurements. *Forensic Science International*, **18**, 189–194.

Grove, D.M. (1984) The statistical interpretation of refractive index measurements II: The multiple source problem. *Forensic Science International*, **24**, 173–182.

Gunel, E. and Wearden, S. (1995) Bayesian estimation and testing of gene frequencies. *Theoretical and Applied Genetics*, **91**, 534–543.

Habbema, J.D.F., Hermans, J. and van den Broek, K. (1974) A stepwise discrimination program using density estimation. In *Compstat 1974* (Bruckman, G., ed.), Physica Verlag, Vienna, pp. 100–110.

Hacking, I. (1975) *The Emergence of Probability*, Cambridge University Press, Cambridge.

Harbison, S.A. and Buckleton, J.S. (1998) Applications and extensions of subpopulation theory: a caseworkers guide. *Science & Justice*, **38**, 249–254.

Harbison, S.A., Stanfield, A.M., Buckleton, J.S and Walsh, S.J. (2002) Allele frequencies for four major sub-populations in New Zealand at three STR loci – *HUMTH01, HUMTPOX and CSF1PO. Forensic Science International*, **126**, 258–260.

Harrison, P.H., Lambert, J.A. and Zoro, J.A. (1985) A survey of glass fragments recovered from clothing of persons suspected of involvement in crime. *Forensic Science International*, **27**, 171–187.

Harvey, W., Butler, O., Furness, J. and Laird, R. (1968) The Biggar murder: dental, medical, police and legal aspects. *Journal of the Forensic Science Society*, **8**, 155–219.

Hicks, T., Monard Sermier, F., Goldmann, T., Brunelle, A., Champod, C. and Margot, P. (2003) The classification and discrimination of glass fragments using non-destructive energy dispersive X-ray microfluorescence. *Forensic Science International*, **137**, 107–118.

Hicks, T.N. (2004) *De l'interprétation des fragments de verre en sciences forensiques*. Doctoral thesis, Ecole des Sciences Criminelles, Lausanne, Switzerland.

Hilton, O. (1995) The relationship of mathematical probability to the handwriting identification problem. *International Journal of Forensic Document Examiners*, **1**, 224–229.

Hoffmann, K. (1991) Statistical evaluation of the evidential value of human hairs possibly coming from multiple sources. *Journal of Forensic Sciences*, **36**, 1053–1058.

Hoggart, C.J., Walker, S.G. and Smith, A.F.M. (2003) Bivariate kurtotic distributions of garment fibre data. *Applied Statistics*, **52**, 323–335.

Holden, C. (1997) DNA fingerprinting comes of age. *Science*, **278**, 1407.

Hu, Y.-Q. and Fung, W.K. (2003) Evaluating forensic DNA mixtures with contributors of different structures ethnic origins: a computer software. *International Journal of Legal Medicine*, **117**, 248–249.

HUGIN Lite (2001) version 5.7, free demonstration version available at http://www.hugin.dk

Hummel, K. (1971) Biostatistical opinion of parentage based upon the results of blood group tests. In *Biostatistische Abstammungsbegutachtung mit Blutgruppenbefunden* (P. Schmidt, ed.), Gustav Fisher, Stuttgart. (Quoted in *Family Law Quarterly*, 1976, **10**, 262.)

Hummel, K. (1983) Selection of gene frequency tables. In *Inclusion Probabilities in Parentage Testing* (R.H. Walker, ed.), American Association of Blood Banks, Arlington, VA, pp. 231–243.

Ihaka, R. and Gentleman, R. (1996) R: a language for data analysis and graphics. *Journal of Computational and Graphical Statistics*, **5**, 299–314.

Iman, K.I. and Rudin, N. (2001) *Principles and Practice of Criminalistics – The Profession of Forensic Science*. CRC Press, Boca Raton, FL.

Intergovernmental Panel on Climate Change (2001) *Report of Working Group I*. Available at http://www.ipcc.ch/pub/spm22-01.pdf

Izenman, A.J. (2000a) Statistical issues in the application of the Federal sentencing guidelines in drug, pornography and fraud cases. In *Statistical Science in the Courtroom* (Gastwirth, J.L., ed.), Springer-Verlag, New York, pp. 25–50.

Izenman, A.J. (2000b) Introduction to two views on the *Shonubi* case. In *Statistical Science in the Courtroom* (Gastwirth, J.L., ed.), Springer-Verlag, New York, pp. 393–403.

Izenman, A.J. (2000c) Assessing the statistical evidence in the *Shonubi* case. In *Statistical Science in the Courtroom* (Gastwirth, J.L., ed.), Springer-Verlag, New York, pp. 415–433.

Izenman, A.J. (2001) Statistical and legal aspects of the forensic study of illicit drugs. *Statistical Science*, **16**, 35–57.

Izenman, A.J. (2003) Sentencing illicit drug traffickers: how do the courts handle random sampling issues? *International Statistical Review*, **71**, 535–556.

Jackson, G. (2000) The scientist and the scales of justice. *Science & Justice*, **40**, 81–85.

Jaynes, E.T. (2003) *Probability Theory*, Cambridge University Press, Cambridge.

Jeffrey, R.C. (1975) Probability and falsification: critique of the Popper program. *Synthèse*, **30**, 95–117.

Jeffreys, A.J., Wilson, V. and Morton, D.B. (1987) DNA fingerprints of dogs and cats. *Animal Genetics*, **18**, 1–15.

Jeffreys, H. (1983) *Theory of Probability*, 3rd edition, Clarendon Press, Oxford.

Jensen, F.V. (2001) *Bayesian Networks and Decision Graphs*, Springer-Verlag, New York.

Johnson, R.E. and Peterson, J. (1999) HLA-DQA1 and polymarker locus allele frequencies for Chicago, Illinois, USA. *Journal of Forensic Sciences*, **44**, 1097.

Jolliffe, I.T. (1986) *Principal Component Analysis*, Springer-Verlag, New York.

Jones, D.A. (1972) Blood samples: probability of discrimination. *Journal of the Forensic Science Society*, **12**, 355–359.

Jordan, M.I. (ed.) (1999) *Learning in Graphical Models*. MIT Press, Cambridge, MA.

Kadane, J.B. and Schum, D.A. (1996) *A Probabilistic Analysis of the Sacco and Vanzetti Evidence*. John Wiley & Sons, Inc., New York.

Kahneman, D., Slovic, P. and Tversky, A. (eds) (1982) *Judgment under Uncertainty: Heuristics and Biases*, Cambridge University Press, Cambridge.

Kass, R.E. and Raftery, A.E. (1995) Bayes factors. *Journal of the American Statistical Association*, **90**, 773–795.

Katterwe, H. (2002a) Comments/objections to reproaches of Forensic Science Service and University of Lausanne. *Information Bulletin for Shoeprint/Toolmark Examiners*, **8**(1), 25–30.

Katterwe, H. (2002b) Comments of Horst Katterwe to the Article of F. Taroni and J.Buckleton *Information Bulletin for Shoeprint/Toolmark Examiners*, **8**(3), 16–20.

Katterwe, H. (2003) True or false. *Information Bulletin for Shoeprint/Toolmark Examiners*, **9**(2), 18–25.

Kaye, D.H. (1979) The laws of probability and the law of the land. *University of Chicago Law Review*, **47**, 34–56.

Kaye, D.H. (1986) Quantifying probative value. *Boston University Law Review*, **66**, 761–766.

Kaye, D.H. (1987) Apples and oranges: confidence coefficients versus the burden of persuasion. *Cornell Law Review*, **73**, 54–77.

Kaye, D.H. (1989) The probability of an ultimate issue: the strange cases of paternity testing. *Iowa Law Review*, **75**, 75–109.

Kaye, D.H. (1993a) *Proceedings of the Second International Conference on Forensic Statistics*, Arizona State University, Center for the Study of Law, Science and Technology, Tempe, AZ. Selected papers included in *Jurimetrics Journal*, **34**(1), 1–115.

Kaye, D.H. (1993b) DNA evidence: probability, population genetics and the courts. *Harvard Journal of Law and Technology*, **7**, 101–172.

Kaye, D.H. (1997a) DNA, NAS, NRC, DAB, RFLP, PCR, and more: an introduction to the symposium on the 1996 NRC report on forensic DNA evidence. *Jurimetrics Journal*, **37**, 395–404.

Kaye, D.H. (1997b) DNA identification in criminal cases: some lingering and emerging evidentiary issues. In *Proceedings of the 7th International Symposium on Human Identification*, Madison, WI, Promega Corporation, pp. 12–25.

Kaye, D.H. and Aickin, M. (1986) *Statistical Methods in Discrimination Litigation*, Marcel Dekker, New York.

Kaye D.H and Koehler, J.J. (1991) Can jurors understand probabilistic evidence? *Journal of the Royal Statistical Society, Series A*, **154**, 21–39.

Kaye, D.H. and Koehler, J.J. (2003) The misquantification of probative value. *Law and Human Behaviour*, **27**, 645–659.

Kaye D.H and Sensabaugh, G.F. (2000) Reference guide on DNA evidence. In *Reference Manual on Scientific Evidence* (Cecil, J. ed.), Federal Judicial Center, Washington, DC, pp. 485–576.

Kendall, M.G. and Buckland, W.R. (1982) *A Dictionary of Statistical Terms*, 4th edition, Longman, London.

Kennedy, R.B., Pressman, I.S., Chen, S., Petersen, P.H. and Pressman, A.E. (2003) Statistical analysis of barefoot impressions. *Journal of Forensic Sciences*, **48**, 55–63.

Kind, S.S. (1994) Crime investigation and the criminal trial: a three chapter paradigm of evidence. *Journal of the Forensic Science Society*, **34**, 155–164.

Kind, S.S., Wigmore, R., Whitehead, P.H. and Loxley, D.S. (1979) Terminology in forensic science. *Journal of the Forensic Science Society*, **19**, 189–192.

Kingston, C.R. (1964) Probabilistic analysis of partial fingerprint patterns. D.Crim. dissertation, University of California, Berkeley.

Kingston, C.R. (1965a) Applications of probability theory in criminalistics. *Journal of the American Statistical Association*, **60**, 70–80.

Kingston, C.R. (1965b) Applications of probability theory in criminalistics – II. *Journal of the American Statistical Association*, **60**, 1028–1034.

Kingston, C.R. (1966) Probability and legal proceedings. *Journal of Criminal Law, Criminology and Police Science*, **57**, 93–98.

Kingston, C.R. (1970) The law of probabilities and the credibility of witness and evidence. *Journal of Forensic Sciences*, **15**, 18–27.

Kingston, C.R. (1988) Discussion of 'A critical analysis of quantitative fingerprint individuality models'. *Journal of Forensic Sciences*, **33**, 9–11.

Kingston, C.R. and Kirk, P.L. (1964) The use of statistics in criminalistics. *Journal of Criminal Law, Criminology and Police Science*, **55**, 514–521.

Kirk, P.L. (1963) The ontogeny of criminalistics. *Journal of Criminal Law, Criminology and Police Science*, **54**, 235–238.

Kirk, P.L. and Kingston, C.R. (1964) Evidence evaluation and problems in general criminalistics. Presented at the Sixteenth Annual Meeting of the American Academy of Forensic Sciences, Chicago.

Knowles, R. (2000) The new (non-numeric) fingerprint evidence standard. *Science & Justice*, **40**, 120–121.

Koehler, J.J. (1992) Probabilities in the courtroom: an evaluation of the objection and policies. In *Handbook of Psychology and Law* (Kagehiro, D.K. and Laufer, W.S., eds), Springer-Verlag, New York, pp. 167–184.

Koehler, J.J. (1993a) Error and exaggeration in the presentation of DNA evidence at trial. *Jurimetrics Journal*, **34**, 21–39.

Koehler, J.J. (1993b) DNA matches and statistics: important questions, surprising answers. *Judicature*, **76**, 222–229.

Koehler, J.J. (1995) The random match probability in DNA evidence: irrelevant and prejudicial? *Jurimetrics Journal*, **35**, 201–219.

Koehler, J.J. (1996) On conveying the probative value of DNA evidence: frequencies, likelihood ratios, and error rates. *University of Colorado Law Review*, **67**, 859–886.

Koehler, J.J. (1997a) One in millions, billions, and trillions: lessons from People v. Collins (1968) for People v. Simpson (1995). *Journal of Legal Education*, **47**(2), 214–223 at 219.

Koehler, J.J. (1997b) Why DNA likelihood ratios should account for error (even when a National Research Council report says they should not). *Jurimetrics Journal*, **37**, 425–437.

Koehler, J.J. (2001a) The psychology of numbers in the courtroom: how to make DNA match statistics seem impressive or insufficient. *Southern California Law Review*, **74**, 1275–1306.

Koehler, J.J. (2001b) When are people persuaded by DNA match statistics? *Law and Human Behaviour*, **25**, 493–513.

Koehler, J.J., Chia, A. and Lindsey, J.S. (1995) The random match probability (RMP) in DNA evidence: irrelevant and prejudicial? *Jurimetrics Journal*, **35**, 201–219.

Koons, R.D. and Buscaglia, J. (1999a) The forensic significance of glass composition and refractive index measurements. *Journal of Forensic Sciences*, **44**, 496–503.

Koons, R.D. and Buscaglia, J. (1999b) Authors' response to Curran *et al.* (1999a). *Journal of Forensic Sciences,* **44**, 1326–1328.

Koons, R.D. and Buscaglia, J. (2002) Interpretation of glass composition measurements: the effects of match criteria on discrimination capability. *Journal of Forensic Sciences,* **47**, 505–512.

Kuo, M. (1982) Linking a bloodstain to a missing person by genetic inheritance. *Journal of Forensic Sciences,* **27**, 438–444.

Kwan, Q.Y. (1977) *Inference of identity of source,* Doctor of Criminology thesis, University of California, Berkeley.

Lambert, J.A. and Evett, I.W. (1984) The refractive index distribution of control glass samples examined by the forensic science laboratories in the United Kingdom. *Forensic Science International,* **26**, 1–23.

Lambert, J.A., Satterthwaite, M.J. and Harrison, P.H. (1995) A survey of glass fragments recovered from clothing of persons suspected of involvement in crime. *Science & Justice,* **35**, 273–281.

Lange, K. (1995) Applications of the Dirichlet distribution to forensic match probabilities. In *Human Identification: The Use of DNA Markers* (Weir, B.S., ed.), Kluwer Academic, Dordrecht, pp. 107–117.

Laplace, Marquis de (1886) Essai philosophique sur les probabilités. *Introduction à la theorie analytique des probabilités, Oeuvres Complètes de Laplace,* Vol.7, Gauthier-Villars, Paris.

Lau, L. and Beveridge, A.D. (1997) The frequency of occurrence of paint and glass on the clothing of high school students. *Journal of the Canadian Society of Forensic Science,* **30**, 233–240.

Lauritzen, S.L. and Mortera, J. (2002) Bounding the number of contributors to mixed DNA stains. *Forensic Science International,* **130**, 125–126.

Lee, J.W., Lee, H.-S., Park, M. and Hwang, J.-J. (1999) Paternity probability when a relative of the father is an alleged father. *Science & Justice,* **39**, 223–230.

Lee, J.W., Lee, H.-S., Han, G.-R. and Hwang, J.-J. (2000) Motherless case in paternity testing. *Forensic Science International,* **114**, 57–65.

Lee, J.W., Lee, H.-S. and Hwang, J.-J. (2002) Statistical analysis for estimating heterogeneity of the Korean population in DNA typing using STR loci. *International Journal of Legal Medicine,* **116**, 153–160.

Lee, P.M. (2004) *Bayesian Statistics: An Introduction,* 3rd edition, Arnold, London.

Lempert, R. (1977) Modelling relevance. *Michigan Law Review,* **89**, 1021–1057.

Lempert, R. (1991) Some caveats concerning DNA as criminal identification evidence: with thanks to the Reverend Bayes. *Cardozo Law Review,* **13**, 303–341.

Lempert, R. (1993) The suspect population and DNA identification. *Jurimetrics Journal,* **34**, 1–7.

Lempert, R. (1997) After the DNA wars: skirmishing with NRC II. *Jurimetrics Journal,* **37**, 439–468.

Lenth, R.V. (1986) On identification by probability. *Journal of the Forensic Science Society,* **26**, 197–213.

Leonard, T. (2000) *A Course in Categorical Data Analysis.* Chapman & Hall/CRC, Boca Raton, FL.

Leonard, T. and Hsu, J.S.J. (1999) *Bayesian Methods,* Cambridge University Press, Cambridge.

Levitt, T.S. and Blackmond Laskey, K. (2001) Computational inference for evidential reasoning in support of judicial proof. *Cardozo Law Review,* **22**, 1691–1731.

Lewontin, R.C. (1993) Which population? (Letter). *American Journal of Human Genetics*, **52**, 205.

Liao, X.H., Lau, T.S., Ngan, K.F.N. and Wang, J. (2002) Deduction of paternity index from DNA mixture. *Forensic Science International*, **128**, 105–107.

Lindley, D.V. (1957) A statistical paradox. *Biometrika*, **44**, 187–192. (Comments by M.S. Bartlett and M.G. Kendall appear in **45**, 533–534.)

Lindley, D.V. (1977a) A problem in forensic science. *Biometrika*, **64**, 207–213.

Lindley, D.V. (1977b) Probability and the law. *The Statistician*, **26**, 203–212.

Lindley, D.V. (1980) L.J. Savage – his work in probability and statistics. *Annals of Statistics*, **8**, 1–24.

Lindley, D.V. (1987) The probability approach to the treatment of uncertainty in artificial intelligence and expert systems. *Statistical Science*, **2**, 17–24.

Lindley, D.V. (1991) Probability. In *The Use of Statistics in Forensic Science* (Aitken, C.G.G. and Stoney, D.A., eds), Ellis Horwood, Chichester, pp. 27–50.

Lindley, D.V. (1998) *Making Decisions*, 2nd edition, John Wiley & Sons, Ltd, London.

Lindley, D.V. and Eggleston, R. (1983) The problem of missing evidence. *Law Quarterly Review*, **99**, 86–99.

Lindley, D.V. and Scott, W.F. (1995) *New Cambridge Statistical Tables*, 2nd edition, Cambridge University Press, Cambridge.

Locard, E. (1914) La preuve judiciaire par les empreintes digitales. *Archives d'Anthropologie Criminelle, de Médecine Légale et de Psychologie Normale et Pathologique*, **28**, 321–348.

Locard, E. (1920) *L'enquête criminelle et les méthodes scientifiques*, Flammarion, Paris.

Locard, E. (1929) L'analyse des poussières en criminalistique. *Revue Internationale de Criminalistique*, September, 176–249.

Louis, T.A. (1981) Confidence intervals for a binomial parameter after observing no successes. *American Statistician*, **35**, 154.

Lyon, T.D. and Koehler, J.J. (1996) The relevance ratio: evaluating the probative value of expert testimony in child sexual abuse cases. *Cornell Law Review*, **82**, 43–78.

Mallows, C. (1998) The zeroth problem. *American Statistician*, **52**, 1–9.

Mardia, K.V., Kent, J.T. and Bibby, J.M. (1979) *Multivariate Analysis*, Academic Press, London.

Massonnet, G. and Stoecklein, W. (1999) Identification of organic pigments in coatings: applications to red automotive topcoats. Part III: Raman spectroscopy (NIR FT-Raman). *Science & Justice*, **39**, 181–187.

Matthews, R. (1994) Improving the odds on justice. *New Scientist*, 16 April, p. 12.

McDermott, S.D. and Willis, S.M. (1997) A survey of the evidential value of paint transfer evidence. *Journal of Forensic Sciences*, **42**, 1012–1018.

McDermott, S.D., Willis, S.M. and McCullough, J.P. (1999) The evidential value of paint. Part II: A Bayesian approach. *Journal of Forensic Sciences*, **44**, 263–269.

McQuillan, J. and Edgar, K. (1992) A survey of the distribution of glass on clothing. *Journal of the Forensic Science Society*, **32**, 333–348.

Meester, R. and Sjerps, M. (2003) The evidential value in the DNA database search controversy and the two-stain problem. *Biometrics*, **59**, 727–732.

Meier, P. and Zabell, S. (1980) Benjamin Peirce and the Howland will. *Journal of the American Statistical Association*, **75**, 497–506.

Mellen, B.G. (2000) A likelihood approach to DNA evidence. In *Statistical Science in the Courtroom* (Gastwirth, J.L., ed.), Springer-Verlag, New York.

Mellen, B.G. and Royall, R.M. (1997) Measuring the strength of deoxyribonucleic acid evidence, and probabilities of strong implicating evidence. *Journal of the Royal Statistical Society, Series A*, **160**, 305–320.

Meuwly, D. (2001) *Reconnaissance de locuteurs en sciences forensiques: l'apport d'une approche automatique.* Doctoral thesis, Institut de Police Scientifique et de Criminologie, University of Lausanne, Switzerland.

Meuwly, D. and Drygajlo, A. (2001) Forensic speaker recognition based on a Bayesian framework and Gaussian mixture modelling (GMM). In *Proceedings of the 2001 Speaker Odyssey Recognition Workshop.* 18–22 June 2000, Crete, Greece, pp. 145–150.

Miller, L.S. (1987) Procedural bias in forensic science examinations of human hair. *Law and Human Behavior*, **11**, 157–163.

Mode, E.B. (1963) Probability and criminalistics. *Journal of the American Statistical Association*, **58**, 628–640.

Moran, B. (2001) Toolmark criteria for identification: pattern match, CMS or Bayesian? *Interface*, **28**, 9–10.

Moran, B. (2002) A report on the AFTE theory of identification and range of conclusions for tool mark identification and resulting approaches to casework. *AFTE (Association of Firearm and Toolmark Examiners) Journal*, **34**, 227–235.

Moran, B. (2003) Comments and clarification of responses from a member of the AFTE 2001 criteria for identification of toolmarks discussion panel. *AFTE (Association of Firearm and Toolmark Examiners) Journal*, **35**, 55–65.

Moras, C. (1906) *L'Affaire Dreyfus: les débats de la Cour de Cassation (15 juin 1906 – 12 juillet 1906).* Société Nouvelle de Librairie et d'Édition, Paris.

Morgan, J.P., Chaganty, N.R., Dahiya, R.C. and Doviak, M.J. (1991) Let's make a deal: the player's dilemma. *American Statistician*, **45**, 284–287.

Mortera, J, Dawid, A.P. and Lauritzen, S.L. (2003) Probabilistic expert systems for DNA mixture profiling. *Theoretical Population Biology*, **63**, 191–205.

Mosteller, F. and Wallace, D.L. (1963) Inference in an authorship problem. *Journal of the American Statistical Association*, **58**, 275–309.

Mosteller, F. and Wallace, D.L. (1984) *Applied Bayesian and Classical Inference: The Case of The Federalist Papers*, Springer-Verlag, New York.

National Research Council (1992) *DNA Technology in Forensic Science*, National Academies Press, Washington, DC.

National Research Council (1996) *The Evaluation of Forensic DNA Evidence*, National Academies Press, Washington, DC.

Nichols, R.A. and Balding, D. (1991) Effect of population structure on DNA fingerprint analysis in forensic science. *Heredity*, **66**, 297–302.

Nichols, R.G. (1997) Firearm and toolmark identification criteria: a review of the literature. *Journal of Forensic Sciences*, **42**, 466–474.

Nichols, R.G. (2003) Firearm and toolmark identification criteria: a review of the literature, Part II. *Journal of Forensic Sciences*, **48**, 318–327.

Ogino, C. and Gregonis, D.J. (1981) Indirect typing of a victim's blood using paternity testing. Presentation before the California Association of Criminalists 57th Semi-annual Seminar, Pasadena, CA.

Ogle, R.R. (1991) Discussion of 'Further evaluation of probabilities in human scalp hair comparisons' (Wickenheiser and Hepworth, 1990). *Journal of Forensic Sciences*, **36**, 971–973.

Olkin, I. (1958) The evaluation of physical evidence and the identity problem by means of statistical probabilities. *General Scientific Session of the American Academy of Forensic Sciences*, Cleveland, Ohio, U.S.A.

Osterburg, J.W. and Bloomington, S.A. (1964) An inquiry into the nature of proof 'The Identity of Fingerprints'. *Journal of Forensic Sciences*, **9**, 413–427.

Owen, D.B. (1962) *Handbook of Statistical Tables*, Addison-Wesley Reading, MA.

Owen, G.W. and Smalldon, K.W. (1975) Blood and semen stains on outer clothing and shoes not related to crime: report of a survey using presumptive tests. *Journal of Forensic Sciences*, **20**, 391–403.

Palmer, R. and Chinherende, V. (1996) A target fibre study using cinema and car seats as recipient items. *Journal of the Forensic Science Society*, **41**, 802–803.

Pankanti, S., Prabhakar, S. and Jain, A.K. (2002) On the individuality of fingerprints. *IEEE Transactions on Pattern Analysis and Machine Intelligence*, **24**, 1010–1025.

Parker, J.B. (1966) A statistical treatment of identification problems. *Forensic Science Society Journal*, **6**, 33–39.

Parker, J.B. (1967) The mathematical evaluation of numerical evidence. *Forensic Science Society Journal*, **7**, 134–144.

Parker, J.B. and Holford, A. (1968) Optimum test statistics with particular reference to a forensic science problem. *Applied Statistics*, **17**, 237–251.

Peabody, A.J., Oxborough, R.J., Cage, P.E. and Evett, I.W. (1983) The discrimination of cat and dog hairs. *Journal of the Forensic Science Society*, **23**, 121–129.

Pearl, J. (1988) *Probabilistic Reasoning in Intelligent Systems: Networks of Plausible Inference*, Morgan Kaufmann, San Mateo, CA.

Pearson, E.F., May, R.W. and Dabbs, M.G.D. (1971) Glass and paint fragments found in men's outer clothing – report of a survey. *Journal of Forensic Sciences*, **16**, 283–300.

Pearson, E.S. and Hartley, H.O. (eds) (1966) *Biometrika Tables for Statisticians, Volume 1*, Cambridge University Press, Cambridge.

Pearson, E.S. and Kendall, M.G. (eds) (1970) *Studies in the History of Statistics and Probability*, Charles Griffin, London.

Peirce, C.S. (1878) The probability of induction. *Popular Science Monthly* Reprinted in *The World of Mathematics*, Volume 2 (Newman, J.R., ed.), Simon and Schuster, New York, (1956), pp. 1341–1354.

Petterd, C.I., Hamshere, J., Stewart, S., Brinch, K., Masi, T. and Roux, C. (2001) Glass particles in the clothing of members of the public in south-eastern Australia – a survey. *Forensic Science International*, **116**, 193–198.

Piattelli-Palmarini, M. (1994) *Inevitable Illusions*. John Wiley & Sons, Inc. New York.

Poincaré, H. (1896) Calcul des Probabilités. Leçons professées pendant le deuxième semestre 1893–1894. In *Calcul des Probabilités* (G. Carré, ed.), Paris.

Poincaré, H. (1912) *Calcul des Probabilités*, Gauthier-Villars, Paris.

Poincaré, H. (1992) *La Science et l'hypothèse*. Editions de la Bohème, Paris.

Poisson, S.D. (1837) *Recherches sur la probabilité des jugements en matière criminelle et en matière civile, précédées des règles générales du calcul des probabilités*, Bachelier, Paris.

Pounds, C.A. and Smalldon, K.W. (1978) The distribution of glass fragments in front of a broken window and the transfer of fragments to individuals standing nearby. *Journal of the Forensic Science Society*, **18**, 197–203.

Puch, R.O. and Smith, J.Q. (2002) FINDS: a training package to assess forensic fibre evidence. *Advances in Artificial Intelligence* (Coella, C.A.C., Albornoz, A. de, Sucar, L.E. and Battistuti, O.S., eds) Springer-Verlag, Berlin, pp. 420–429.

Rabinovitch, N.L. (1969) Studies in the history of probability and statistics XXII: Probability in the Talmud. *Biometrika*, **56**, 437–441.

Rabinovitch, N.L. (1973) *Probability and Statistical Inference in Ancient and Medieval Jewish Literature*, University of Toronto Press, Toronto.

Race, R.R., Sanger, R., Lawler, S.D. and Bertinshaw, D. (1949) The inheritance of the *MNS* blood groups: a second series of families. *Heredity*, **3**, 205–213.

Rahne, E., Joseph, L. and Gyorkos, T.W. (2000) Bayesian sample size determination for estimating binomial parameters from data subject to misclassification. *Applied Statistics*, **49**, 119–128.

Ramsey, F.P. (1931) Truth and probability. In *The Foundations of Mathematics and Other Logical Essays* (R.B. Braithwaite ed.), Routledge & Kegan Paul, London.

Redmayne, M. (1995) Doubts and burdens: DNA evidence, probability and the courts. *Criminal Law Review*, 464–482.

Redmayne, M. (1997) Presenting probabilities in court: the DNA experience. *International Journal of Evidence and Proof*, **4**, 187–214.

Redmayne, M. (2002) Appeals to Reason. *Modern Law Review*, **65**, 19–35.

Reinstein, R.S. (1996) Comment. In *Convicted by Juries, Exonerated by Science: Case Studies in the Use of DNA Evidence to Establish Innocence after Trial* (Connors, E., Lundregan, T., Miller, N. and McEwen, T., eds), US Department of Justice, Washington, DC.

Robertson, B. and Vignaux, G.A. (1991) Extending the conversation about Bayes. *Cardozo Law Review*, **13**, 629–645.

Robertson, B. and Vignaux, G.A. (1992) Unhelpful evidence in paternity cases. *New Zealand Law Journal*, **9**, 315–317.

Robertson, B. and Vignaux, G.A. (1993a) Probability – the logic of the law. *Oxford Journal of Legal Studies*, **13**, 457–478.

Robertson, B. and Vignaux, G.A. (1993b) Biology, logic, statistics and criminal justice. *The Criminal Lawyer*, **35**.

Robertson, B. and Vignaux, G.A. (1993c) Taking fact analysis seriously. *Michigan Law Review*, **91**, 1442–1464.

Robertson, B. and Vignaux, G.A. (1994) Crime investigation and the criminal trial. *Journal of the Forensic Science Society*, **34**, 270.

Robertson, B. and Vignaux, G.A. (1995a) *Interpreting Evidence: Evaluating Forensic Science in the Courtroom*, John Wiley & Sons, Ltd, Chichester.

Robertson B. and Vignaux G.A.(1995b) DNA evidence: wrong answers or wrong questions? In *Human Identification: The use of DNA Markers* (Weir, B.S., ed.), Kluwer Academic, Dordrecht, pp. 145–152.

Robertson, B. and Vignaux, G.A. (1998) Explaining evidence logically. *New Law Journal*, **148**, 159–162.

Robinson, N., Taroni, F., Saugy, M., Ayotte, C., Mangin, P. and Dvorak, J. (2001) Detection of nandrolone metabolites in urine after a football game in professional and amateur players: a Bayesian comparison. *Forensic Science International*, **122**, 130–135.

Rose, P.J. (2002) *Forensic Speaker Identification*. Taylor & Francis, London.

Rose, P.J. (2003) The technical comparison of forensic voice samples. In *Expert Evidence* (Freckleton, I. and Selby, H., eds), Thomson Lawbook Company, Sydney, Chapter 99.

Roux, C. and Margot, P. (1997) An attempt to assess relevance of textile fibres recovered from car seats. *Science & Justice*, **37**, 225–230.

Roux, C., Chable, J. and Margot, P. (1996) Fibre transfer experiments onto car seats. *Science & Justice*, **36**, 143–151.

Roux, C., Langdon, S., Waight, D. and Robertson, J. (1999) The transfer and persistence of automotive carpet fibres on shoe soles. *Science & Justice*, **39**, 239–251.

Roux, C., Kirk, R., Benson, S., Van Haren, T. and Petterd, C.I. (2001) Glass particles in footwear of members of the public in south-eastern Australia – a survey. *Forensic Science International*, **116**, 149–156.

Royall, R. (1997) *Statistical Evidence: A Likelihood Paradigm*, Chapman & Hall, London.

Royall, R. (2000) On the probability of observing misleading statistical evidence. *Journal of the American Statistical Association*, **95**, 760–780.

Rudin, M. and Inman, K. (2003) *An Introduction to Forensic DNA Analysis*, 2nd edition, CRC Press, Boca Raton, FL.

Ryan, B.F., Joiner, B.L. and Ryan, T.A. (2000) *MINITAB Handbook*, Fourth edition, Brooks Cole, Pacific Grove, CA (http://www.minitab.com/).

Ryland, S.G., Kopec, R.J. and Somerville, P.N. (1981) The evidential value of automobile paint chips. Part II: The frequency of occurrence of topcoat colours. *Journal of Forensic Sciences*, **26**, 64–74.

Sacco, N. (1969) *The Sacco–Vanzetti Case: Transcript of the Record of the Trial of Nicola Sacco and Bartolomeo Vanzetti in the Courts of Massachusetts and Subsequent Proceedings, 1920–7*, 6 vols, P.P. Appel, Mamaroneck, NY.

Saks, M.J. and Koehler, J.J. (1991) What DNA 'fingerprinting' can teach the law about the rest of forensic science. *Cardozo Law Review*, **13**, 361–372.

Salmon, D. and Salmon, C. (1980) Blood groups and genetic markers polymorphisms and probability of paternity. *Transfusion*, **20**, 684–694.

Salter, M.T. and Cook, R. (1996) Transfer of fibres to head hair, their persistence and retrieval. *Forensic Science International*, **81**, 211–221.

Savage, L.J. (1954) *The Foundations of Statistics*, Dover, New York.

Schum, D.A. (1994) *Evidential Foundations of Probabilistic Reasoning*, John Wiley & Sons, Inc. New York.

Schum, D.A. (1999) Inference Networks and the Evaluation of Evidence: Alternative Analyses. In *Uncertainty in Artificial Intelligence: Proceedings of the Fifteenth Conference* (Laskey, K. Prade, H., eds), Morgan Kaufmann Publishers, San Francisco, pp. 575–584.

Schum, D.A. (2000) Singular evidence and probabilistic reasoning in judicial proof. In *Harmonisation in Forensic Expertise* (Nijboer, J.F. and Sprangers, W.J.J.M., eds), Thela Thesis, Leiden, The Netherlands, pp. 587–603.

Scientific Sleuthing Newsletter (1988) Hair analysis. *Scientific Sleuthing Newsletter*, **12**(3), 4, 6.

Scott, A.J. and Knott, M. (1974) A cluster analysis method for grouping means in the analysis of variance. *Biometrics*, **30**, 507–512.

Seheult, A. (1978) On a problem in forensic science. *Biometrika*, **65**, 646–648.

Selvin, S. (1975) On the Monty Hall problem. *American Statistician*, **29**, 134.

Shafer, G. (1976) *A Mathematical Theory of Evidence*, Princeton University Press, Princeton, NJ.

Shafer, G. (1978) Non-additive probabilities in the work of Bernoulli and Lambert. *Archive for History of Exact Sciences*, **19**, 309–370.

Shafer, G. (1982) Lindley's paradox (with discussion). *Journal of the American Statistical Association*, **77**, 325–351.

Shannon, C.E. (1948) A mathematical theory of communication. *Bell System Technical Journal*, **27**, 379–423.

Shannon, C.R. (1984) *Royal Commission concerning the conviction of Edward Charles Splatt*, Woolman, D.J., Government Printer, South Australia, Australia.

Sheynin, O.B. (1974) On the prehistory of the theory of probability. *Archive for History of Exact Sciences*, **12**, 97–141.

Shoemaker, J.S. Painter, I.S., and Weir, B.S. (1999) Bayesian statistics in genetics: a guide for the uninitiated. *Trends in Genetics*, **15**, 354–358.

Siegel, J.A. (1997) Evidential value of textile fiber – transfer and persistence of fibers. *Forensic Science Review*, **9**, 81–96.

Silverman, B.W. (1986) *Density Estimation*, Chapman & Hall, London.

Simon, R.J. and Mahan, L. (1971) Quantifying burdens of proof. *Law and Society Review*, **5**, 319–330.

Simons, A.A. (1997) Technical Working Group on Friction Ridge Analysis, Study and Technology (TWGFAST) Guidelines. *Journal of Forensic Identification*, **48**, 147–162.

Simpson, E.H. (1949) Measures of diversity. *Nature*, **163**, 688.

Sinha, S.K. (ed.) (2003) Y-chromosome: genetics, analysis, and application in forensic science. *Forensic Science Review*, **15**(2), 77–201.

Sjerps, M. and Kloosterman, A.D. (1999) On the consequences of DNA profile mismatches for close relatives of an excluded suspect. *International Journal of Legal Medicine*, **112**, 176–180.

Smalldon, K.W. and Moffat, A.C. (1973) The calculation of discriminating power for a series of correlated attributes. *Journal of the Forensic Science Society*, **13**, 291–295.

Smeeton, N.C. and Adcock, C.J. (eds) (1997) Sample size determination – Special issue. *The Statistician*, **46**(2).

Smith, J.A.L. and Budowle, B. (1998) Source identification of body fluid stains using DNA profiling. In *Proceedings from the Second European Symposium of Human Identification*, Promega Corporation, Innsbruck, Austria, pp. 89–90.

Smith, R.L. and Charrow, R.P. (1975) Upper and lower bounds for the probability of guilt based on circumstantial evidence. *Journal of the American Statistical Association*, **70**, 555–560.

Souder, W. (1934/1935) The merits of scientific evidence. *Journal of the American Institute of Criminal Law and Criminology*, **25**, 683–684.

Srihari, S.N., Cha, S.-H., Arora, H. and Lee, S. (2002) Individuality of handwriting. *Journal of Forensic Sciences*, **47**, 856–872.

Stockmarr, A. (1999) Likelihood ratios for evaluating DNA evidence when the suspect is found through a database search. *Biometrics*, **55**, 671–677.

Stockton, A. and Day, S. (2001) Bayes, handwriting and science. In *Proceedings of the 59th Annual ASQDE Meeting – Handwriting & Technology: at the Crossroads*. Des Moines, Iowa, U.S.A., pp. 1–10.

Stoney, D.A. (1984a) Evaluation of associative evidence: choosing the relevant question. *Journal of the Forensic Science Society*, **24**, 473–482.

Stoney, D.A. (1984b) Statistics applicable to the inference of a victim's blood type from familial testing. *Journal of the Forensic Science Society*, **24**, 9–22.

Stoney, D.A. (1991a) Transfer evidence. In *The Use of Statistics in Forensic Science* (Aitken, C.G.G and Stoney, D.A., eds), Ellis Horwood, Chichester, pp. 107–138.

Stoney, D.A. (1991b) What made us think we could individualise using statistics? *Journal of the Forensic Science Society*, **31**, 197–199.

Stoney, D.A. (1992) Reporting of highly individual genetic typing results: a practical approach. *Journal of Forensic Sciences*, **37**, 373–386.

Stoney, D.A. (1994) Relaxation of the assumption of relevance and an application to one-trace and two-trace problems. *Journal of the Forensic Science Society*, **34**, 17–21.

Stoney, D A. (2001) Measurement of fingerprint individuality. In *Advances in Fingerprint Technology* (Lee, H.C. and Gaensslen, R.E., eds), CRC Press, Boca Raton, FL pp. 327–387.

Stoney, D.A. and Thornton, J.I. (1986) A critical analysis of quantitative fingerprint individuality models. *Journal of Forensic Sciences*, **33**, 11–13.

Taroni, F. and Aitken, C.G.G. (1998a) Probabilités et preuve par l'ADN dans les affaires civiles et criminelles. Questions de la cour et réponses fallacieuses des experts. *Revue Pénale Suisse*, **116**, 291–313.

Taroni, F. and Aitken, C.G.G. (1998b) Probabilistic reasoning in the law, part 1: Assessment of probabilities and explanation of the value of DNA evidence. *Science & Justice*, **38**, 165–177.

Taroni, F. and Aitken, C.G.G. (1998c) Probabilistic reasoning in the law, part 2: Assessment of probabilities and explanation of the value of trace evidence other than DNA. *Science & Justice*, **38**, 179–188.

Taroni, F. and Aitken, C.G.G. (1999a) The likelihood approach to compare populations: a study on DNA evidence and pitfalls of intuitions. *Science & Justice*, **39**, 213–222.

Taroni, F. and Aitken, C.G.G. (1999b) DNA evidence, probabilistic evaluation and collaborative tests. *Forensic Science International*, **108**, 121–143.

Taroni, F. and Aitken, C.G.G. (2000) Fibres evidence, probabilistic evaluation and collaborative test. Letter to the Editor. *Forensic Science International*, **114**, 45–47.

Taroni, F. and Buckleton, J. (2002) Likelihood ratio as a relevant and logical approach to assess the value of shoeprint evidence. *Information Bulletin for Shoeprint/Toolmark Examiners*, **8**(2), 15–25.

Taroni, F. and Champod, C. (1994) Forensic medicine, P.C.R. and Bayesian approach. *Journal of Medical Genetics*, **31**, 896–898.

Taroni, F. and Mangin, P. (1999) *La preuve ADN, les probabilitēs, les experts and les juristes. Necessitē de developpement et de communication*. Final report, 1115-054002/98, to Swiss National Foundation.

Taroni, F. and Margot, P. (2000) Fingerprint evidence evaluation: is it really so different to other evidence types? (Letter to the Editor). *Science & Justice*, **40**, 277–280.

Taroni, F. and Margot, P. (2001) General comments on the scale of conclusions in shoemarks – the need for a logical framework. *Information Bulletin for Shoeprint/Toolmark Examiners*, **7**(2), 37–41.

Taroni, F., Champod, C., and Margot, P. (1996) Statistics: a future in tool marks comparison? *AFTE (Association of Firearm and Toolmark Examiners) Journal*, **28**, 222–232.

Taroni, F., Champod, C. and Margot, P. (1998) Forerunners of Bayesianism in early forensic science. *Jurimetrics Journal*, **38**, 183–200.

Taroni, F., Aitken, C.G.G. and Garbolino, P. (2001) De Finetti's subjectivism, the assessment of probabilities, and the evaluation of evidence: a commentary for forensic scientists. *Science & Justice*, **41**, 145–150.

Taroni, F., Lambert, J.A., Fereday, L. and Werrett, D.J. (2002) Evaluation and presentation of forensic DNA evidence in European laboratories. *Science & Justice*, **42**, 21–28.

Taroni, F., Biedermann, A., Garbolino, P. and Aitken, C.G.G. (2004) A general approach to Bayesian networks for the interpretation of evidence. *Forensic Science International*, **139**, 5–16.

Thagard, P. (2003) Why wasn't O.J. convicted? Emotional coherence in legal inference. *Cognition and Emotion*, **17**, 361–383.

Thanasoulias, N.C., Parisis, N.A. and Evmiridis, N.P. (2003) Multivariate chemometrics for the forensic discrimination of blue ball-point pen inks based on their Vis spectra. *Forensic Science International*, **138**, 75–84.

Thompson, W.C. (1989) Are juries competent to evaluate statistical evidence? *Law and Contemporary Problems*, **52**, 9–41.

Thompson, W.C. (1993) Evaluating the admissibility of a new genetic identification test: lessons from the DNA war. *Journal of Criminal Law and Criminology*, **84**, 22–104.

Thompson W.C. (1995) Subjective interpretation, laboratory error and the value of forensic DNA evidence: three cases studies. *Genetica*, **96**, 153–68.

Thompson, W.C. (1997) Accepting lower standards: the National Research Council's second report on forensic DNA evidence. *Jurimetrics Journal*, **37**, 405–424.

Thompson W.C. and Ford S. (1989) DNA typing: acceptance and weight of the new genetic identification tests. *Virginia Law Review*, **75**, 45–108.

Thompson, W.C. and Schumann, E.L. (1987) Interpretation of statistical evidence in criminal trials. The prosecutor's fallacy and the defence attorney's fallacy. *Law and Human Behaviour*, **11**, 167–187.

Thompson, W.C, Taroni, F. and Aitken, C.G.G. (2003) How the probability of a false positive affects the value of DNA evidence. *Journal of Forensic Sciences*, **48**, 47–54.

Thompson, Y. and Williams, R. (1991) Blood group frequencies of the population of Trinidad and Tobago, West Indies. *Journal of the Forensic Science Society*, **31**, 441–447.

Tillers, P. (2001) Artificial intelligence and judicial proof. *Cardozo Law Review*, **22**, 1433–1851.

Tippett, C.F., Emerson, V.J., Fereday, M.J., Lawton, F. and Lampert, S.M. (1968) The evidential value of the comparison of paint flakes from sources other than vehicles. *Journal of the Forensic Science Society*, **8**, 61–65.

Tribe, L. (1971) Trial by mathematics: precision and ritual in the legal process. *Harvard Law Review*, **84**, 1329–1393.

Triggs, C.M. and Buckleton, J.S. (2002) Logical implications of applying the principles of population genetics to the interpretation of DNA profiling evidence. *Forensic Science International*, **128**, 108–114.

Triggs, C.M. and Buckleton, J.S. (2003) The two-trace problem re-examined. *Science & Justice*, **43**, 127–134.

Triggs, C.M., Curran, J.M., Buckleton, J.S. and Walsh, K.A.J. (1997) The grouping problem in forensic glass analysis: a divisive approach. *Forensic Science International*, **85**, 1–14.

Triggs, C.M., Harbison, S.A. and Buckleton, J.S. (2000) The calculation of DNA match probabilities in mixed race populations. *Science & Justice*, **40**, 33–38.

Tryhorn, F.G. (1935) The assessment of circumstantial scientific evidence. *Police Journal*, **8**, 401–411.

Tversky, A. and Kahneman, D. (1974) Judgement under uncertainty: heuristics and biases. *Science*, **185**, 1124–1131.

Tzidony, D. and Ravreboy, M. (1992) A statistical approach to drug sampling: a case study. *Journal of Forensic Sciences*, **37**, 1541–1549.

United Nations (1998) *Recommended Methods for Testing Opium, Morphine and Heroin.* Manual for use by National Drug Testing Laboratories. United Nations, New York.

United States Supreme Court (1995) Schlup v. Delo, United States Reports, **513**, 298–322.

Venables, W.M. and Ripley, B.D. (2002) *Modern Applied Statistics with S-Plus*, 4th edition, Springer-Verlag, New York.

Vicard, P. and Dawid, A.P. (2003) Estimating mutation rates from paternity data. In *Atti del Convegno Modelli complessi e metodi computazionali intensivi per la stima e la previsione*, Università Cà Foscari, Venezia, Italy, pp. 415–418.

Vito, G.F. and Latessa, E.J. (1989) *Statistical Applications in Criminal Justice*, Sage, London.

Wakefield, J.C., Skene, A.M., Smith, A.F.M. and Evett, I.W. (1991) The evaluation of fibre transfer evidence in forensic science: a case study in statistical modelling. *Applied Statistics*, **40**, 461–476.

Walsh, K.A.J. and Buckleton, J.S. (1986) On the problem of assessing the evidential value of glass fragments embedded in footwear. *Journal of the Forensic Science Society*, **26**, 55–60.

Walsh, K.A.J. and Buckleton, J.S. (1988) A discussion of the law of mutual independence and its application to blood group frequency. *Journal of the Forensic Science Society*, **28**, 95–98.

Walsh, K.A.J. and Buckleton, J.S. (1991) Calculating the frequency of occurrence of a blood type for a 'random man'. *Journal of the Forensic Science Society*, **31**, 49–58.

Walsh, K.A.J. and Buckleton, J.S. (1994) Assessing prior probabilities considering geography. *Journal of the Forensic Science Society*, **34**, 47–51.

Walsh, K.A.J., Buckleton, J.S. and Triggs, C.M. (1996) A practical example of the interpretation of glass evidence. *Science & Justice*, **36**, 213–218.

Weir, B.S. (1992) Population genetics in the forensic DNA debate. *Proceedings of the National Academy of Science USA*, **89**, 11654–11659.

Weir, B.S. (1995) DNA statistics in the Simpson matter. *Nature Genetics*, **11**, 365–368.

Weir, B.S. (1996a) *Genetic Data Analysis II*, Sinauer Associates, Sunderland, MA.

Weir, B.S. (1996b) Presenting DNA statistics in court. In *Proceedings of the 6th International Symposium on Human Identification*, Promega Corporation, Madison, WI, pp. 128–136.

Weir, B.S. (1998) The coancestry coefficient. In *Proceedings of the 8th International Symposium on Human Identification*, Promega Corporation, Madison, WI, pp. 87–91.

Weir, B.S. (2000a) Statistical analysis. In *Encyclopedia of Forensic Sciences* (Siegel, J.A., Saukko, P.J. and Knupfer, G.C., eds), Academic Press, San Diego, CA, pp. 545–550.

Weir, B.S. (2000b) The consequences of defending DNA statistics. In *Statistical Science in the Courtroom* (Gastwirth, J.L., ed.), Springer-Verlag, New York, pp. 86–97.

Weir, B.S. (2001a) Forensics. In *Handbook of Statistical Genetics* (Balding, D.J., Bishop, M. and Cannings, C., eds), John Wiley & Sons, Ltd, Chichester, pp. 721–739.

Weir, B.S. (2001b) DNA match and profile probabilities – Comment on Budowle *et al.* (2000) and Fung and Hu (2000a). *Forensic Science Communications*, **3**. Available at http://www.fbi.gov/hq/lab/fsc/backissu/jan2001/weir.htm

Weir, B.S. and Evett, I.W. (1992) Whose DNA? *American Journal of Human Genetics*, **50**, 869.

Weir, B.S. and Evett, I.W. (1993) Reply to Lewontin (1993) (Letter). *American Journal of Human Genetics*, **52**, 206.

Weir, B.S. and Hill, W.G. (1993) Population genetics of DNA profiles. *Journal of the Forensic Science Society*, **33**, 218–225.

Weir, B.S., Triggs, C.M., Starling, L., Stowell, L.I., Walsh, K.A.J. and Buckleton, J. (1997) Interpreting DNA mixtures. *Journal of Forensic Sciences*, **42**, 213–222.

Weiss, C. (2003) Expressing scientific uncertainty. *Law, Probability and Risk*, **2**, 25–46.

Welch, B.L. (1937) The significance of the difference between two means when the population means are unequal. *Biometrika*, **29**, 350–362.

Wickenheiser, R.A. and Hepworth, D.G. (1990) Further evaluation of probabilities in human scalp hair comparisons. *Journal of Forensic Sciences*, **35**, 1323–1329.

Wickenheiser, R.A. and Hepworth, D.G. (1991) Authors' response. *Journal of Forensic Sciences*, **36**, 973–976.

Wigmore, J. (1937) *The Science of Proof: as Given by Logic, Psychology and General Experience and Illustrated in Judicial Trials*, 3rd edition, Little, Brown, Boston.

Wooley, J.R. (1991) A response to Lander: the Courtroom perspective. *American Journal of Human Genetics*, **49**, 892–893.

Wright, S. (1922) Coefficients of inbreeding and relationship. *Amererican Naturalist*, **56**, 330–338.

Wright, S. (1951) The genetical structure of populations. *Annals of Eugenics*, **15**, 323–354.

Wright, S. (1965) The interpretation of population structure by F-statistics with special regard to systems of mating. *Evolution*, **19**, 395–420.

Yellin, J. (1979) Book review of '*Evidence, proof and probability, 1st edition*', Eggleston, R. (1978) Weidenfeld and Nicolson, London. *Journal of Economic Literature*, 583.

Zabell, S. (1976) Book review of '*Probability and statistical inference in ancient and medieval Jewish literature*', Rabinovitch, N.L. (1973), University of Toronto Press, Toronto, Canada. *Journal of the American Statistical Association*, **71**, 996–998.

Zeisel, H. and Kaye, D.H. (1997) *Prove It with Figures*, Springer-Verlag, New York, pp. 216–217.

Notation

The Greek and Roman alphabets provide a large choice of letters to be used for mathematical notation. Despite this large choice, some letters, such as x, are used in this book to mean more than one thing. It is hoped that no letter or symbol is asked to mean more than one thing at the same time and that the list which follows will help readers to know what each letter or symbol does mean at any particular point. Chapter or section references are given for the first or main use of many of the letters or symbols.

... : three dots, written on the line, indicate 'and so on in sequence to'. Thus $x_1, \ldots, x_5$ can be read as 'x_1 and so on in sequence to x_5' and is short-hand for the sequence x_1, x_2, x_3, x_4, x_5. Usually, the last subscript is a general one such as n so that a sequence of n items would be written as $x_1, \ldots, x_n$.

$\cdots$: three dots, in raised position, indicate 'a repeat of the operation immediately before and after the dots'. Thus $x_1 + \cdots + x_5$ is short-hand for '$x_1 + x_2 + x_3 + x_4 + x_5$'. Similarly $x_1 \times \cdots \times x_5$ is short-hand for '$x_1 \times x_2 \times x_3 \times x_4 \times x_5$'. Usually the last subscript is a general one such as n, so that a sum or product of n items would be written as $x_1 + \cdots + x_n$ or $x_1 \times \cdots \times x_n$. Also, the symbol $\times$ is often omitted and $x_1 \times \cdots \times x_n$ written as $x_1 \cdots x_n$.

$\sum$: the sum of terms following the symbol. For example, $\sum_{i=1}^{n} x_i$ denotes the sum of $x_1, \ldots, x_n$ $(x_1 + \cdots + x_n)$; Section 2.3.2.

$\prod$: the product of terms following the symbol. For example, $\prod_{i=1}^{n} Pr(S_i)$ denotes the product of the probabilities $Pr(S_1), \ldots, Pr(S_n)$ $(Pr(S_1) \cdots Pr(S_n))$; Section 4.5.4.

$^-$: to be read as 'the opposite of' or the 'complement of', thus if M denotes male, $\bar{M}$ denotes female; Section 3.1.1.

$^-$: to be read as 'the mean of', thus $\bar{x}$ is the mean of a set of measurements $x_1, \ldots, x_n$; Section 2.4.1.

$\equiv$: to be read as 'as equivalent to'. For example, if M denotes male and F denotes female, then $M \equiv \bar{F}$ and $F \equiv \bar{M}$.

Statistics and the Evaluation of Evidence for Forensic Scientists: Second Edition
C.G.G. Aitken and F. Taroni © 2004 John Wiley & Sons, Ltd   ISBN: 0-470-84367-5

$\gg$:   to be read as 'is very much greater than' (in contrast to $>$ which is simply 'is greater than'); Section 4.6.2.

$\propto$:   to be read as 'is proportional to'. For example, this is often used in Bayesian analysis where the distribution of the random variable is taken to be proportional to an expression involving only terms in the random variable and omitting other terms which are needed to ensure the distribution is a probability distribution (i.e., has a total probability of 1). Use of such a notation eases the algebraic manipulations associated with Bayesian inference; Section 7.4.

$(\mathbf{X} \mid \boldsymbol{\theta}, \Sigma) \sim \mathbf{N}(\boldsymbol{\theta}, \Sigma)$:   multivariate random variable $\mathbf{X}$ has a Normal distribution with mean vector $\boldsymbol{\theta}$ and covariance matrix Σ; Section 2.4.6.

$\mid x \mid$:   for x a number, the absolute value of x; if $x > 0, \mid x \mid = x,$; if $x < 0, \mid x \mid = -x$; for example, $\mid 6 \mid = 6, \mid -6 \mid = 6$.

$\mid \Sigma \mid$:   determinant of the matrix Σ.

α :   a prior parameter for the beta distribution; Section 2.4.4.

$b_{g,\mathbf{m}}$:   probability of g groups of sizes $(m_1, \ldots, m_g) = \mathbf{m}$ being found; Section 10.5.

β :   a prior parameter for the beta distribution; Section 2.4.4.

$B(\alpha, \beta)$:   the normalising constant for a beta distribution; Section 2.4.4.

$B(\alpha_1, \ldots, \alpha_k)$:   the normalising constant for a Dirichlet distribution; Section 2.4.5.

$C_X(Y)$:   the probability a person of blood group Y innocently bears a bloodstain of blood group X; Section 8.3.3.

Γ :   analytical result from the inspection of trace evidence.

γ :   frequency of Γ in a relevant population; a parameter.

$\Gamma(x+1)$:   the gamma function; Section 2.4.4.

E :   quality or measurements of evidential material; Chapter 1.

   E_c :   quality or measurements of evidential material of source form, also denoted x; Section 1.6.1.

   E_s :   quality or measurements of evidential material of receptor form, also denoted y; Section 1.6.1.

   Ev :   the totality of the evidence, equals (M, E); Section 1.6.1.

$E(\theta_i)$:   the mean of the variable θ_i, also known as the *expectation*; Section 2.4.5.

$f_{t,z-1}\{.\}$:   the probability density function of the t-distribution with $z-1$ degrees of freedom; Section 2.4.3.

g :   number of groups; Section 12.2.

H_d :   the proposition of the defence; Section 3.1.1.

H_p :   the proposition of the prosection; Section 3.1.1.

$\log_e$:   logarithm to base e.

$\log_{10}$:   logarithm to base 10.

M :   evidential material; Section 1.6.1.

   M_c :   evidential material of source form; Section 1.6.1.

   M_s :   evidential material of receptor form; Section 1.6.1.

m :   number of items inspected; Section 6.1.

n :   number of items not inspected; Section 6.2.2.

n :   number of groups transferred between two objects; Section 12.2.1.

N :   consignment size ($= m + n$); Section 6.1.

$N(\theta, \sigma^2)$:   Normal distribution of mean θ and standard deviation σ.

ν :   degrees of freedom; Section 2.4.3

O :   odds; Section 3.1.

p :   probability of transfer of material to the suspect from the crime scene or from the suspect to the crime scene, persisting and being recovered if the suspect were innocent; Section 9.5.4.

p_i :   probability of presence of $i (\geq 0)$ groups of material on the suspect; Section 10.5.4.

Pr :   probability.

Q :   random variable corresponding to quantity to be estimated; Section 6.3.

q :   equals $1 - p$, where p is the relative frequency in a sample.

q :   quantity to be estimated; Section 6.3.

ρ :   population correlation coefficient; Section 2.4.6.

R :   number of items in the consignment which are illicit; Section 6.2.2.

r :   the probability of relevance: some, all or none of the transferred material may be present for innocent reasons (e.g., reasons unassociated with the offender) and some, none or all for guilty reasons (e.g., reasons associated with the offender); some of the material is selected for analysis. If the selected material is part of that which was there for guilty reasons then it is defined as relevant; Section 9.5.3.

s :   standard deviation of a sample or of measured items; Section 2.2.

s_l :   probability that a group of fragments found on members of a population is large; Section 10.5.4.

σ :   standard deviation of a population; Section 2.2.

$\sum$:   covariance matrix; Section 2.4.6.

$t_\nu(P)$:   the $100P\%$ point of the t-distribution with ν degrees of freedom; Section 2.4.3.

t_n :   probability of transfer of $n (\geq 0)$ items of material to the suspect from the crime scene or from the suspect to the crime scene, persisting and being recovered if H_p is true. Section 10.5.4.

t_W :   the numerator of Student's t-density (10.23); Section 10.6.

t'_n :   probability of transfer of $n (\geq 0)$ items of material to the offender from the crime scene or from the offender to the crime scene, persisting and being recovered if H_d is true; Section 12.2.1.

θ :   probability of at least one match of evidence of a given frequency with an identified individual in a population of individuals unrelated to the identified individual and of finite size; Section 3.3.5.

θ :   mean of a Normal distribution; Section 2.4.2.

θ :   parameter of a probability distribution, for a prior distribution it is treated as a variable; Sections 2.4.4 and 2.4.5.

θ :   proportion of the consignment which contains illicit items; Section 6.2.1.

θ :   co-ancestry coefficient F_{ST}; Section 13.6.

θ_0 :   lower bound for the proportion of the consignment which contains illicit items; Section 6.2.1.

V :   the value of evidence, the likelihood ratio; Section 3.5.1.

$V_s(x)$:   the value of the evidence, the likelihood ratio for x versus s; Section 13.9.

w_j :   the weight of the contents of the jth item not examined which is illicit; Section 6.3.2.

$\bar{w}$:   mean weight of items not inspected which are illicit; Section 6.3.2.

x :   measurement on source or control material; Sections 1.4 and 10.1.

x_i :   the weight of the contents of the ith item examined which is illicit; Section 6.3.2.

$\bar{x}$:   mean weight of inspected items which are illicit; Section 6.3.2.

$\mathbf{x}_{ij}$:   background multivariate data for sample j in group i, $i = 1, \ldots, m$, $j = 1, \ldots, n$; Section 7.3.8.

$x!$: x factorial; when x is a positive integer, the product of x with all integers less than it and greater than zero, $x(x-1)(x-2)\cdots 2 \cdot 1$; conventionally $0! = 1$; Section 2.3.1.

y :   measurement on receptor or recovered material; Sections 1.4 and 10.1.

y :   number of items not inspected which are illicit; this number is unknown and modelled by a beta-binomial distribution; Section 6.2.2.

$\mathbf{y}_1$:   multivariate source or control data; Chapter 11.

$\mathbf{y}_2$:   multivariate receptor or recovered data; Chapter 11.

z_i :   a member of the background data for univariate data, $i = 1, \ldots, k$; Section 10.3.

z :   number of items inspected which are found to be illicit; $z \leq m$; Section 6.2.1.

$\binom{n}{x}$:   the binomial coefficient, the number of ways in which x ($0 \leq x \leq n$) items may be chosen from n ($\geq x$) in which no attention is paid to ordering; equals $n!/\{x!(n-x)!\}$; Section 2.3.3.

Cases

The *Belhaven and Stenton Peerage Case*, 1 App. Cas. 279, p. 208.

Commonwealth of Massachusetts v. Nicola Sacco and Bartolomeo Vanzetti, 1921, pp. 5, 429.

Daubert v. Merrell Dow Pharmaceuticals Inc., 509 U.S. 579, 1993, pp. 197–198.

Johannes Pruijsen v. H.M. Customs & Excise, Crown Court, Chelmsford, UK, 30 July 1998, p. 108.

Kumho Tire Co. Ltd. v. Carmichael, 526 U.S. 137, 1999, p. 198.

People v. Collins, 68 Cal 2d 319, 438. P.2d 33, 36, 66 Cal. Rptr. 497 (Cal, 1968), pp. 126–128, 149, 210, 212.

R. v. Adams, D.J., 1997, 2 Cr. App. Rep. 4679, pp. 78, 79, 208, 211, 213.

R. v. Clark, 2003, EWCA Crim 1020, 2003, All ER (D) 223 (Apr), CA and 2000, All ER (D) 1219, CA, pp. 79, 128, 211–215.

R. v. Dallagher, 2003, 1 Crim. App. R 195, p. 222.

R. v. Deen, Court of Appeal, Criminal Division, 21 December 1993, pp. 79, 91.

R. v. Doheny and Adams, 1 Cr. App. R. 369, 375, 1997, pp. 79, 206.

R. v. Gordon, M., Court of Appeal, 22 November 1993, 22 April and 26 May 1994, p. 91.

R. v. Kempster, 2003, EWCA Crim 3555, p. 222.

R. v. Lashley, CA 9903890 Y3, 8 February 2000, p. 211.

R. v. Montella, 1 NZLR High Court, 1992, 63–68, p. 92.

R. v. Smith, CA 9904098 W3, 8 February 2000, p. 211.

Re the Paternity of M.J.B. : T.A.T., 144 Wis. 2d 638; 425 N.W. 2d 404, 1988, pp. 315–316.

Rise v. State of Oregon, n. 93-35521, D.C. No. CV-91-06456-MRH, 9th Cir. 1995, p. 414.

Ross v. State of Indiana, Indiana Court of Appeal, 13 May 1996, pp. 82, 91.

Ross v. State, B14-90-00659, Tex. App., 13 February 1992, pp. 85, 92.

State v. Klindt, District Court of Scott County, Iowa, Case number 115422, 1968, pp. 248–249.

State of New Jersey v. J.M. Spann, 130 N.J. 484; 617 A. 2d 247, 1993, p. 316.

State of Vermont v. T. Streich, 658 A.2d 38, 1995, p. 82.

Statistics and the Evaluation of Evidence for Forensic Scientists: Second Edition
C.G.G. Aitken and F. Taroni © 2004 John Wiley & Sons, Ltd   ISBN: 0-470-84367-5

U.S. v. Jakobetz, 955 F. 2d 786 2nd Cir. 1992, p. 92.

U.S. v. Llera Plaza, 188 F. Supp. 2d 549, E.D. Pa. 2002, vacating 179 F. Supp. 2d 492, E.D. Pa. 2002, p. 228.

U.S. v. Pirre, 927, F. 2d 694, 2nd Cir., 1991, p. 192.

U.S. v. Shonubi: Shonubi V: 962 F.Supp.370 (E.D.N.Y. 1997); *Shonubi IV*: 103 F.3d 1085 (2d Cir. 1997); *Shonubi III*: 895 F.Supp. 460 (E.D.N.Y. 1995); *Shonubi II*: 998 F.2d 84 (2d Cir. 1993); *Shonubi I*: 802 F.Supp. 859 (E.D.N.Y. 1992), p. 179.

Wike v. State, 596 So 2nd 1020, Fla S. Ct., 1992, p. 81.

Wilson v. Maryland, Court of Appeals of Maryland, 370 Md 191, 803 A. 2d 1034, 4 August 2002, pp. 79, 128.

Author Index

Statistics and the Evaluation of Evidence for Forensic Scientists: Second Edition
C.G.G. Aitken and F. Taroni © 2004 John Wiley & Sons, Ltd   ISBN: 0-470-84367-5

Subject Index

Statistics in Practice

Human and Biological Sciences

Brown and Prescott – Applied Mixed Models in Medicine
Ellenberg, Fleming and DeMets – Data Monitoring Committees in Clinical Trials: A Practical Perspective
Lawson, Browne and Vidal Rodeiro – Disease Mapping With WinBUGS and MLwiN
Lui – Statistical Estimation of Epidemiological Risk
*Marubini and Valsecchi – Analysing Survival Data from Clinical Trials and Observation Studies
Parmigiani – Modeling in Medical Decision Making: A Bayesian Approach
Senn – Cross-over Trials in Clinical Research, Second Edition
Senn – Statistical Issues in Drug Development
Spiegelhalter, Abrams and Myles – Bayesian Approaches to Clinical Trials and Health-Care Evaluation
Whitehead – Design and Analysis of Sequential Clinical Trials, Revised Second Edition
Whitehead – Meta-Analysis of Controlled Clinical Trials

Earth and Environmental Sciences

Buck, Cavanagh and Litton – Bayesian Approach to Interpreting Archaeological Data
Glasbey and Horgan – Image Analysis in the Biological Sciences
Webster and Oliver – Geostatistics for Environmental Scientists

Industry, Commerce and Finance

Aitken and Taroni – Statistics and the Evaluation of Evidence for Forensic Scientists, Second Edition
Lehtonen and Pahkinen – Practical Methods for Design and Analysis of Complex Surveys, Second Edition
Ohser and Mücklich – Statistical Analysis of Microstructures in Materials Science

*Now available in paperback